The Nucleus

Volume 2

For other titles published in this series, go to
www.springer.com/series/7651

The Nucleus

Volume 2: Chromatin, Transcription, Envelope, Proteins, Dynamics, and Imaging

Ronald Hancock

Editor

Laval University Cancer Research Centre, Hôtel-Dieu Hospital,
Québec, Canada

 Humana Press

Editor
Ronald Hancock
Laval University Cancer Research Centre
Hôtel-Dieu Hospital,
Québec, Canada

Series Editor
John M. Walker
University of Hertfordshire
Hatfield, Herts., UK

ISBN: 978-1-60327-460-9   e-ISBN: 978-1-60327-461-6
ISSN: 1064-3745   e-ISSN: 1940-6029
DOI:10.1007/978-1-60327-461-6

Library of Congress Control Number: 2008929443

Cover illustration: Figure 2, Chapter 15, "Detection of the Nuclear Poly (ADP-ribose)-Metabolizing Enzymes and Activities in Response to DNA Damage," by Jean-Christophe Amé, Antoinette Hakmé, Delphine Quenet, Elise Fouquerel, Françoise Dantzer, and Valérie Schreiber.

Printed on acid-free paper

9 8 7 6 5 4 3 2 1

springer.com

Preface

As noted in the preface to *The Nucleus: Volume 1: Nuclei and Subnuclear Components*, it was a pleasure to have the efficient and generous collaboration of the contributors. For many of them, the good personal and scientific contacts that facilitated this project owe much to Wilhelm Bernhard's ideals of life and science and to the atmosphere of the "Wilhelm Bernhard Nuclear Workshop" that perpetuates them. Numerous other protocols of interest to those who work with nuclei are available in previous volumes of the Methods in Molecular Biology and other series, as well as online. The present volumes attempt to draw attention to and foster interest in less well-explored and emerging areas, and to offer a more global perspective on nuclear biology. While the subjects presented inevitably reflect to some extent the interests of the editor, the emphasis on imaging methods in *The Nucleus: Volume 2: Chromatin, Transcription, Envelope, Proteins, Dynamics, and Imaging* can be justified plausibly by the major contributions that imaging has made in recent years to our understanding of the nucleus. The help of Stan Fakan in identifying contributors to this volume, of Joanna for congenial working conditions, and of David Casey of Humana Press, who prepared these volumes with meticulous expertise, is gratefully acknowledged.

Québec   Ronald Hancock
December 2007

Contents

Preface... v

Contributors xi

List of Color Plates.. xvii

Part I   Physical and Mechanical Properties

1   **Physical Properties of the Nucleus Studied
    by Micropipette Aspiration** 3
    Amy C. Rowat

2   **Mechanical Properties of Interphase Nuclei Probed
    by Cellular Strain Application** 13
    Jan Lammerding and Richard T. Lee

Part II   Chromatin and Transcription

3   **Gene Expression in Polytene Nuclei**....................................... 29
    Petra Björk and Lars Wieslander

4   **Electron Microscope Visualization of RNA Transcription
    and Processing in *Saccharomyces cerevisiae*
    by Miller Chromatin Spreading** 55
    Yvonne N. Osheim, Sarah L. French, Martha L. Sikes,
    and Ann L. Beyer

5   **Combing Genomic DNA for Structural and Functional Studies** 71
    Catherine Schurra and Aaron Bensimon

6   **Using Molecular Beacons to Study Dispersal of mRNPs
    from the Gene Locus** 91
    Patrick T. C. van den Bogaard and Sanjay Tyagi

7   Mapping *Cis*- and *Trans*- Chromatin Interaction Networks
    Using Chromosome Conformation Capture (3C) 105
    Adriana Miele and Job Dekker

8   Recognition Imaging of Chromatin and Chromatin-Remodeling
    Complexes in the Atomic Force Microscope 123
    Dennis Lohr, Hongda Wang, Ralph Bash, and Stuart M. Lindsay

9   Using Cells Encapsulated in Agarose Microbeads to Analyse
    Nuclear Structure and Functions 139
    Dean Jackson

Part III   The Nuclear Envelope

10 Investigation of Nuclear Envelope Structure
    and Passive Permeability 161
    Victor Shahin, Yvonne Ludwig, and Hans Oberleithner

11 Reconstitution of Nuclear Import in Permeabilized Cells 181
    Aurelia Cassany and Larry Gerace

12 Nuclear Envelope Formation In Vitro: A Sea Urchin Egg
    Cell-Free System 207
    Richard D. Byrne, Vanessa Zhendre, Banafshé Larijani,
    and Dominic L. Poccia

Part IV   Modifications of Nuclear Proteins

13 Detection and Analysis of (O-linked ß-*N*-Acetylglucosamine)-
    Modified Proteins 227
    Natasha E. Zachara

14 Detection of Sumoylated Proteins 255
    Ok-Kyong Park-Sarge and Kevin D. Sarge

15 Detection of the Nuclear Poly(ADP-ribose)-Metabolizing
    Enzymes and Activities in Response to DNA Damage 267
    Jean-Christophe Amé, Antoinette Hakmé, Delphine Quenet,
    Elise Fouquerel, Françoise Dantzer, and Valérie Schreiber

16 Purification and Analysis of Variant and Modified Histones
    Using 2D PAGE 285
    George R. Green and Duc P. Do

17 Quantification of Redox Conditions in the Nucleus 303
   Young-Mi Go, Jan Pohl, and Dean P. Jones

Part V Protein Dynamics in the Nucleus

**18 Fluorescence Correlation Spectroscopy to Assess the Mobility
    of Nuclear Proteins** 321
    Stefanie Weidtkamp-Peters, Klaus Weisshart, Lars Schmiedeberg,
    and Peter Hemmerich

**19 Single Molecule Tracking for Studying Nucleocytoplasmic
    Transport and Intranuclear Dynamics** 343
    Jan Peter Siebrasse and Ulrich Kubitscheck

**20 Fluorescence Recovery After Photobleaching (FRAP)
    to Study Nuclear Protein Dynamics in Living Cells** 363
    Martin E. van Royen, Pascal Farla, Karin A. Mattern, Bart Geverts,
    Jan Trapman, and Adriaan B. Houtsmuller

Part VI Imaging Methods

**21 Nanosizing by Spatially Modulated Illumination (SMI)
    Microscopy and Applications to the Nucleus** 389
    Udo J. Birk, David Baddeley, and Christoph Cremer

**22 Visualisation of RNA by Electron Microscopic
    In Situ Hybridisation** 403
    Jacques Rouquette, Karl-Henning Kalland, and Stanislav Fakan

23 Electron Spectroscopic Imaging of the Nuclear Landscape 415
    Kashif Ahmed, Ren Li, and David P. Bazett-Jones

**24 Cryoelectron Microscopy of Vitreous Sections: A Step Further
    Towards the Native State** 425
    Cedric Bouchet-Marquis and Stanislav Fakan

Index 441

Contributors

Kashif Ahmed
Program in Genetics and Genome Biology, The Hospital
for Sick Children, Toronto, Ontario, Canada

Jean-Christophe Amé
Département Intégrité du Génome, UMR 7175-LCI du CNRS,
École Supérieure de Biotechnologie de Strasbourg, Illkirch, France

David Baddeley
Kirchhoff Institut für Physik, Universität Heidelberg, Heidelberg, Germany

Ralph Bash
Department of Chemistry and Biochemistry, Biodesign Institute,
and Department of Physics and Astronomy, Arizona State University,
Tempe, AZ, USA

David P. Bazett-Jones
Program in Genetics and Genome Biology, The Hospital
for Sick Children, Toronto, Ontario, Canada

Aaron Bensimon
Genomic Vision, Paris Biotech Santé, Paris, France

Ann L. Beyer
Department of Microbiology, University of Virginia Health System,
Charlottesville, VA, USA

Udo J. Birk
Kirchhoff Institut für Physik, Universität Heidelberg, Heidelberg, Germany

Petra Björk
Department of Molecular Biology and Functional Genomics,
Stockholm University, Stockholm, Sweden

Patrick T. C. van den Bogaard
Public Health Research Institute, New Jersey Medical School,
University of Medicine and Dentistry of New Jersey, Newark, NJ, USA

Cedric Bouchet-Marquis
Centre of Electron Microscopy, University of Lausanne, Lausanne, Switzerland

Richard D. Byrne
Cell Biophysics Laboratory, Cancer Research UK (CR-UK),
London Research Institute (LRI), London, UK

Aurelia Cassany
Department of Cell and Molecular Biology, The Scripps Research Institute,
La Jolla, CA, USA

Christoph Cremer
Kirchhoff Institut für Physik, Universität Heidelberg, Heidelberg, Germany

Françoise Dantzer
Département Intégrité du Génome, UMR 7175-LCI du CNRS,
École Supérieure de Biotechnologie de Strasbourg, Illkirch, France

Job Dekker
Program in Gene Function and Expression and Department of Biochemistry
and Molecular Pharmacology, University of Massachusetts Medical School,
Worcester, MA, USA

Duc P. Do
Department of Pharmaceutical Sciences, Mercer University
College of Pharmacy and Health Sciences, Atlanta, GA, USA

Stanislav Fakan
Centre of Electron Microscopy, University of Lausanne, Lausanne, Switzerland

Pascal Farla
Department of Pathology, Josephine Nefkens Institute,
Erasmus University Medical Centre, Rotterdam, The Netherlands

Elise Fouquerel
Département Intégrité du Génome, UMR 7175-LC1 du CNRS,
École Supérieure de Biotechnologie de Strasbourg, Illkirch, France

Sarah L. French
Department of Microbiology, University of Virginia Health System,
Charlottesville, VA, USA

Larry Gerace
Department of Cell and Molecular Biology, The Scripps Research Institute,
La Jolla, CA, USA

Bart Geverts
Department of Pathology, Josephine Nefkens Institute,
Erasmus University Medical Centre, Rotterdam, The Netherlands

Young-Mi Go
Department of Medicine, Emory University, Atlanta, GA, USA

George R. Green
Department of Pharmaceutical Sciences, Mercer University
College of Pharmacy and Health Sciences, Atlanta, GA, USA

Ronald Hancock
Laval University Cancer Research Centre, Hôtel-Dieu Hospital, Québec, Canada

Antionette Hakmé
Département Intégrité du Génome, UMR 7175-LCI du CNRS,
École Supérieure de Biotechnologie de Strasbourg, Illkirch, France

Peter Hemmerich
Leibniz Institute for Age Research, Fritz Lipmann Institute, Jena, Germany

Adriaan B. Houtsmuller
Department of Pathology, Josephine Nefkens Institute,
Erasmus University Medical Centre, Rotterdam, The Netherlands

Dean Jackson
Faculty of Life Sciences, MIB, University of Manchester, Manchester, UK

Dean P. Jones
Division of Pulmonary Medicine, Department of Medicine,
Emory University, Atlanta, GA, USA

Karl-Henning Kalland
Centre for Research in Virology, University of Bergen,
The Gade Institute, Bergen, Norway

Ulrich Kubitscheck
Institute for Physical and Theoretical Chemistry, Department
of Biophysical Chemistry, University of Bonn, Bonn, Germany

Jan Lammerding
Brigham and Women's Hospital/Harvard Medical School,
Department of Medicine, Cardiovascular Division, Cambridge, MA, USA

Banafshé Larijani
Cell Biophysics Laboratory, Cancer Research UK (CR-UK),
London Research Institute (LRI), London, UK

Richard T. Lee
Brigham and Women's Hospital/Harvard Medical School,
Department of Medicine, Cardiovascular Division, Cambridge, MA, USA

Ren Li
Program in Genetics and Genome Biology, The Hospital
for Sick Children, Toronto, Ontario, Canada

Stuart M. Lindsay
Department of Chemistry and Biochemistry, Biodesign Institute,
and Department of Physics and Astronomy, Arizona State University,
Tempe, AZ, USA

Dennis Lohr
Department of Chemistry and Biochemistry, Arizona State University,
Tempe, AZ, USA

Yvonne Ludwig
Institute of Physiology II, University of Münster, Münster, Germany

Karin A. Mattern
Department of Pathology, Josephine Nefkens Institute, Erasmus
University Medical Centre, Rotterdam, The Netherlands

Adriana Miele
Program in Gene Function and Expression and Department of Biochemistry
and Molecular Pharmacology, University of Massachusetts Medical School,
Worcester, MA, USA

Hans Oberleithner
Institute of Physiology II, University of Münster, Münster, Germany

Yvonne N. Osheim
Department of Microbiology, University of Virginia Health System,
Charlottesville, VA, USA

Ok-Kyong Park-Sarge
Department of Physiology, University of Kentucky, Lexington, KY, USA

Dominic L. Poccia
Department of Biology, Amherst College, Amherst, MA, USA

Jan Pohl
Microchemical and Proteomics Facility, Department of Medicine,
Emory University, Atlanta, GA, USA

Delphine Quenet
Département Intégrité du Génome, UMR 7175-LCI du CNRS,
École Supérieure de Biotechnologie de Strasbourg, Illkirch, France

Jacques Rouquette
Centre of Electron Microscopy, University of Lausanne, Lausanne, Switzerland

Amy C. Rowat
Department of Physics/Division of Engineering and Applied Science,
Harvard University, Engineering Sciences Laboratory, Cambridge, MA, USA

Martin E. van Royen
Department of Pathology, Josephine Nefkens Institute,
Erasmus University Medical Centre, Rotterdam, The Netherlands

Kevin D. Sarge
Department of Molecular and Cellular Biochemistry, University of Kentucky,
Lexington, KY, USA

Lars Schmiedeberg
Wellcome Trust Center for Cell Biology, Institute of Cell and
Molecular Biology, University of Edinburgh, Edinburgh, Scotland, UK

Valérie Schreiber
Département Intégrité du Génome, UMR 7175-LCI du CNRS,
École Supérieure de Biotechnologie de Strasbourg, Illkirch, France

Catherine Schurra
Institut Pasteur, Unité de Stabilité des Génomes, Paris, France

Victor Shahin
Institute of Physiology II, University of Münster, Münster, Germany

Jan Peter Siebrasse
Institute for Physical and Theoretical Chemistry, Department of
Biophysical Chemistry, University of Bonn, Bonn, Germany

Martha L. Sikes
Department of Microbiology, University of Virginia
Health System, Charlottesville, VA, USA

Jan Trapman
Department of Pathology, Josephine Nefkens Institute,
Erasmus University Medical Centre, Rotterdam, The Netherlands

Sanjay Tyagi
Public Health Research Institute, New Jersey Medical School,
University of Medicine and Dentistry of New Jersey, Newark, NJ, USA

Hongda Wang
Biodesign Institute and Department of Physics and Astronomy,
Arizona State University, Tempe, AZ, USA

Stefanie Weidtkamp-Peters
Leibniz Institute for Age Research, Fritz Lipmann Institute, Jena, Germany

Klaus Weisshart
Carl Zeiss Jena, Jena, Germany

Lars Wieslander
Department of Molecular Biology and Functional Genomics,
Stockholm University, Stockholm, Sweden

Natasha E. Zachara
The Department of Biological Chemistry, The Johns Hopkins
University School of Medicine, Baltimore, MD, USA

Vanessa Zhendre
Cell Biophysics Laboratory, Cancer Research UK (CR-UK),
London Research Institute (LRI), London, UK

List of Color Plates

The images listed below appear in the color insert.

Color Plate 1   *Fig. 1, Chapter 1*. Images of nuclei deformed by micropipette aspiration. *See* complete caption on p. 8.

Color Plate 2   *Fig. 1, Chapter 3*. Immunofluorescent labelling of splicing factors in an intact polytene nucleus and in an isolated polytene chromosome. *See* complete caption on p. 32.

Color Plate 3   *Fig. 2, Chapter 5*. Images of fluorescent replication signals visualised at ×40 magnification. *See* complete caption on p. 74.

Color Plate 4   *Fig. 4, Chapter 6*. Tracks of a few of the mRNP particles in a nucleus overlaid on an image of chromatin density. *See* complete caption on p. 101.

Color Plate 5   *Fig. 4, Chapter 8*. Examples of Recognition Imaging. *See* complete caption on p. 134.

Color Plate 6   *Fig. 5, Chapter 12*. Binding of membrane vesicle fractions MV1 and MV2beta to decondensed sea urchin sperm chromatin. *See* complete caption on p. 216.

Color Plate 7   *Fig. 2, Chapter 15*. Detection of PARP-1, PARP-2, and PARG by immunofluorescence microscopy. *See* complete caption on p. 276.

Color Plate 8   *Fig. 4, Chapter 15*. Detection of PAR by immunofluorescence microscopy. *See* complete caption on p. 280.

Color Plate 9   *Fig. 2, Chapter 18*. Subnuclear structures. *See* complete caption on p. 323.

Color Plate 10   *Fig. 4, Chapter 18*. FCS analysis procedure. *See* complete caption on p. 325.

Color Plate 11   *Fig. 8, Chapter 18*. Fitting an autocorrelation function to model equations. *See* complete caption on p. 335.

Color Plate 12   *Fig. 2, Chapter 19*. Import of single molecules of human NTF2/p10 (a 14-kDa protein implicated in NLS-mediated nuclear import) in digitonin-permeabilised HeLa cells stably expressing POM121-GFP, a GFP conjugate of the nucleoporin POM121. *See* complete caption on p. 353.

Color Plate 13   *Fig. 4, Chapter 21.* The results as saved by the command "save-fis_rep." *See* complete caption on p. 398.

Color Plate 14   *Fig. 1, Chapter 23.* Correlative and fluorescence micrographs of a 70-nm-thick section of an SK-N-SH neuroblastoma cell nucleus. *See* complete caption on p. 417.

Part I
Physical and Mechanical Properties

Chapter 1
Physical Properties of the Nucleus Studied by Micropipette Aspiration

Amy C. Rowat

Keywords Nuclear mechanics; Micropipette aspiration; Confocal imaged microdeformation

Abstract Understanding the physical properties of the cell nucleus is critical for developing a deeper understanding of nuclear structure and organization as well as how mechanical forces induce changes in gene expression. We use micropipette aspiration to induce large, local deformations in the nucleus, and microscopy to image nuclear shape as well as the response of fluorescently labeled components in the inner nucleus (chromatin and nucleoli) and the nuclear envelope (lamins and membranes). By monitoring the response of nuclear structures to these deformations, we gain insights into the material properties of the nucleus. Here we describe the experimental protocols for micropipette aspiration of nuclei in living cells as well as isolated nuclei. In addition to confocal imaging, deformed nuclei can be imaged by brightfield or epifluorescence microscopy.

1 Introduction

Many fundamental biochemical processes, including DNA replication and gene transcription, are compartmentalized within the cell nucleus. In addition to these biochemical functions, the nucleus is an organelle within the cell that is subject to mechanical forces transmitted by the surrounding cytoskeleton. Despite evidence that physical forces induce changes in gene expression (*1*), the mechanical properties of the cell nucleus are not well understood. Understanding the interplay between gene expression and physical environment is critical for a more complete knowledge of phenomena from differentiation to aging. The mechanical properties of the nucleus are also of interest in the context of disease: increasing evidence suggests that nuclear structure and stability are altered in laminopathies, disorders associated with mutations in genes encoding nuclear envelope proteins (reviewed in refs. (*2–4*)). The mechanism of these tissue-specific disorders is not completely understood, but one hypothesis proposes that nuclei with decreased mechanical stability become damaged in load-bearing tissues, thus giving rise to disease.

Table 1.1 Components of isolated nuclei and nuclei in living cells are labeled with fluorescent probes and visualized subsequent to micropipette deformation

Structure	Fluorescent probe	Method
Lamins	GFP-Lamin A	Transfection
Membranes	Membrane dye (DiIC18)	Incubation with isolated nuclei
Nuclear pore complexes	Anti-nucleoporin 62 + secondary antibody	Indirect immunofluorescence
DNA	SYTOX orange	Incubation with isolated nuclei
Nucleoli	CFP-NLS-NLS-NLS	Transfection

Transfection of genes encoding fluorophore-tagged proteins is an effective way to stain nuclei in living cells. Isolated nuclei can be additionally stained by indirect immunofluorescence and/or by incubation with fluorescent probes, including DNA dyes and membrane stains (Fig. 1.1)

CFP, cyan fluorescent protein

Mechanical properties of nuclei can be determined experimentally: techniques to investigate how the nucleus responds to mechanical forces involve deforming the cell or nucleus, and monitoring the resultant response *(4)*. Micropipette aspiration is one such technique that facilitates large, local deformations of cells and nuclei; characterizing the response of nuclei to such external forces yields insights into material properties of nuclei *(5–8)*.

We couple micropipette aspiration to brightfield as well as confocal microscopy to probe and characterize nuclear physical properties under deformation in three dimensions. To gain detailed information about the response of particular nuclear structures to mechanical perturbation, we label specific nuclear components in isolated nuclei and in living cells, deform nuclei by micropipette aspiration, and image the resulting deformations. These studies reveal that the nuclear envelope behaves as a solid–elastic shell *(7)*, that nuclear membranes are fluid in nature, and that loss of specific nuclear envelope proteins alters nuclear envelope elastic properties *(8)* (*see* Table 1.1).

2   Materials

2.1   Cell Culture and Transfection

1. Dulbecco's Modified Eagle Medium (DMEM) containing 10% fetal bovine serum (FBS), 1% glutamate, and 1% penicillin/streptomycin (BioWhittaker, Cambrex, East Rutherford, NJ, USA).
2. Carrier DNA: herring testes DNA (Clontech, San Jose, CA, USA).
3. Enhanced GFP conjugated to lamin A construct (eGFP-LamA) (a generous gift from D. Shumaker and R.D. Goldman) *(9)*.

4. Construct of CFP conjugated to three nuclear localization sequences (CFP-NLS-NLS-NLS) (Clontech).
5. $CaPO_4$ Transfection kit (Invitrogen, Carlsbad, CA, USA).

2.2 Preparation of Isolated Nuclei

All salts, creatine phosphokinase, and creatine phosphate are obtained from Sigma (St. Louis, MO, USA). Stocks are stored at 4°C unless otherwise specified. Note that $MgCl_2$ is very hygroscopic, so it is best to buy it in small vials and not to leave open for long periods of time (*see* **Note 1**).

1. Hypotonic buffer: 10 mM HEPES-KOH (pH ~7.9), 10 mM KCl, 1.5 mM $MgCl_2$, 0.5 mM DTT (add fresh from a frozen stock solution), protease inhibitor (complete tablets, EDTA-free; Roche, Penzburg, Germany) (1 tablet/50 mL buffer, shake to dissolve). Prepare fresh just prior to isolation.
2. Dounce tissue homogenizer (7 mL) (Wheaton Scientific, Millville, NJ, USA).
3. Refrigerated laboratory centrifuge.
4. Isolated nuclei are resuspended in a physiological buffer whose ionic composition mimics that of the cytoplasm [9] (*see* **Note 2**). Stock solutions are prepared: 1 M KCl, 1 M Na_2ATP, 1 M DTT, and 0.2 M Na_2HPO_4 and stored at 4°C (1 M KCl and 0.2 M Na_2HPO_4) or at −20°C (1 M Na_2ATP and 1 M DTT). An energy regeneration system is added of creatine phosphokinase (obtain in powder form, store at −20°C, and add just prior to resuspension) and creatine phosphate (1 M, store in 1-mL aliquots at −20°C). The final concentrations of components are described in **Section 3.2.4.**
5. SYTOX Orange (Molecular Probes, Eugene, OR, USA).
6. DiIC18 (1,1'-dioctadecyl-3,3,3',3'-tetramethylindocarbocyanine perchlorate) (Molecular Probes).
7. Anti-nucleoporin p62 IgG, 250 μg/mL (BD Transduction Laboratories, San Diego, USA).
8. Cy3-conjugated goat anti-mouse IgG, 0.75 mg/mL (Jackson ImmunoResearch Laboratories, West Grove, PA, USA).

2.3 Micropipette Aspiration

2.3.1 Preparing Pipettes

1. Glass capillaries, 1-mm diameter (World Precision Instruments, Sarasota, FL, USA).
2. Pipette puller (Sutter Instruments Co., Novato, CA, USA).
3. Microforge (MF-900, Narishige, Japan).

4. Silanization solution II (~2% dimethyldichlorosilane in 1,1,1-trichloroethane; Sigma-Aldrich, Denmark).
5. 1-mL plastic syringe fitted with a 97-mm long, 28-gauge backfiller (Microfil, World Precision Instruments, Sarasota, FL, USA).

2.3.2 Micropipette Setup

1. Micropipette setup:
   (a) A home-built micropipette setup consisting of a manometer, a pressure transducer, and micromanipulators or *x*-*y* translation stage *(11, 12)* or
   (b) An Eppendorf micropipette system, e.g., Injectman and Cell-Tram Air.

2. Microscope:
   (a) An inverted microscope (Zeiss Axiovert S100, Zeiss LD Achroplan ×40/0.65; Zeiss, Göttingen, Germany) or
   (b) A confocal microscope (Axiovert 200 M laser scanning confocal model 510 equipped with a META polychromatic multichannel detector and Apochromat 403/1.2 W corrected water immersion objective (Zeiss).

3. Charge-coupled device (CCD) camera (XC-85000E Donpisha; Sony, Japan) connected to a computer via a frame grabber (Sigma-SLC; Matrix Vision, Oppenweiler, Germany).

2.3.3 Aspirating Isolated Nuclei

1. Two-well chamber with a borosilicate coverslip bottom (0.17-mm thick) (Nalge Nunc, Slangerup, DK).

2.3.4 Aspirating Nuclei in Intact, Adherent Cells

1. CO_2-independent medium: Minimum Essential Medium without phenol red containing 10% FBS, 1% glutamate, 1% sodium pyruvate, 1% penicillin/streptomycin, and 1% 1 M HEPES (Invitrogen).
2. Objective heater (Bioptech, Butler, PA, USA).
3. Coverslips (20×0.17 mm) (LaCon, Staig, Germany).

3 Methods

3.1 Cell Culture and Transfection

Cells are grown at 37°C. Transfection is performed with a twofold excess of carrier DNA into HeLa/MEF cells at 70–80% confluency using $CaPO_4$ and cells are then plated on coverslips. This procedure is used to transfect HeLa and MEF cells with eGFP-Lamin A, as well as HeLa cells with CFP-NLS-NLS-NLS, and mutant lamin proteins (*see* **Note 3**).

3.2 *Preparation of Isolated Nuclei*

1. Steps 3–6 are carried out on ice, so an ice bucket must be prepared in advance. Both the hypotonic buffer and Dounce homogenizer are cooled on ice before use and the homogenizer is rinsed with hypotonic buffer.
2. Cells are grown to 90% confluency and are harvested by trypsinization, washed three times with PBS, and resuspended in 5 mL of hypotonic buffer. The resuspended cells are incubated on ice for 8–10 min or until the plasma membranes are sufficiently swollen, as observed using phase contrast microscopy. When swollen and looking like they are about to burst, the cells are transferred to the Dounce homogenizer.
3. After ten strokes of the tight pestle in a Dounce homogenizer, 2–3 μL of the nuclear suspension is placed on a microscope slide for inspection: nuclei should appear clean, without too many external membranes. If swollen cells are still visible, a few more strokes of Douncing is required to release the nuclei (*see* **Note 1**).
4. Nuclei are collected by centrifugation (~200×g, 4°C) for 5 min, forming a nuclei-enriched pellet. The supernatant is collected and stock solutions (*see* **Section 2.2.4**) are added to give final concentrations of 130 mM KCl, 1.5 mM MgCl$_2$, 10 mM Na$_2$HPO$_4$, 1 mM Na$_2$ATP, and 1 mM DTT, pH 7.4. Adjusting the concentration of the supernatant rather than using fresh buffer maintains all soluble factors that may be released during nuclear isolation. An energy regeneration system (0.1 mg/mL creatine phosphokinase and 5 mM creatine phosphate) is also added to the supernatant and the nuclei-enriched pellet is resuspended in 1 mL of this physiological buffer. A good nuclear isolation for one 10-cm and one 15-cm plate (*see* **Note 1**) typically yields 10^7 nuclei/mL.

3.2.1 Staining of Isolated Nuclei

1. Nuclear pore complexes (indirect immunofluorescence). A volume of 600 μL of isolated nuclei, resuspended in "physiological" buffer, are incubated with 1.5 μg of p62 antibody for 15 min. After centrifugation (100×g), the supernatant is removed and the nuclear pellet is resuspended in physiological buffer and incubated for 15 min with 4.5 μg of secondary antibody (Cy3-conjugated goat anti-mouse IgG). Subsequent centrifugation and resuspension in physiological buffer results in nuclei with labeled nuclear pore complexes and minimal background fluorescence.
2. Membranes. Membranes of isolated nuclei are labeled by incubation with DiIC18 at a final concentration of 0.01 μM in physiological buffer for 15 min. Nuclei that are not cleanly isolated and remain associated with external

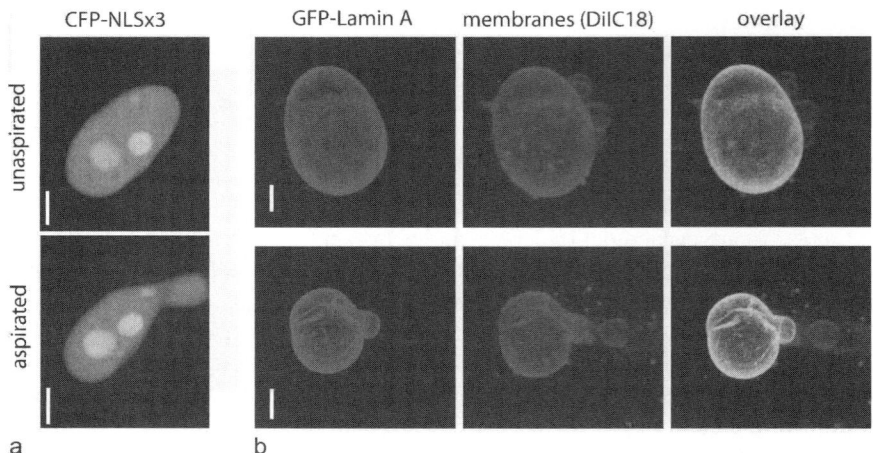

Fig. 1.1 Images of nuclei deformed by micropipette aspiration. Shown here are examples of nuclei both isolated from and in living cells with different stained components. **a** A nucleus in an intact HeLa cell transiently expressing CFP-NLSx3. This fluorophore preferentially localizes to nucleoli. **b** Nucleus isolated from a HeLa cell. The nuclear lamina is visualized in the nuclei of cells transiently expressing GFP-Lamin A. Nuclear membranes of isolated nuclei are stained with the lipophilic probe, DilC18. Reprinted with permission from [8]. To view this figure in color, see COLOR PLATE 1

      membranes from, for example, the endoplasmic reticulum, are easily identified because the membrane staining extends beyond the periphery of the nucleus itself; such nuclei should be excluded from analysis. *See* Fig. 1.1b for images of membrane-stained nuclei.

3. Nucleic Acids. SYTOX Orange is used to stain nucleic acids by incubation with isolated nuclei at a final concentration of 0.1 μM for 15 min prior to imaging (*see Note 4*).

3.3   Micropipette Aspiration

3.3.1   Preparing Pipettes

1. Pipettes are pulled from glass capillaries using a pipette puller and are forged to ensure a flat tip with an inner diameter of 1.5–5.0 μm.
2. After pulling and forging the capillaries, the micropipettes are dipped in silanization solution to prevent sticking of biological material to the surface and are then incubated in an oven at 90°C for 120 min. After cooling to room temperature they are inspected on the microforger for clogging that can result from the silanization treatment. The bottom coverslip of the chamber is also silanized.
3. Pipettes are backfilled with physiological buffer using the 1-mL syringe fitted with a backfiller.

3.3.2   Micropipette Setup

The microscope and micropipette setup are mounted on an air-cushioned table that dampens vibrations, ensuring the pipette remains stationary during image acquisition. The pipette is secured in place so that applying an aspiration pressure does not cause it to move. Micromanipulators are used to position the micropipette and pressures are applied to nuclei through the micropipette. Using a custom-built manometer system and a pressure transducer *(11, 12)*, the aspiration pressure is measured to range from 1 to 7 kPa. Note that pressure transducers are necessary for quantitative studies of deformation as a function of aspiration pressure.

3.3.3   Aspirating Isolated Nuclei

1. Isolated nuclei are visualized at room temperature (25°C) in a chamber that allows for micropipette access: the side of a two-well borosilicate chamber is carefully excised using a razor blade (Fig. 1.2a). A thermostated chamber may also be used to regulate temperature (Fig. 1.2b).
2. A large drop of buffer (~500 µL) is placed in one well of the two-well borosilicate chamber and the micropipette tip is placed inside the drop (Fig. 1.2a).
3. A volume of ~100 µL of resuspended nuclei are pipetted into the drop. Nuclei settle to the bottom of the chamber over several min.
4. The pipette is lowered to the bottom of the chamber into the same focal plane as the nuclei. With the micropipette fixed in position, the chamber is translated until the desired nucleus is located. The micropipette is manipulated so that the edge is as close to the nucleus as possible and an aspiration pressure is applied. Pressures on the order of 1 kPa induce initial deformation of isolated nuclei: the nucleus is drawn toward the mouth of the pipette by the negative pressure and into the pipette, forming a nuclear projection (tongue) inside the pipette. During pipette manipulation, dust and cell debris may block the orifice; if these cannot be removed (see **Note 5**) a fresh pipette must be mounted.
5. To avoid interactions between substrate and nucleus during measurement, the pipette is raised in the vertical direction. Deformed nuclei are imaged either by brightfield, epifluorescence, or confocal microscopy (*see* **Note 5**). Measurements are typically performed either by applying increasing aspiration pressure and analyzing the increase of the tongue length, or by analyzing the tongue length as a function of time at constant pressure.
6. After an aspiration measurement is complete, the nucleus is ejected from the pipette by applying a positive pressure. Gentle tapping of the pipette with a fine object such as a screwdriver or pen can help to dislodge nuclei from the pipette. Larger aggregates of nuclei and/or unwanted particles can be removed by withdrawing the pipette from the drop of buffer and reinserting it through the air-liquid interface. In the case of persistent clogging, it is necessary to load a fresh pipette. Buffer and nuclei should be replaced after no more than 15 min to avoid marked changes in buffer conditions (*see* **Note 6**).

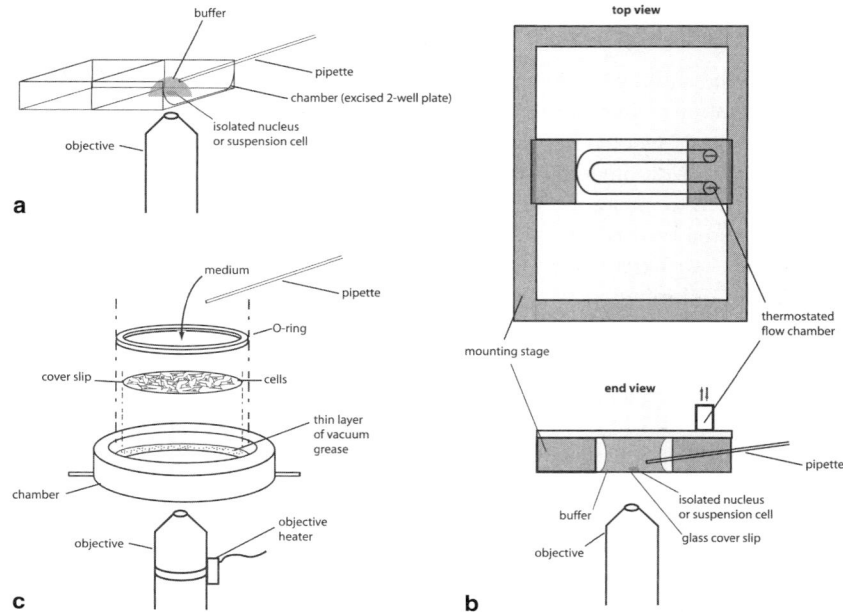

Fig. 1.2 Schematic illustrations of the micropipette aspiration setup. **a** A chamber for isolated nuclei or suspension cells is constructed by excising the side of a two-well chamber with a 0.17-mm-thick coverslip bottom. Nuclei are placed in a drop of buffer, and the pipette enters into the drop. **b** A thermostated chamber constructed of rust-free steel for isolated nuclei or suspended cells is constructed by affixing a glass coverslip to the bottom of the chamber with a small amount of vacuum grease. Buffer is inserted in between the coverslip and top of the chamber. Temperature is controlled by water flowing (*red arrows*) through channels on the upper side of the chamber. **c** Adherent cells plated on a coverslip are immersed in CO_2-independent medium. Nuclei are deformed and imaged at 37°C

3.3.4   Aspirating Nuclei in Intact, Adherent Cells

1. Prior to imaging, the coverslip on which the cells are cultured is mounted in the custom-designed chamber (Fig. 1.2c). A thin layer of vacuum grease is applied to the circumference of the chamber before affixing the coverslip. A rubber O-ring constitutes the third layer and seals the chamber, facilitating the immersion of the cells in CO_2-independent medium. The objective heater is set to 38.5°C, allowing for imaging the cells at 37°C.

2. The pipette is placed in the center of the field of view (aligned with the center of the light beam) and is carefully lowered into the same focal plane as the cells. With the micropipette fixed in position slightly above the cells, the stage is moved in the *x-y* plane until a desirable nucleus is located.

3. The pipette is brought to the cell surface adjacent to the nucleus and aspiration pressure is applied, causing the nucleus to deform into the micropipette. Nuclei are imaged either by brightfield, epifluorescence, or confocal microscopy (*see* **Note 5**).

4. After image acquisition and measurement of a given nucleus, the nucleus can be expelled from the pipette by applying a positive pressure while moving the pipette up and away from the cell. The same pipette can be used to aspirate more than one nucleus; however, the coverslip of cells should be replaced every 15 min or so to avoid excessive buffer evaporation (*see* **Note 6**).

3.4 *Image Analysis*

To interpret images of deformed nuclei, quantitative image analysis methods must be developed. Such analyses can be performed using image-processing software such as ImageJ (National Institutes of Health, Bethesda, MD, USA) or Zeiss LSM 510 v.3.0 software. For example, to characterize the mechanical properties of the nuclear envelope and inner nucleus, we have analyzed both changes in nuclear shape and subnuclear body position upon deformation, as well as fluorescence intensity distribution of labeled nuclear components (*see* **Note 4**) *(7, 8)*. Labeling nuclear structures with different fluorophores and imaging the deformations of nuclear components provides a powerful method to probe the physical properties of nuclei both isolated, and in intact, living cells. These methods provide a set of basic tools to further our understanding of the physical nature of the cell nucleus, and how the nucleus responds to its mechanical environment.

4   Notes

Here we add further notes on problems encountered during micropipette aspiration of nuclei.
1. Nuclear isolation. Isolation by Dounce homogenization works best with a large number of nuclei. To isolate a sufficient quantity of nuclei while conserving the amount of DNA used for transfection, only one 10-cm plate is transfected; cells from this small plate harvested together with cells from a 15-cm plate are pooled for nuclear isolation. The success of nuclear isolation and the "cleanliness" of isolated nuclei varies: when not successful, extranuclear membranes remain associated with the nucleus. Unsuccessful nuclear isolation may also result from altered magnesium concentration of the hypotonic buffer.
2. Buffer conditions for isolated nuclei. The physical properties and morphology of isolated nuclei are extremely sensitive to buffer conditions (unpublished observations) *(5)*, suggesting that salts induce changes in the structure (e.g., chromatin packing) *(13)* and mechanical properties of the inner nucleus; buffers should thus be chosen with care.
3. Extension to other systems. This protocol can be adapted for other cell types including mouse embryo fibroblasts as well as isolated plant cell nuclei.

4. Quantitative imaging. SYTOX orange stains nucleic acids very effectively, but is not appropriate for quantitative intensity analysis of DNA concentration. For quantitative analysis of chromatin density, PicoGreen (Molecular Probes) would be a more appropriate choice of dye.
5. Extension to confocal microscopy. Accommodating the confocal microscope setup for micropipette aspiration requires the design of alternative "open" chambers that allow for the entry of the micropipette as nearly horizontal as possible.
6. Timescale. Due to buffer evaporation from the open chamber, it is best to limit the experimental timescale to 10–15 min. Fresh buffer can be added during the experiment, but will alter buffer conditions.

Acknowledgments Many thanks to J.H. Ipsen for critical discussions. Thanks also to D.K. Shumaker and R.D. Goldman, Northwestern University, Chicago, USA for generously providing the GFP-Lam A construct as well as to J.S. Andersen, Y.W. Lam, and J. Lammerding for helpful advice. This work was supported by the Danish National Research Foundation and a NSERC Julie Payette Scholarship (ACR). ACR is a Human Frontiers Science Program Cross-disciplinary fellow.

References

1. Garcia-Cardena, G., Comander, J., Anderson. K.R., Blackman, B.R., and Gimbrone, M.A. Jr. (2001) Biomechanical activation of vascular endothelium as a determinant of its functional phenotype. *Proc. Natl. Acad. Sci. USA* **98**, 4478–4485.
2. Mattout, A., Dechat, T., Adam, S.A., Goldman, R.D., and Gruenbaum, Y. (2006) Nuclear lamins, diseases and aging. *Curr. Opin. Cell Biol.* **18**, 335–341.
3. Broers, J.L, Ramaekers, F.C., Bonne, G., Yaou, R.B., and Hutchison, C.J. (2006) Nuclear lamins: laminopathies and their role in premature ageing. *Physiol. Rev.* **86**, 967–1008.
4. Rowat, A.C., Lammerding, J., Herrmann, H., and Aebi, U. (2008) Towards an integrated understanding of the structure and mechanics of the cell nucleus. *Bioessays* **30**, 226–236
5. Dahl, K.N., Engler, A.J., Pajerowski, J.D., and Discher, D.E. (2005) Power-law rheology of isolated nuclei with deformation mapping of nuclear substructures. *Biophys. J.* **89**, 2855–2864.
6. Guilak, F., Tedrow, J.R., and Burgkart, R. (2000) Viscoelastic properties of the cell nucleus. *Biochem. Biophys. Res. Commun.* **269**, 781–786.
7. Rowat, A.C., Foster, L.J., Nielsen, M.M., Weiss, M., and Ipsen, J.H. (2005) Characterization of the elastic properties of the nuclear envelope. *J. R. Soc. Interface* **2**, 63–69.
8. Rowat, A.C., Lammerding, J., and Ipsen, J.H. (2006) Mechanical properties of the cell nucleus and the effect of emerin deficiency. *Biophys. J.* **91**, 1–16.
9. Moir, R.D., Yoon, M., Khuon, S., and Goldman, RMD. (2000) Nuclear lamins A and B1: different pathways of assembly during nuclear envelope formation in living cells. *J. Cell Biol.* **151**, 1155–1168.
10. Jackson, D.A., Yuan, J., and Cook, P.R. (1988) A gentle method for preparing cyto- and nucleo-skeletons and associated chromatin. *J. Cell Sci.* **90**, 365–378.
11. Henriksen, J.R. and Ipsen, J.H. (2004) Measurement of membrane elasticity by micro-pipette aspiration. *Eur. Phys. J. E Soft Matter* **14**, 149–167.
12. Evans, E. and Needham, D. (1987) Physical properties of surfactant bilayer membranes: thermal transitions, elasticity, rigidity, cohesion, and colloidal interactions. *J. Phys. Chem.* **91**, 4219–4228.
13. Bojanowski, K. and Ingber, D.E. (1998) Ionic control of chromosome architecture in living and permeabilized cells. *Exp. Cell Res.* **244**, 286–294.

Chapter 2
Mechanical Properties of Interphase Nuclei Probed by Cellular Strain Application

Jan Lammerding and Richard T. Lee

Keywords Nucleus; Nuclear stiffness; Lamin; Muscular dystrophy; Cell mechanics; Strain

Abstract The mechanical properties of the interphase nucleus have important implications for cellular function and can reflect changes in nuclear envelope structure and/or chromatin organization. Mutations in the nuclear envelope proteins lamin A and C cause several human diseases, such as Emery–Dreifuss muscular dystrophy, and dramatic changes in nuclear stiffness have been reported in cells from lamin A/C–deficient mice. We have developed a cellular strain technique to measure nuclear stiffness in intact, adherent cells and have applied this experimental method to fibroblasts from mouse models of Emery–Dreifuss muscular dystrophy and to skin fibroblasts from laminopathy patients and healthy control subjects. The experimental protocol is based on measuring induced nuclear deformations in cells plated on a flexible silicone substrate; the nuclear stiffness can subsequently be inferred from the ratio of induced nuclear strain to the applied membrane strain. These experiments reveal that lamins A and C are important determinants of nuclear stiffness and that lamin mutations associated with muscular dystrophies and other laminopathies often result in disturbed nuclear stiffness that could contribute to the tissue-specific disease phenotypes.

1   Introduction

Mutations in the nuclear envelope proteins lamin A and C cause a variety of human diseases that include Emery-Dreifuss muscular dystrophy and limb-girdle muscular dystrophy (reviewed in refs. *(1, 2)*). Although the mechanisms leading to these tissue-specific phenotypes remain unclear, it is thought that decreased nuclear stability could contribute to the muscular phenotypes in these diseases *(1, 3)*. Because the nucleus is surrounded by the cytoskeleton, direct measurements of its physical properties are inherently difficult. Micropipette aspiration of single nuclei allows direct measurements of nuclear mechanics (see Chap. 1 by Rowat in this volume), but is limited by the risks that the nuclear isolation procedure might damage nuclei (especially in mutant cells) and/or alter nuclear or chromatin structure depending on the buffer conditions. Here, we describe an alternative method in which cells plated on transparent silicone membranes are subjected to uniform strain. The applied substrate strain is transmitted to the cytoskeleton through focal adhesion complexes, resulting in cytoskeletal strain that closely matches the applied membrane strain (*4*), while the stiffer nucleus deforms less. Applied membrane strain and induced nuclear strain are quantified based on phase contrast and fluorescence images of fluorescently labeled nuclei taken before, during, and after strain application. An advantage of this technique is that it probes nuclear mechanics in the normal cellular environment without having to isolate the nucleus, thus preserving the normal nuclear and cytoskeletal architecture; furthermore, the techniques resemble physiological load application as found in many tissues such as muscle or blood vessel walls. However, the experiments only provide information on the nuclear stiffness relative to the surrounding cytoskeleton, and the induced nuclear deformation depends on cell adhesion to a substrate and intracellular force transmission through the cytoskeleton. We have adapted this technique to measure the contribution of specific nuclear envelope proteins to nuclear stiffness and fragility, and are currently studying the effect of specific lamin mutations (*3, 5, 6*).

2   Materials

2.1   Preparation of Silicone Membrane Dishes

1. Components for nine strain dishes, each consisting of a bottomless dish and a plastic O-ring to hold the silicone membrane (*see* **Note 1** and Fig. 2.1). The strain dishes are autoclavable for sterile cell culture conditions. The cell culture dishes are equipped with threads and a wide collar on the outside to fit securely into the dish-holder plate of the strain device in a predetermined position.

2. Silicone membrane: 0.005-inch-thick silicone sheeting in 12×12-inch sheets (Gloss/Gloss non-reinforced silicone sheeting; Specialty Manufacturing Inc., Saginaw, MI, USA).

3. Scalpel or razor blade and 12-inch ruler to cut silicone sheets.

4. Human plasma fibronectin (Invitrogen, Carlsbad, CA, USA) is dissolved in sterile water at 1 mg/mL and stored in aliquots of 500 μL at −20°C. For frequent use, keep one aliquot at 4°C and use it within ~4 weeks.

5. Phosphate-buffered saline solution (PBS) without calcium and magnesium. Final 1× buffer is prepared from 10× concentrate (Invitrogen) adjusted to pH 7.1, autoclaved, and stored at 4°C.

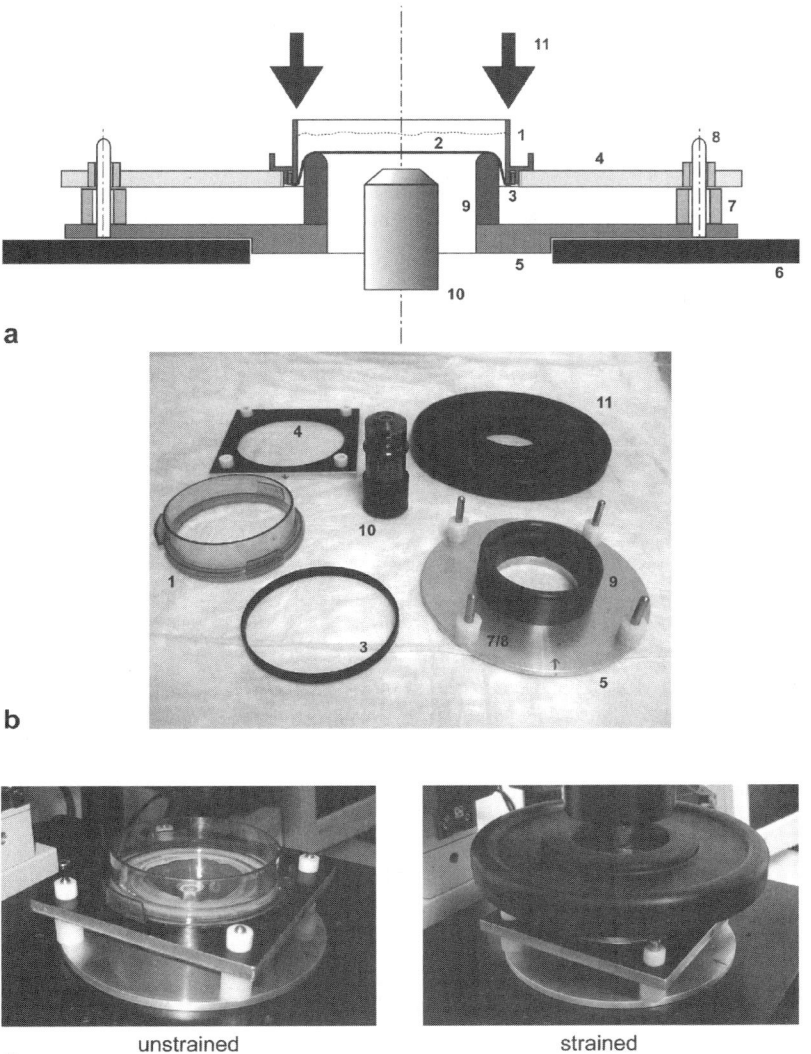

Fig. 2.1 Overview of the strain device. **a** Schematic cross-section. The strain dish (*1*) contains cells plated on a transparent silicone membrane (*2*) held in place by a plastic O-ring (*3*) wedged into a groove of the strain dish. The strain dish is mounted onto the dish-holder plate (*4*) and carefully placed on the base plate (*5*) that fits firmly onto the microscope stage (*6*). Nylon spacers (*7*) limit the vertical displacement of the dish-holder plate and thus the applied membrane strain. Polytetrafluoroethylene bearings in the dish-holder plate and vertical pins (*8*) on the base plate provide alignment and stabilization of the movable plate relative to the stationary platen (*9*). A central bore in the base plate enables access of the microscope objective (*10*) mounted on an extension tube. Pushing down the dish/dish-holder plate with a weight plate (*11*) results in uniform membrane strain. **b** Select components of the strain device, using the same labels as above. **c** Photos of strain device in the resting condition (*left*) and during strain application (*right*). The weight to push down the dish-holder plate is visible on top of the strain dish

2.2   Cell Culture and Plating Cells

1. Appropriate cell culture medium for the desired cell type, e.g., Dulbecco's Modified Eagle Medium (DMEM; Invitrogen) supplemented with 10% fetal calf serum and antibiotics for mouse embryo fibroblasts.
2. Hanks buffered saline solution (HBSS) without calcium and magnesium. Final 1× buffer is prepared from 10× concentrate (Invitrogen), adjusted to pH 7.1, sterile filtered or autoclaved, and stored at 4°C. HBSS+ with calcium and magnesium is also required in **Section 3.3.1.**
3. Solution of 0.05% (w/v) trypsin and ethylenediamine tetraacetic acid (Trypsin-EDTA; Invitrogen).
4. Phosphate-buffered saline solution (PBS) without calcium and magnesium: see **Section 2.1.5.**
5. 10-cm style untreated polystyrene dishes (Corning, Acton, MA, USA). These dishes are used to carry the silicone membrane dishes and to keep their contents sterile when taken outside the hood.

2.3   Strain Experiments

1. Inverted microscope (e.g., IX-70, Olympus, Center Valley, PA, USA) with a digital camera suited for fluorescence microscopy (e.g., CoolSNAP HQ, Roper Scientific, Tucson, AZ, USA) and appropriate image acquisition software (e.g., IPLab, ImagePro, etc.).
2. High-power objective (at least ×40 magnification) appropriate for the inverted microscope used in the experiments. We use a ×60 phase contrast objective (Olympus LCPlanFl ×60, 0.70NA Ph2).
3. Microscope objective extension tube (ThorLabs, Newton, NJ, USA) to compensate for the raised sample plane of the strain device (~20 mm above the normal sample position).
4. Strain device consisting of microscope platform, spacers, and dish-holder plate.
5. 2.5-pound rubberized weight plate with 2-inch central bore (Olympia Sports, Westbrook, ME, USA).
6. Chemically inert, silicone-impermeant grease (Braycote 804; Castrol, Irvine, CA, USA).
7. HBSS without calcium and magnesium (*see* **Section 2.2.2**).
8. Cell-permeable Hoechst 33342 nucleic acid stain (Invitrogen/Molecular Probes): 10 mg/mL solution in water, stored at 4°C.

2.4   Analysis

1. Image analysis software. We use MATLAB (Mathworks, Natick, MA, USA) to analyze nuclear and membrane strain based on TIFF images taken during the experiments, but other image analysis software should be equally suitable.

2. Spreadsheet program. We use Microsoft Excel for the final data analysis to exclude contracted and damaged cells and for the statistical analysis of the results.

3 Methods

The experiments are carried out on a custom-made strain system mounted on an inverted epifluorescence microscope. Cells are plated on custom-made plastic strain dishes with a silicone membrane serving as the cell culture substrate that are described in Fig. 2.1 and elsewhere *(7)*. The strain dish is then mounted on a strain device consisting of a base plate that fits onto the microscope, a movable dish holder plate that can slide up and down on four guidance pins, and a separate "weight plate" to apply a load (see 11 in Fig. 2.1). The base is made of an aluminum plate that securely fits onto the microscope stage. A low-friction poly-tetrafluoroethylene-impregnated Delrin platen is located in the center of the base and serves to apply biaxial strain to a silicone membrane when the membrane is pressed down over the platen. The platen has a large central bore to accommodate the microscope objective mounted on an objective extension tube. Vertical steel pins positioned at each corner of the plate are used to align the dish-holder plate with the base plate. The dish-holder plate contains four polytetrafluoroethylene bearings, one in each corner, to allow precise alignment with the vertical guidance pins from the base plate and to keep the dish/plate assembly in a horizontal position during strain application. In the resting position, the strain dish rests with the silicone membrane on the central platen. To apply biaxial strain, weights are used to press down the strain dish mounted on the dish-holder plate, resulting in a homogeneous and uniform biaxial strain field in the central section of the silicone membrane. Friction between the platen and the membrane is minimized by application of chemically inert, silicone-impermeant grease. The maximal membrane strain is limited by nylon spacers placed on the vertical alignment pins, effectively limiting the vertical displacement of the dish.

Images of cells and fluorescently labeled nuclei are acquired before, during, and after strain application and subsequently analyzed to compute the normalized nuclear strain, defined as the ratio of induced nuclear strain to applied membrane strain. The analysis is carried out using custom-written MATLAB software and is based on calculating the scaling factor of the linear affine image transformation (i.e., image transformation that allows for translation, rotation, and scaling) that best fits manually selected control points from the pre-strain image to the corresponding positions in the full-strain image *(3)*. Measuring nuclear strain based on the displacements of distinct markers is more reliable than measurements that rely on changes in feature length or cross-sectional area, because the apparent dimensions can vary with the fluorescence intensity, whereas the centroid position of small distinct features is not affected as long as the corresponding features can be identified.

3.1 Preparation of Silicone Membrane Dishes

3.1.1 Assembly of Strain Dishes

1. Thoroughly clean the plastic components of the strain dishes with 70% alcohol and subsequently soak in hot water for at least 30 min; rinse with deionized water and then air-dry. Prepare nine sets of dishes for each 12×12-inch silicone membrane (*see* **Note 2** on how to reuse dishes).
2. Cut out nine 4×4-inch pieces from a single silicone sheet using a ruler and a razor blade.
3. Carefully place a 4×4-inch silicone membrane piece on top of one of the plastic rings (rounded side facing up), making sure not to have any creases or folds in the membrane.
4. Concentrically place the plastic strain dish on top of the ring and push down over the membrane and plastic ring, thereby fixing the silicone membrane in the groove between the ring and the strain dish.
5. Cut the access membrane with scissors (*see* **Note 3**).
6. Rinse the assembled strain dish under hot water, followed by a quick rinse with deionized water.
7. Place the assembled strain dishes in an autoclave bag and autoclave with a dry cycle for 30 min.

3.1.2 Fibronectin-Coating Silicone Membranes

1. Remove the sterilized strain dishes from the autoclave bag inside a sterile cell culture hood. For better image quality during the strain experiments, wipe down the outside of the silicone membrane with a 70% alcohol sterilization pad. Place the strain dish inside an inverted, 10-cm cell culture dish, which allows easy movement of the dishes outside the cell culture hood while maintaining a sterile interior.
2. Mark a small dot on the center of the bottom of the silicone membrane (outside) with a fine-tip marker. This ink spot will serve as a reference frame during the strain experiments and will help in locating the cells throughout the experiments as the membrane moves during the strain application.
3. Prepare fibronectin solution (*see* **Note 4**). Fill a 50-mL tube with PBS for 1–4 dishes (each dish requires 11 mL of PBS), then add the appropriate amount of fibronectin stock solution for a final concentration of 2 μg/mL. Briefly invert or vortex to mix, then transfer 11 mL of fibronectin solution into each dish.
4. Incubate the dishes with fibronectin solution overnight at 4°C.
5. Before plating cells in strain dishes, aspirate off the fibronectin solution, rinse dishes once with PBS (~5 mL per dish), then fill each dish with ~10 mL of growth media.

3.2 Plating Cells

1. Once the cells cultured in a T25 flask reach confluence, rinse them once with HBSS and detach them with trypsin/EDTA (*see* **Note 5**).
2. Inactivate trypsin by addition of growth medium with FCS (use 1–2× the amount of trypsin), collect in a 50-mL tube, and spin down the cells for 5 min at 380×*g*.
3. Resuspend the cell pellet in 3–5 mL of growth media and transfer an appropriate amount of cell suspension into a strain dish containing 10 mL of growth media, then gently pipette up and down to mix (*see* **Note 6**).
4. Transfer the dishes into the cell culture incubator and incubate for 48 h before use for nuclear strain experiments (*see* **Note 7**).

3.3 Strain Experiments

3.3.1 Strain Experiments

1. Mount the high-power objective on the objective extension tube on the inverted fluorescence microscope; bring the objective into the lowest possible position to avoid contact with the silicone membrane when mounting the strain dish.
2. Place the base of the strain device on the microscope stage.
3. Apply the desired amount of spacers on the strain device.
4. Set up the CCD camera, image acquisition software, and microscope.
5. Incubate the cells in the strain dish with growth media containing 1 μg/mL of Hoechst 33342 nuclear stain for 15 min at 37°C.
6. Aspirate off the medium and replace with 20 mL of HBSS+ (*see* **Note 8**).
7. Carefully apply grease to the perimeter of the bottom of the silicone membrane, making sure to leave the center of the membrane free to allow for unobstructed imaging. In addition, apply a small amount of grease to the strain device platen.
8. Tightly screw the strain dish into the dish holder plate and carefully mount the dish-holder plate with the strain dish onto the strain device on the microscope.
9. Adjust the microscope focus and stage position to locate the dot marked in the center of the membrane through the eyepiece; the dot will serve as the starting point for all image acquisitions and aids in locating the same cells as the membrane initially moves during strain application.
10. Starting from the dot, move the microscope stage to locate a field of view containing well-spread cell(s) with centrally located nuclei.
11. Acquire a phase contrast image and a fluorescence image (blue channel) of the first section (*see* **Note 9**).
12. Move the microscope stage to a new field with well-spread cells (*see* **Note 10**). Acquire phase contrast and fluorescence images of this field of view and repeat the procedure for three to six additional sections. These images are saved as the pre-strain images. The pre-strain image acquisition process (and all subsequent

stages) should be completed in less then 10 min to avoid remodeling of the cells when strain is applied later on.

13. Move the stage back to locate the starting point (ink dot). If using a water or oil-immersion objective, retract the objective to avoid pushing the membrane over the objective. Slowly place the weight plate on the top of the dish-holder plate until it rests firmly on the spacers, resulting in reproducible membrane strain.

14. Adjust the microscope focus and locate the ink dot on the silicone membrane, then find the initial starting section based on the phase contrast images taken at the pre-strain stage.

15. Acquire phase contrast and fluorescence images of the strained cells as described above, carefully adjusting the microscope stage and focus to precisely match the nuclear cross-section imaged at the pre-strain stage. Once again, this process should not exceed 10 min to avoid active remodeling and adaptation of the cell to the strained substrate.

16. After all the corresponding images have been acquired, move the microscope stage back to the starting point. Carefully remove the weight from the dish-holder plate to allow the silicone membrane to relax (*see* **Note 11**).

17. Adjust the microscope focus and locate the ink dot again. Subsequently, acquire phase contrast and fluorescence images of the post-strain cells, carefully adjusting the microscope stage and focus to match the images taken at pre-strain and full-strain.

18. Carefully remove the dish from the strain device and repeat the entire procedure with additional dishes if necessary. Add bleach to the strain dishes to disinfect them and *see* **Notes** for **Section 3.1.1** for reusing the strain dishes for new experiments.

3.4 Analysis of Results

3.4.1 Membrane Strain Measurements

1. Select pairs of control points in the corresponding image sections taken before and during strain application, as shown in Fig. 2.2. For accurate measurements of membrane strain, select three to eight image features that are located directly on the silicone membrane and that are distributed throughout the field of view (*see* **Note 12**).

Compute the linear conformal spatial transformation matrix based on the pairs of control points selected above (*see* **Note 13**). Each control point p_i in the pre-strain image can be represented in its spatial coordinates (x_i, y_i) and written as one row of the control point matrix u as:

$$u = \begin{pmatrix} x_1 & y_1 & 1 \\ x_2 & y_2 & 1 \\ \dots & \dots & \dots \end{pmatrix}$$

(Eq. 2.1)

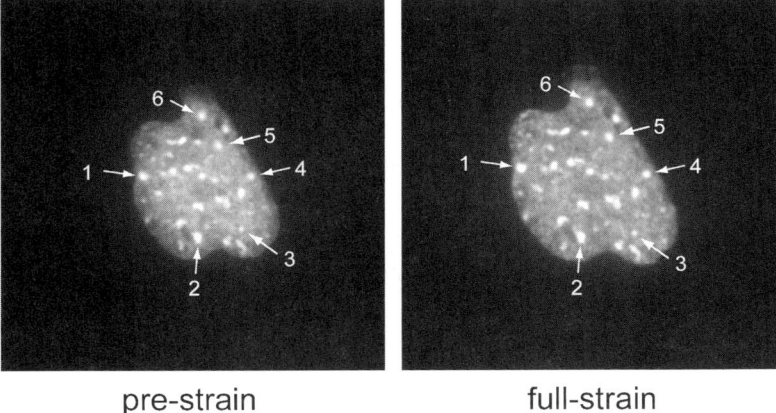

pre-strain   full-strain

Fig. 2.2 Example of control points selected for nuclear strain analysis. Shown is a pair of images of a lamin A/C-deficient mouse embryo fibroblast nucleus with fluorescently labeled chromatin before (*left*) and during (*right*) strain application. The nucleus is visibly enlarged as a result of the strain application. Distinct spots of high chromatin density are apparent in the fluorescence images, and six of these speckles are used as control points for the strain calculations (*#1–6*). These pairs of control points of matching image features are subsequently used to compute the best-fit affine linear conformal transformation matrix

The corresponding positions of control points in the matching image taken during strain application (full-strain image), represented by the matrix u', can be computed by applying the transformation matrix T to the control point matrix u of the pre-strain image:

$$u' = uT, \tag{Eq. 2.2}$$

where T is the spatial transformation matrix of the form:

$$T = \begin{pmatrix} a_{11} & a_{12} & 0 \\ a_{21} & a_{22} & 0 \\ a_{31} & a_{32} & 1 \end{pmatrix}. \tag{Eq. 2.3}$$

The angle of rotation α and the scaling factor s can then be determined from the least-square fit solution of Eq. 2.2 to the control point matrix u^* of the actual coordinates of the control points in the full-strain image, while taking advantage of the restriction that $a_{11} = a_{22}$ and $a_{12} = -a_{21}$ (*see* **Note 14**):

$$\tan \alpha = \frac{a_{12}}{a_{11}}$$

$$s = \frac{a_{11}}{\cos \alpha}. \tag{Eq. 2.4}$$

In the case of biaxial strain application, rotations are generally negligible, i.e., $\alpha \to 0$ and thus $\cos\alpha \approx 1$, reducing the expression for the scaling factor s to:

$$s = a_{11} \qquad \qquad \text{(Eq. 2.5)}$$

See **Note 14** for details. The scaling factor s is related to the membrane stain in percent strain by:

$$Strain \text{ [percent]} = (s-1) \times 100 \qquad \qquad \text{(Eq. 2.6)}$$

3. To validate the solution to the transformation matrix, apply the transformation matrix to the pre-strain image (see Eq. 2.2) and overlay the resulting image with the actual full-strain image. This procedure will also reveal any nonuniform deformations and partial cell detachment.
4. Subsequently, repeat this procedure for each pair of image sections taken during the strain application. Also use the procedure for the combination of pre-strain and post-strain images to make sure that the membrane has returned to baseline after the experiments.
5. Compute the induced nuclear strain for each cell following the same procedure, using bright chromatin spots within a single nucleus in the pre-strain and full-strain images as control points (*see* **Note 12**).
6. For each nucleus, determine the residual strain after strain application by selecting control points based on images taken before and after the strain application. The residual strain yields important information on potential cellular damage during the strain application.
7. Compute the normalized nuclear strain for each nucleus as the ratio of induced nuclear strain to applied membrane strain. Normalized nuclear strain data are pooled from at least three independent experiments (each containing measurements on ~5–10 nuclei) and compared with other cell or treatment groups by statistical analysis (*see* **Note 15**). Increased normalized nuclear strain indicates reduced nuclear stiffness.
8. Because the normalized nuclear strain is a function of the ratio of nuclear stiffness to cytoskeletal stiffness, it is recommended that cytoskeletal structure and function be compared between different experimental groups by, for example, magnetic bead microrheology *(3, 8)* or at least by immunofluorescence labeling of cytoskeletal components.

4   Notes

1. We use a custom-made strain device in our experiments. A similar setup can be reproduced based on the design described here. Similar systems might also be available commercially, but one should make sure that they apply uniform strain across the image field. Our particular strain device can apply up to 35% mem-

brane strain, but we routinely perform nuclear strain experiments with membrane strains of only 5%. Larger strain magnitudes improve the precision of the nuclear strain measurements because they generally result in larger induced nuclear deformation, but excessive membrane strain can cause cell detachment and structural damage. Because cellular sensitivity to strain varies with cell type, we recommend performing initial experiments with increasing strain magnitudes while carefully monitoring cell appearance during and after the strain application.

2. Assembled strain dishes can be reused after the experiments. Disinfect the interior of the cell dish with 10% bleach, then rinse repeatedly under hot water. Fill the inside with deionized water and incubate at room temperature for at least 24 h, occasionally changing the water. Subsequently, rinse the inside of the dishes again with hot water, and remove the grease on the outside (bottom side) of the strain dish with Kim-wipes and 70% alcohol, because residual grease might impair the optical transparency in subsequent experiments. Also, clean the inside of the silicone membrane with 70% ethanol to remove residual cell and extracellular matrix debris. Finally, rinse the dish with deionized water and autoclave with a dry cycle for 30 min.

3. Excess membrane material can interfere with screwing the strain dish into the strain device. Cutting away excess membrane works best by keeping one of the scissor blades in contact with the strain dish during the cutting process and slowly rotating the dish around.

4. Depending on the cell type used, other extracellular matrix proteins such as laminin, collagen, or gelatin can be substituted. It is recommended that a suitable concentration be established for each protein and cell type combination when beginning work with new cell types.

5. The experimental technique is limited to adherent cells that spread well on the appropriately coated substrate without detaching during strain application. Nonadherent or weakly adherent cells are not suitable for these experiments. A confluent T25 flask of human or mouse fibroblasts is generally sufficient for one to three strain dishes.

6. The number of cells per dish will depend on the specific cell type. In general, it is best to plate cells at an intermediate cell density, because experiments should be performed in subconfluent (~30–60% confluence) cell layers to avoid cell-to-cell interactions, while too low cell densities can negatively affect cell viability. For human and mouse fibroblasts, we found that ~1,000–2,000 cells/cm^2 is a good starting cell density. The optimal cell density is best determined by initially plating cells at a range of densities. Once the optimal density has been determined, it is often not necessary to count cells for each experiment but instead to use the appropriate fraction of the cell suspension, e.g., 0.5 mL of 4-mL cell suspension harvested from a confluent T25 flask.

7. Cells should become sufficiently attached for experiments after overnight incubation and can be used once they reach the appropriate cell density. For fast-growing cells, we recommend incubating cells for 48 h with serum-free media supplemented with insulin, transferrin, and selenium to minimize the number of

mitotic cells because the experiment is designed to measure physical properties of interphase nuclei.

8. Generally, experiments are performed at room temperature, and the HBSS should be equilibrated to room temperature before use. However, temperature-sensitive cells might require working at 37°C, which can be achieved by using a microscope stage incubation chamber or a perfusion system. In this case, HBSS should be prewarmed to 37°C.

9. In the phase contrast images it is best to focus on the membrane and the cell outline, because this can be subsequently used to compute the membrane strain and to detect potential cell detachment; in the fluorescence images of the nucleus, focus on the center cross-section of the nucleus; optionally, acquire 3D image stacks of the entire nucleus, but this can lead to photobleaching. Optionally, acquire images in other color channels if additional fluorescent markers are used.

10. It is easiest to locate the imaged cells again by moving in only one direction from the initial starting point, or, alternatively, moving along the perimeter of the ink dot. In order to avoid excessive photobleaching, avoid overlap of subsequently imaged sections.

11. In order to make sure that the silicone membrane completely returns to the pre-strain state, it can help to gently lift up the dish-holder plate a little and then gently lower it onto the platen.

12. The silicone membrane generally contains sufficient intrinsic distinct marks (e.g., impurities or debris) visible in the phase contrast images that can be used as control points to calculate the applied membrane strain. Alternatively, additional membrane markers consisting of markings with a felt-tip pen or fluorescently labeled polystyrene beads absorbed to the silicone membrane can be used. For the nuclear strain determination, bright fluorescent chromatin speckles serve as excellent markers in mouse embryo fibroblasts. If measurements of cytoskeletal strain are required, these can be obtained by using endocytosed small, fluorescently labeled polystyrene beads (Molecular Probes-Invitrogen) as control points. The 0.2-μm beads are small enough to be internalized by the cell, yet not so small that they diffuse freely through the cytoskeletal network *(4)*.

13. Determining the four independent transformation matrix entries ($a_{11} = a_{22}$; $a_{12} = -a_{21}$, a_{31}, a_{32}) requires a minimum of two pairs of control points. In our experiments, we generally use 4–15 pairs of control points to reduce the influence of a single control point on the overall results, and we improve the localization of the manually selected control points with a normalized cross-correlation algorithm (function *cpcorr* in the Matlab Image Processing toolbox). In this case, the transformation matrix is computed based on the least-square fit solution for all control points.

14. The spatial transformation matrix T is a combination of scaling, rotation, and translation image transformations, so that Eq. 2.2 can also be expressed as:

$$u' = \begin{pmatrix} x_1 & y_1 & 1 \\ x_2 & y_2 & 1 \\ \cdots & \cdots & \cdots \end{pmatrix} \begin{pmatrix} s & 0 & 0 \\ 0 & s & 0 \\ 0 & 0 & 1 \end{pmatrix} \begin{pmatrix} \cos\alpha & \sin\alpha & 0 \\ -\sin\alpha & \cos\alpha & 0 \\ dx & dy & 1 \end{pmatrix}, \qquad \text{(Eq. 2.7)}$$

where s is the scaling constant, α the degree of rotation, and dx and dy represent the translation in x and y, respectively. For each control point (x_i, y_i) in the pre-strain image and its corresponding point (x'_i, y'_i) in the strain image, this can be expressed as

$$(x'_i \ y'_i \ 1) = (x_i \ y_i \ 1) \begin{pmatrix} s\cos\alpha & s\sin\alpha & 0 \\ -s\sin\alpha & s\cos\alpha & 0 \\ dx & dy & 1 \end{pmatrix}. \qquad \text{(Eq. 2.8)}$$

Comparing Eq. 2.5 with Eqs. 2.2 and 2.3 reveals the relationship between the general transformation matrix entries a_{ij} and the linear conformal spatial transformation parameters s, α, dx, and dy:

$$\begin{aligned} a_{11} &= a_{22} = s\cos\alpha \\ a_{12} &= -a_{21} = s\sin\alpha \\ a_{31} &= dx \\ a_{32} &= dy \end{aligned} \qquad \text{(Eq. 2.9)}$$

or

$$\begin{aligned} \tan\alpha &= \frac{a_{12}}{a_{11}} \\ s &= \frac{a_{11}}{\cos\alpha}. \end{aligned} \qquad \text{(Eq. 2.10)}$$

15. Because cytoskeletal damage and partial cell detachment during strain application generally result in collapse and apparent shrinkage of nuclei, we exclude any nuclei from the analysis that have a nuclear scaling factor of less than 0.99 (corresponding to a 1% size reduction for the transformation between pre- and post-strain images).

References

1. Burke, B. and C.L. Stewart (2006) The laminopathies: the functional architecture of the nucleus and its contribution to disease. *Annu. Rev. Genomics Hum. Genet.* **7**, 369–405.
2. Capell, B.C. and F.S. Collins (2006) Human laminopathies: nuclei gone genetically awry. *Nat. Rev. Genet.* **7**, 940–952.

3. Lammerding, J., Schulze, P.C., Takahashi, T., Kozlov, S., Sullivan, T., Kamm, R.D., Stewart, C.L., and Lee, R.T. (2004) Lamin A/C deficiency causes defective nuclear mechanics and mechanotransduction. *J. Clin. Invest.* **113**, 370–378.
4. Caille, N., Y. Tardy, and J.J. Meister (1998) Assessment of strain field in endothelial cells subjected to uniaxial deformation of their substrate. *Ann. Biomed. Eng.* **26**, 409–416.
5. Lammerding, J., Fong, L.G., Ji, J.Y., Reue, K., Stewart, C.L., Young, S.G., and Lee, R.T. (2006) Lamins A and C but not lamin B1 regulate nuclear mechanics. *J. Biol. Chem.* **281**, 25768–25780.
6. Lammerding, J., Hsiao, J., Schulze, P.C., Kozlov, S., Stewart, C.L. and Lee, R.T. (2005) Abnormal nuclear shape and impaired mechanotransduction in emerin-deficient cells. *J. Cell Biol.* **170**, 781–791.
7. Cheng, G.C., Briggs, W.H., Gerson, D.S., Libby, P., Grodzinsky, A.J., Gray, M.L., and Lee, R.T. (1997) Mechanical strain tightly controls fibroblast growth factor-2 release from cultured human vascular smooth muscle cells. *Circ. Res.* **80**, 28–36.
8. Bausch, A.R., Ziemann, F., Boulbitch, A.A., Jacobson, K., and Sackmann, E. (1998) Local measurements of viscoelastic parameters of adherent cell surfaces by magnetic bead microrheometry. *Biophys. J.* **75**, 2038–2049.

Part II
Chromatin and Transcription

Chapter 3
Gene Expression in Polytene Nuclei

Petra Björk and Lars Wieslander

Keywords Polytene chromosomes; Balbiani ring genes; Chromatin; Gene expression; transcription; mRNA processing; Nucleocytoplasmic export; Immunolabelling; Electron microscopy

Abstract Gene expression in eukaryotic cells is a multi-step process. Many of the steps are both co-ordinated and quality controlled. For example, transcription is closely coupled to pre-messenger RNA (mRNA)–protein assembly, pre-mRNA processing, surveillance of the correct synthesis of messenger ribonucleoprotein (mRNP), and export. The coordination appears to be exerted through dynamic interactions between components of the transcription, processing, surveillance, and export machineries. Our knowledge is so far incomplete about these molecular interactions and where in the nucleus they take place. It is therefore essential to analyze the intranuclear steps of gene expression in vivo. Polytene nuclei are exceptionally large and contain chromosomes and individual genes that can be structurally analyzed in situ during ongoing transcription. Furthermore, they contain gene-specific pre-mRNPs/mRNPs that can be visualised and analyzed as they are synthesised on the gene and then followed on their path to the cytoplasm. We describe methods for investigating the structure and composition of active chromatin and gene-specific pre-mRNPs/mRNPs in the context of analyses of gene expression processes in the nuclei of polytene cells.

1   Introduction

Polytene nuclei, compared with diploid nuclei, lend themselves more easily to in vivo analyses of individual genes in action. The focus of this chapter is to present methods that can be used to investigate the structure and composition of active genes and their pre-mRNP/mRNP products. In addition, polytene nuclei make studies possible of where and when the various molecular machines involved in synthesis, processing, and export of mRNPs operate in the nucleus. The methods are based on experience from studies in *Chironomus tentans*, and they are presented in the context of the experimental possibilities offered by polytene nuclei for analysis of gene expression.

R. Hancock (ed.) *The Nucleus: Volume 2: Chromatin, Transcription, Envelope,*
Proteins, Dynamics, and Imaging,
© Humana Press 2008

1.1 Polytene Nuclei, Polytene Chromosomes, and the Balbiani Ring Genes

The exceptional spatial resolution that can be obtained in polytene nuclei is the result of the amplification and organization of the chromosomes. Polytenization is a process that uncouples cell division from chromosome replication and cell growth. In holometabolic species, such as *Drosophila melanogaster* and *C. tentans*, many tissues serve only during early stages in development and are broken down during metamorphosis and replaced by the adult tissues. For example, salivary glands develop and are active only during the larval stage. After reaching a small number of cells during development of the *C. tentans* salivary gland containing approximately 40 cells, each cell continues to grow in size and replicate its chromosomes, but does not divide. Each cell becomes very large and the cell nucleus can reach a diameter of $70\,\mu m$, i.e., about ten times that of nuclei in mammalian cells. After 13 chromosome replications, each chromosome will consist of 8,000 chromosome copies, attached to each other along their length in perfect register to form the polytene chromosome. Also the homologous chromosomes are attached to each other and the number of polytene chromosomes equals the haploid chromosome number.

In a polytene nucleus of *C. tentans*, the four polytene chromosomes are intermingled, but they are separate from each other and can be individually identified. Two of the chromosomes contain one nucleolar organiser region each and, around these, the nucleoli form as distinct round volumes. The polytene chromosomes are surrounded by the interchromatin space. In mammalian cell nuclei, the interchromatin space forms a complex intranuclear network. In polytene nuclei, the interchromatin is also found in between the chromosomes, but mainly forms large open spaces.

Polytene nuclei have a different organization than mammalian cell nuclei and it is appropriate to ask to what extent knowledge obtained from polytene nuclei is generally valid. It is our opinion that that knowledge is generally valid, although the different types of organization may emphasise different aspects of the nuclear processes. This opinion is based on the following facts. First, the polytene nucleus develops from a diploid nucleus and the biological processes are the same. Second, all known molecular components that are involved in nuclear processes in polytene nuclei are homologues to mammalian counterparts. Third, in several instances observations have been made in polytene nuclei, while technical problems have obscured and delayed the same observation in mammalian cell nuclei. One example is that spliceosome assembly occurs co-transcriptionally. A second example is that mRNP particles move away from the site of synthesis in a non-directional manner consistent with diffusion. These observations have later been substantiated in mammalian cell nuclei.

In *C. tentans* a family of genes, the Balbiani ring gene family, encode secretory proteins that are produced in large quantities by the salivary gland cells *(1, 2)*. Four of the genes within this family form extremely large puffs, called Balbiani rings

(BR). We focus on the BR genes because they provide possibilities to analyze spatial- and time-resolved aspects of gene expression in situ on endogenous genes. Many of the protocols and analytical strategies can be used in *D. melanogaster*. Several of the experimental possibilities are however particularly favourable for the BR genes. This is because of the combination of the length of the BR genes, their exon–intron structure, the high intensity of transcription, and the large size of the BR mRNPs. So far it has not been possible to study an individual *D. melanogaster* gene product in such detail from the gene to the nuclear pore complex (NPC). A limitation of the BR gene system is the difficulty of interfering with steps of gene expression by producing mutants and knocking down mRNA levels. So far, this has not been done successfully. To interfere with gene expression processes, it is possible to perform microinjection experiments (**Section 3.5**), and/or treatment with different drugs.

1.2 Analysis of Transcription—Dynamics of Active Chromatin

So far, it has not been possible to describe the detailed structure of endogenous transcribing mammalian genes in situ. In contrast, activation of genes in polytene chromosomes is morphologically visible as unfolding of the chromatin, both at the light and the electron microscope (EM) level (**Sections 3.1–3.4**), allowing detailed ultrastructural analysis. In a typical mammalian cell, chromosome territories and chromatin-free space each occupy approximately half the volume of the nucleus. In the chromatin territories, the DNA is organised into various chromatin fibers, for example, the 10-nm and 30-nm fibers, but also thicker fibers. These fibers build higher-order structures, probably loops of different lengths (reviewed in ref. *(3)*). Studies in mammalian cells rely mainly on biochemical analyses and immunofluorescence studies that do not allow resolution of structural details for an individual gene. Activation of genes, involving chromatin modification and unfolding, is generally coupled to the formation of chromatin loops. It has for example been shown that protein–protein contacts can result in looping out of the chromatin between the contact points in the β-globin locus in mammals *(4)*, in the hsp70 genes in *D. melanogaster (5, 6)*, and in yeast genes *(7)*. Looping mediates long-range interactions at the base of the loop that can result in significant influences on transcription by enhancing interactions between transcription factors and/or insulating the active chromatin from neighbouring silent chromatin (reviewed in refs. *(8, 9)*).

The possibility of observing a transcribing gene in detail in polytene nuclei depends on several factors, most notably the transcription initiation rate and the length of the unfolded gene. In favourable cases, such as the BR genes, the transcribing gene can be visualised and analysed in considerable detail. Many thousands of copies of a gene are unfolded. This is an advantage when locating specific proteins or protein modifications by immunolabelling at the level of the light microscope, because the signal is amplified (**Sections 3.1** and **3.3**) (Fig. 3.1a, b).

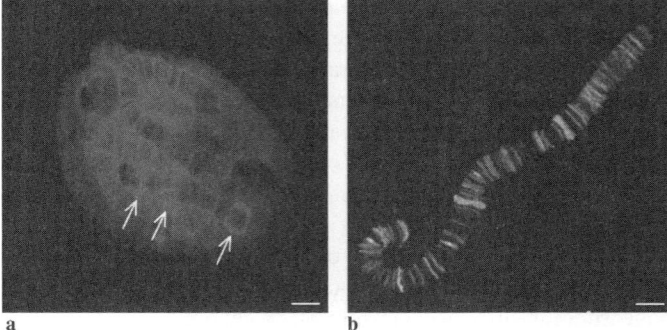

Fig. 3.1 Immunofluorescent labelling of splicing factors in an intact polytene nucleus and in an isolated polytene chromosome. **a** Specific gene loci along the chromosomes are labelled as seen in a confocal section through a nucleus (**Section 3.1**). The three *arrows* point at the BR1, BR2, and BR3 gene loci in chromosome IV, all intensely labelled. Labelling is also detected in the interchromatin space surrounding the chromosomes. **b** Chromosome I was isolated and double labelled with antibodies directed against two splicing factors (**Section 3.3**). Bars, 10 μm. To view this figure in color, see COLOR PLATE 2

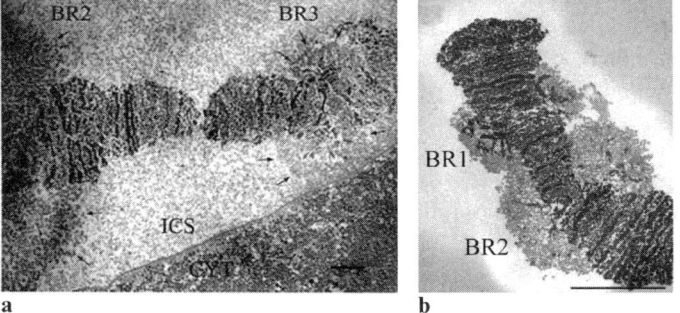

Fig. 3.2 Electron micrographs of chromosome IV. **a** A section through an intact cell (**Section 3.2**) shows part of chromosome IV, with the transcriptionally active BR2 and BR3 genes. The surrounding interchromatin space (*ICS*) and part of the cytoplasm (*CYT*) are separated by the nuclear membrane. **b** Part of an isolated chromosome IV (**Section 3.4**) is shown with the active BR1 and BR2 genes. Bars, 10 μm. **a** Reproduced from Fig. 2 in ref. (*27*)

Analyses of polytene chromosomes give a complete view of the whole genome (*10*). At the EM level (**Sections 3.2** and **3.4**), a section through a gene locus contains parts of many gene copies, but not a complete gene (Fig. 3.2), and to follow an individual gene from its beginning to its end requires that reconstructions are made from serial sections.

In the ~40-kb long BR1 (or BR2) gene, the transcribing chromatin forms a 3–4-μm-long relaxed loop, starting and ending at compact chromatin ((*11*); S. Masich and B. Daneholt, personal communication). Between 100 and 200 transcripts are simultaneously synthesised on this gene loop. The spacing of RNA polymerase II along the gene varies naturally, but can also be experimentally influenced *(12, 13)*. The DNA in the unfolded loop is marginally longer than the transcribed part of the gene. At the 5′ end of the gene, about 500 bp of DNA forms an extended chromatin fiber protruding from more compact chromatin to the first RNA polymerase. At the 3′ end, about 3 kb of DNA extends as a 10-nm chromatin fiber from the last pre-mRNP to more compact chromatin *(14)*.

In the most highly transcribed BR gene, the overall compaction of the DNA in the loop is approximately 2.5 times, showing that the chromatin fiber is extensively unfolded. The structure of the chromatin fiber between two RNA polymerases varies, reflecting different degrees of transcription-dependent chromatin reorganization *(15)*. When very active, the distance between polymerases is about 200 bp and the fiber width is about 5 nm, consistent with nuclease digestion data showing that nucleosomes are disrupted *(16)*. If the distance between polymerases is longer, 10 nm and even 30 nm fibers are formed. If transcription is turned off, the BR gene rapidly folds into a supercoiled thick fiber loop *(17)*. It is likely that also in mammalian nuclei, the structure of the transcribing chromatin within loops is variable and reflects the density of polymerases.

The potential for analysing the dynamic structure of the transcribing chromatin in polytene nuclei in situ has not yet been fully explored. Immunolabelling of proteins and histone modifications should allow analysis of the composition and specific modification of the chromatin in relation to the different phases of transcription *(18)*. Reconstruction in 3D at the EM level should provide high-resolution structures and detailed localization of specific epitopes in these structures. Such studies can be combined with methods to turn off the BR genes and then turn them on again. Structural and cytological analyses in situ of the individual genes can also be complemented with biochemical methods such as Chip analyses *(19)* to merge information at different levels.

1.3   Analysis of Transcription—Synthesis of the RNP Fiber

Transcription takes place in many specific gene loci along the polytene chromosome on chromatin loops protruding from the chromosome. Transcription factors *(20)* and chromatin-modifying factors *(21)* accumulate in these active loci in a transcription-dependent manner. At the light microscope level, transcription in diploid nuclei is observed in many foci, and it has been suggested that when genes are activated they move to such foci *(22)*. The transcription foci are located in perichromatin regions, since EM data show that nascent RNA in mammalian nuclei is essentially found only in the perichromatin region (reviewed in ref. *(3)*), corresponding structurally to where the active loops in polytene chromosomes are. Available data from polytene chromosomes suggest that RNA polymerases are recruited to the gene loci and move along the active chromatin loop, concomitant with structural reorganization of the chromatin *(23, 24)*. If this is also the case in mammalian nuclei, or if transcription is organised in a different way, remains to be established.

One main advantage of polytene nuclei is that in particularly favourable genes, such as the BR genes, individual pre-mRNP/mRNP complexes can be analyzed from their site of synthesis to their passage through the NPCs at the nuclear membrane *(24)*. This relies on EM methods that permit identification of the specific genes as well as the specific mRNP complexes in the interchromatin space. For this purpose, we find it convenient to combine EM analyses of cryosections through the polytene nucleus (**Section 3.2**) (Fig. 3.2a) with EM analyses of isolated polytene chromosomes (**Section 3.4**) (Fig. 3.2b). In both cases, the morphological analyses can be combined with immunodetection of specific proteins by gold-conjugated antibodies.

Several observations have been made in situ on the BR genes that are most likely of importance for eukaryotic genes in general. The first is that the growing RNA immediately associates with proteins to form an RNP complex. All nascent eukaryotic pre-mRNAs are therefore likely to be RNP complexes. The second observation is that the RNP complex changes its structure as it grows longer during transcription *(25)*; reconstructions in 3D have identified an approximately 7-nm fiber as the basic element that is gradually folded into reproducible higher-order structures along the BR1 and BR2 genes *(26)*. Transcripts on short eukaryotic genes are therefore likely to form fibers and transcripts on long genes are likely to form more complex, ring-like structures. The third observation is that, even though the basic pre-mRNP element may be a thin fiber, nascent pre-mRNPs on different genes are likely to have different overall structures. We have observed that the structure of nascent BR3 pre-mRNPs with many introns is different from that of nascent BR1 or BR2 pre-mRNPs with few introns and long exons *(27)*. The co-transcriptional assembly of spliceosomes and the structural dynamics of the spliceosomes during the splicing reaction give rise to structural heterogeneity of the pre-mRNPs along the gene. The presence or absence of exon–junction complexes and association of the pre-mRNA with different combinations of hnRNP proteins should also contribute to structural variations. We therefore predict that pre-mRNP/mRNP complexes from different eukaryotic genes will look different.

1.4   Analysis of the Dynamic Composition of pre-mRNP/mRNP

It is essential to understand the composition of pre-mRNPs/mRNPs and how the composition changes during the different phases of gene expression in the nucleus. This is difficult to achieve in diploid cell nuclei. In polytene nuclei, it can be determined if a particular protein associates with defined pre-mRNPs/mRNPs and when and where the association occurs. Immunolabelling of polytene chromosomes allows genome-wide analyses of the association of specific proteins with pre-mRNPs at defined genes. This can be performed with chromosomes in fixed intact nuclei (**Section 3.1**; Fig. 3.1a), or better with isolated chromosomes (**Section 3.3**; Fig. 3.1b). For example, heterogeneous nuclear RNP (hnRNP) A/B proteins bind to pre-mRNPs from almost all genes *(28)*, while other proteins bind to pre-mRNPs from a subset of genes *(29–31)*.

The transcribing loops of the BR genes furthermore allow detailed analyses of when defined molecules associate with a gene-specific pre-mRNP during the transcription process (**Sections 3.2** and **3.4**). Some proteins associate at the 5′ end of the pre-mRNPs early during transcription and remain bound *(32, 33)*, while other proteins bind at many sites on the pre-mRNPs continuously throughout transcript elongation *(29, 34)*. Yet other proteins, and also small nuclear RNPs (snRNPs), associate with and subsequently leave the pre-mRNPs during transcription (*(35)*; Björk et al., manuscript in preparation).

It is important to note that immunolocalization at the EM level has a high, but limited resolution (**Sections 3.2** and **3.4**). The combined length of the primary and secondary gold-conjugated antibodies sets the resolution at approximately 20–30 nm, depending on the size of the gold marker. This is not a problem when the labelled object is large. However, a more detailed localization of an epitope requires more sophisticated methodology. Precise localization would be needed if, for example, the position of a protein within a pre-mRNP complex is studied or if the structural relationship between components involved in co-ordination of transcription elongation, processing, and surveillance are analyzed. To reach higher resolution, a combination of immunolabelling with 3D EM reconstruction methods can be used *(36)*. Using this combination of methods, the electron-dense connection (the antibodies) between the gold label and the epitope in a pre-mRNP complex could be observed in the reconstruction *(27)* (see Fig. 3.3e).

In general, surprisingly many proteins already associate with mRNAs during transcription. Most of the processing events seem to take place at the genes and these processes are likely to be responsible for the major changes in the composition of the mRNPs. Leaving the gene, the mRNPs appear to be largely ready for export and even for cytoplasmic events, since some proteins accompany the mRNPs into the cytoplasm. Some changes in composition do however take place during export. The polytene nucleus and the BR genes make it possible to follow the composition of the gene-specific mRNPs during intranuclear transport, using immunolabelling at the EM level (**Section 3.2**). Some proteins associate with the BR mRNPs during transcription and leave them at the basket of the NPC *(28, 37)*, while others become associated with the mRNPs only close to or at the NPC *(38)*.

1.5   Analysis of Co-transcriptional pre-mRNP Maturation Processes

In polytene nuclei it is possible to determine where the different pre-mRNA processing events take place and to study the in vivo structures in which they occur. This allows a detailed view of the co-ordination of the processes involved in gene expression.

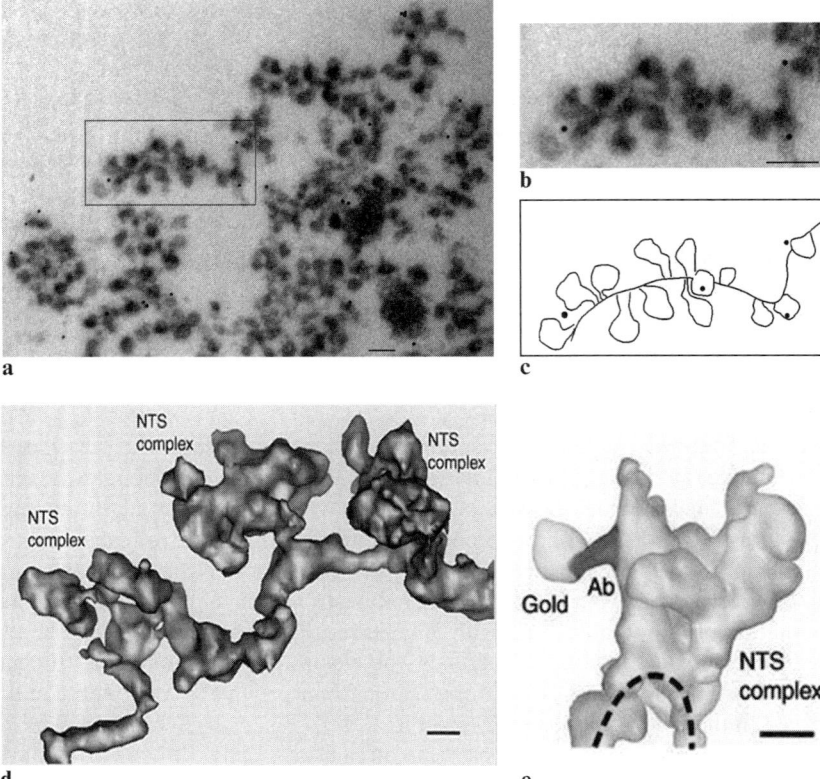

Fig. 3.3 Immunoelectron microscopy on an active BR2 gene. **a** Segments of unfolded BR2 gene loops with growing pre-mRNPs are shown (**Section 3.4**). The image is a higher magnification of a region from an isolated chromosome IV. The area within the rectangle is shown in higher magnification in (**b**) and in a schematic drawing in (**c**). The thin line traces the transcribed chromatin fiber. The contours of the growing pre-mRNP complexes are also shown. The *black dots* are gold particles conjugated with the secondary antibodies. Four of the pre-mRNPs are labelled. Bars, 100 nm. **d** A three-dimensional reconstruction of a part of a transcribing BR3 gene is shown. The ~11 nm chromatin fiber with three protruding pre-mRNP complexes are reconstructed. These complexes contain the pre-mRNP, the RNA polymerase II, and splicing factors, and are called nascent transcript and splicing complexes (*NTS complexes*). **e** A three-dimensional reconstruction of one NTS complex is shown, labelled with a primary antibody directed against the U2 snRNP-specific U2 B″ protein and a secondary antibody conjugated with a gold particle. The gold and the electron-dense connection to the NTS complex indicate where the epitope is located. Bars, 10 nm. **d** and **e** Reproduced from Figs. 3B and 4C in ref. *(27)*

Capping is known to occur early during transcription. In accordance, immunolocalization showed that also CBP20, one of the nuclear cap-binding proteins, binds to the pre-mRNP early during transcription *(32)*. It is expected that the cap-binding complex interacts with both splicing *(39)* and export machineries *(40)*.

Immunolabelling of splicing factors such as serine-arginine-rich (SR) proteins and snRNPs in intact cell nuclei (**Section 3.1**) and on isolated chromosomes (**Section 3.3**) revealed that these factors are recruited to genes in a transcription-dependent manner *(41)*. Polytene nuclei and the BR genes have made it possible to directly demonstrate that splicing is mainly a co-transcriptional process *(42)*. The polytene nucleus is large enough to permit microdissection of nuclear compartments and subsequent extraction of the RNA (**Section 3.6**). For further experimental details of microdissection, see ref. *(43)*. Even the BRs can be isolated by microdissection, and therefore it is uniquely possible to directly obtain the growing BR pre-mRNAs and determine their exon–intron structure. These studies showed that introns are excised during transcription and that the position of the intron in relation to the remaining time of transcription determines if splicing is a completely co-transcriptional process or if it is initiated co-transcriptionally and completed after transcription *(44)*. In agreement, immunolocalization of snRNPs (**Section 3.2**) showed that snRNPs associate with and leave the pre-mRNPs during transcription *(35)*.

Data from mammalian cells strongly suggest co-ordination between pre-mRNP processing and transcription (reviewed in ref. *(45)*). The BR genes allow direct demonstration that the splicing and the transcript-elongation machineries are coupled in time and space, but the molecular details are not known. Using immunolabelling and 3D reconstruction, it was shown that a complex containing elongating RNA polymerase II, the pre-mRNP, and snRNPs is attached to the transcribing chromatin *(27)*. Furthermore, these transcription-splicing complexes change structure during the splicing reaction; estimations of molecular masses strongly indicated that splicing factors are repeatedly recruited to and dissociate from the pre-mRNP during transcript elongation.

Direct analysis of BR1 and BR2 pre-mRNAs, obtained by microdissection, supports a model in which transcription termination, 3′ end processing, and splicing of the last intron are co-ordinated *(46)*. This agrees with data from yeast mutants suggesting that 3′ end processing, transcription termination, and assembly of export-competent mRNPs are dependent on each other (reviewed in ref. *(47)*). The structural basis for such a co-ordination of events at the 3′ end of genes is not yet studied in vivo. The pre-mRNPs at the end of the BR1 and BR2 genes, where these events take place, can however be identified in the EM (**Sections 3.2** and **3.4**) making structural studies possible. The BR genes have considerable potential for further in vivo studies of pre-mRNA processing. This requires that appropriate antibodies will be available, for example targeting mRNP surveillance and export complexes. Development of methods for knocking down expression of specific genes in the salivary gland cells would also be important. If reconstructions in 3D will be possible at even greater resolution, extensive in situ structural information about the processing events and their co-ordination with each other and with transcription can be obtained.

1.6   Analysis of Intranuclear Movement of mRNP

In yeast, active genes interact with components of the NPC *(48–50)*. Such interactions could be important for transcription activation and could also facilitate export of mRNPs. In mammalian cells and even more so in polytene cells, the size of the nucleus and the chromosome organization make it unlikely that such NPC contacts take place continuously. It remains possible that contacts are transient, maybe establishing gene activity during differentiation. Alternatively, such interactions could involve molecules associated with both the NPCs and intranuclear regions.

There is no clear evidence that positioning of active genes is important for export of mRNPs in mammalian or polytene nuclei. Insight into how mRNPs move from the gene to the NPCs has been gained in polytene nuclei. Such studies require that specific gene products can be identified in the cell nucleus as they move from the gene. In one approach, fluorescent in situ hybridization was used to detect mRNAs from a highly transcribed gene in nuclei of *D. melanogaster* salivary gland cells *(51)*. In a second approach, the size and defined structure of mRNPs from the BR1 and 2 genes were taken advantage of *(52)*. It is possible to identify these gene specific mRNP particles in the interchromatin space, and these properties were combined with incorporation of BrUTP into the BR mRNA in a short pulse. Immunodetection of BrUTP in the newly synthesised BR mRNP particles (**Section 3.2**) revealed how these mRNPs move as they are released from the gene. Both approaches showed that mRNPs moved away from the gene in all directions and that the kinetics of this movement was consistent with diffusion. In diploid nuclei, direct studies of the intranuclear movement of poly(A) RNA *(53)* and subsequently of single mRNPs *(54)* confirm that mRNPs move by diffusion inside the interchromatin space.

It should be pointed out that immunolocalization in polytene nuclei allows detection of specific molecules that may associate with or dissociate from the mRNPs in the interchromatin space. In the case of BR mRNPs, it is further possible to detect details in their movement; for example it has been described that these mRNPs can interact with structures in the interchromatin space *(55)*. With improved methods for labelling mRNPs and fluorescence microscopy, it is also likely that movements of the mRNPs in vivo can be described.

1.7 Analysis of Nuclear–Cytoplasmic Export of mRNP

In polytene nuclei, the absence of chromatin at the nuclear membrane gives a clear view of the NPCs (**Section 3.2**; Fig. 3.4).

In combination with the large and structurally defined BR mRNPs, this allows detailed structural studies of how mRNPs dock and translocate through NPCs (reviewed in ref. *(24)*). The BR mRNPs first interact with the basket of the NPC, an interaction that may be mediated by RAE1 (Gle2p), which binds to BR mRNPs only at this stage *(38)*. According to studies in yeast, interactions with Mlp proteins (Tpr in higher eukaryotes) represent a quality control mechanism that prevents unprocessed or incorrectly packaged pre-mRNPs from being exported *(56)*. After docking, the BR mRNP translocates with its 5′ end first through the NPC channel *(57)*. During translocation and exit into the cytoplasm, the BR mRNPs unfold and presumably lose some proteins (Fig. 3.4). Unfolding and

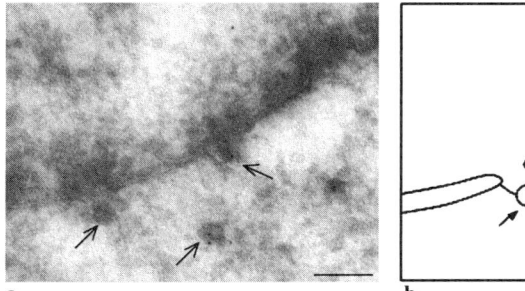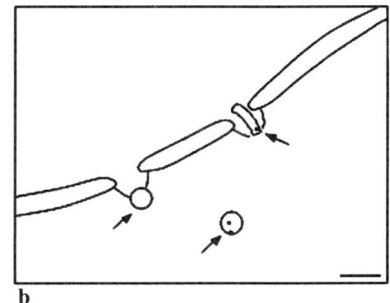

Fig. 3.4 Immunoelectron microscopy on BR1 and BR2 mRNPs in intact nuclei (**Section 3.2**). **a** Three mRNPs are indicated by *arrows*. The mRNP in the middle is in the interchromatin space and labelled with two gold particles. The mRNP to the left is docked to the basket at the nuclear side of a NPC. The mRNP to the right is being exported to the cytoplasm through the NPC and is labelled with one gold particle. During translocation through the NPC, the mRNP is partially unfolded. **b** The mRNPs, the nuclear membrane, and the gold particles are shown schematically. Bars, 100 nm

possibly translocation are dependent on the DEAD-box helicase Dbp5, when activated by Gle1 and bound to the fibrils of the NPC (reviewed in ref. *(58)*). Dbp5 was shown to associate with the BR mRNPs already co-transcriptionally *(33)*, underlining that transcription, assembly of mRNP, and export are co-ordinated. The unfolding of the BR mRNP during exit at the cytoplasmic side of the NPC results in an extended mRNP, with which ribosomes rapidly become associated. In accordance, it has been demonstrated that a translation-initiation factor, eIF4H, binds to the exiting mRNPs *(59)*.

2   Materials

2.1   Biological Material

1. Animals: *C. tentans* is cultured as previously described *(60)*. Salivary glands, manually isolated from fourth instar larvae, are used in all protocols.
2. Primary antibodies: monoclonal or polyclonal, directed against your protein of interest (*see* **Note 1**).
3. Secondary antibodies: fluorophore-labelled for detection in the fluorescence microscope (DakoCytomation Denmark A/S, Glostrup, Denmark), and gold-conjugated for EM detection (Jackson ImmunoResearch, West Grave, PA, USA).

2.2   Immunofluorescent Labelling of Nuclei in Intact Salivary Gland Cells

2.2.1   Buffers and Reagents

1. Phosphate-buffered saline (PBS): 137 mM NaCl, 10 mM Na$_2$HPO$_4$, 1.7 mM K$_2$HPO$_4$, and 2.7 mM KCl, pH 7.4.
2. Fixation buffer: 4% (w/v) paraformaldehyde in PBS.
3. Permeabilization buffer: 0.2% (w/v) lauryl sulfate sodium salt (SDS) in PBS.
4. Blocking buffer: PBS supplemented with 5% (w/v) nonfat dry milk and 5% (w/v) bovine serum albumin (BSA).
5. Antibody dilution buffer: 0.5% (w/v) milk powder and 0.5% (w/v) BSA in PBS.
6. Mounting medium: Vectashield (Vector Laboratories, Burlingame, CA, USA).

2.2.2   Fluorescence Microscope

We use an Axioplan II microscope or a LSM 510 Meta Laser Scanning microscope (Carl Zeiss, Jena, Germany).

2.3   Immunoelectron Microscopy on Nuclei in Intact Salivary Gland Cells

2.3.1   Buffers and Reagents

1. PBS.
2. Fixation buffer: 4% (w/v) paraformaldehyde and 0.1% (v/v) glutaraldehyde in PBS.
3. Cryoprotection solution: 2.3 M sucrose in PBS.
4. PBS containing 0.1 M glycine (PBSG).
5. Blocking solution: 10% (v/v) fetal calf serum (FCS) in PBS.
6. Antibody dilution buffer: 5% FCS in PBS.
7. Contrasting solution for cryosections: 2% (w/v) uranyl acetate (UAc) in double-distilled (dd) H$_2$O.
8. Polyvinyl alcohol (PVA) solution: 4% (w/v) PVA (9–10 kDa, 80% hydrolyzed, Sigma-Aldrich) and 0.3% (w/v) UAc in ddH$_2$O.

2.3.2   Electron Microscopy

1. Nickel grids coated with formvar and carbon (Agar Scientific, Stansted, Essex, England).

2. Cryoultramicrotome: Leica Ultracut UCT + Leica EM FCS (Leica Mikrosysteme, Vienna, Austria).
3. Transmission EM: we view sections in a Tecnai G2 Spirit BioTwin (FEI, Hillsboro, OR, USA) and record images with a Gatan US 1000P camera (Gatan, Pleasanton, CA, USA).

2.4 Immunofluorescent Labelling and Immunoelectron Microscopy on Isolated Polytene Chromosomes

2.4.1 Buffers and Reagents

1. TKM buffer: 10 mM triethanolamine-HCl, pH 7.0, 100 mM KCl, and 1 mM MgCl$_2$. Store in aliquots at −20°C.
2. Fixation buffers: 2% or 4% (w/v) paraformaldehyde in TKM or 2% (v/v) glutaraldehyde in TKM, as indicated in the protocol.
3. Permeabilization buffer: 0.025% or 2% (w/v) NP40 in TKM.
4. Blocking buffer: TKM supplemented with 2% (w/v) BSA.
5. Antibody dilution buffer: 0.5% (w/v) BSA in TKM.
6. Washing solutions: TKM or TKM containing 0.01% (v/v) Tween-20 as indicated.
7. Mounting medium: Vectashield.
8. Dehydration solutions: 90%, 95% (v/v), or absolute (abs.) ethanol as indicated.
9. Plastic embedding mixture: prepared using the components of an R1031 Agar 100 resin kit (Agar Scientific). Mix 24 g of Agar 100 resin (47% v/v), 10 g of dodecenyl succinic anhydride (DDSA) (20%), 16 g of methyl nadic anhydride (MNA) (31%), and 1 g of N-benzyldimethylamine (BDMA) (2%).
10. Contrasting and washing solutions for sections: 2% (w/v) UAc in 50% (v/v) ethanol, and 50% ethanol.

2.4.2 Glassware

1. Micropipettes: made from sodium glass tubes by manually drawing out a tip using a Bunsen burner. The tip is checked for correct diameter (~200 μm).
2. Micro-culture slides with two wells (Menzel-Gläser, Braunschweig, Germany) for initial fixation and washing of glands (**steps 1–5**).
3. Glass slides with eight wells (Erie Scientific, Portsmouth, NH, USA) for all other incubations and for embedding in plastic.
4. Before use, the micropipettes and the glass slides used for incubations and subsequent plastic embedding are siliconised (five layers) using SERVA silicon solution (SERVA, Heidelberg, Germany).

2.4.3 Fluorescence Microscopy

As in **Section 2.2.2.**

2.4.4 Electron Microscopy

1. Microtome: we section plastic-embedded samples using a Leica Ultracut UCT (Leica Mikrosysteme).
2. Electron microscope: we view sections and record images as in **Section 2.3.2.**

2.5   *Microinjection into Nuclei and Cytoplasm of Salivary Gland Cells*

2.5.1 Buffers and Reagents

1. Basic ZO medium (originally described in ref. *(61)*) (order number 992145; National Veterinary Institute, Uppsala, Sweden): $0.1\,\mu g/mL$ of insulin, $2.5\,\mu g/mL$ of fungizone, $1\,mM$ L-glutamine, $50\,U/mL$ of penicillin, $50\,\mu g/mL$ of streptomycin, $1\,mg/mL$ of bacto-peptone, and 2% (v/v) FCS.
2. ZO medium containing 0.4% (w/v) trypan blue.
3. Paraffin oil (Paraffin liquid for spectroscopy, Merck, Darmstadt, Germany).

2.5.2 Microinjection Equipment

1. InjectMan NI 2 coupled to a FemtoJet (Eppendorf, Hamburg, Germany) and to a Zeiss Axiovert microscope (Carl Zeiss, Jena, Germany).
2. Injection needles: Eppendorf Femtotips II.

2.5.3 Fluorescence Microscopy

As in **Section 2.2.2.**

2.5.4 Electron Microscopy

As in **Sections 2.3.2** and **2.4.4.**

2.6   *Isolation of RNA from Microdissected Nuclear Compartments*

2.6.1 Buffers and Reagents

1. Fixation: 70% ethanol.

2. Dehydration and storage: ethanol:glycerol (1:1 v/v).
3. Paraffin oil.
4. Extraction buffer: 20 mM Tris-HCl, pH 7.4, 1 mM EDTA, 0.5% SDS, and 0.5 mg/mL of proteinase K (Roche Diagnostics, Indianapolis, IN, USA).

2.6.2 Equipment

1. Cover slips: 6×32 mm (Menzel-Gläser).
2. Glass chamber: overall dimensions: 40×70×5 mm. The central part, 25-mm long, is 2-mm thick (produced by a local glass workshop).
3. Needles: made from sodium glass rods with a length of about 8 cm and diameter of 2 mm. One end is first drawn out in the flame of a Bunsen burner to a thinner rod, about 3-cm long and 0.5-mm in diameter. The final tip is made at the end of the thin rod with a microforge (de Fonbrune F-1000; Helmut Saur Laborbedarf, Reutlingen, Germany).
4. Micromanipulator: De Fonbrune precision pneumatic (Harvard Apparatus, Edenbridge, UK).

3 Methods

In our view, the rationale for using polytene nuclei for studies of nuclear functions is the unique possibility to study the structure of active genes and their RNP products in vivo. Combined with immunological methods, the composition of these structures can be determined. Furthermore, the high resolution in the polytene nucleus allows determination of when and where components interact and processes take place. The methods presented here therefore aim at combining structure, composition, and spatial relationships in polytene nuclei for analyses of gene expression processes. We describe methods for the study of the localization of components in intact polytene nuclei and along the chromosomes in the light microscope and in the EM. We also describe methods for influencing nuclear processes by microinjection. Finally, we describe a method for isolating nuclear compartments by microdissection. The protocols have been worked out for studies of BR genes and BR mRNPs in *C. tentans* but they are also applicable to *D. melanogaster*, for example. The methods described are the results of work in the laboratories of B. Daneholt and N. Visa and in our own laboratory.

3.1 Immunofluorescent Labelling of Nuclei in Intact Salivary Gland Cells

1. Salivary glands are dissected from *C. tentans* fourth instar larvae and transferred to PBS buffer.

2. Fix the glands in 4% paraformaldehyde in PBS for 15 min on ice.
3. Wash the glands 3× 5 min in PBS.
4. Permeabilise the glands in PBS containing 0.2% SDS for 10 min at room temperature (*see* **Note 2**).
5. Wash the glands 3× 10 min in PBS.
6. Block the cells in PBS containing 5% milk and 5% BSA for 1 h (*see* **Note 3**).
7. Incubate the cells for 1 h with the primary antibody diluted in PBS containing 0.5% milk and 0.5% BSA (*see* **Note 4**).
8. Wash the cells 3× 10 min in PBS.
9. Incubate the cells with the secondary antibody diluted in PBS containing 0.5% milk and 0.5% BSA, for 1 h in the dark.
10. Wash the glands 3× 10 min in PBS.
11. Mount the preparation in a drop of Vectashield and add a cover slip. The cover slip can be sealed with rubber cement or nail polish. The preparation can now be viewed in the fluorescence microscope.

3.2   *Immunoelectron Microscopy of Nuclei in Intact Salivary Gland Cells*

1. Salivary glands are dissected from *C. tentans* fourth instar larvae and transferred to PBS buffer.
2. Fix the glands in 4% paraformaldehyde and 0.1% glutaraldehyde in PBS for 20–25 min at room temperature (*see* **Note 5**).
3. Wash the glands 3× 5 min with PBS.
4. Infiltrate the glands with 2.3 M sucrose in PBS for 40 min to 1 h at room temperature.
5. Put the sample on a specimen holder for the cryoultramicrotome. Remove most of the excess sucrose using a filter paper, and then freeze the mounted sample in liquid nitrogen (*see* **Note 6**).
6. Mount the sample in the cryoultramicrotome.
7. Cut ~70-nm cryosections (*see* **Note 7**), pick up on drops of 2.3 M sucrose in PBS, and mount on nickel grids coated with formvar and carbon.
8. Rinse the grids once in PBSG and then incubate in PBSG for 5–10 min to quench free aldehyde groups (*see* **Note 8**).
9. Block the sections in 10% FCS in PBS for 40 min.
10. Incubate the sections for 1 h with primary antibody diluted in 5% FCS in PBS (*see* **Note 9**).
11. Wash 4× with PBS for a total of 10 min.
12. Incubate the sections for 40 min with secondary antibody diluted in 5% FCS in PBS.
13. Wash with PBS for a total of 10 min.
14. Wash 4× with ddH$_2$O for a total of 10 min.
15. Stain the sections with 2% UAc in ddH$_2$O for 2–5 min.

16. Incubate sections in a drop of PVA solution for at least 3 min. Pick up grids from the drops in a loop. Remove excess PVA solution by stroking the edge of the loop and grid against a filter paper (tilted to an angle of 45 degrees). Air-dry the grids. Remove the grids from the loop with fine tweezers.

17. Examine the preparations in a transmission electron microscope at 80 kV. The distribution of immunogold labelling on chromatin, in the interchromatin region, at the nuclear membrane, and specifically at the NPCs can be analysed. We analyze the composition of the BR mRNPs, components in close association of the mRNPs, and the localization in situ. Quantification of labelling of BR mRNPs along the genes can be performed as described below (**Section 3.4, step 25**).

3.3 *Immunofluorescent Labelling of Isolated Polytene Chromosomes*

1. Dissect salivary glands from *C. tentans* fourth instar larvae and transfer the glands to TKM buffer in a siliconised micro-culture glass slide (*see* **Note 10**).

2. Prefix the glands in 2% paraformaldehyde in TKM buffer for 90 sec at room temperature (*see* **Note 11**).

3. Wash the glands 3× 1 min with TKM buffer.

4. Keep the glands in TKM containing 2% NP40 for 30 sec at room temperature and then transfer them to TKM containing 0.025% NP40.

5. On a cold stage (+4°C) under a dissection microscope, chromosomes are isolated by repeatedly pipetting the glands through a siliconised glass micropipette with a diameter of 200 µm (*see* **Note 12**).

6. Transfer the released chromosomes, by a micropipette, to a drop of TKM placed in a well on a glass slide. Allow the chromosomes to settle on the glass surface and attach.

7. Wash the chromosomes 3× 5 min each with TKM (*see* **Note 13**).

8. Fix the chromosomes in 4% paraformaldehyde in TKM for 30 min (*see* **Note 14**).

9. Wash the samples 3× 5 min with TKM (see **Note 15**).

10. Block the chromosomes in 2% BSA in TKM for 30 min.

11. Add the primary antibody diluted in 0.5% BSA in TKM to the preparations and incubate for 1 h (*see* **Note 16**).

12. Wash the chromosomes 3× 5 min in TKM containing 0.01% Tween-20.

13. Incubate the preparations with the secondary, fluorescent-labelled antibody, diluted in 0.5% BSA in TKM, for 1 h (*see* **Note 17**).

14. Wash the chromosomes 3× 5 min with TKM.

15. Mount the chromosomes in a drop of Vectashield and apply a cover slip. The cover slip can be sealed with rubber cement or nail polish. The chromosomes can now be viewed in a fluorescence microscope.

3.4   Immunoelectron Microscopy on Isolated Polytene Chromosomes

1–12. The same steps as in **Section 3.3**, except that all glass slides are siliconised.

13. Incubate the preparations with gold-conjugated secondary antibody, diluted in 0.5% BSA in TKM, for 1 h.

14. Wash the chromosomes 3× 5 min with TKM.

15. Postfix the chromosomes in 2% glutaraldehyde in TKM for 1 h.

16. Wash the samples 3× 5 min with TKM.

17. Dehydrate the chromosomes, first 2× 5 min in 90% ethanol, then 2× 5 min in 95% ethanol, and finally 4× 10 min in abs. ethanol (*see* **Note 18**).

18. Embed the chromosomes in Agar 100 plastic mixture at room temperature (*see* **Note 19**), starting with a mix of three parts abs. ethanol and one part Agar 100 plastic mixture, overnight. The following day, incubate in a mix of one part abs. ethanol and one part Agar 100 plastic mixture for 2 h with at least one change of the solution. The embedding continues with a mix of one part of abs. ethanol and three parts of Agar 100 plastic mixture for 1 h, and finally in pure Agar 100 plastic mixture for 5 h with one to two changes.

19. Let the preparations polymerise at 60°C for 2 days (*see* **Note 20**).

20. Remove the plastic-embedded chromosomes from the slide by putting the slide on a metal surface cooled by liquid nitrogen or dry ice (*see* **Note 21**).

21. Cut the plastic samples containing the chromosomes to a block to fit on plastic supports (plastic capsules) for the ultramicrotome and glue thereon (*see* **Note 22**).

22. Trim the sides of the block and mount the sample in an ultramicrotome.

23. Collect sections of ~70-nm thickness (*see* **Note 23**) and up on Cu grids (pre-washed in acetone, 70% ethanol, and air-dried).

24. Stain the sections with 2% (w/v) UAc in 50% ethanol for 10 min and wash in 50% ethanol.

25. Examine the preparations in a transmission electron microscope at 80 kV. Quantitative analysis can be carried out by recording images of BR1 and BR2 gene loci. The number of gold particles in the proximal, middle, and distal regions of the BR genes can be counted and presented as percentage of immunogold labelling in each region. The position of the immunogold label on the BR mRNP can also be analysed.

3.5   Microinjection into Nuclei and Cytoplasm of Salivary Gland Cells

1. Centrifuge solutions for injection at 5,600×g for 1 h at 4°C before the injection needles are filled.

2. Carefully dissect a salivary gland from *C. tentans* fourth instar larvae, cover the gland with hemolymph or ZO medium, and check the gland for possible damage (*see* **Note 24**).

3. Move the gland with a pipette into ~30-µL drop of medium and place the gland in a small plastic petri dish. Cover the drop with paraffin oil (*see* **Note 25**).

4. Remove most of the medium surrounding the gland, and inject either into the nucleus or into the cytoplasm. For the nucleus, we use an injection pressure of 800 hPa and an injection time of 1.5 sec, corresponding to about 0.01 nL injected. For the cytoplasm, larger volumes are injected (*see* **Note 26**). Document the gland morphology and the cells that have been injected.

5. After injection, add 30 µL of hemolymph containing the desired material (*see* **Note 26**) and incubate the glands for an appropriate time (*see* **Note 27**).

6. Analyse the injected glands according to any of the protocols for immunolabelling (intact glands or isolated chromosomes) at the light microscope or the EM level.

3.6   *Isolation of RNA from Microdissected Nuclear Compartments*

1. Dissect salivary glands from *C. tentans* fourth instar larvae and transfer the glands to a drop of hemolymph on a 6×32-mm cover slip, one gland per drop.

2. Carefully move the glands with dissection needles within the drop to stretch out the glands and avoid folded regions.

3. Carefully remove the hemolymph from the gland with a pipette, and immediately put the cover slip into a vial (we use a scintillation vial) filled with cold 70% ethanol on ice.

4. Change the 70% ethanol after 30 min and incubate on ice for a further 30 min.

5. Move the cover slip with the glands into a new container containing cold ethanol:glycerol (1:1) and keep on ice for at least 1 h. If necessary, the glands can be stored for several weeks in the vial at −20°C.

6. Put the cover slip on a glass slide, and in the dissection microscope move the glands with dissection needles to a new narrow cover slip. Make sure that a small volume of ethanol:glycerol surrounds each gland. Stretch out the gland if necessary.

7. Place the cover slip on a glass chamber with the glands facing downwards. The glass chamber has a central thinner part, about 2-mm deep, that will be filled with paraffin oil, and the cover slip bridges the thin part of the chamber.

8. Fill the space under the cover slip with paraffin oil. The glands will be surrounded with paraffin oil, preventing evaporation of the drop of ethanol:glycerol around the gland. The volume of this drop has to be small enough such that the gland does not float around during dissection.

9. Put the glass chamber on the stage of an ordinary light microscope equipped with phase contrast. View the glands through the cover slip.

10. Attach glass needles (**Section 2.6.2**) to the micromanipulator and move the needles into the paraffin oil space below the glands (*see* **Note 28**).

11. Under visual control in the microscope, raise the needles toward the gland. The tip of one needle is used to hold the gland, and the other needle is used to dissect the gland cells. It is best to keep the two needles close to each other.

12. The nucleus and the chromosomes are well seen in phase contrast. Complete nuclei can be isolated, moved away from the gland, and opened with the needles. The chromosomes can be cleaned from visible interchromatin material with the needles, and individual chromosomes and the interchromatin material can be isolated and put at separate positions on the cover slip. Finally, specific chromosomal regions, for example, a BR locus, can be cut out from the chromosome.

13. Collect and pick up isolated compartments, for example, chromosomes, BRs, and interchromatin material, on the tip of a needle. Move the needle out of the paraffin oil and put the needle into about 100 μL of extraction buffer in an Eppendorf tube. After 2 min remove the needle and visually check it to ensure that the material does not remain on its tip (*see* **Note 29**).

14. Perform RNA extraction for a total of 30 min at room temperature. Add NaCl to a final concentration of 0.1 M and add 5 μg of *E. coli* transfer RNA (tRNA). Extract once with phenol, followed by extraction with chloroform once. Add ethanol (2.5 volumes) to precipitate the RNA.

15. The RNA is now ready for analyses, for example by RT-PCR (*see* **Note 30**).

4 Notes

1. Both monoclonal and polyclonal antibodies can be used. However, time must be spent to evaluate each antibody for specificity by Western blotting and by labelling of chromosomes and cells, for the appropriate dilution, and for optimal incubation time and temperature. Some antibodies work well enough for labelling isolated chromosomes at the light microscope level, but less well for immuno EM. In that case, it can be worth trying different fixation conditions.

2. The glands can also be permeabilised in Triton X-100, a milder treatment. Each antibody has to be tested for optimum permeabilisation method and time. It is advisable to study a number of cells in each gland and several glands to determine the optimal and reproducible labelling pattern.

3. Blocking and incubation with antibodies are performed in a humid chamber at room temperature.

4. Some antibodies will work better if the incubation is done overnight at +4°C followed by 1 h at room temperature.

5. An alternative fixation is to use 4% paraformaldehyde in PBS for 1 h at room temperature or overnight at +4°C. The material will be less fixed, but this may provide better conditions for the immunoreaction.

6. To improve the grip of the sample to the specimen holder, the top of the holder should be roughened using a file or sandpaper. The holder is then washed in ethanol before use. Some sucrose should be left to glue the sample to the

specimen holder. The mounted samples are stored in cryotubes in a liquid nitrogen storage system.

7. When starting with a new sample, semi-thin sections (~500 nm) are collected at −80°C, stained with 0.5% (w/v) toluidine blue, and examined in the light microscope until you find the position of the cell nucleus. From there, ultra-thin sections are collected at −105°C. Before finishing sectioning, a few ~300-nm sections are collected, stained, and examined in the light microscope to check the position in the nucleus.

8. All incubation steps in the immunolabelling reaction are performed on drops on Parafilm in a Petri dish, at room temperature.

9. To economise on antibody, it is possible to use volumes of 10 μL or even less. For the washing steps, the volumes should be large (100–200 μL).

10. An alternative approach is to perform immunolabelling of chromosomes prepared by squashing salivary gland cells, for example according to the protocol described in ref. *(10)*, although in our hands isolated chromosomes perform more consistently.

11. It is convenient to have drops of the various solutions ready on siliconised micro culture slides. The glands are moved with the help of dissection needles.

12. The glands are rigid and can be difficult to suck into the pipette. If they are cut into smaller pieces (with dissection needles or a scalpel) prior to pipetting, the release of individual chromosomes is more efficient.

13. The chromosomes can be treated with RNase at this stage. This is often performed to determine if immunolabelling is dependent on intact RNA.

14. The fixation, blocking, and incubations with antibodies are performed in a humid chamber at room temperature.

15. At this stage, the chromosomes can be covered by a drop of glycerol/TKM (87%/13% v/v) and stored at −20°C. Upon use, they are washed in TKM to remove the glycerol.

16. Some antibodies will work better if the incubation is done overnight at +4°C.

17. The chromosomes can also be labelled with 6-nm gold-conjugated secondary antibodies and the labelling enhanced with immunogold silver enhancement solution (IntenSEM; GE Health Care, Arlington Heights, IL, USA) according to the manufacturer. In our hands, this is sometimes more sensitive than fluorescent labelling.

18. All incubations before the dehydration (**Section 3.4, step 17**) are performed at room temperature in a humid chamber.

19. Dehydration and embedding in Agar 100 plastic mixture are performed in a chamber (for example, a Petri dish) in the presence of silica gel to minimise humidity, and all steps are at room temperature. The beaker with magnet, tubes, and pipette tips for making the Agar 100 plastic mixture have to be first incubated at 80°C for at least 15 min to minimise humidity in the samples. The ready-made mixture can be stored at +4°C overnight in the glass beaker for use the next day, covered with Parafilm and put in a vessel containing silica gel with a tight lid. Before use the next day, the mixture should adopt room temperature, which takes at least 1 h.

20. After polymerisation, the position of the chromosomes is carefully marked before removing the sample from the slide. It can be difficult to see the chromosomes in the plastic, and the plastic has to be cut and mounted on a plastic capsule and trimmed before sectioning, so it is important to know the exact position of the chromosomes.

21. If the plastic block does not separate from the slide, it may be necessary to use alternating heat (boiling water) and cold (liquid nitrogen or dry ice). Be careful not to scratch or damage the surface near the chromosomes, because then it will not be possible to see the chromosomes or to section the sample.

22. The sample has to be glued onto a plastic support to mount it in the holder of the ultramicrotome. We use a gelatin capsule filled with polymerised Agar 100 mixture. A super fixing glue should be used (e.g., Super Attak; Loctite, Henkel, Düsseldorf, Germany).

23. Because the chromosomes are located very close to the front surface of the block, we do not trim the surface and collect sections from the start of the sectioning.

24. To check that the cells have not been damaged during dissection, the gland is incubated in ZO medium containing 0.4% trypan blue for 30 sec. Healthy cells will exclude the dye, but if the cell membranes have been damaged the dye will diffuse into the cells. If any cell in the gland is damaged, the gland is discarded. A healthy gland is rinsed in ZO medium before further processing.

25. During injection and subsequent incubations, the gland can be surrounded by either hemolymph (which is carefully bled from four to five other larvae) or ZO medium. The surrounding oil prevents evaporation and oxidation of the hemolymph.

26. Antibodies, oligopeptides, isotopes, or drugs can be injected. The concentration of antibodies should generally be higher than 1 mg/mL and the solution should be free of sodium azide. According to our experience, PBS or water can be used but not Tris buffers or solutions containing DMSO. Appropriate control solutions (antibodies or buffer) can be injected into control cells in the same gland. Uninjected cells in the same gland also serve as controls.

27. In our experience, the glands are healthy for at least 2–3 h in hemolymph.

28. The drawn-out tip of the needle should be slightly bent, and it should not be too long, otherwise it breaks more easily. When the needles are put into the micromanipulator, the two tips should be close to each other, directed upward, and in the focal plane of the microscope. The needles are then lowered, and positioned under the gland by moving the glass chamber. The needles are then raised toward the gland under visual control.

29. The isolated material, surrounded by a small amount of ethanol:glycerol, sticks well to the tip of a needle. Care must be taken when the needle is moved out from the paraffin oil; the critical step when the tip of the needle just leaves the oil should be done slowly and under visual control.

30. If RNA is extracted from chromosomal material, it is best to treat the extracted nucleic acids with DNase (RNase-free) to remove DNA in the sample.

Acknowledgments We are grateful for helpful comments from Prof. B. Daneholt and Prof. N. Visa. We also acknowledge the assistance of K. Bernholm.

References

1. Case, S.T. and Wieslander, L. (1992) Secretory proteins of Chironomus salivary glands: structural motifs and assembly characteristics of a novel biopolymer. *Results Probl. Cell Differ.* **19**, 187–226.
2. Wieslander, L. (1994) The Balbiani ring multigene family: coding repetitive sequences and evolution of a tissue-specific cell function. *Progr. Nucleic Acid Res. Mol. Biol.* **48**, 275–313.
3. Cremer, T., Cremer, M., Dietzel, S., Müller, S., Solovei, I., and Fakan, S. (2006) Chromosome territories—a functional nuclear landscape. *Curr. Opin. Cell Biol.* **18**, 307–316.
4. Carter, D., Chakalova, L., Osbourne, C.S., Dai, Y.-F., and Fraser, P. (2002) Longe-range chromatin regulatory interactions in vivo. *Nature Genet.* **32**, 623–626.
5. Udvardy, A., Maine, E., and Schedl, P. (1985) The 87A7 chromomere. Identification of novel chromatin structures flanking the heat shock locus that may define the boundaries of higher order domains. *J. Mol. Biol.* **185**, 341–358.
6. Blanton, J., Gaszner, M., and Schedl, P. (2003) Protein:protein interactions and the pairing of boundary elements in vivo. *Genes Dev.* **17**, 664–675.
7. O'Sullivan J.M., Tan-Wong S.M., Morillon, A., Lee, B., Coles, J., Mellor, J., and Proudfoot, N.J. *Nat. Genet.* **36**, 1014–1018.
8. Saiz, L. and Vilar J.M.G. (2006) DNA looping: the consequences and its control. *Curr. Opin. Cell Biol.* **16**, 344–350.
9. Chambeyron, S. and Bickmore, W.A. (2004) Does looping and clustering in the nucleus regulate gene expression? *Curr. Opin. Cell Biol.* **16**, 256–262.
10. Eissenberg, J.C. (2006) Functional genomics of histone modification and non-histone chromosomal proteins using the polytene chromosomes of Drosophila. *Methods* **40**, 360–364.
11. Andersson, K., Björkroth, B., and Daneholt, B. (1980) The in situ structure of the active 75 S RNA genes in Balbiani rings of *Chironomus tentans*. *Exp. Cell Res.* **130**, 313–326.
12. Beermann, W. (1973) Directed changes in the pattern of Balbiani ring puffing in Chironomus: effects of a sugar treatment. *Chromosoma* **41**, 297–326.
13. Egyházi, E. (1975) Inhibition of Balbiani ring RNA synthesis at the initiation level. *Proc. Natl. Acad. Sci. USA* **72**, 947–950.
14. Ericsson, C., Mehlin, H., Björkroth, B., Lamb, M.M., and Daneholt, B. (1989) The ultrastructure of upstream and downstream regions of an active Balbiani ring gene. *Cell* **56**, 631–639.
15. Björkroth, B., Ericsson, C., Lamb, M.M., and Daneholt, B. (1988) Structure of the chromatin axis during transcription. *Chromosoma* **96**, 333–340.
16. Belikov, S., Paulsson, G., and Wieslander, L. (1998) Promoter regions of four Balbiani ring genes in *Chironomus tentans* exhibit a common salivary gland-specific chromatin organisation, which is dependent of the rate of transcriptional initiation. *Mol. Gen. Genet.* **258**, 420–426.
17. Andersson, K., Björkroth, B., and Daneholt, B. (1984) Packaging of a specific gene into higher order structures following repression of RNA synthesis. *J. Cell Biol.* **98**, 1296–1303.
18. Ericsson, C., Grossbach, U., Björkroth, B., and Daneholt, B. (1990) Presence of histone H1 on an active Balbiani ring gene. *Cell* **60**, 73–83.
19. Ringrose, L., Ehret, H., and Paro, R. (2004) Distinct contributions of histone H3 lysine 9 and 27 methylation to locus-specific stability of polycomb complexes. *Mol. Cell* **16**, 641–653.
20. Westwood, J.T., Clos, J., and Wu, C. (1991) Stress-induced oligomerization and chromosomal relocalization of heat-shock factor. *Nature* **353**, 822–827.

21. Sjölinder, M., Björk, P., Söderberg, E., Sabri, N., Farrants, A.K., and Visa, N. (2005) The growing pre-mRNA recruits actin and chromatin-modifying factors to transcriptionally active genes. *Genes Dev.* **19**, 1871–1874.

22. Osborne, C.S., Chakalova, L., Brown, K.E., Carter, D., Horton, A., Debrand, E., Goyenechea, B., Mitchell, J.A., Lopes, S., Reik, W., and Fraser, P. (2004) Active genes dynamically colocalize to shared sites of ongoing transcription. *Nat. Genet.* **10**, 1065–1071.

23. Daneholt, B. (1992) The transcribed template and the transcription loop in Balbiani rings. *Cell Biol. Int. Rep.* **16**, 709–715.

24. Daneholt, B. (2001) Assembly and transport of a premessenger RNP particle. *Proc. Natl. Acad. Sci. USA* **98**, 7012–7017.

25. Skoglund, U., Andersson, K., Björkroth, B., Lamb, M.M., and Daneholt, B. (1983) Visualization of the formation and transport of a specific hnRNP particle. *Cell*, **34**, 847–855.

26. Lönnroth, A., Alexciev, K., Mehlin, H., Wurtz, T., Skoglund, U., and Daneholt, B. (1992) Demonstration of a 7 nm RNP fiber as the basic structural element in a premessenger RNP particle. *Exp. Cell Res.* **199**, 292–296.

27. Wetterberg, I., Zhao, J., Masich, S., Wieslander, L., and Skoglund, U. (2001) In situ transcription and splicing in the Balbiani ring 3 gene. *EMBO J.* **20**, 2564–2574.

28. Björk, P., Wetterberg-Strandh, I., Baurén, G., and Wieslander, L. (2006) *Chironomus tentans* repressor splicing factor represses SR protein function locally on pre-mRNA exons and is displaced at correct splice sites. *Mol. Biol. Cell* **17**, 32–42.

29. Wurtz, T, Kiseleva, E., Nacheva, G., Alzhanova-Ericsson, A.T., Rosen, A., and Daneholt, B. (1996) Identification of two RNA-binding proteins in Balbiani ring premessenger ribonucleoprotein granules and presence of these proteins in specific subsets of heterogeneous nuclear ribonucleoprotein particles. *Mol. Cell. Biol.* **16**, 1425–1435.

30. Sun, X., Zhao, J., Kylberg, K., Soop, T., Palka, K., Sonnhammer, E., Visa, N., Alzhanova-Ericsson, A.T., and Daneholt, B. (2004) Conspicuous accumulation of transcription elongator repressor hrp130/CA150 on the intron-rich Balbiani ring 3 gene. *Chromosoma* **113**, 244–257.

31. Singh, O.P., Visa, N., Wieslander, L., and Daneholt, B. (2006) A specific SR protein binds preferentially to the secretory protein gene transcripts in salivary glands of *Chironomus tentans*. *Chromosoma* **115**, 449–458.

32. Visa, N., Izaurralde, E., Ferreira, J., Daneholt, B., and Mattaj, I.W. (1996) A nuclear cap-binding complex binds Balbiani ring pre-mRNA co-transcriptionally and accompanies the ribonucleoprotein particle during nuclear export. *J. Cell Biol.* **133**, 5–14.

33. Zhao, J., Jin, S.B., Björkroth, B., Wieslander, L., and Daneholt, B. (2002) The mRNA export factor Dbp5 is associated with Balbiani ring mRNP from gene to cytoplasm. *EMBO J.* **21**, 1177–1187.

34. Alzhanova-Ericsson, A.T., Sun, X., Visa, N., Kiseleva, E., Wurtz, T., and Daneholt, B. (1996) A protein of the SR family of splicing factors binds extensively to exonic Balbiani ring pre-mRNA and accompanies the RNA from the gene to the nuclear pore. *Genes Dev.* **10**, 2881–2893.

35. Kiseleva, E., Wurtz, T., Visa, N., and Daneholt, B. (1994) Assembly and disassembly of spliceosomes along a specific pre-messenger RNP fiber. *EMBO J.* **13**, 6052–6061.

36. Skoglund, U., Öfverstedt, L.-G., and Daneholt, B. (1998) Procedures for three-dimensional reconstruction from thin sections with electron tomography. In: RNP particles, splicing and autoimmune diseases, Springer lab manual, J. Schenkel ed., Springer Verlag Berlin, Heidelberg, New York, pp 72–94.

37. Sun, X., Alzhanova-Ericsson, A.T., Visa, N., Aissouni, Y., Zhao, J., and Daneholt, B. (1998) The hrp23 protein in the Balbiani ring pre-mRNP particles is released just before or at the binding to the nuclear pore complex. *J. Cell Biol.* **142**, 1167–1180.

38. Sabri, N. and Visa, N. (2000) The Ct-RAE1 protein interacts with Balbiani ring RNP particles at the nuclear pore. *RNA* **6**, 1597–1609.

39. Gournemann, J., Kotovic, K.M., Hujer, K., and Neugebauer, K.M. (2005) Cotranscriptional spliceosome assembly occurs in a stepwise fashion and requires the cap binding complex. *Mol. Cell* **19**, 53–63.
40. Cheng, H., Dufu, K., Lee, C.-S., Hsu, J.L., Dias, A., and Reed, R. (2006) Human mRNA export machinery recruited to the 5′ end of mRNA. *Cell* **127**, 1389–1400.
41. Baurén, G., Jiang, W.O., Bernholm, K., Gu, F., and Wieslander, L. (1996) Demonstration of a dynamic transcription-dependent organization of pre-mRNA splicing factors in polytene nuclei. *J. Cell Biol.* **135**, 929–941.
42. Baurén, G. and Wieslander, L. (1994) Splicing of Balbiani ring 1 gene pre-mRNA occurs simultaneously with transcription. *Cell* **76**, 183–192.
43. Lambert, B. and Daneholt, B. (1975) Microanalysis of RNA from defined cellular components. *Methods Cell Biol.* **10**, 17–47.
44. Wetterberg, I., Baurén, G., and Wieslander, L. (1996) The intranuclear site of excision of each intron in the Balbiani ring 3 pre-mRNA is influenced by the time remaining to transcription termination and different excision efficiencies for the various introns. *RNA* **2**, 641–651.
45. Bentley, D. (2005) Rules of engagement: co-transcriptional recruitment of pre-mRNA processing factors. *Curr. Opin. Cell Biol.* **17**, 251–256.
46. Baurén, G., Belikov, S., and Wieslander, L. (1998) Transcriptional termination in the Balbiani ring 1 gene is closely coupled to 3′-end formation and excision of the 3′-terminal intron. *Genes Dev.* **12**, 2759–2769.
47. Aguilera, A. (2005) Co-transcriptional mRNP assembly: from the DNA to the nuclear pore. *Curr. Opin. Cell Biol.* **17**, 242–250.
48. Rodriguez-Navarro, S., Fischer, T., Luo, M.J., Antunez, O., Brettschneider, S., Lechner, J., Perez-Ortin, J.E., Reed, R., and Hurt, E. (2004) Sus1, a functional component of the SAGA histone acetylase complex and the nuclear pore-associated mRNA export machinery. *Cell* **116**, 75–86.
49. Casolari, J.M., Brown, C.R., Drubin, D.A., Rando, O.J., and Silver, P.A. (2005) Developmentally induced changes in transcriptional program alter spatial organization across chromosomes. *Genes Dev.* **19**, 1188–1198.
50. Schmid, M., Arib, G., Laemmli, C., Nishikawa, J., Durussel, T., and Laemmli, U. K. (2006) Nup-PI: The nucleopore-promoter interaction of genes in yeast. *Mol. Cell* **21**, 379–391.
51. Zachar, Z., Kramer, J., Mims, I.P., and Bingham, P.M. (1993) Evidence for channelled diffusion of pre-mRNAs during nuclear RNA transport in metazoans. *J. Cell Biol.* **121**, 729–742.
52. Singh, O.P., Björkroth, B., Masich, S., Wieslander, L., and Daneholt, B. (1999) The intranuclear movement of Balbiani ring premessenger ribonucleoprotein particles. *Exp. Cell Res.* **25**, 135–146.
53. Politz, J.C., Tuft, R.A., Pederson, T., and Singer, R.H. (1999) Movement of nuclear poly(A) RNA throughout the interchromatin space in living cells. *Curr. Biol.* **9**, 285–291.
54. Shav-Tal, Y., Darzacq, X., Shenoy, S.M., Fusco, D., Janicki, M., Spector, D.L., and Singer, R.H. (2004) Dynamics of single mRNPs in nuclei of living cells. *Science* **304**, 1797–1800.
55. Miralles, F., Öfverstedt, L.G., Sabri, N., Aissouni, Y., Hellman, U., Skoglund, U., and Visa, N. (2000) Electron tomography reveals posttranscriptional binding of pre-mRNPs to specific fibers in the nucleoplasm. *J. Cell Biol.* **148**, 271–282.
56. Vinciguerra, P. and Stutz, F. (2004) mRNA export: an assembly line from genes to nuclear pores. *Curr. Opin. Cell Biol.* **16**, 285–292.
57. Mehlin, H., Daneholt, B., and Skoglund, U. (1992) Translocation of a specific premessenger ribonucleoprotein particle through the nuclear pore studied with electron microscope tomography. *Cell* **69**, 605–613.
58. Cole, C.N. and Scarcelli, J.J. (2006) Transport of messenger RNA from the nucleus to the cytoplasm. *Curr. Opin. Cell Biol.* **18**, 299–306.

59. Björk, P., Baurén, G., Gelius, B., Wrange, Ö., and Wieslander, L. (2003) The *Chironomus tentans* translation initiation factor eIF4H is present in the nucleus but does not bind to mRNA until the mRNA reaches the cytoplasm. *J. Cell Sci.* **116**, 4521–4532.
60. Meyer, B., Mähr, R., Eppenberger, H.M., and Lezzi, M. (1983). The activity of Balbiani rings 1 and 2 in salivary glands of *Chironomus tentans* larvae under different modes of development and after pilocarpine treatment. *Develop. Biol.* **98**, 265–277.
61. Wyss, C. (1982) *Chironomus tentans* epithelial cell line sensitive to ecdysteroids, juvenile hormone, insulin and heat shock. *Exp. Cell Res.* **139**, 309–319.

Chapter 4
Electron Microscope Visualization of RNA Transcription and Processing in *Saccharomyces cerevisiae* by Miller Chromatin Spreading

Yvonne N. Osheim, Sarah L. French, Martha L. Sikes, and Ann L. Beyer

Keywords Transcription; Ribosomal RNA; RNA processing; Electron microscopy; Yeast

Abstract The Miller chromatin spreading technique for electron microscopic visualization of gently dispersed interphase chromatin has proven extremely valuable for analysis of genetic activities in vivo. It provides a unique view of transcription and RNA processing at the level of individual active genes. The budding yeast *Saccharomyces cerevisiae* has also been an invaluable model system for geneticists and molecular biologists. In this chapter, we describe methods for applying the Miller chromatin-spreading method to *Saccharomyces cerevisiae*. This allows one to use electron microscopic visualization of a gene of interest to study effects of specific mutations on gene activity. We are applying the method to study transcription and processing of ribosomal RNA.

1 Introduction

The Miller chromatin spreading method, developed over 35 years ago by Oscar L. Miller, Jr. and colleagues, allows visualization of chromatin-associated genetic events by electron microscopy (EM) *(1)*. The method uses hypotonic lysis of the cells of choice, allowing the chromatin to be released from the nucleus and to slowly disperse. The dispersed chromatin is centrifuged onto a carbon-coated EM grid, resulting in a loosened two-dimensional array of chromatin. Nascent transcripts on the chromatin are retained and they, too, are loosened in structure. However, the basic nucleosomal structure of the chromatin and the basic ribonucleoprotein (RNP) structure of the nascent transcripts are maintained. One can identify actively transcribing genes by the RNP fibers extending from the chromatin backbone; these transcripts increase in length with increasing distance from the presumed promoter. Formation of RNP particles and secondary structures can be observed on the growing transcripts. In the past we have used the Miller chromatin spreading method mainly with amphibian oocytes and Drosophila tissues to analyze genes transcribed

R. Hancock (ed.) *The Nucleus: Volume 2: Chromatin, Transcription, Envelope, Proteins, Dynamics, and Imaging,*
© Humana Press 2008

by RNA polymerase I and II (Pol I and II) *(2, 3)*, including analysis of splicing *(4)* and 3′ end processing of nascent Pol II transcripts *(5)*, and also to study DNA replication and amplification *(6, 7)*. Previous articles describe our detailed methods for chromatin spreading of amphibian oocytes and various Drosophila tissues *(8–10)*, cells that are particularly amenable to analysis by this method due to such features as amplified ribosomal RNA (rRNA) genes, small genomes, and tractability to mechanical removal or permeabilization of cell membranes. This chapter focuses on spreading of the budding yeast *Saccharomyces cerevisiae*.

Saccharomyces cerevisiae (hereafter called yeast) is an ideal organism for analysis by Miller spreading because it grows quickly and easily, has a small genome, and can be manipulated genetically to express, or not express, a variety of genes. Over the years, several attempts were made to spread yeast cells with limited success. Recently, we have developed a method that works routinely, resulting in efficient and rapid release of yeast chromatin from growing cells *(11)*. To date we have focused our analysis on the yeast genes encoding rRNA, which are transcribed by RNA Pol I. They are present in 150–200 tandem copies at a single chromosomal locus and make up about 10% of the yeast genome. Their abundance and tandem repetition make the rRNA genes easy to detect in a chromatin spread, as shown in the low magnification view of a lysed yeast cell in Fig. 4.1a, and also at higher magnification in Fig. 4.1b.

We use two slightly different methods for spreading yeast chromatin. One is the traditional Miller spreading method that utilizes a small amount of detergent in water that has been adjusted to pH 9. This method (Method 1) removes more of the proteins from the chromatin and from the nascent transcripts, allowing the chromatin to disperse more freely. An example of a single rRNA gene visualized by this method is shown in Fig. 4.2a. We frequently use this method when the experiment calls for counting the number of active rRNA genes in a given stretch of chromatin or counting the number of polymerase molecules transcribing individual rRNA genes.

In Method 2, the cells are lysed in the same manner but then diluted into a solution containing 11 m*M* KCl. This small amount of salt allows retention of more nucleosomes on the chromatin and more proteins on the nascent transcripts. The chromatin does not disperse quite as well as in Method 1, but one typically sees many more yeast cells on an EM grid because the chromatin masses are more concentrated. The image in Fig. 4.1 shows a chromatin mass of this sort, with the inset in Fig. 4.1c showing a portion of nucleosomal chromatin from this cell. Figure 4.2b, c shows single rRNA genes visualized by this method, with different methods of contrast enhancement, as will be discussed.

In yeast cells, we have used the chromatin spreading method to study aspects of transcriptional regulation by Pol I (e.g., **refs.** *(11, 12)*) and also to study the structure and co-transcriptional processing of nascent rRNA transcripts (e.g., **refs.** *(13, 14)*). Visualization of genes by this method allows one to infer a temporal series of RNA processing events, due to the multiple transcripts per gene and the fact that each transcript is slightly older or younger than its neighboring transcripts. Nascent rRNA transcripts are bound by a characteristic "terminal knob," a particle that is thought to include U3 small nucleolar RNA (snoRNA) *(2, 13)* and has been seen at

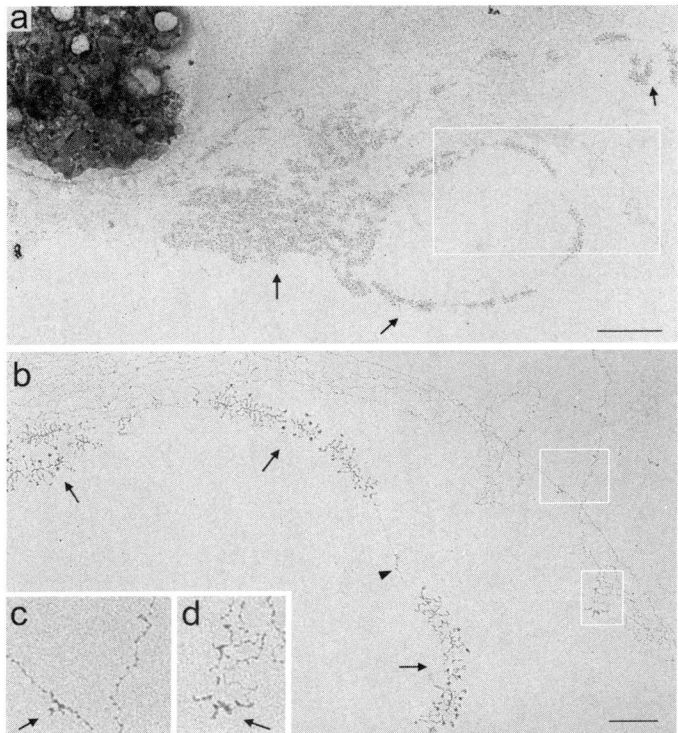

Fig. 4.1 Overview of chromatin released from a lysed yeast cell by Miller spreading. **a** Low-magnification view of a lysed cell ghost (large mottled dark object in upper left corner) with chromatin spilling out to the right and below. The tandemly repeated and transcriptionally active rRNA genes are especially noticeable in the chromatin mass due to their relatively dark appearance, which in turn is due to multiple nascent RNA transcripts in a RNP complex. *Arrows* indicate regions that correspond to the dispersed nucleolus containing these rRNA genes. Nonnucleolar regions of the chromatin mass (upper middle region of the panel) appear lighter due to the relative paucity of transcripts. Bar, 2 μm. **b** Higher-magnification view of the *white-boxed* region from (**a**), showing a few tandem repeats of active rRNA genes (*black arrows*) with some strands of non-nucleolar chromatin to the right. An active 5 S rRNA gene is also visible in the spacer between the large rRNA genes (*arrowhead*). Bar, 0.5 μm. (**c, d**) Higher-magnification views of the *white-boxed* regions in (**b**), showing pre-mRNA genes transcribed by pol II (*arrows*) and also showing nucleosomal chromatin structure

the 5′ end of rRNA transcripts from all eukaryotes visualized by this method (Fig. 4.2a–c). In yeast, the terminal knob enlarges co-transcriptionally to encompass the 18S rRNA coding region, and in this form, the terminal knob is called the small subunit (SSU) processome. The SSU processome has been isolated biochemically and corresponds to a complex RNP particle that is essential for production of small subunit rRNA *(13)*. In yeast, we found that the SSU processome is often cleaved

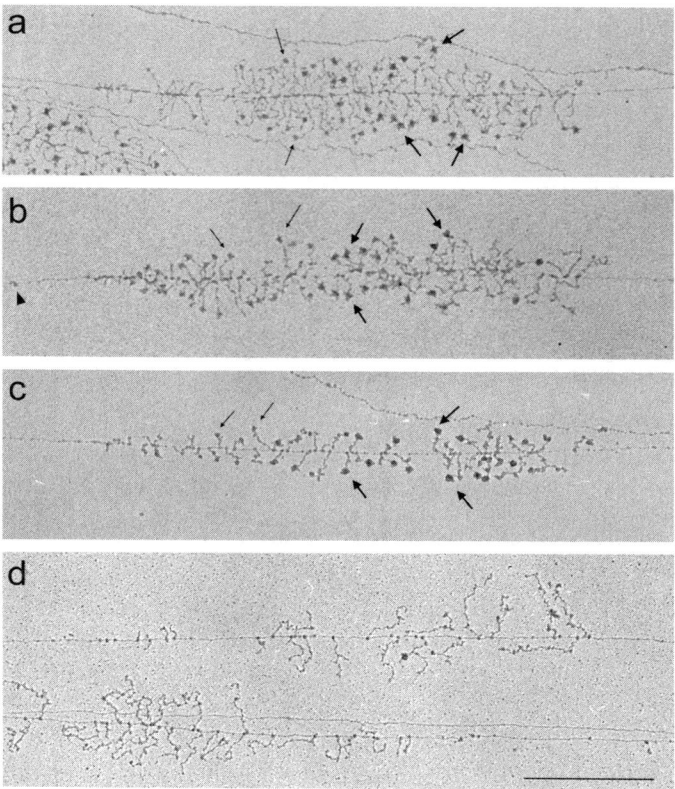

Fig. 4.2 Representative EMs of yeast rRNA genes obtained using different spreading conditions. In each case, the gene promoter is to the left and the terminator to the right. Shown at the same magnification; bar, 0.4 μm. **a** Method 1 conditions (typical Miller spread conditions), unshadowed image. Characteristic structures on the nascent transcripts include 5′ terminal knobs (*small arrows*) and SSU processomes (*larger arrows*), as discussed in the text. Many of the transcripts in the last quarter of yeast rRNA genes are missing terminal knobs because they have been cleaved in an early step of rRNA processing (to separate small and large subunit RNAs). **b** Method 2 conditions (slightly increased salt concentration during lysis), unshadowed image. Note that the transcripts in this gene are somewhat more proteinaceous and compacted as compared with the transcripts on the gene in (**a**). *Arrows* as in (**a**). *Arrowhead* indicates an active 5 S rRNA gene. **c** Method 2 conditions plus rotary shadowing with platinum (see text). Note that contrast is improved when grids are shadowed, but ultrastructural details are lost. For example, note the difference in shape between the SSU processomes (*large arrows*) in (**b**) and (**c**): protrusions are visible in the unshadowed image but not in the shadowed image. **d** Method 1 conditions using a yeast strain in which the single gene for U3 snoRNA is under the control of a repressible promoter. Shown are two rRNA genes 3 h after repression of U3 RNA synthesis (with the gene at the bottom in opposite 5′ to 3′ orientation relative to the gene at top). Note the absence of terminal knobs, SSU processomes, and cleaved transcripts, as seen in the control rRNA genes above

off the transcripts while the transcript is nascent *(14)* (Fig. 4.2a–c). By spreading cells with genes of interest under the control of repressible promoters, one can see the effect of depleting a specific protein or RNA on both the rRNA transcription levels and nascent transcript structure (e.g., **refs.** *(14)*). For example, the rRNA genes in Fig. 4.2d are from yeast cells that have been depleted of U3 snoRNA. Note the absence of 5′ terminal knobs on these transcripts as compared with the control genes in Fig. 4.2a–c *(13)*.

Yeast chromatin spreads also allow analysis of the 5 S rRNA genes transcribed by RNA polymerase III, since the 5 S genes are located in the intergenic spacers between each of the large rRNA genes and so can be easily located (Figs. 4.1b, 4.2b, arrowheads). Active 5 S genes are seen to be engaged by one to three polymerases. Their transcripts are only 132-nt long and thus are too short to be seen protruding from the polymerases. These very short genes have not previously been identifiable in Miller spreads.

Genes presumed to be transcribed by Pol II, encoding pre-messenger RNAs (mRNAs), are also routinely seen in yeast chromatin spreads. As expected in yeast cells, Pol II genes encoding pre-mRNA are shorter than Pol I genes encoding pre-rRNA, but not as short as Pol III genes. Examples are boxed in Fig. 4.1b, and shown at higher magnifications in Fig. 4.1c, d (arrows). Most often, yeast Pol II genes will exhibit only a single transcript, though occasionally they will be more highly transcribed. On rare occasions, short intron loops are seen near the 5′ ends of these transcripts (not shown). We have not yet identified specific Pol II genes in yeast chromatin spreads, though this should be feasible with genetic manipulation, such as by placing the gene of interest on an extrachromosomal plasmid.

2   Materials

2.1   Growing Yeast

1. Growth medium: appropriate for the yeast strain of interest, such as YP+dextrose (YPD) or YP+galactose (YPgal). YP medium is 1% (w/v) yeast extract (BD Bacto; VWR, West Chester, PA, USA; cat. DF 0886-17), 2% (w/v) peptone (BD Bacto; VWR; cat. DF 0118-17), and 2% (w/v) dextrose (BD Bacto; VWR; cat. 90000-908) or galactose (Sigma-Aldrich, St. Louis, MO, USA; cat. G0625). Make as sterile 2× solutions and dilute 1:1 with either sterile distilled water (dH$_2$O) or with 2 M sorbitol before use.
2. 2 M sorbitol solution: filter-sterilize and mix 1:1 with 2× growth medium as required.

2.2   Lysing Yeast

1. 30°C water bath.
2. Rocking platform.
3. Zymolyase 20T (USBiological, Swampscott, MA, USA; cat. Z1000).
4. Nalgene filters: 115-mL size, 0.2 μm, disposable (Nalgene, Rochester, NY, USA; cat. 120-0200). The solutions for chromatin spreading are filtered through a filter that has been extensively rinsed with dH$_2$O to free it of residual detergents. The filters (labeled for a particular solution) are reused over many months, since the goal is not sterility but rather removing particulate material.
5. Water: for all the following solutions, ChromAR water is used (Mallinckrodt Baker, Phillipsburg, NJ, USA; cat. 6795-10). It is important that the water quality be reproducible and that it be of very low ionic strength.
6. Buffer solution pH 10 (Fisher Scientific, Pittsburgh, PA, USA; cat. SB116-500): this is a potassium carbonate–potassium borate–potassium hydroxide buffer for standardizing pH meters that we use to adjust the pH of the following solutions without adding a lot of ions. For this reason, the pH of the spreading solutions is not stable, so all must be made fresh daily. We normally make 100 mL each of the next four solutions.
7. Triton X-100 solution: 0.025% v/v Triton X-100 (Acros scintillation grade; Fisher Scientific) in dH$_2$O, pH adjusted to 9.0–9.1 with Fisher pH 10 buffer for Method 1. For Method 2, the pH is adjusted to 9.25. Both solutions are filtered and made fresh daily.
8. Sucrose–formalin solution: 0.1 M sucrose, 10% (v/v) formaldehyde (e.g., Sigma-Aldrich; cat. F-1635 supplied as a 37% solution) in dH$_2$O, pH adjusted to 8.8 with Fisher pH 10 buffer, and filtered. It is important that no more than 20–25 drops of pH 10 buffer (using a Pasteur pipette) are needed to achieve the desired pH (*see* **Note 1**).
9. Photo-Flo 200 solution: 0.06% (v/v) Photo-Flo 200 (Eastman Kodak, Rochester, NY, USA) in dH$_2$O, adjusted to pH 8.8 with Fisher pH 10 buffer and filtered. Make fresh daily.
10. For Method 2: dH$_2$O adjusted to pH 9.0 with Fisher pH 10 buffer and filtered. Make fresh daily.
11. For Method 2: 0.1 M KCl adjusted to pH 7.0 with Fisher pH 10 buffer and filtered. Make fresh daily.
12. For Method 2: 11 mM KCl solution made by an 8:1 dilution of dH$_2$O (pH 9.0):0.1 M KCl (pH 7.0).
13. Plastic petri dishes: 35×10 mm (e.g., Falcon #35-1008; BD Biosciences, San Jose, CA, USA).

2.3   Depositing Yeast Lysates on EM Grids

1. EM grids, carbon film covered: we routinely prepare our own carbon film (*see* **Note 2**), depositing it on 300-mesh copper grids (Electron Microscopy Sciences,

Hatfield, PA, USA; cat. 0300-Cu). Commercially available carbon film-covered grids can be used, but we have no experience with these.

2. Microcentrifugation chambers: used to contain the chromatin lysate, allowing the lysate to be centrifuged down onto the grid, which is situated at the bottom of the chamber. Chambers are made by cutting a ~2-cm diameter clear Plexiglas rod into disks ~6-mm high. A 4-mm diameter hole is drilled almost all the way through the center of the disk, leaving only a very thin layer of undrilled Plexiglas at the bottom. The sides and bottom of the hole must be milled smooth, so that the grid can lie flat and the film on the grid does not break.

3. Centrifuge: Sorvall RC-4 with HB-4 rotor (DuPont, Wilmington, DE, USA) or similar.

4. Centrifuge adaptors for microcentrifugation chambers: standard laboratory bottle caps with an outer diameter of ~2.8 cm and height of ~1.8 cm are used.

5. Tweezers: anti-capillary, self-closing (Dumont pattern N4; Ernest Fullam, Latham, NY, USA; cat. 14140). It is useful to have two tweezers so that two grids can be manipulated simultaneously.

6. Forceps: 10–12 inches long.

7. Ross optical lens tissue (Fisher Scientific; cat. 09-740).

8. Bibulous paper (VWR; cat. 28511-007).

9. Round cover slips: 18-mm diameter (e.g., Fisher Scientific).

10. Plastic petri dishes, 100×15 mm (e.g., Fisher Scientific).

11. Round filter paper: 90-mm diameter (e.g., Whatman, Clifton, NJ, USA; cat. 1001 090).

2.4   Contrast Enhancement

1. Phosphotungstic acid (PTA) stain: 4% (w/v) stock solution in dH_2O. We have used PTA from Sigma-Aldrich, Electron Microscopy Sciences, Polysciences (Warrington, PA, USA), and Mallinkrodt (Hazelwood, MO, USA) with similar results. After filtering, store in a stoppered brown bottle and use over the course of a year. *PTA is a heavy metal acid which can irritate skin, eyes, and respiratory tract.*

2. Uranyl acetate stain: 4% (w/v) stock solution in dH_2O. We have used the suppliers listed in **Section 2.4.1**. After filtering, store in a stoppered brown bottle and use over the course of a year. A yellow precipitate will form in the bottom of the bottle, and care should be taken not to disturb this precipitate when removing stain from the bottle. *Uranyl acetate is a radioactive heavy metal salt and should be handled accordingly.*

3. 95% ethanol.

4. Microtest tissue culture plates: 96-well (Falcon; BD Biosciences; cat. 35-3072).

5. Small glass petri dish filled with 95% ethanol.

6. Plastic wash bottles: one containing 95% ethanol and the other dH_2O, and waste receptacles for each bottle.

2.5   Examining the Grids

1. Transmission EM: we use a JEOL100CX.

3   Methods

3.1   Growing Yeast

1. On the morning of the day before an experiment, inoculate a single yeast colony from a fresh plate into a 100-mL Erlenmeyer flask containing 5 mL of the appropriate growth medium and grow at the appropriate temperature (usually 30°C) with shaking.
2. In the late afternoon, measure the OD_{600} and, based on the doubling time determined previously, calculate the dilution appropriate to reach an OD_{600} of 0.4 the next morning. Dilute the cells into the appropriate growth medium plus $1 M$ sorbitol, which protects the yeast from osmotic shock during zymolyase treatment; we grow cells for extended periods in sorbitol so that they are fully adapted to this medium before lysis. Cells transferred to sorbitol medium usually undergo a lag before recommencing rapid growth, and to compensate for the variable length of this lag we generally make two additional dilutions, one with 3× the calculated amount of yeast cells and the second with 10× the calculated amount. This helps to ensure that we will have cells at the appropriate density to spread the next day (OD_{600} between 0.4 and 0.6).
3. The following morning, determine the OD_{600} of the various dilutions and use the culture that is at the appropriate density. If all the cultures have grown to an OD_{600} beyond 0.6, we dilute to an OD_{600} of 0.1–0.2 and let the culture grow to the appropriate density (*see* **Note 3**). The best chromatin spreads are obtained from cells that have not grown beyond mid-log phase (OD_{600} between 0.4 and 0.6), so we routinely use this growth range unless the experiment calls for cells at a different stage (*see* **Note 4**).

3.2   Lysing Yeast

1. Measure 5 mg of zymolyase into a 15-mL conical disposable centrifuge tube and add 200 μL of the appropriate sorbitol growth medium. Vortex the tube to dissolve the zymolyase and put the tube in the 30°C water bath for a few minutes to equilibrate the temperature.
2. Working rapidly, add 1 mL of yeast cell culture in mid-log phase to the tube containing zymolyase, vortex, and put the tube in the 30°C water bath. Shake the tube continuously by hand for 4 min with occasional vortexing, allowing the

zymolyase to just begin to digest the yeast cell wall so that hypotonic lysis will occur in steps 4–5.

3. Transfer the yeast to an Eppendorf tube and spin at full speed (15,000×*g*) in a microcentrifuge for 6 sec, just enough time to result in a soft pellet of yeast cells. Invert the tube and shake out the supernatant; it is important to get all of the medium out of the tube so that it does not contribute to the ionic strength of the lysate. To achieve this, wipe the inside of the tube with a Kim-wipe that has been twisted into a sharp-pointed, thin shape, being careful not to touch the pellet.

4. For Method 1, add slowly 1 mL of 0.025% Triton, pH 9.1 to the pellet.

   For Method 2, add slowly 1 mL of 0.025% Triton, pH 9.25 to the pellet.

In both cases, one should see cells being detached from the pellet incrementally rather than having the entire pellet detach from the tube at once.

5. For Method 1, the same 1-mL micropipette and tip can be used to transfer the yeast–Triton suspension to a small flask containing an additional 3 mL of 0.025% Triton, pH 9.1, solution. For Method 2, transfer the suspension to a small flask containing 6 mL of 11 m*M* KCl solution.

In both cases, as the yeast hits the additional volume, one should see the solution become cloudy. The flask is then swirled and the same micropipette and tip is used to pipette the solution ~10× to help with lysis and dispersal. After an additional swirl, remove 1 mL to a 35×10-mm plastic petri dish, cover the flask and petri dish, and put both on a rocker platform at medium speed (*see* **Note 5**).

6. After 30 to 45 min of gentle rocking to further loosen the chromatin, add 1/10 volume of sucrose–formalin solution to each container and allow the chromatin to disperse for an additional 15 min to 1 h or so.

3.3 *Depositing the Yeast Lysate on EM Grids*

1. Place carbon film-covered EM grids film-side up in a small glass petri dish filled with 95% ethanol, and allow them to sit for ~1 min to make them hydrophilic. We usually prepare eight grids at a time (*see* **Note 6**).

2. Fill eight grid chambers with sucrose–formalin solution by inserting a filled Pasteur pipette to the bottom of each chamber and releasing the solution. The goal is to avoid bubbles in the chamber, since then the grid will not lie flat on the bottom. Chambers are slightly overfilled such that the surface is convex.

3. Remove the grids one by one from the ethanol with tweezers (holding onto the edge of the grid) and hold the grids over a wash beaker. Pipette 1–2 Pasteur pipette volumes of sucrose–formalin solution slowly onto the tweezers, allowing the solution to run gently down over the grid, rinsing off *all* the ethanol (if the ethanol is not rinsed off the chromatin will precipitate). Hold the wet grid film-side up over the central cavity of a grid chamber inside the rounded-up liquid volume, and release it. It should gently float to the bottom of the chamber and lie flat.

4. After rinsed grids have been placed in the bottom of each of the eight chambers, pull off the excess amount of sucrose–formalin solution over each grid with a Pasteur pipette, to leave the chamber about half full of sucrose–formalin, which acts as a sucrose cushion to keep cell debris off the grid.

5. Using a 200-µL micropipette, gently layer ~65 µL of the yeast cell lysate from the small petri dish over the sucrose–formalin cushion above the grid so that the chamber is slightly overfilled. A round, 18-mm cover slip is then dropped in place over the chamber, and is sealed by the excess liquid.

6. Blot the cover slip's surface by pressing down with a folded sheet of bibulous paper, and place the grid chamber in the appropriate-sized bottle cap, which is used as a centrifuge adaptor. The placement in the bottle cap is most easily achieved by using an old tweezers stretched so that the tines are on opposite sides of the grid chamber, which can then be gently lowered into the bottle cap. A second filled grid chamber can be positioned over the first in the same bottle cap. The remaining six chambers, also filled and sealed, are placed into three more bottle caps (*see* **Note 7**).

7. Using the 10- or 12-inch forceps, place the bottle cap adapters containing two grid chambers each into the swinging buckets of an HB-4 or similar centrifuge rotor. The tines of the forceps are inserted on opposite sides of the grid chamber and on the inside surface of the bottle cap, with outward pressure exerted. With practice, one can use this method to gently lower the bottle cap to the bottom of the bucket and also to remove it after the centrifugation is finished. For Method 1, spin grid chambers at ~7,000×*g* for 5 min with the brake off. For Method 2, spin at ~1,600×*g* for 5 min with the brake off. In addition to having the brake off, it is important when starting the centrifuge that the speed dial be turned up slowly and smoothly so that the chromatin is not sheared or stretched and is minimally disturbed.

8. After centrifugation, remove the bottle caps from the buckets and the grid chambers from the bottle caps, using old tweezers. Slide off the cover slips and add one or two drops of sucrose–formalin solution over the center cavity until it is slightly overfilled. The grid chamber is then inverted and tapped gently on its side. The grid should float to the bottom of the drop of sucrose–formalin hanging from the chamber and can then be carefully grasped on an edge with anticapillary tweezers and dipped into a beaker of Photo-Flo solution for 30 sec.

9. The grid is then carefully blotted with lens tissue that has been folded in half; the tissue should touch *only* the edge of the grid and the region between the tines of the tweezers. From the dull side of the grid, one can watch the drying process. The Photo-Flo reduces surface tension and allows for more uniform drying (*see* **Note 8.**) Grids can be stained immediately or accumulated for later staining. We normally put the grids of a particular yeast strain on filter paper in a 100-mm petri dish that has been labeled appropriately, and store them at room temperature.

3.4   Contrast Enhancement

1. Grids are stained both with PTA to preferentially stain proteins, and with uranyl acetate (UA) to stain nucleic acids (*see* **Note 9**). Staining is done in 96-well microtiter plates. To stain eight grids: using a Pasteur pipette, nine drops of 95% ethanol are added to each of eight wells in the first vertical row of a microtiter plate. Three drops of PTA stock stain solution are added to each of the top four wells and mixed; three drops of UA stock stain solution are added to each of the bottom four wells and mixed. *Gloves should be worn and all washes and pipettes saved for proper disposal. Pipettes can be rinsed with 95% ethanol from a wash bottle and stored in 15-mL culture tubes, which are appropriately labeled and re-used over many months.*

2. A grid is picked up on the edge with anticapillary tweezers and submerged for 30 sec in the first well containing PTA. It is important that the grid be completely submerged, but not so far that it touches the bottom of the well, which will result in bending the grid. The grid is then rinsed with one Pasteur pipette volume of 95% ethanol by running the ethanol down the tines of the tweezers over the grid. The grid is then blotted on the edge with a small piece of filter paper to remove excess ethanol and immediately submerged in the first well of UA for 1 min. The grid is rinsed as before with one to two pipette volumes of ethanol and dried with lens tissue. Each well of stain can be used for staining two grids (*see* **Note 10**). The grids can then be viewed in the EM, and genes will appear with contrast similar to that shown in Fig. 4.2a, b.

3. If one wants additional contrast, one can evaporate a thin layer of platinum over the chromatin and genes will appear similar to that in Fig. 4.2c (*see* **Note 11**). There are many different vacuum evaporators that can be used for platinum shadowing; the two essential items are that the platinum is evaporated at a 7–8-degree angle relative to the grids, and that the grids are spinning at about 30–60 rpm during platinum deposition. A light coating of platinum (~15–20 Å) is preferable, since too much platinum can ruin a grid and if the shadow is too light, one can always add more later. It is also critical that the grids be flat during the shadowing procedure so that platinum deposition is uniform and reaches all areas of the grid.

3.5   Examining the Grids

1. We view the grids in a JEOL100CX transmission EM at 60 kV using a 40-μm objective aperture. After determining that there are lysed yeast cells on the grid, we scan row by row at a low magnification of ~×6,000. One can easily see the yeast ghosts without using the binoculars, and the rRNA genes frequently extend out from the ghosts. We normally take a low magnification (×4,000–6,000)

micrograph of the chromatin field and then higher magnification images (~×14,000) of genes or regions of interest.

2. We scan all grids, even if some are not useful. The quality of the chromatin spread and the amount of chromatin material often varies from grid to grid, even though the same lysate from the same petri dish is centrifuged onto all eight grids (*see* **Notes 12** and **13**).

4   Notes

1. If more than 20–25 drops of pH 10 buffer are needed to achieve the desired pH in the sucrose–formalin solution, this suggests that the free acid level in the formaldehyde is too high. The acidity of the formalin will increase with age, but we find that the primary cause of high free acid is the variability of this parameter during the manufacturing process. Before purchasing, we contact various companies (e.g., Sigma-Aldrich) and choose a bottle from the lot that contains the lowest amount of free acid.

2. We make carbon film-covered grids using 300-mesh copper grids that have a shiny and a dull side. The film is deposited on the shiny side. Making high-quality film that is thin yet strong enough to withstand chromatin deposition is a technically demanding endeavor. Many different protocols are available and are beyond the scope of this chapter. The method we use is described below, and a somewhat more detailed version, including schematics, was previously published (8). In brief, we dip clean, warmed microscope slides into a 40% glycerol solution and let the glycerol drain evenly from the slides. The glycerol is then wiped from the back of the slide and the slides are put on a piece of filter paper in a vacuum evaporator containing a carbon source. The evaporator is pumped down and the grids are glow-discharged at maximum current for 1.5–2 min. The evaporator is then evacuated and the current increased until the carbon rods begin to evaporate. The amount of carbon deposited is determined by monitoring the color of the filter paper under the slides; a light beige color results in a very thin film, which is the goal. A large black-coated Pyrex dish is then filled to overflowing with distilled water. A small beaker with a thin mesh screen laid across its top is submerged in the water. One of the carbon-coated slides is then held at a 45-degree angle with the carbon-glycerin side up and gently lowered into the water. The carbon film will float on the surface of the water and the glycerol will dissolve. Using two forceps, the carbon film is maneuvered so that it is out of the way while grids are being placed on the mesh screen, shiny side up. The grids are dipped in 95% ethanol before being submerged; one attempts to replicate with the grids the shape of the piece of carbon film. The carbon film is then gently moved into position so that it is directly over the grids and the water is aspirated, allowing the film to gently settle over the grids. The grids are left on the wire mesh to dry overnight and sorted the next day for their degree

of coverage, which can be ascertained by looking with reflected light at the dull side of the grid held at a 30–40-degree angle under a dissecting microscope. The sorted grids can be stored in the refrigerator and used over the course of a year.

3. If the yeast have not grown as well as expected or if the OD is much higher than expected, we examine the yeast culture in an inverted optical microscope after spotting 20 µL on a slide. One can determine whether most of the yeast cells are budded, as expected for a rapidly growing strain, and also if the solution is contaminated. If the culture is contaminated, throw it out and start over again.

4. Although we usually prepare cells in mid-log phase after growth at 30°C, the experiments sometimes call for examining cells after they leave log phase or, for temperature-sensitive mutations, after growth at 37°C. In both of these situations, the cells are refractory to lysis, and it is more difficult to get optimal dispersal.

5. It is critical that this procedure (**Section 3.2**, **steps 2–5**) be done very rapidly, especially if one is interested in transcriptional regulation. If the cells are allowed to linger in conditions in which growth continues but is slowed (e.g., at room temperature instead of 30°C) transcription initiation can be down-regulated, resulting in 5′ ends of rRNA genes that are not completely loaded with Pol I molecules. Therefore, one wants to go from transcribing cells to lysed cells as rapidly and with as little perturbation as possible. Once the cells are lysed in a large volume of low ionic strength liquid, biological processes cease and transcription and replication intermediates appear to arrest in place.

6. We usually prepare eight grids at a time from a particular yeast strain lysate. All eight grid chambers are centrifuged simultaneously, and are treated identically as far as possible. It is good practice to have eight grids, because a variety of pitfalls can occur: grids can be accidentally crunched at any of the steps, the film can break or tear, and some grids may have little material on them, especially with Method 1. The remaining lysate, both in the petri dish and the unused lysate in the Erlenmeyer flask, can be covered with Parafilm and put in the refrigerator overnight. The next day, if one discovers when examining the grids that there is not sufficient material or that most of the film is broken one can make additional grids from the refrigerated lysates. This works best if one is examining transcriptional regulation (using Method 1), since the number of transcripts per gene remains stable over time in these conditions. We do not use saved lysates when examining the structure of nascent transcripts (using Method 2), since there can be changes in the degree of protein removal from RNA over time.

7. It is easier to do experiments involving several different strains or conditions if one has sufficient grid chambers so that cleaning is not necessary between preparations. We recommend a minimum of 32 grid chambers, allowing spreads from four different conditions.

8. Up until this point it is extremely important that the grid not dry out at any of the steps.

9. Before staining, one can examine the grids by putting them on a slide and looking with an inverted optical microscope. One should see lysed yeast cells on at
   least some of the grid squares. One can also ascertain how much film is broken.
   If one sees no yeast cells or very little intact film, there will be little or no chromatin on the grid and therefore no point in staining it. One can also put 20-μL
   drops of the lysate on a slide, add a cover slip, and look in the inverted optical
   microscope. Lysed cells become a dark gray and one can frequently see a halo
   of chromatin emanating from them, suggesting a good lysis has occurred.
   Unlysed cells have a bright, birefringent appearance; their presence in a preparation frequently results in more broken film.

10. Staining is very tedious, and to speed it up we stain two grids simultaneously
    by holding one tweezers with a grid in each hand and staining in two adjacent
    wells. In fact, after becoming adept with grasping grids on the edge with the
    tweezers, two grids can be processed simultaneously in all of the steps.

11. The platinum shadow makes viewing the grids much easier, but it also obscures
    some of the ultrastructural details as can be seen by comparing yeast rRNA
    genes in Fig. 4.2b (unshadowed, Method 2) with Fig. 4.2c (shadowed, Method
    2). We suggest looking at the grids first without shadowing with platinum, and
    only shadowing when necessary or when ultrastructural detail is not relevant to
    the goals of the experiment.

12. A good spread will result in a low density of chromatin lying across the grid in
    a relaxed, nonstretched manner. Transcripts will be unwound and lie on either
    side of the chromatin template; they should have reproducible and distinct RNP
    structures. One has to be patient in searching for such regions on a grid. Grids
    will often exhibit regions with stretched chromatin. Transcript structure will
    occasionally be rather muddy, or look "precipitated," as if adjacent structures of
    a transcript coalesced. Too much salt in the lysate can cause this, for example,
    if not all of the growth medium was removed from the Eppendorf tube after
    the yeast was pelleted. Likewise, chromatin will frequently not be dispersed
    enough to see distinct genes. Rather, the darkly stained Pol I backbones and
    SSU processomes will be apparent over a chromatin background. Frequently
    yeast cells are flocculent so they tend to stick together, and in these strains, one
    often sees clumps of yeast on a grid square with chromatin emanating from the
    outer periphery. This material on the periphery often contains useful views of
    active genes.

13. We are often interested in analyzing rRNA genes from strains that have been
    depleted of a particular protein thought to have a role in ribosome biogenesis.
    Generally, depleted strains are more difficult to spread. They do not grow as
    well and may be more refractory to lysing. Depending on the protein depleted,
    transcription levels can go up or down and the morphologies of the nascent
    rRNA transcripts can change; when components of the SSU processome are
    depleted, nascent transcripts lack the distinctive SSU processomes and do not
    undergo nascent transcript cleavage (Fig. 4.2d). It is critical to always do a con-
    trol (nondepleted) spread at the same time as the experimental (depleted), using
    the same solutions and grids. If one is examining a question concerning gene
    regulation, and is depleting a protein whose gene is regulated by a gal-inducible

promoter (and thus is switching from galactose to glucose media), the controls should also include the parental strain after the same media switch because rRNA transcription is up-regulated in glucose.

Acknowledgments The authors thank Dr. Oscar L. Miller, Jr., who trained all of us in the Miller chromatin spreading method. The writing of this article was supported by the National Science Foundation (NSF) grant MCB448171. Research in the authors' lab is supported by grants from the NSF and the National Institutes of Health (NIH).

References

1. Miller, O.L. Jr. and Beatty, B.R. (1969) Visualization of nucleolar genes. *Science* **164**, 955–957.
2. Mougey, E.B., O'Reilly, M., Osheim, Y., Miller, O.L. Jr., Beyer, A., and Sollner-Webb, B. (1993) The terminal balls characteristic of eukaryotic rRNA transcription units in chromatin spreads are rRNA processing complexes. *Genes Dev.* **7**, 1609–1619.
3. Osheim, Y.N., Proudfoot, N.J., and Beyer, A.L. (1999) EM visualization of transcription by RNA polymerase II: downstream termination requires a poly(A) signal but not transcript cleavage. *Mol. Cell* **3**, 379–387.
4. Beyer, A.L. and Osheim, Y.N. (1988) Splice site selection, rate of splicing, and alternative splicing on nascent transcripts. *Genes Dev.* **2**, 754–765.
5. Osheim, Y.N., Sikes, M.L., and Beyer, A.L. (2002) EM visualization of Pol II genes in Drosophila: most genes terminate without prior 3′ end cleavage of nascent transcripts. *Chromosoma* **111**, 1–12.
6. French, S. (1992) Consequences of replication fork movement through transcription units in vivo. *Science* **258**, 1362–1365.
7. Osheim, Y.N., Miller, O.L., Jr., and Beyer, A.L. (1988) Visualization of Drosophila mela-nogaster chorion genes undergoing amplification. *Mol. Cell. Biol.* **8**, 2811–2821.
8. Osheim, Y.N. and Beyer, A.L. (1989) Electron microscopy of ribonucleoprotein complexes on nascent RNA using Miller chromatin spreading method. *Meth. Enzym.* **180**, 481–509.
9. Beyer, A.L., Sikes, M., and Osheim, Y. (1994) EM methods for visualization of genetic activity from disrupted nuclei. *Meth. Cell Biol.* **44**, 613–630.
10. Osheim, Y.N. and Beyer, A.L. (1998) EM visualization of transcriptionally active genes after injection into Xenopus oocyte nuclei. *Meth. Cell Biol.* **53**, 471–496.
11. French, S.L., Osheim, Y.N., Cioci, F., Nomura, M., and Beyer, A.L. (2003) In exponentially growing Saccharomyces cerevisiae cells, rRNA synthesis is determined by the summed RNA polymerase I loading rate rather than by the number of active genes. *Mol. Cell. Biol.* **23**, 1558–1568.
12. Schneider, D.A., French, S.L., Osheim, Y.N., Bailey, A.O., Vu, L., Dodd, J., Yates, J.R., Beyer, A.L., and Nomura, M. (2006) RNA polymerase II elongation factors Spt4p and Spt5p play roles in transcription elongation by RNA polymerase I and rRNA processing. *Proc. Natl. Acad. Sci. USA* **103**, 12707–12712.
13. Dragon, F., Gallagher, J.E., Compagnone-Post, P.A., Mitchell, B.M., Porwancher, K.A., Wehner, K.A., Wormsley, S., Settlage, R.E., Shabanowitz, J., Osheim, Y., Beyer, A.L., Hunt, D.F., and Baserga, S.J. (2002) A large nucleolar U3 ribonucleoprotein required for 18 S ribosomal RNA biogenesis. *Nature* **417**, 967–970.
14. Osheim, Y.N., French, S.L., Keck, K.M., Champion, E.A., Spasov, K., Dragon, F., Baserga, S.J., and Beyer, A.L. (2004) Pre-18 S ribosomal RNA is structurally compacted into the SSU processome prior to being cleaved from nascent transcripts in Saccharomyces cerevisiae. *Mol. Cell* **16**, 943–954.

Chapter 5
Combing Genomic DNA for Structural and Functional Studies

Catherine Schurra and Aaron Bensimon

Keywords Molecular combing; Fluorescent hybridisation; FH; DNA replication; Genomic structure analysis; Stretching

Abstract Molecular combing is a process whereby single DNA molecules bind by their extremities to a silanised surface and are then uniformly stretched and aligned by a receding air/water interface *(1)*. This method, with a high resolution ranging from a few kilobases to megabases, has many applications in the field of molecular cytogenetics, allowing structural and functional analysis at the genome level. Here we describe protocols for preparing DNA for combing and for the use of fluorescent hybridisation (FH) applied to combed DNA to conduct physical mapping or genomic structural analysis. We also present the methodology for visualising and studying DNA replication using combed DNA.

1 Introduction

Genomic instability, a major feature of cancer cells, has important implications for cell cycle regulation, replication, differentiation, and apoptosis. This instability is mainly characterised by changes in chromosomal structure such as rearrangements, translocations, deletions, amplifications, and/or modification of the replication pattern. Many molecular cytogenetic techniques used to investigate chromosomal aberrations which employ fluorescence hybridisation (FH) of DNA molecules, e.g., on metaphase spreads, are based on condensed DNA. The spatial resolution of these methods is limited to 1–5 Mb.

Attempts to improve the resolution by decondensing DNA have included the use of lysed cells, for example, interphase cell nuclei *(2)*, a human–hamster hybrid cell line treated with an inhibitor of topoisomerase II and alkaline buffer *(3)*, or chromatin released with formamide or NaOH/ethanol solutions *(4, 5)*, as well as methods using direct visual hybridisation (DIRVISH) on stretched and deproteinised DNA *(6, 7)*.

For methods that employ physical stretching of DNA, the resolution is a few kilobases, but the drawback of these methods is the uncontrolled and variable stretching of the DNA molecules. Consequently, external standard controls have to be used in parallel to calibrate the measurements. These methods are also limited by the low number of DNA molecules available for observation, making quantitative and statistically significant results difficult to obtain.

Molecular combing provides a reliable tool to resolve some of the problems mentioned above. Molecular combing involves two steps: first, specific anchorage of the DNA molecules at one or both extremities under precise physico-chemical conditions on a glass surface coated with silane, which has exposed vinylic $(-CH=CH_2)$ end groups; and second, extension of the DNA molecules by the movement of a receding air/water interface. The stretching of DNA is uniform and reproducible, yielding a very precise extension factor of 2 Kb/μm, which may vary depending upon the treatment used to coat the surface (8) but is independent of the size and sequence of the DNA molecules. The bond between silane and the DNA end is strong enough to withstand the capillary forces (>160 pN) that stretch the molecule to 150% of its crystallographic length. Since stretching occurs in the immediate vicinity of the meniscus, all molecules are uniformly stretched. The specific binding at the ends of DNA molecules is probably due to partial unwinding of the double helix at the extremities. The pH-induced denaturation of the DNA ends should expose the hydrophobic part of the nucleotide bases and thus promote interaction with the silanised surface. Many attempts have been made to specifically anchor DNA to a solid surface, but most of these approaches have involved modification of the ends of the molecules. Molecular combing, on the other hand, presents the advantage that no chemical modification of the extremities is required; the molecules bind spontaneously to the coated glass surface. The average length of the combed molecules is 250–500 Kb, and occasionally molecules of more than 1 Mb are obtained. Typically, 20 to 100 genomes can be stretched on a single 22×22-mm silanised surface, making quantitative results feasible.

Purity of DNA is a critical feature of the combing process. Preparation begins by embedding cells in agarose plugs to protect DNA from shearing. The DNA is then deproteinised and the agarose is removed by digestion. The extracted DNA is transferred to a buffered solution and placed in a Teflon reservoir, where a silanised coverslip is inserted by a motorised translation stage. The DNA, in a random coil conformation in solution, binds through one or both extremities to the surface. After a few minutes the surface is removed vertically from the solution at a constant speed (300 μm/s) to stretch the DNA molecules bound on the surface (Fig. 5.1). After combing, the DNA is irreversibly fixed to the surface and provides a target for FH. Sequence-specific probes are labelled with haptens such as biotin or digoxygenin and hybridised to the combed genomic DNA. The hybridised regions are detected with several layers of affinity molecules coupled to different fluorochromes, visualised as linear fluorescent segments with a fluorescence microscope, and recorded with a high-resolution digital camera.

Molecular combing has found many applications in the molecular cytogenetics field: FH of probes onto combed DNA allows direct physical mapping of their respective positions along the stretched molecules. The reliability of FH on combed DNA has been shown by the refined physical mapping of the Ca^{2+}-activated neutral

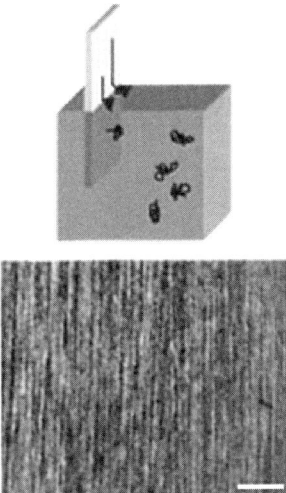

Fig. 5.1 The mechanism of combing. Randomly coiled DNA molecules in solution in the binding buffer (0.5 M MES, pH 5.5) bind at one of their extremities to the silanised surface of a coverslip. This surface is then slowly removed from the solution by the lifting apparatus (*upper panel*). The interaction of the hydrophobic surface and the liquid solution of DNA forms a receding meniscus as the coverslip rises, and this stretches the DNA molecules (*lower panel*). Bar, 20 μm

protease 3 gene, the detection of subtle deletions or rearrangements that could not be detected by classical methods such as for tuberous sclerosis (9) and for micro-deletions in the *BRCA1* gene (10), the determination of the number of gene copies in an amplified 21 trisomy region (11), and the elucidation of the structures of amplicons (12) and of highly repeated regions such as human ribosomal DNA (rDNA) arrays displaying non-canonical structures (13).

Additionally, DNA combing can be applied to the functional study of the replication process. Classical techniques to identify origins of replication are competitive polymerase chain reaction (PCR) and 2D-gel electrophoresis, which identifies replication intermediates as bubbles or replicating forks, but these methods provide only partial information. Molecular combing offers the capacity to localise and quantify the firing of several origins on the DNA molecule. For this technique, growing cells are first differentially labelled with halogenated thymidine analogs. Consequently, replicating DNA can be visualised by immunofluorescence techniques (14). This allows the visualisation of replication origins as well as the measurement of parameters of replication such as fork velocities, inter-origin distances, fork densities, the timing of fired origins, and certain anomalies such as the presence of fork barriers or over-replication (Fig. 5.2). When coupled to FH, it is possible to localise and map fired origins on specific regions of the genome such as rDNA arrays (15). These parameters provide information about the replication

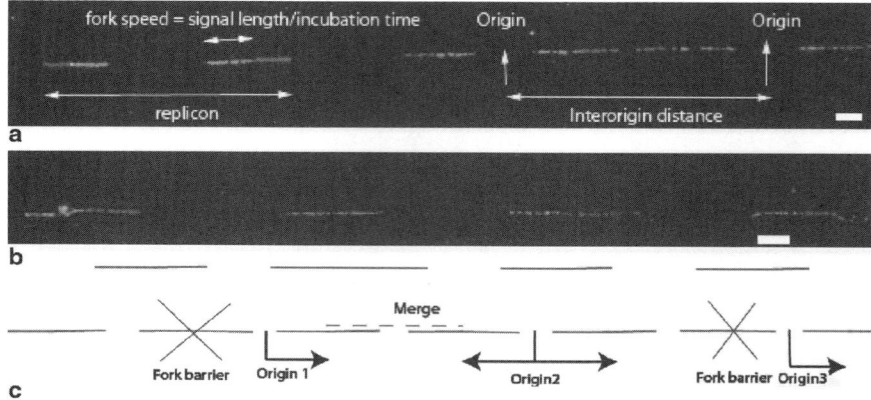

Fig. 5.2 Images of fluorescent replication signals visualised at ×40 magnification. Bars, 10 μm. **a** Three replicons localised on the same DNA molecule, which was not labelled and is not visible; only the replication signals from the incorporated IdU and CldU are visualised. A replicon is defined as a region replicated from a single fired origin. The distance between two activated origins can be measured when initiation takes place before adding the first label (IdU); in the present case the origin is localised precisely at the middle of the gap between the two green segments. The replication fork speed is measured by dividing the length of the replicated region by the incubation time with the label (IdU or CldU). **b** Another replication signal. Example of fork barriers. **c** Interpretation of signals. The first signal with two contiguous green-red segments is followed by a green segment where a red-green was expected. This means that a fork barrier occurs at origin 1: it is a unidirectional origin. The green signal alone is a merge of outgoing forks coming from origins 1 and 2. Origin 2 is bidirectional and represents the majority of cases. Origin 3 is unidirectional with another fork barrier. To view this figure in color, see COLOR PLATE 3

profile of the cell line and are particularly useful to screen novel anti-cancer drugs and to target a lead compound that would re-establish a normal replication program in treated cancer cells. The technique should be useful for discovery of more specific drugs that are less toxic for normal tissues and so avoid heavy chemotherapeutic treatment.

Molecular combing may also be applied to the study of interactions between DNA and proteins. Binding of the DnaA protein has been detected on combed genomic DNA from *E. coli* (*16*); however, it is not known if the protein is bound in a sequence-specific manner. Transcription on combed DNA has also been observed, but only when the extension was close to the normal length; no transcription occurred on DNA overstretched to ~150% of the molecule's normal contour length (*17*).

Molecular combing offers a very powerful tool for wide genome studies that could complement other cytogenetic and molecular biology techniques to better understand chromosomal aberrations. This chapter describes, with all relevant practical details, the individual steps required to prepare DNA for combing, and the different processes applied to combed DNA for replication and/or genomic structural analysis using methods of FH validated to fit the molecular combing tool.

2   Materials

2.1   Cell Culture and Replication Labelling

1. 5-chloro-2′-deoxyuridine (CldU) and 5-iodo-2′-deoxyuridine (IdU) (Sigma-Aldrich, Saint-Quentin Fallavier, France).
2. Foetal bovine serum (Invitrogen, Cergy Pontoise, France).
3. RPMI 1640 (BioWhittaker, Fontenay-sous-bois, France).
4. MEM Earl's salts and L-glutamine (Invitrogen).
5. Trypsin-EDTA (BioWhittaker).

2.2   Preparation of Agarose Blocks for Combing

1. LMP agarose 1.5% (w/v) in phosphate-buffered saline (PBS): dissolve 1.5 g of LMP agarose powder (Tebu-Bio, Le Perray en Yvelines, France; cat. 50080E) in 100 mL of PBS in a microwave oven. Store at +4°C for up to 3 months.
2. Proteinase K (20 mg/mL) (Eurobio, Courtaboeuf, France; cat. GEXPRK01.B5).
3. Plug molds (Bio-Rad, Marnes-la-Coquette, France; cat. 1703713).
4. N-lauroylsarcosine sodium salt (sarcosyl), EDTA, Tris-Cl, and phosphate-buffered saline (PBS) (all from Sigma Aldrich).
5. β-agarase I (New England Biolabs; Ozyme, Saint Quentin Yvelines, France; cat. M0392S).
6. ESP solution: prepare a fresh solution with 0.1 g of sarcosyl, 1 mL of proteinase K (20 mg/mL), and 9 mL of 500 mM EDTA, pH 8.0. Put at 50°C to dissolve the sarcosyl and activate the enzyme.
7. 0.5 M EDTA, pH 8.0. Autoclave, can be stored for at least 1 year at room temperature.
8. TE: 10 mM Tris-HCl 1 mM EDTA, pH 8.0. Autoclave, can be stored for 2 months at room temperature.

2.3   Silanisation (see Note 1)

1. Coverslips (Erie Scientific Company; VWR, Fontenay-sous-Bois, France; cat. 631-0302).
2. 7-octenyltrichlorosilane (Interchim, Montlucon, France; cat. SI 06708.0). *Use gloves, glasses and a chemical hood to manipulate silane, which is very volatile and toxic*. To neutralise its toxicity, add water and wait a few minutes until the silane is polymerised and can be then handled without a chemical hood. Use a hood to handle solvents (chloroform, toluene, pirana solution, and ammonium hydroxide).

3. Teflon reservoirs and glass vessel for silanisation: *see* **Note 1**. Ceramic 12-place coverslip racks (Thomas Scientific, Swedesboro, NJ, USA; cat. EA 8542-E40).
4. NH_4OH solution (ACS reagent quality, 28.0–30.0% as NH_3; Sigma Aldrich; cat. 221228). *Handle under a chemical hood and with gloves and glasses.*
5. 33% H_2O_2 (VWR; cat. 23622.298).
6. Toluene (for analysis, VWR; cat. 28676.297). *Handle under a chemical hood.*
7. Ethanol (EtOH) (Sigma-Aldrich; cat. E-7023).
8. Syringe for injection of silane: 100 µL (Roucaire, Courtaboeuf, France; cat. 0030081030).
9. Oxygen and argon (Air Liquide, France). *Use a chemical hood to avoid enriching the air with these gases.*

2.4 Combing

1. YOYO-1 iodide (Molecular Probes; Invitrogen, Cergy Pontoise, France; cat. Y3601). *Use gloves: YOYO-1 is a DNA intercalant and may be a carcinogen.*
2. 0.5 *M* 2-morpholinoethanesulfonic acid (MES), pH 6.5: dissolve MES (Sigma-Aldrich; cat. M-2933) in distilled water, adjust the pH with NaOH, and autoclave, can be stored for 1–2 months at +4°C.
3. Superglue, cyanoacrylate (Henkel, Boulogne Billancourt, France; Radiospares cat. 379–8946).
4. Combing apparatus (*see* Fig. 5.3 and **Note 1**).

2.5 Labelling of Probes

1. BioPrime DNA labelling kit (Invitrogen; cat. 18094-011).
2. Dig 11-dUTP (Roche Diagnostics, Meylan, France; cat. 11 558 706 910).
3. 100 m*M* (25 µmoles) dATP, dCTP, dGTP, and dTTP (MP Biomedicals, Illkirch, France; cat. NTATP 100, NTCTP100, NTGTP100, and NTTTP100, respectively).

2.6 Checking the Quality of Probes

1. Anti-digoxygenin-AP (Roche; cat. 1 093 274).
2. Digoxygenin-labelled control DNA (Roche; cat. 1 585 738).
3. Alkaline phosphatase substrate kit IV (Vector Labs; ABCYS, Paris, France; cat. SK-5400). This includes the chromogenic substrate (BCIP and NBT).
4. Hybond N+ membranes (Amersham, Les Ulis, France; cat. RPN82B).
5. Alkaline phosphatase-conjugated streptavidin (Jackson ImmunoResearch, West Grove, PA, USA; cat. 016-050-084).

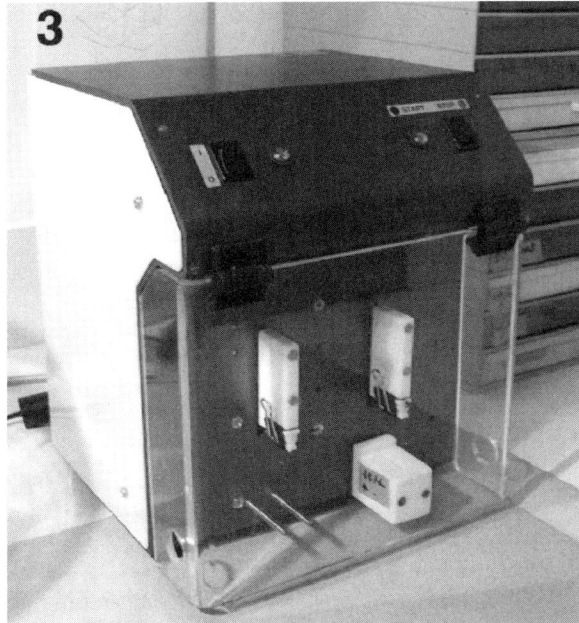

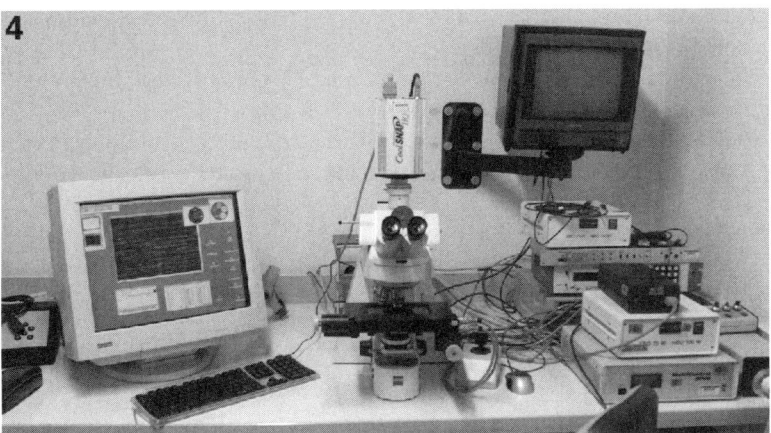

Figs. 5.3 and **5.4** Combing apparatus and microscope station

2.7   *Fluorescent Hybridisation*

1. Deionised formamide (Qbiogene, Illkirch, France; cat. FORMD002). *Use a chemical hood to manipulate formamide because it is a teratogen.*
2. Blocking-Aid (Molecular Probes).
3. Herring sperm DNA (Promega, Charbonnières, France).

4. Vectashield mounting medium for fluorescence (Vector Labs; AbCys cat. H-1000).
5. 20× SSC (Interchim, Montluçon, France; cat. 0804-4 L).
6. Tween 20, NaCl and sodium acetate (NaAc) (cat. S2889) (Sigma Aldrich).
7. Human cot-1 DNA (Invitrogen; cat.15279-011).
8. Post-hybridisation solution: formamide 50%/2× SSC.
9. Hybridisation solution: 50% formamide, 2× SSC, 0.01 M NaCl, and 0.5% Tween 20 in Blocking-Aid. Do not use dextran sulphate because it causes noise on hydrophobic surfaces.
10. Denaturation solution: 0.05 M NaOH, 0.1 M NaCl.

2.8 Antibodies for FH and Replication

1. FITC-labelled mouse anti-BrdU and mouse anti-BrdU (Beckton Dickinson, Le Pont-de-Claix, France; cat. 347583 and 347580, respectively).
2. Rat anti-BrdU (AbD Serotec, Cergy Saint-Christophe, France; cat. OBT0030).
3. Alexa 488-labelled, highly cross-absorbed goat anti-mouse IgG (Molecular Probes; cat. A11029).
4. Streptavidin labelled with Alexa 594 (cat. S-32356), Alexa 488, Alexa 350, and Alexa 750 (Molecular Probes).
5. Alexa 350-labelled goat anti-mouse IgG (cat. A21049) and donkey anti-goat IgG (cat. A 21081) (Molecular Probes).
6. Alexa 594-labelled donkey anti-rat IgG (cat. A 21209) (Molecular Probes).
7. Biotinylated rabbit anti-streptavidin (Rockland Immunochemicals, Gilbertsville, PA, USA; cat. 200–4695).
8. Alexa 488-labelled donkey anti-sheep IgG (Molecular Probes; cat. A11015).
9. FITC-labelled sheep anti-digoxygenin (Roche; cat. 1207741).
10. FITC-labelled rabbit anti-donkey (Rockland; cat. 616–4202).

Antibodies are resuspended following the manufacturers instructions, and aliquoted and stored at −20°C unless +4°C is specified. When thawed, aliquots are stored for no more than 1 week in the refrigerator.

2.9 Visualisation of Fluorescent Signals

A microscope station (Fig. 5.4) is required to observe the fluorescent signals and to acquire images. This consists of a microscope (e.g., Axioplan 2; Zeiss), a UV lamp (XBO 75 W HBO 100 W; Zeiss), a motorised stage (MultiControl 2000; ITK, Lahnau, Germany), and a CCD camera (Photometrics HQ Coolsnap; Roper Scientific, Evry, France). Filter sets consist of excitation, emission, and dichroic filters (XF03 for blue fluorescence, XF100-2 for green, XF102-2 for red, and

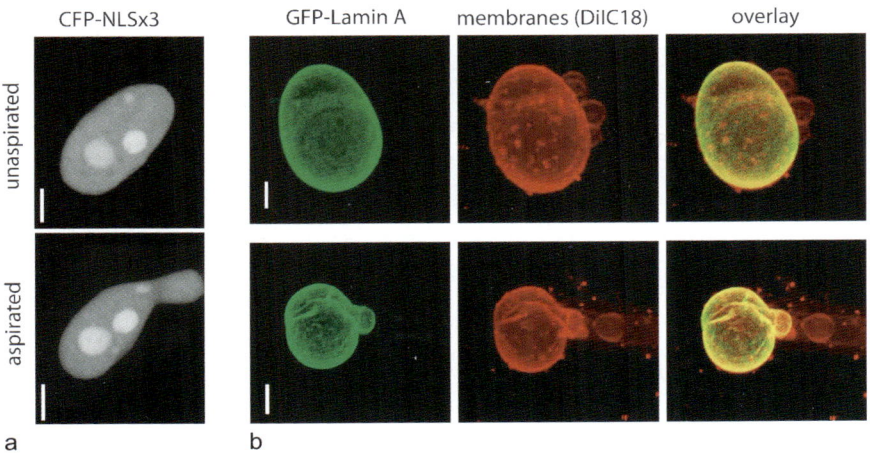

COLOR PLATE 1. **Fig. 1.1** Images of nuclei deformed by micropipette aspiration. *See* complete caption on p. 8

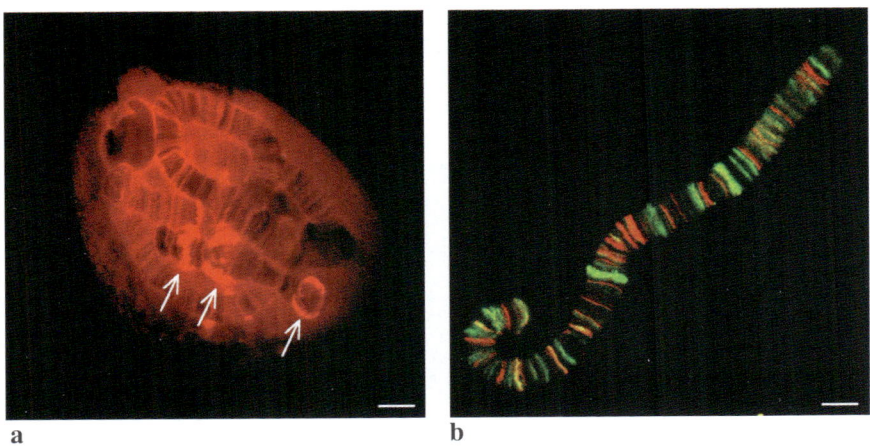

COLOR PLATE 2. **Fig. 3.1** Immunofluorescent labelling of splicing factors in an intact polytene nucleus and in an isolated polytene chromosome. *See* complete caption on p. 32

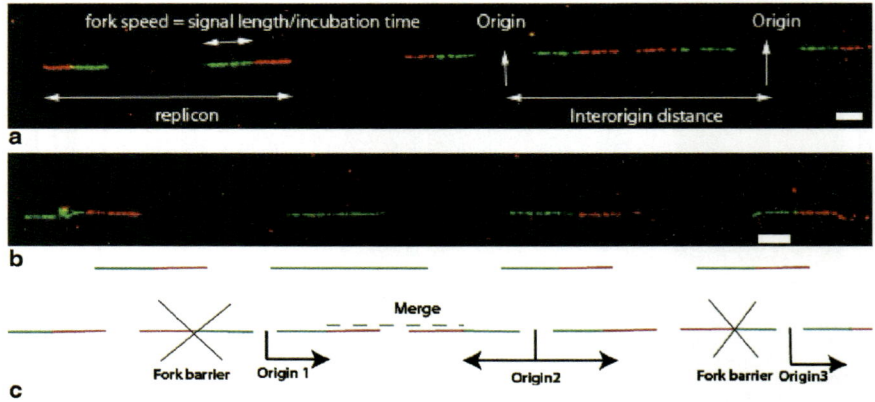

a

b

Merge

Fork barrier   Origin 1   Origin2   Fork barrier   Origin3

c

COLOR PLATE 3. **Fig. 5.2** Images of fluorescent replication signals visualised at ×40 magnification. *See* Complete caption on p. 74

COLOR PLATE 4. **Fig. 6.4** Tracks of a few of the mRNP particles in a nucleus overlaid on an image of chromatin density. *See* Complete caption on p. 101

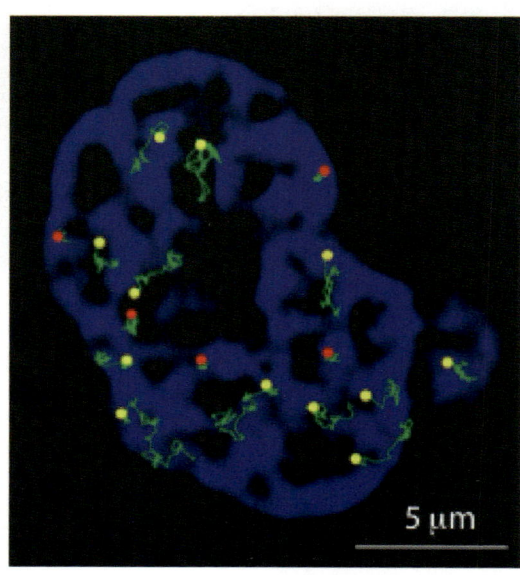

5 μm

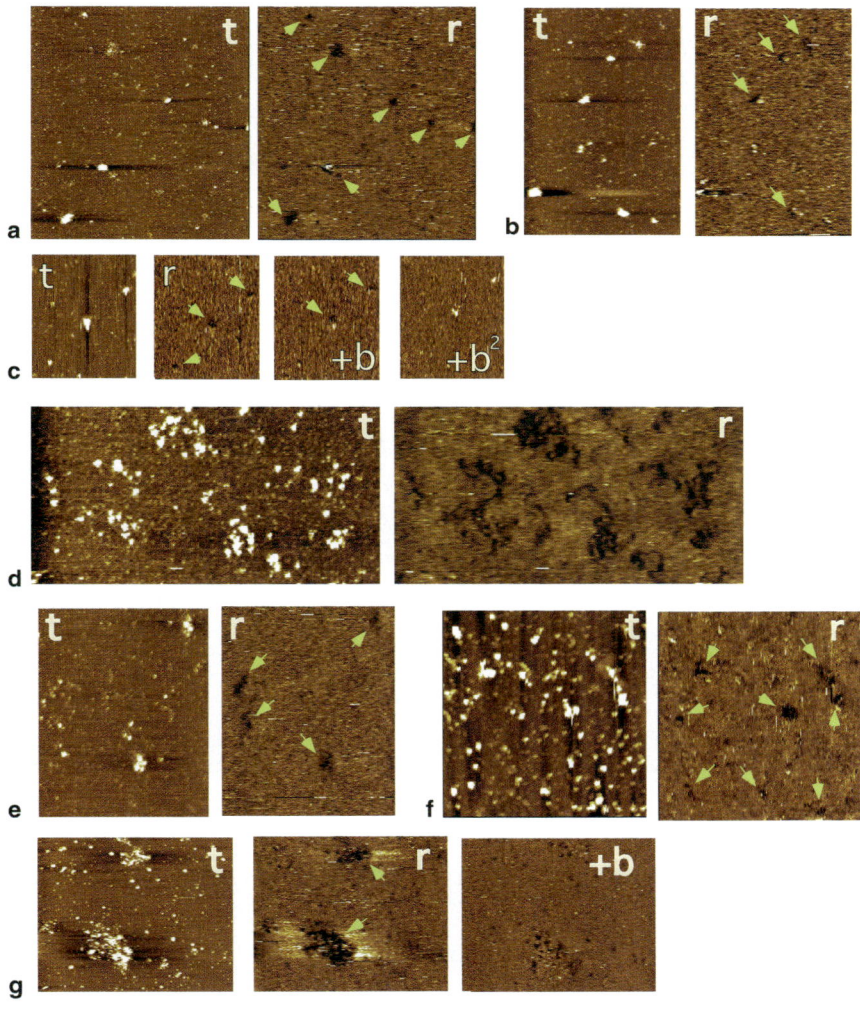

Color Plate 5. **Fig. 8.4** Examples of Recognition Imaging. *See* complete caption on p. 134

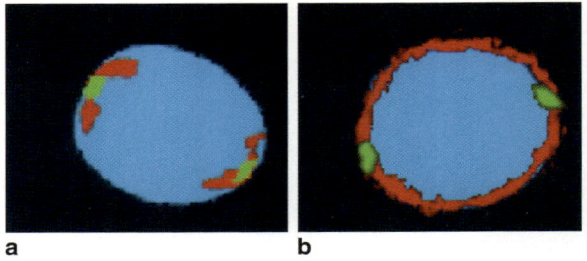

CeLOR PLATE 6. **Fig. 12.5** Binding of membrane vesicle fractions MV1 and MV2beta to decondensed sea urchin sperm chromatin. *See* complete caption on p. 216

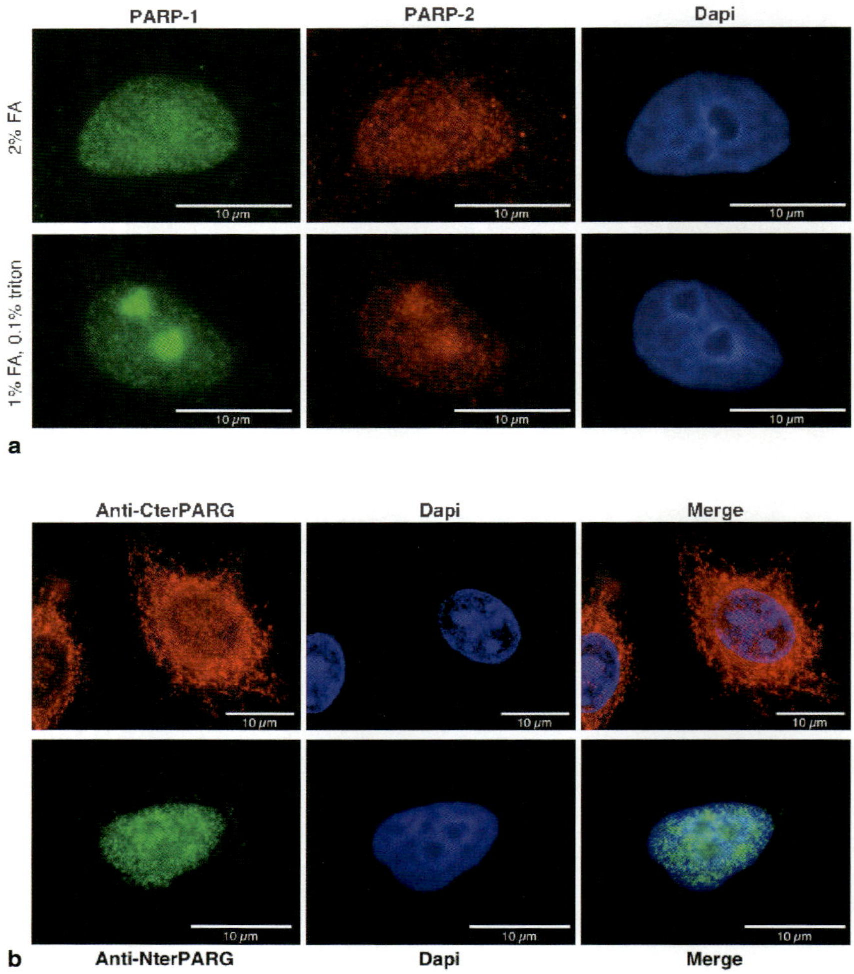

PARP-1	PARP-2	Dapi

2% FA

1% FA, 0.1% triton

a

Anti-CterPARG	Dapi	Merge

b   Anti-NterPARG   Dapi   Merge

CeLOR PLATE 7. **Fig. 15.2** Detection of PARP-1, PARP-2, and PARG by immunofluorescence microscopy. *See* complete caption on p. 276

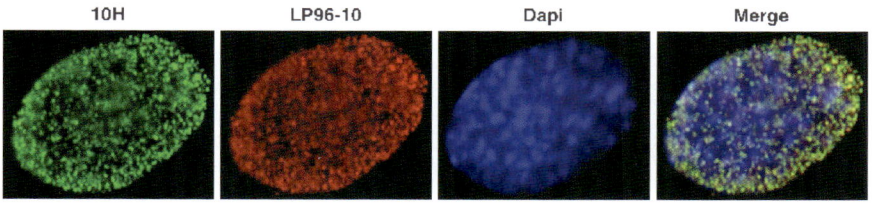

COLOR PLATE 8. **Fig. 15.4** Detection of PAR by immunofluorescence microscopy. *See* complete caption on p. 280

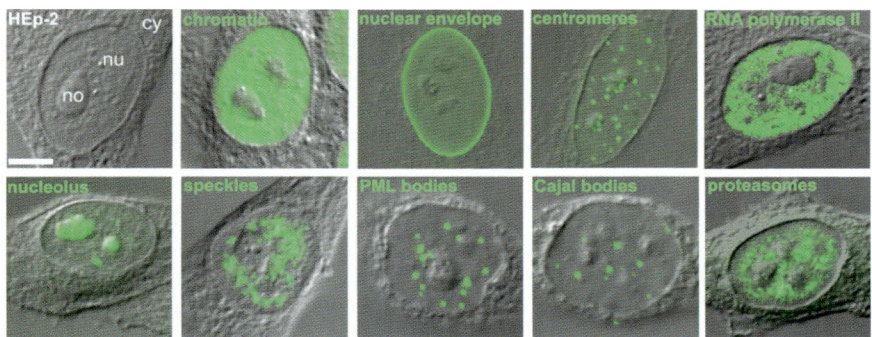

COLOR PLATE 9. **Fig. 18.2** Subnuclear structures. *See* complete caption on p. 323

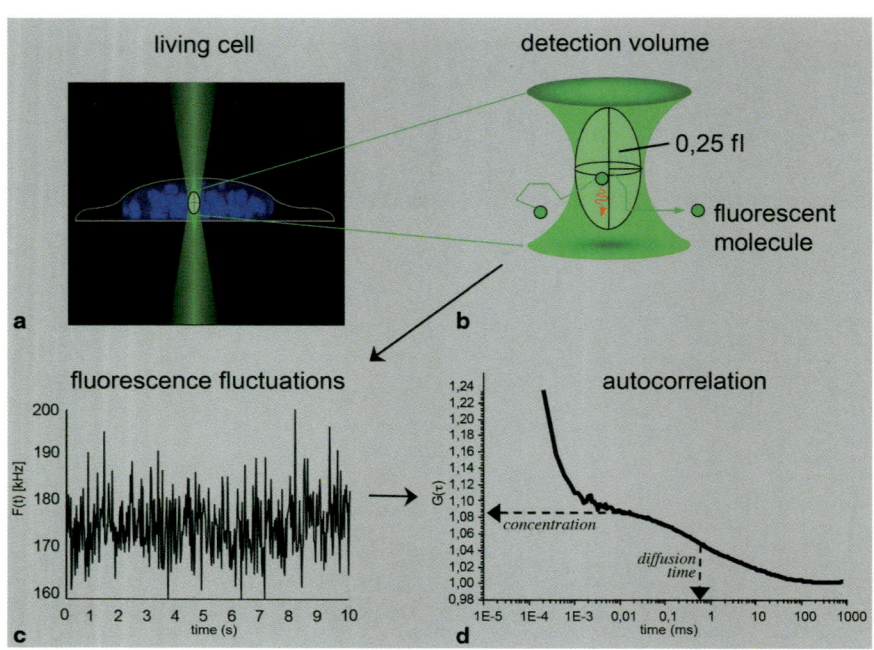

COLOR PLATE 10. **Fig. 18.4** FCS analysis procedure. *See* complete caption on p. 325

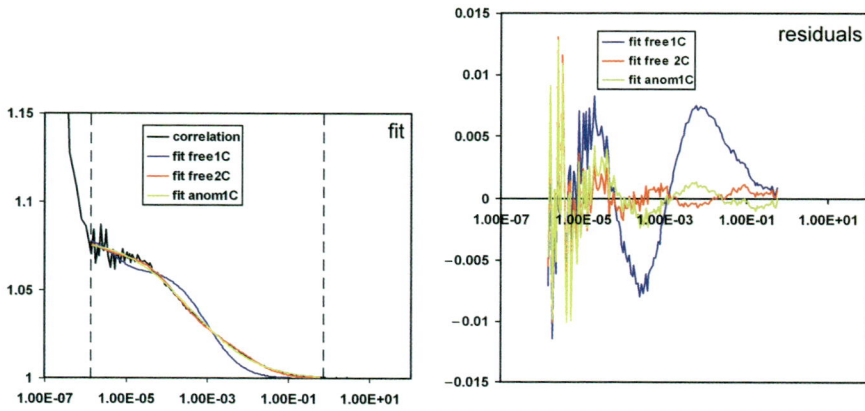

COLOR PLATE 11. **Fig. 18.8** Fitting an autocorrelation function to model equations. *See* complete caption on p. 335

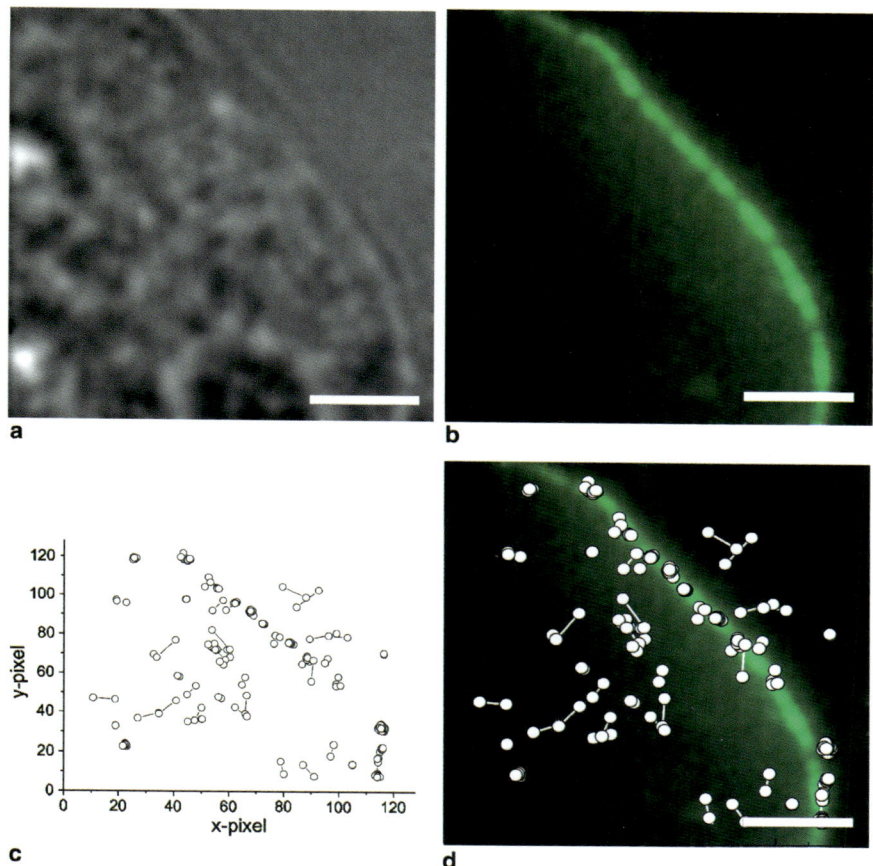

COLOR PLATE 12. **Fig. 19.2** Import of single molecules of human NTF2/p10 (a 14-kDa protein implicated in NLS-mediated nuclear import) in digitonin-permeabilised HeLa cells stably expressing POM121-GFP, a GFP conjugate of the nucleoporin POM121. *See* complete caption on p. 353

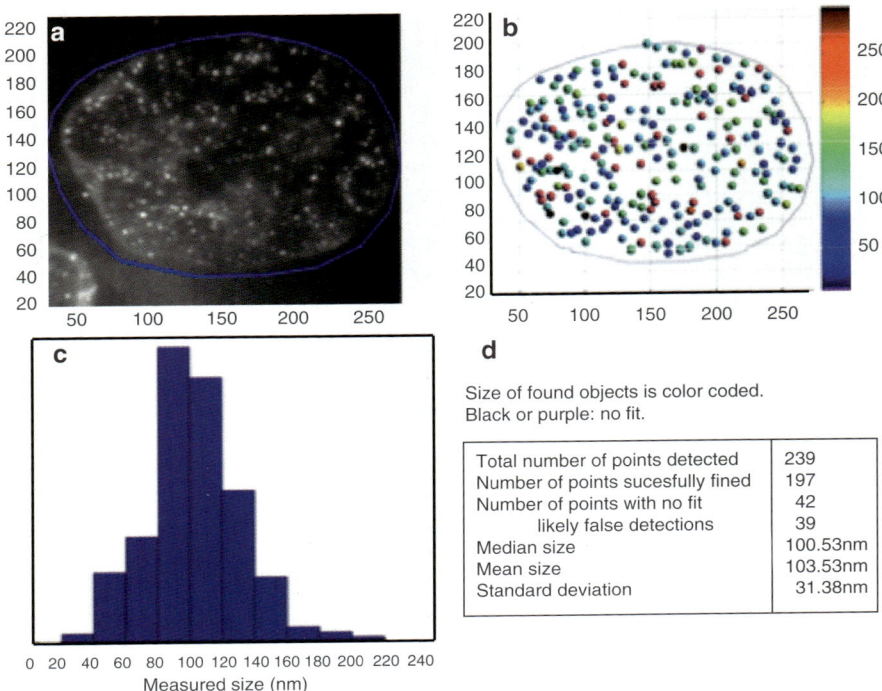

Size of found objects is color coded.
Black or purple: no fit.

Total number of points detected	239
Number of points sucesfully fined	197
Number of points with no fit	42
likely false detections	39
Median size	100.53nm
Mean size	103.53nm
Standard deviation	31.38nm

Measured size (nm)

COLOR PLATE 13. **Fig. 21.4** The results as saved by the command "savefis_rep." *See* complete caption on p. 398

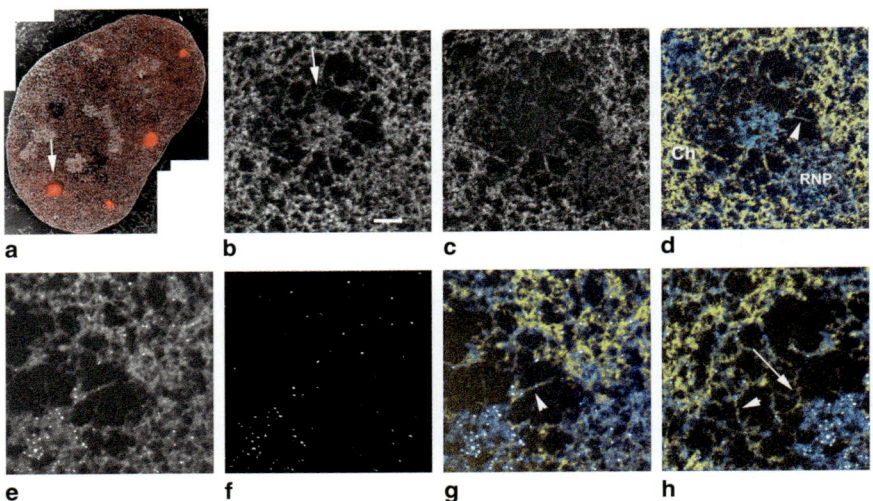

COLOR PLATE 14. **Fig. 23.1** Correlative and fluorescence micrographs of a 70-nm-thick section of an SK-N-SH neuroblastoma cell nucleus. *See* complete caption on p. 417

XF143-2 for infrared; Omega Optical, Brattleboro, VT, USA). These are all connected together with a computer. The program SmartCapture 2.1 (Digital Scientific, Cambridge, UK) controls the camera, the motorised filter wheel and objective(s), and the automated microscope stage. This software allows the acquisition and storage of images, and home-made software (Jmeasure 2.2.4) allows length measurements of FH and replication signals.

3 Methods

3.1 *Labelling Cells for Studies of DNA Replication*

It is essential to label the cells when they are growing exponentially and asynchronously; only a few percent of the cells are in a replicating state. Sequential labelling with two halogenated nucleotides allows analysis of the spatio-temporal profile of replication at the genome level of cells treated with compounds that influence replication compared with control cells. The protocol varies depending on the cell type (suspension culture or adherent cells like fibroblasts). The species of origin of the cells is important, because some cells like murine lymphoblastoid cells have problems incorporating IdU but not BrdU and CldU, for unknown reasons.

3.1.1 Adherent Cells

1. Add an appropriate volume of IdU stock solution to give a $100\,\mu M$ final concentration in the culture medium, or for a better result remove the medium and add fresh pre-warmed medium containing $100\,\mu M$ IdU at 37°C. Incubate the cells at 37°C for the appropriate time depending on the replication speed, typically 30 min. Use a timer (*see* **Note 2**).
2. At t = 30 min, remove the medium and wash the cells twice with 1× PBS.
3. Add medium pre-warmed at 37°C containing $100\,\mu M$ CldU, and incubate the cells at 37°C for the same time as that used for IdU.
4. A t = 60 min, wash the cells twice in 1× PBS.
5. Trypsinize the cells for 2–3 min at 37°C, stop the trypsinisation by adding serum.
6. Centrifuge the cells at room temperature.

3.1.2 Cells in Suspension Culture

1. Add IdU to $100\,\mu M$ final concentration and mix the culture flask very gently to avoid stressing the cells. Incubate for a precise time at 37°C.
2. At t = 30 min, without washing (*see* **Note 3**), add CldU to $100\,\mu M$ final concentration and incubate the culture further for the same time.
3. At t = 60 min, centrifuge the cells at room temperature.

3.2 Preparation of Genomic DNA in Agarose Blocks

1. Melt a bottle of 1.5% LMP agarose in PBS in a microwave oven and keep molten in a 50°C water bath.
2. Pre-heat ESP solution for at least 1 h at 50°C to activate proteinase K.
3. Prepare plug molds sealed with adhesive tape on the bottom to avoid leaking.
4. Pre-chill 50 mL of sterile 1× PBS in an ice bucket (see **Note 4**).
5. Switch on a water bath at 37°C.
6. Start from **step 6, Section 3.1.1** or step 3, **Section 3.1.2**, depending on the type of cells, and remove the supernatant.
7. Resuspend the cells thoroughly in cold 1× PBS to give 500,000 cells/50 μL (see **Note 5**).
8. Place at 37°C for 3 min (see **Note 6**).
9. Add an equal volume of molten 1.5% LMP agarose.
10. Vortex for 5 sec and immediately distribute a 90-μL aliquot into each well of the plug mold.
11. Put the plug mold in the refrigerator for at least 30 min to allow the block to solidify.
12. Remove the tape from the plug molds and eject the agarose block into a 15-mL plastic tube by tapping the mold vigorously on the tube, or by pushing the block with the small plastic strip provided with the plug molds.
13. Fill the tube with ESP solution to just cover the blocks, i.e., 250 μL/block.
14. Incubate the tubes overnight at 50°C for deproteinisation.
15. The following day, remove the ESP solution using a spatula or gauze to retain the block. Avoid pipetting, because the blocks can break.
16. Fill the tube with at least 10 mL of 0.5 M EDTA, pH 8.0. Blocks can be stored for months (even for years) at +4°C if necessary.

3.3 Treatment of Agarose Blocks for Combing DNA

1. Transfer a block from EDTA solution to a 50-mL (or at least 15-mL) conical tube, fill the tube with TE, and wash the block for 1 h at room temperature on a slowly rotating wheel.
2. Change the TE solution and wash again for 1 h in the same conditions. Washing removes residual cell fragments and enzymes.
3. Remove the TE solution, put the block in a 2-mL Eppendorf tube with 1.5 mL of 0.1 M MES, pH 6.5, and place the tube in a 68°C water bath for 20 min to melt the agarose.
4. Place the tube for 5 min at 42°C (see **Note 7**).
5. Add 1–2 μL of β-agarase I, mix the tube very gently by inverting it slowly once or twice, and incubate overnight at 42°C to allow the enzyme to digest the agarose (see **Note 8**).
6. The following day, pour the DNA solution very gently into a 1.5-mL Teflon reservoir (see **Note 9**) and place it in the combing apparatus.

3.4 Silanisation of Glass Surfaces (see Fig. 5.5)

The silanisation is a critical step, and must be very carefully done to avoid dust on the glass coverslips, which could hinder silanisation. Start with very clean coverslips and select the best ones. The silanisation cell must be cleaned with pirana solution (*see* **Section 3.4.2.1**) when it becomes dirty to remove silane residues, and then rinsed very thoroughly. The solvents can be filtered after use for one or two times maximum. Exchange the bottle of silane for a new one before finishing it, because when opened very often the silane (initially under argon) becomes less efficient.

3.4.1 Pre-pirana Washing

1. For both pre- and post-pirana washing, the coverslips (22×22 mm) are placed on the ceramic racks in a large beaker that contains the appropriate solution, which is then placed in a bath sonicator. First wash 3× with Milli-Q water.

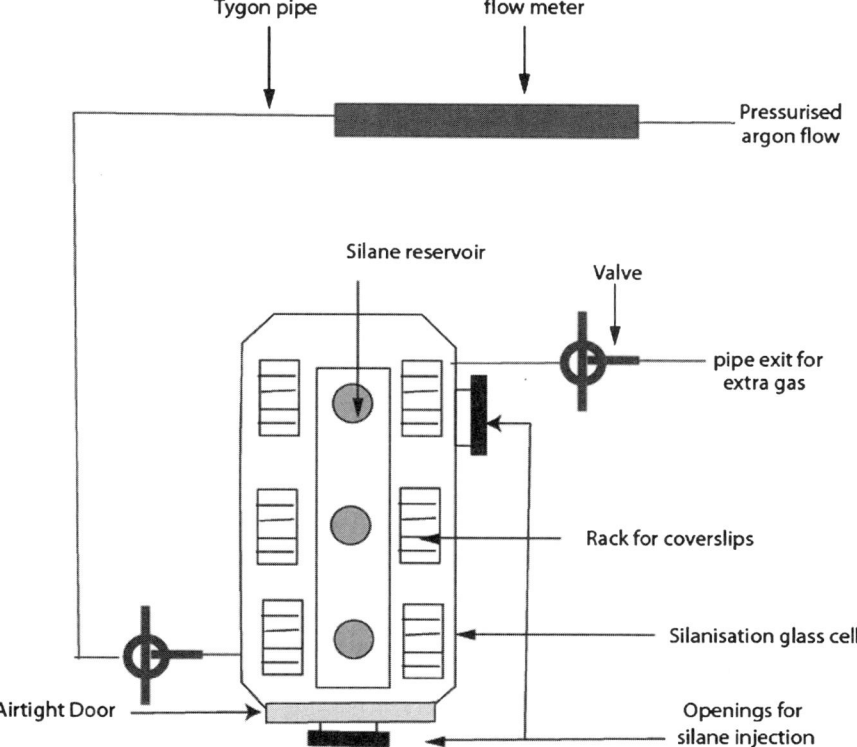

Fig. 5.5 Silanisation device. An aerial view with the glass treatment vessel connected to tubes for gas exit and for pressurised argon flow monitored through the flow meter. The airtight door of the glass cell allows introduction of racks with coverslips and of three reservoirs for silane (made with aluminium foil) on a Teflon rack. The openings allow the injection of silane with a micro-syringe

2. Sonicate the coverslips in ethanol for 10 sec and repeat this step five times.
3. Sonicate in chloroform for 10 sec and repeat this step five times.
4. Rinse the coverslips in ethanol three times.

3.4.2 Washing with Pirana Solution

Work under a chemical hood using gloves and glasses.

1. Prepare pirana solution: 390 mL of NH_4OH solution (**Section 2.3.4**) and 360 mL of 33% H_2O_2 (**Section 2.3.5**).
2. Place the racks with coverslips in a beaker containing pirana solution, and heat it on a heating block with boiling for about 30 min.
3. Wash five times in Milli-Q water.

3.4.3 Post-pirana Washing

1. Sonicate the coverslips in 6% HCl for 5× 15 sec.
2. Rinse five times in Milli-Q water.

3.4.4 Pre-silanisation Washing

1. Sonicate in ethanol 5× 10 sec.
2. Sonicate in ethanol/toluene (1:1) 5× 10 sec.
3. Sonicate in toluene 5× 10 sec.
4. Dry the coverslips very well with a stream of oxygen or argon.

3.4.5 Silanisation

1. Install the racks containing the coverslips and the three silane reservoirs in aluminium foil in the middle of the glass reactor chamber and close it hermetically. Dry the coverslips with a flux of argon (900 mL/min) for 15 min.
2. Inject 100 µL of 7-octenyltrichlorosilane in each reservoir and leave it sealed overnight.

3.4.6 Post-silanisation Washing

1. Dry the reactor with an argon stream (900 mL/min) for 15 min.
2. Open the reactor chamber and put the silane reservoir into tap water to polymerise the silane.
3. Sonicate the coverslips in toluene 5× 10 secs.
4. Sonicate in ethanol/toluene (1:1) 5× 10 secs.
5. Sonicate in ethanol 5× 10 secs.

6. Dry the coverslips very well with an oxygen or argon stream and store them in a vacuum to avoid dust.

3.5 Molecular Combing of DNA

1. Place a silanised coverslip on the coverslip holder of the combing apparatus and push the start button to allow the coverslip to dip into the DNA solution in the reservoir.
2. Incubate for 5 min to allow binding of DNA molecules to the silanised surface.
3. The coverslip is removed automatically at a constant speed of 300 μm/s (see **Note 10**).
4. To check the quality of combing, dip the combed coverslip again into a Teflon reservoir containing 1.5 mL of 0.5 M MES, pH 5.5, with 2 μL of YOYO-I and incubate for 1 min. Then remove the coverslip by pushing the stop and start buttons.
5. Glue the coverslip onto a glass slide with Superglue (does not autofluoresce).
6. Put a drop of Vectashield antifading medium and cover with a non-silanised coverslip.
7. Check the density, integrity, and linearity of the DNA molecules with a fluorescence microscope (this coverslip is sacrificed and cannot be used for replication or FH).
8. The other coverslips are combed in a blind manner without YOYO-I staining, glued on slides, and baked overnight at 60°C to bind the DNA irreversibly to the surface.

3.6 Labelling Probes

3.6.1 Labelling with Biotin-14-dCTP

The random primers, Klenow fragment, and dNTP mix are provided in the Bioprime kit.

1. Place between 500 ng and 1 μg of template DNA in a 1.5-mL Eppendorf tube containing 20 μL of 2.5× random primers.
2. Adjust the volume to 44 μL with sterile distilled water.
3. Put the tube in boiling water for 8 min to denature DNA.
4. Immediately place the tube in icy water for 5 min to keep the DNA in a denatured state.
5. Then add 5 μL of the dNTP mix containing biotin dCTP and 1 μL of Klenow fragment.
6. Mix well and spin briefly to collect the reaction mix at the bottom of the tube.
7. Incubate the tube at 37°C overnight (see **Note 11**).
8. The following day, add 2 μL of 0.5 M EDTA, pH 8.0 to stop the reaction.
9. If the probes are not used the same day for FH, they can be stored indefinitely at −20°C. There is no need to aliquot them and they can be thawed and re-frozen without damage.

3.6.2   Labelling with dig-11dUTP

Only the Klenow fragment and the random primers in the kit are used.

1. Mix the template DNA and primers, adjust the volume to 42 μL with sterile distilled water and denature as in Section **3.6.1, steps 1–4.**
2. Add 2 μL of 1 μ*M* dig-dUTP and 5 μL of a nucleotide mixture containing 90.5 μL of sterile water, 1 μL each of 1 μ*M* dATP, 1 μ*M* dGTP, and 1μ*M* dCTP, and 6.5 μL of 0.1 μ*M* dTTP (*see* **Note 12**). Add 1 μL of Klenow fragment.
3. Mix well and spin briefly.
4. Incubate the tube at 37°C overnight.
5. The following day, add 2 μL of 0.5 *M* EDTA, pH 8.0 to stop the reaction.

3.6.3   Checking Probes by Electrophoresis

Aliquots (4 μL) of labelled probes mixed with 2 μL of glycerol/gel loading buffer containing 0.2% Orange G, and a DNA marker ladder, are electrophoresed in a 0.6% agarose gel in 1× TBE for 1 h at 50 V to estimate the size of the labelled probes. A smear or band 300 basepairs in length is expected.

3.6.4   Evaluation of the Incorporation of Biotin or Digoxygenin and of the Quantity of Probe

To test digoxygenated probes, spot 1 μL of a series of five successive 1:3 dilutions starting from 5 ng DNA onto a Hybond N+ membrane disc; and for biotin probes, 1 μL of a series of five successive 1:2 dilutions starting at 250 pg DNA. For standards, a digoxygenin-labelled DNA at 5 ng/μL is used; since there is no commercially available biotin-labelled DNA we use highly purified lambda DNA which we biotin label ourselves; we find that the labelling is reproducible enough to serve as a standard. The same serial dilutions of the corresponding control DNA are spotted on the membrane disc next to the experimental samples.

1. After 10 min to allow the DNA to bind to the membrane, place the membrane disc in a Petri dish and cover with 0.4 N NaOH to denature the DNA. Incubate for 20 min at room temperature with slow agitation.
2. Neutralise with five washes in SSC. Check the pH of the last wash with pH paper; it should be about 7.5.
3. Remove the 5× SSC, add 6% BSA in 1× PBS, and incubate for 1 h at 37°C to block non-specific binding of proteins.
4. Add 10 μL of streptavidin–alkaline phosphatase (for biotin-labelled probes) or anti-digoxygenin–alkaline phosphatase (for digoxygenin-labelled probes) at 1 mg/mL. Incubate for 30 min at 37°C.
5. Wash thoroughly with 250 mL of 1× PBS for 30 min to eliminate unbound reagents.
6. Prepare the chromogenic substrate solution provided with the alkaline phosphatase substrate kit IV.

7. For one membrane disc, put 7.5 mL of 0.1 M Tris-Cl, pH 9.5 and add three drops each of NBT, BCIP, and MgCl$_2$.
8. Let the reaction develop for about 5 min at room temperature.
9. Stop by rinsing with 100 mM EDTA.
10. The labelling is considered efficient if the spot intensities are similar to those of the corresponding standard DNA.

3.7 FH and/or Replication

Depending on the question to be asked, process for FH with two labelled probes (biotin and digoxygenin) to detect two colors of signal and to obtain a precise physical map of a region of interest. Alternatively, one may process for FH and replication together to study replication in a defined local region by detecting one or two labelled FH probes together with one or two colors for the replication signal. Figure 5.6 summarises the different steps for studying FH and replication processes,

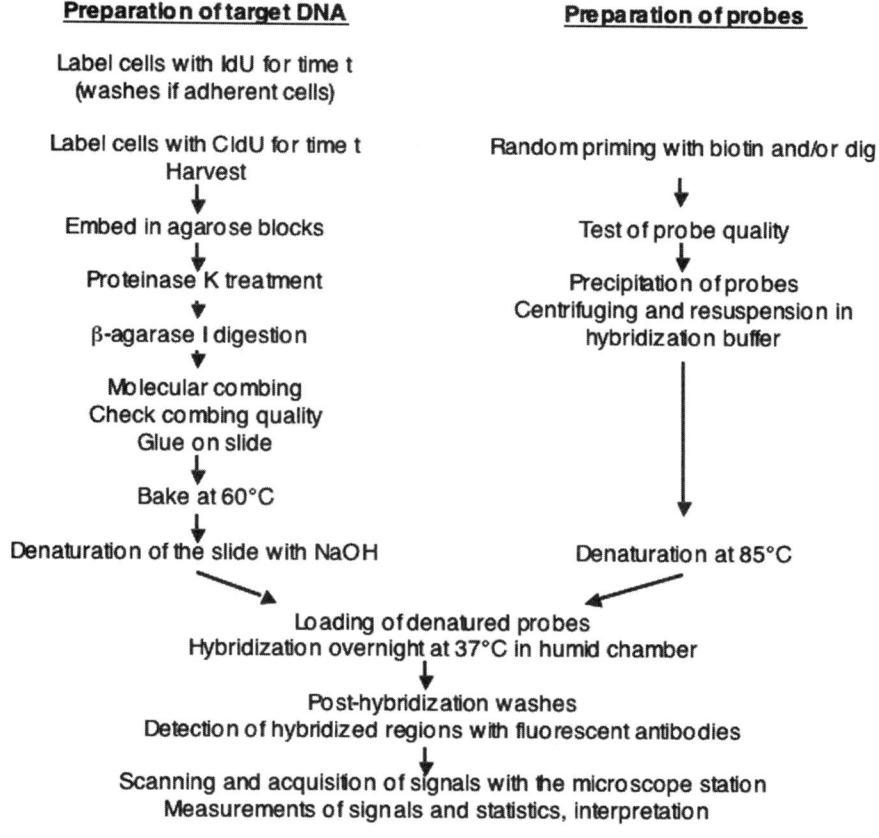

Preparation of target DNA

Label cells with IdU for time t
(washes if adherent cells)

Label cells with CldU for time t
Harvest
↓
Embed in agarose blocks
↓
Proteinase K treatment
↓
β-agarase I digestion
↓
Molecular combing
Check combing quality
Glue on slide
↓
Bake at 60°C
↓
Denaturation of the slide with NaOH

Preparation of probes

Random priming with biotin and/or dig
↓
Test of probe quality
↓
Precipitation of probes
Centrifuging and resuspension in
hybridization buffer
↓
Denaturation at 85°C

Loading of denatured probes
Hybridization overnight at 37°C in humid chamber
↓
Post-hybridization washes
Detection of hybridized regions with fluorescent antibodies
↓
Scanning and acquisition of signals with the microscope station
Measurements of signals and statistics, interpretation

Fig. 5.6 The major steps of FH on combed DNA coupled to visualisation of replication

and some possibilities of immunofluorescent detection for studying FH and/or replication are described in Table 5.1, but this list is not exhaustive.

3.7.1 Slide Preparation for FH Alone or Together with Replication

1. Place slides (from **Section 3.5**) in a Coplin jar with $0.05\,N$ NaOH and $0.1\,M$ NaCl for 15 min at room temperature and agitate slowly to denature the combed DNA (*see* **Note 13**).
2. Rinse 3× in cold PBS to neutralise the NaOH, and immerse for 3 min each in 70%, 90%, and 100% ethanol (*see* **Note 14**).
3. Air-dry the slides at room temperature.

3.7.2 Slide Preparation for Replication Alone

1. As in **Section 3.7.1.**

3.7.3 Probe Preparation for FH

1. In a 1.5-mL Eppendorf tube, place 1 µg of labelled probe(s), five times the total amount of cot-1 DNA if necessary (*see* **Note 15**), 1 µg of herring sperm DNA, 0.1 volume of $3\,M$ Na acetate, pH 5.2, and 3 volumes of 100% ETOH at $-20°C$.
2. Put at $-80°C$ for 30 min to precipitate.
3. Centrifuge for 30 min at $15,000 \times g$ at 4°C.
4. Remove the supernatant and dry the pellet in a Speed Vac for 5 min (*see* **Note 16**).
5. Re-suspend the pellet in 20 µL of hybridisation buffer and incubate for at least 30 min in a 37°C water bath.
6. Denature the probes for 5 min in a water bath at 85°C.
7. Load the denatured probe immediately onto the denatured slide and cover with a non-silanised coverslip (*see* **Note 17**). Incubate the slide overnight at 37°C in a humid chamber.

3.8 Post-hybridisation Steps for FH Alone or Associated with Studies of Replication

1. Remove the coverslip gently and wash three times with 50% formamide/2× SSC for 5 min each at room temperature in a Coplin jar.
2. Wash 3× 5 min each in 2× SSC at room temperature.

Table 5.1 Examples of detection schemes

Layer	1	2	3	4
Case 1. FH and replication: biotin and digoxygenin probes with two replication signals (*see* Note 19)				
Digoxygenin (green)	Sheep anti-digoxygenin FITC, 1:25	Donkey anti-sheep Alexa 488, 1:25	—	—
Biotin (infrared)	Streptavidin–Alexa 750, 1:25	Rabbit anti-streptavidin–biotin, 1:25	Streptavidin–Alexa 750, 1:25	—
IdU (blue)	—	—	Mouse anti-BrdU, 2:5	Goat anti-mouse Alexa 350, 1:25
CldU (red)	—	—	Rat anti-BrdU, 1:5	Donkey anti-rat Alexa 594, 1:25
Case 2. FH and replication: biotin-labelled probe with two replication signals				
Biotin (green)	Streptavidin–Alexa 488, 1:25	Rabbit anti-streptavidin–biotin, 1:50	Streptavidin–Alexa 488, 1:25	—
IdU (blue)	Mouse anti-BrdU, 2:5	Goat anti-mouse Alexa 350, 1:25	Donkey anti-goat Alexa 350, 1:25	—
CldU (red)	Rat anti-BrdU, 1:5	Donkey anti-rat Alexa 594, 1:25	—	—
Case 3. FH and replication: biotin- and digoxygenin-labelled probes with one replication signal				
Digoxygenin (green)	Sheep anti-digoxygenin–FITC, 1:50	Donkey anti-sheep Alexa 488, 1:25	Rabbit anti-donkey FITC, 1:50	—
Biotin (blue)	Streptavidin–Alexa 350, 1:25	Rabbit anti- streptavidin–biotin, 1:50	Streptavidin–Alexa 350, 1:25	—
CldU (red)	Rat anti-BrdU, 1:5	Donkey anti-rat Alexa 594, 1:25	—	—
Case 4. FH alone: biotin- and digoxygenin-labelled probes together				
Digoxygenin (green)	Sheep anti-digoxygenin FITC, 1:50	Donkey anti-sheep Alexa 488, 1:25	Rabbit anti-donkey FITC, 1:50	—
Biotin (red)	Streptavidin–Alexa 594, 1:25	Rabbit anti- streptavidin–biotin, 1:50	Streptavidin–Alexa 594, 1:25	—
Case 5. Replication alone: two replication labels with different colors				
IdU (green)	Mouse anti-BrdU–FITC, 2:5	Goat anti-mouse Alexa 488, 1:25	—	—
CldU (red)	Rat anti-BrdU, 1:5	Donkey anti-rat Alexa 594, 1:25	—	—

Colours of signals are indicated in parentheses and antibody dilutions are shown after their name

3.9   Immunofluorescent Steps for All Studies

The combed DNA molecules are usually not counterstained, but if necessary, label the cells with BrdU for 24 h in order to completely cover the S phase and detect the BrdU with an anti-BrdU antibody coupled to a green fluorochrome, the biotin probe with a blue antibody, and the digoxygenin probe with a red antibody. Alternatively, you can use an antibody directed against guanosine or adenine or a chromosome-painting probe.

Antibodies are diluted in blocking solution. Many combinations of antibodies are possible; to avoid cross-reactions, use antibodies that have been highly cross-absorbed against the species in which they were developed, and minimise the number of antibody layers as much as possible.

All incubation steps are for 20 min at 37°C in a humid chamber (*see* **Note 18**). The antibody mix is prepared fresh a few minutes before each step, and ~25 µL is deposited on each slide and covered with a non-silanised coverslip. After each step, the coverslip is removed gently and the slide is washed three times for 3 min each in a Coplin jar filled with 1× PBS under slow agitation at room temperature. The slides are finally mounted in Vectashield medium.

4   Notes

1. The Teflon reservoirs and the Molecular Combing System (combing machine) are now manufactured by Bio-Rad and are commercially available from Genomic Vision. Silanised cover slips will also be commercially available in the next few months. For details, please contact Yamina Ghomchi (yamina. ghomchi@genomicvision.com).
2. The incubation time with modified nucleotides must be very precise, and has to be respected scrupulously to have accurate measurements.
3. Avoid washing the cells because this stops the replication process. In this case, replication signals will have a different pattern since the IdU is not removed when adding the CldU.
4. It is easier to re-suspend cells in cold PBS, since the low temperature inhibits their aggregation. If they do aggregate, the number of cells per block will be variable and this will cause a problem for combing.
5. Measure the residual volume precisely when removing the supernatant, because it may be non-negligible. This step is very tricky since the number of cells per block is very important to obtain a good density for combing; if it is too high the DNA molecules will not be correctly combed because they will form bundles, supercoiled fibers, or non-linearised molecules. If there are too few cells, the low density of combed DNA will not give enough signal for FH, in particular for single-locus targets. For replication, the density needed is lower than for FH. Note that the number of cells given here is only an indication, and should be adapted if the cells are not diploid; for example, since HeLa cells are triploid, use 2/3 of the cell concentration. Counting of cells has to be very precise and

should be done with two different methods in parallel to be accurate. Errors can arise at this step because the cells drop to the bottom of the tube by gravity after homogenisation, or they aggregate causing underestimation when counted in a Coulter counter. The first time, do a range of concentrations near the recommended value. To avoid aggregation of the cells when counting, use a hypotonic solution of 0.9% (v/w) Na citrate and take an aliquot from this solution to count the cells. In less than 5 min, they become a little swollen and free from each other, and counting is then more reliable. Do not leave them for more than 5 min in this solution, or the cell membrane will be disrupted.

6. The cells have to be heated just enough, not to chill the agarose too much but to have time to distribute it in the mold before it solidifies. But do not heat too long, or the cells will bind together again.

7. This step avoids overheating the enzyme.

8. After this step the DNA is in solution and no longer protected by agarose, so do not vortex or pipette the solution.

9. Teflon is a good material for this purpose since it does not bind DNA and can be cleaned easily after use, by boiling water for example. When in solution, DNA can be stored for at least 2 months at +4°C. If the solution is too concentrated, for better combing quality it can help to wait 2–3 weeks and to keep it in the Teflon reservoir covered with Parafilm to avoid evaporation. A faster alternative is to cut the agarose block with a scalpel before the agarose step to reduce the quantity of DNA in the tube.

10. Notice that the DNA will be combed on both faces of the silanised coverslip.

11. For a better yield, incubate overnight.

12. This mixture can be stored at −20°C and thawed and refrozen many times without loss of efficiency.

13. Make this solution fresh, as the NaOH reacts with air to produce CO_2, which could change the pH. Respect the time precisely. This step is critical because the incubation time must be long enough to denature the DNA, but not too long since this could remove silane and the combed DNA from the surface.

14. Cold PBS helps to neutralise the pH and to keep the DNA denatured. Rinse very quickly and check the pH with paper until it goes down to 7.5.

15. The amount of Cot-1 DNA has to be determined empirically. The herring sperm DNA helps DNA precipitation and raises the FH efficiency. The amount of probe can be lowered to at least 200 ng if necessary, and if two labelled probes are used, they can be precipitated together.

16. Do not exceed this time, since if the pellet is too dry it will be difficult to resuspend.

17. Try to denature the slides and the probes in parallel; if the slides are not ready on time, put the denatured probes on ice water.

18. Respect the incubation times scrupulously to avoid background problems. Do not allow the slides to dry at any step, since this will result in background.

19. Note that the biotin and digoxygenin probes are detected together at the same time. Mouse anti-BrdU-FITC from Becton Dickinson recognises BrdU but also cross-reacts with IdU (but not with CldU). Rat anti-BrdU (ImmunologicalsDirect;

Abcys, Paris, France) recognises BrdU and also CldU, but not IdU. Table 5.1 shows a few possibilities: it is not exhaustive. You can imagine other systems depending on what you want to detect.

References

1. Bensimon, A., Simon, A., Chiffaudel, A., Croquette, V., Heslot, F., and Bensimon D. (1994) Alignment and sensitive detection of DNA by a moving interface. *Science* **265**, 2096–2098.
2. Van Dekken H., Pinkel D., Mullikin J., Trask B., Van den Engh G., and Gray J. (1989) Three-dimensional analysis of the organization of human chromosome domains in human and human-hamster hybrid interphase nuclei. *J. Cell Sci.* **94**, 299–306.
3. Heng, H.H., Squire, J., and Tsui, L.C. (1992) High-resolution mapping of mammalian genes by in situ hybridization to free chromatin. *Proc. Natl. Acad. Sci. USA* **89**, 9509–9513.
4. Senger, G., Jones, T.A., Fidlerova, H., Sanseau, P., Trowsdale J., Duff, M., and Sheer, D. (1994) Released chromatin: linearized DNA for high resolution fluorescence in situ hybridization. *Hum. Mol. Genet.* **3**, 1275–1280.
5. Fidlerova, H., Senger, G., Kost, M., Sanseau, P., and Sheer, D. (1994) Two simple procedures for releasing chromatin from routinely fixed cells for fluorescence in situ hybridization. *Cytogenet. Cell Genet.* **65**, 203–205.
6. Parra, I. and Windle, B. (1993) High resolution visual mapping of stretched DNA by fluorescent hybridization. *Nat. Genet.* **5**, 17–21.
7. Heiskanen, M., Karhu, R., Hellsten, E., Peltonen, L., Kallioniemi, O.P., and Palotie, A. (1994) High resolution mapping using fluorescence in situ hybridization to extended DNA fibers prepared from agarose-embedded cells. *Biotechniques* **17**, 928–933.
8. Allemand, J-F., Bensimon, D., Jullien, L., Bensimon, A., and Croquette, V. (1997) pH-Dependent specific binding and combing of DNA. *Biophys. J.* **73**, 2064–2070.
9. Michalet, X., Ekong R., Fougerousse, F., Rousseaux, S., Schurra, C., Hornigold, N., Van Slegtenhorst, M., Wolfe, J., Povey, S., Beckmann, J. S., and Bensimon, A. (1997) Dynamic molecular combing: stretching the whole human genome for high-resolution studies. *Science* **277**, 1518–1523.
10. Gad S., Aurias, A., Puget, N., Mairal, A., Schurra, C., Montagna, M., Pages, S., Caux, V., Mazoyer, S., Bensimon, A., and Stoppa-Lyonnet, D. (2001) Color bar coding the *BRCA1* gene on combed DNA: A useful strategy for detecting large gene rearrangements. *Genes Chromosomes Cancer* **31**, 75–84.
11. Herrick, J., Michalet, X., Conti, C., Schurra, C., and Bensimon, A (2000) Quantifying single gene copy number by measuring fluorescent probe lengths on combed genomic DNA. *Proc. Natl. Acad. Sci. USA* **97**, 222–227.
12. Herrick, J., Conti, C., Teissier, S., Thierry, F., Couturier J., Sastre-Garau, X., Favre, M., Orth, G., and Bensimon, A. (2005) Genomic organization of amplified *MYC* genes suggests distinct mechanisms of amplification in tumorigenesis. *Cancer Res.* **65**, 1174–1179.
13. Caburet, S., Conti, C., Schurra, C., Lebofsky, R., Edelstein, S.J., and Bensimon, A. (2005) Human ribosomal RNA genes arrays display a broad range of palindromic structures. *Genome Res.* **15**, 1079–1085.
14. Herrick, J., Stanislawski P., Hyrien, O., and Bensimon, A. (2000) Replication fork density increases during DNA synthesis in X. laevis egg extracts. *J. Mol. Biol.* **300**, 1133–1142.
15. Lebofsky, R. and Bensimon, A. (2005) DNA replication origin plasticity and perturbed fork progression in human inverted repeats. *Mol. Cell Biol.* **25**, 6789–6797.
16. Herrick, J. and Bensimon, A. (1999) Single molecule analysis of DNA replication. *Biochimie* **81**, 859–871.
17. Gueroui, Z., Place, C., Freyssingeas, E., and Berge, B. (2002) Observation by fluorescence microscopy of transcription on single combed DNA. *Proc. Natl. Acad. Sci. USA* **99**, 6005–6010.

Chapter 6
Using Molecular Beacons to Study Dispersal of mRNPs from the Gene Locus

Patrick T. C. van den Bogaard and Sanjay Tyagi

Keywords Molecular beacons; Single particle tracking; mRNA transport; Nuclear viscosity; Live-cell imaging

Abstract Before leaving the site of transcription, newborn messenger RNAs (mRNAs) become associated with a number of different proteins. How these large messenger ribonucleoprotein (mRNP) complexes then move through the dense nucleoplasm to reach the nuclear periphery has been a fascinating question for the last few years. We have studied the mechanism of this process by tracking individual mRNPs in real time. We were able to track mRNPs at single-molecule resolution because we utilized mRNAs that were engineered to have a sequence motif repeated 96 times in their untranslated region. These mRNAs were visualized with the help of molecular beacons that were specific for the repeated sequence; the binding of 96 molecular beacons to each mRNA molecule rendered them so intensely fluorescent that they were visible as fine fluorescent spots that could be tracked by high-speed video microscopy. In this chapter, we describe the details of the construction of genes containing the tandem repeats, the integration of such genes into the genome of a cell line, the design and testing of molecular beacons, time-lapse imaging of mRNPs, and computer-aided generation and analysis of the tracks of the individual mRNPs. These methods will be useful for studies of other dynamic processes such as mRNA export, splicing, and decay.

1 Introduction

The nuclei of cells of higher eukaryotes have a diameter of about $5\,\mu m$ and yet they house DNA strands with cumulative length of $3\,m$ or more. This large quantity of DNA is tightly associated with histones and other chromatin proteins. Even though the DNA exists in a highly organized state in the interphase nucleus, it creates a highly viscous environment. When mRNAs are produced at a gene locus they get immediately associated with a number proteins (as many as two dozen) that remain bound to them throughout their journey to the ribosomes in the cytoplasm. These

R. Hancock (ed.) *The Nucleus: Volume 2: Chromatin, Transcription, Envelope,*   91
Proteins, Dynamics, and Imaging,
© Humana Press 2008

ribonucleoprotein complexes (RNPs) are much larger than the mRNA in terms of their mass. How these large complexes then travel to the nuclear pore through the dense chromatin has remained an unresolved question for the last few years *(1)*. Assuming that the viscosity of the nucleus would be too high for the mRNP to overcome, early workers proposed that mRNPs are transferred along a chain of receptors until they reach a nuclear pore, expending metabolic energy in the process *(2)*. This solid-state transport model is supported by observations that some mRNA transcripts are distributed along tracks that originate from the locus of the parent gene in fixed nuclei *(3)*. According to an alternate theory called the gene-gating hypothesis, active genes would be situated near nuclear pores and mRNAs would exit the nucleus through the nearest pores *(4)*. The gene-gating hypothesis is supported by observations that certain mRNAs exit from one side of the nucleus *(5)* and that, in yeast, many transcriptionally active gene loci are associated with nuclear pore proteins while quiescent loci, which are normally present near the center of the nucleus, move to the periphery upon induction *(6)*. More recent studies indicate that simple Brownian diffusion is sufficient to account for mRNA transport within the nucleus *(7, 8)*.

We have studied this issue by developing unique reporter genes and oligonucleotide probes that allow imaging and tracking of individual mRNPs in live cells *(8)*. We utilize oligonucleotide probes called molecular beacons that become fluorescent upon hybridization *(9)*. Molecular beacons are hairpin-shaped molecules with an internally quenched fluorophore whose fluorescence is restored when they bind to a target nucleic acid (Fig. 6.1).

The fluorescence of a target-bound molecular beacon is a hundred to a thousand times more intense than that of the free molecular beacon *(9)*. Owing to their stem, the recognition of targets by molecular beacons is so specific that if the target differs even by a single nucleotide the probe does not bind to it. In order to ensure that the molecular beacons will be stable in the cell and that their hybrids will not be digested by ribonuclease H, the molecular beacons can be made from 2′-*O*-methylribonucleotides *(10)*. Many different fluorophores can be utilized to detect

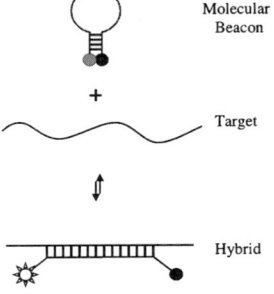

Fig. 6.1 Operation of molecular beacons. On their own, these molecules are nonfluorescent because the stem hybrid keeps the fluorophore close to the quencher. When the probe sequence in the loop hybridizes to its target, forming a rigid double helix, a conformational reorganization occurs that separates the quencher from the fluorophore and leads to an increase in fluorescence

several mRNAs in the same cell. The capacity of molecular beacons to recognize and report the presence of a specific mRNA in the complex cellular environment was demonstrated by imaging the distribution and following the localization process of native *oskar* mRNA from nurse cells to the posterior end of the fruit fly oocyte *(11)*.

Since most genes produce just a few molecules of mRNA, it is desirable to be able to detect and track individual mRNA molecules. The fluorescence emitted by a single fluorophore, though detectable under specialized microscopic conditions, is too faint for tracking individual mRNPs. Since the monitoring of molecular motion requires high-speed video microscopy with exposure times around 100 msec, it is important to have a high fluorescence signal emanating from the target molecules. Therefore, we have developed an approach where multiple target sequences were incorporated in a relatively less important 3'-untranslated region of an mRNA *(8)*. This multimeric target sequence could bind 96 copies of the same molecular beacon, rendering this mRNA so fluorescent that each individual molecule could be detected in the living cell as a diffraction-limited spot. The gene encoding a GFP-mRNA-96-mer was expressed in CHO cells and its mRNA was successfully tracked in the nucleus. Moreover, after export from the nucleus, the mRNAs could be clearly detected in the cytoplasm and their translation gave raise to green fluorescent protein (GFP) fluorescence, indicating that the molecular beacons bound to the mRNA did not hamper its nucleo–cytoplasmic export and subsequent translation.

So far, this approach has been utilized only for the exploration of mRNA dynamics within the nucleus but it should be effective for other contexts in which mRNA dynamics are studied. Other processes include mRNA splicing and maturation, mRNA export from the nucleus to the cytoplasm, mRNA localization in cytoplasm, and mRNA decay.

In this chapter, we describe procedures for making a gene encoding the tandem repeats, making a cell line in which this gene is integrated, design of molecular beacons that will be effective in mRNA imaging, delivery of molecular beacons into the cell, time-lapse imaging of individual mRNA molecule in living cells, and a computer algorithm to obtain and analyze single molecules tracks from the time lapse images.

2   Materials

2.1   Equipment

1. Spectrofluorometer (Photon Technology International, Birmingham, NJ, USA).
2. Spectrofluorometric thermal cycler: 7700 Prism (Applied Biosystems Foster City, CA, USA).
3. Microinjection apparatus (FemtoJet; Brinkmann, Westbury, NY, USA).

4. Inverted fluorescence microscope with heated stage and ×100 oil-immersion objective (Axiovert 200 M; Carl Zeiss MicroImaging, Thornwood, NY, USA).
5. CCD camera cooled to −30°C (CoolSNAP HQ; Photometrics, Tucson, AZ, USA) mounted on the bottom port of the microscope, with image acquisition software.
6. Open culture dish environmental control system with culture dishes coated with conductive material for controlled heating (Delta T4; Bioptechs, Butler, PA, USA).

2.2   Cells and Cell Culture

1. CHO–AA8-Tet-off cell line (Clontech, Mountain View, CA, USA).
2. Eagle's minimal essential medium supplemented with glutamine (Sigma-Aldrich, St. Louis, MO, USA) and 10% TET-System-Aproved FBS (Clontech).
3. Geneticin (G-418; Clontech).
4. Doxycycline (Clontech).

2.3   Plasmids and Cloning

1. Plasmid pGEM-11Zf(+) (Promega, Fitchburg Center, WI, USA).
2. Plasmid pTRE-d2EGFP (Clontech).
3. Competent cells: *Escherichia coli* MAX Efficiency Stbl2 (Invitrogen, Carlsbad, CA, USA).
4. Restriction enzymes.
5. Complementary DNA oligonucleotides for the repeat sequence (IDT, Coralville, IA, USA) (*see* **Section 3.1.2**).

2.4   Reagents and Solutions

1. Nuclease-free water (Ambion, Austin, TX, USA).
2. TE buffer: 1 mM EDTA, 10 mM Tris-HCl, pH 8.0.
3. Phosphate-buffered saline (PBS) with $CaCl_2$ and $MgCl_2$ (Sigma-Aldrich).
4. Molecular Beacon: dissolve molecular beacons for stock solutions in nuclease-free water and store at −20°C. Dilute working solutions in water, keep protected from light, and store at −20°C for up to 1 month. Filter through 0.2-μm filters to avoid clogging of needles. Wash the filters with the same water in which the probes are dissolved beforehand.
5. Oligonucleotide complementary to the molecular beacon: dissolve in TE and store at −20°C.

6. Full-length mRNA to be studied: transcribed in vitro from a plasmid (*see* **Section 3.2.2**).
7. Hybridization buffer: 1 m*M* MgCl$_2$, 20 m*M* Tris-HCl, pH 8.0.
8. Image processing software: MATLAB with the imaging-processing toolbox (MathWorks, Natick, MA, USA).

3   Methods

3.1   *Construction of a Gene Encoding an mRNA with Tandem Repeats*

3.1.1   Choosing the Sequence of Repeats

An important consideration for choosing the nucleotide sequence of the tandem repeats is that the molecular beacon target sequence in the repeat should be accessible to the probe and not occluded by secondary structure. It is possible to predict probe accessibility to some extent using the RNA-folding program m-fold (www. bioinfo.rpi.edu/applications/mfold) (*see* **Note 1** and **ref. (*12*)**). This algorithm presents the user with a number of thermodynamically favorable structures. In practice, none of the predicted individual secondary structures represents the true naturally occurring conformation; however, an analysis of all suboptimal structures is useful in identifying the probe-accessible sites. For that analysis, we look at the *ss-count* and *P-num* parameters to find probable molecular beacon binding targets. The *ss-count* describes the number of times a specific nucleotide is present in single-stranded form among all the predicted structures, while the *P-num* value counts the number of different base pairs that was found for that particular nucleotide when it was not single stranded. The repeated sequence should preferably have a high overall *ss-count* (>50%), since nucleotides with a low *ss-count* are strongly involved in base pairing. However, the *P-num* indicates if these nucleotides with low *ss-count* have a high stability (low *P-num*) or are unstable and not very consequential (high *P-num*). Since any arbitrary sequence can be chosen for the repeat unit, alter your sequence until it has almost no secondary structures. Then repeat the analysis with tandem concatemers of the sequence.

A second consideration in the selection of the repeat sequence is that it should be longer than the molecular beacon target site. If the molecular beacon probe sequence is the same length as the repeat sequence, the fluorophore of each molecular beacon will be quenched by the quencher of the molecular beacon bound at the neighboring site. In order to overcome this problem, we make the repeat unit at least 10 nucleotides (nt) longer than the target. The spacer of 10 nt that this strategy provides places the fluorophore more than 100 Å apart from the quencher, a distance beyond which there is no significant fluorescence resonance energy transfer.

A further consideration is that one arm of the molecular beacon is made to bind the target so that the arms of neighbor molecular beacons do not bind to each other. Below we will describe the construction of the multimer used for the study of mRNA transport in the nucleus as an example *(8)*.

3.1.2   Cloning Tandem Repeats

As an example, we describe the cloning of 96 repeats of the 50-nt-long sequence 5′-CAGGAGTTGTGTTTGTGGACGAAGAGCACCAGCCAGCTGATCGACCT CGA-3′ used in our studies of mRNP diffusion *(8)*. As a repeating unit, this multimer has an overall high *ss-count* as well as a high *P-num* per nucleotide, indicating that many different secondary structures are predicted (*mfold*) but none of them are considered stable. The construction of the direct repeats was carried out by the procedure described previously by Robinett et al. *(13)*. In this procedure, a synthetic DNA insert corresponding to one element of the array is cloned into a plasmid. This insert is then excised and reinserted at the tail end of the monomeric insert, producing a clone containing a dimeric insert. This cycle is repeated with the dimeric clone to obtain a tetrameric insert, and then the cycle is repeated again to obtain the desired length of the insert, which doubles in each cycle.

In our implementation, the repeat sequence contained an *Sal*I site on its 5′ terminus and an *Xho*I and *Bam*HI site respectively on its 3′ terminus (Fig. 6.2).

The procedure relied on the fact that the restriction enzymes *Sal*I and *Xho*I generate overhangs that are compatible with each other, however, the product of ligation cannot be cut again by the same enzymes (*see* **Note 2**). Two strands for the repeat sequence (*see* **Section 2.3.5**) were ordered separately and annealed together to form a double-stranded DNA. The sequences of the strands are TCGACAGGA GTTGTGTTTGTGGACGAAGAGCACCAGCCAGCTG ACTCGAGCCGAGG and GATCCCTCGGCTCGAGTCAGCTGGCTGGTGCTCTTCGTCCACAAAC ACAACTCCTG. The annealed strands were phosphorylated, which resulted in a double-stranded fragment with *Sal*I and *Bam*HI overhangs and an internal *Xho*I site. The vector pGEM-11 was digested with *Sal*I and *Bam*HI and ligated with this synthetic insert. The integrity of the resulting construct (pGEM-1×) was verified by restriction analysis with *Sal*I and *Bam*HI. Due to its small size, the presence of the insert was confirmed by electrophoresis on a 10% polyacrylamide gel against a 100-bp marker.

For dimerization of the insert, a monomeric *Sal*I and *Bam*HI fragment obtained by digestion of pGEM-1× was inserted in the same plasmid digested with *Xho*I and *Bam*HI, yielding pGEM-2×. To create the tetramer repeat, a portion of pGEM-2× was digested with *Sal*I and *Bam*HI and the dimer insert was ligated into the vector fragment (containing the dimer) of another portion of pGEM-2× that was digested with *Xho*I and *Bam*HI, yielding pGEM-4×. Repeating these steps led ultimately to vectors containing a 32-mer and a 64-mer repeat (pGEM-32× and pGEM-64×). In the same way as described above, these two vectors were used to create pGEM-96×. During these cloning steps, special host cells that minimize recombination (MAX Efficiency

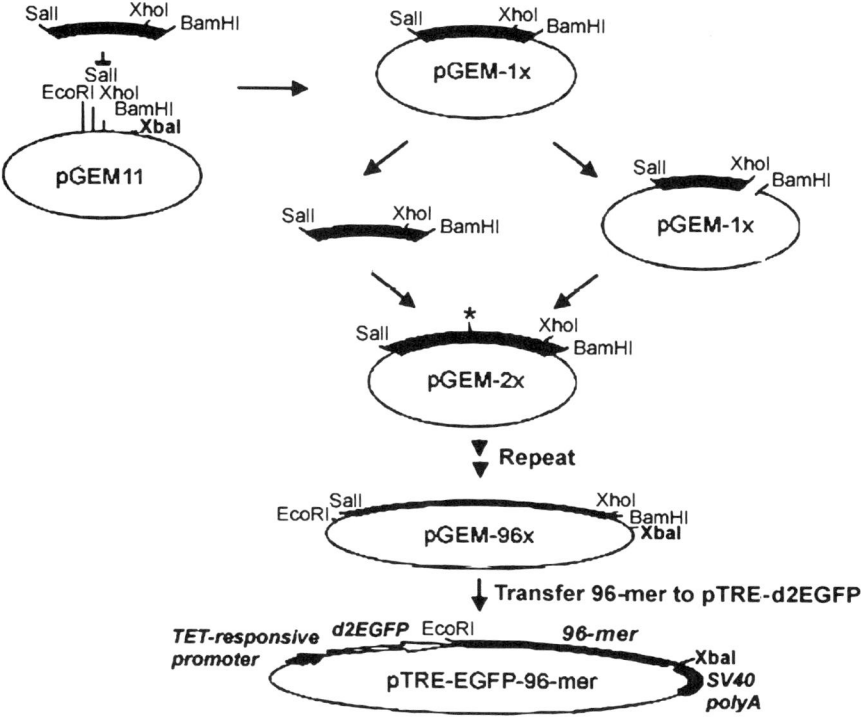

Fig. 6.2 Assembly of the reporter gene with multimeric repeats. After annealing both strands, the basic unit is inserted into pGEM11 digested with restriction enzymes *Sal*I and *Bam*HI, resulting in the vector pGEM-1×. The principle of forming the multimeric repeats is to ligate the new unit digested with *Sal*I and *Bam*HI into the same vector digested with *Xho*I and *Bam*HI, and then to repeat the cloning cycle again until the desired number of repeats is obtained

Stbl2 competent cells) were used to reduce the tendency to deletions and rearrangements of the multimer and that were exclusively grown at 30°C.

The pTRE-GFP-96-mer was created by excision of the 96-mer from pGEM-96× with *Eco*RI and *Xba*I and ligation into pTRE-d2EGFP digested with the same enzymes. Since long tandem repeats are not well tolerated by the host cells, in order to make a preparation of plasmids containing 32 or more repeats, we grow many pure colonies in broth and utilize only those that have maintained the full-length insert.

pTRE-GFP-96-mer was used to transfect the CHO-AA8-Tet-off cell line that contains a stably integrated gene for the tetracycline-controlled Tet-off transactivator. A geneticin (G-418)-resistant clone (CHO-GFP-96-mer) was selected based on GFP expression in the absence of doxycycline. This clone did not express GFP in presence of doxycycline (10 ng/mL).

3.2   Molecular Beacons

3.2.1   Design

For live cell imaging, we normally use molecular beacons with a 2′-O-methyl backbone as these are resistant to cellular nucleases, and their hybrids with RNA are not degraded by ribonuclease H. However, this change in the backbone alters the thermodynamic characteristics of the probe; both the probe–target hybrid and the stem hybrids become stronger (melting at higher temperatures). While the former does not cause any problems, the latter results in probes that have difficulty opening in response to the target so that the kinetics of probe–target hybridization become sluggish. To overcome this problem, a smaller (4–5 nt) and/or less G/C-rich sequence can be chosen for the stem. The length of the probe sequence in the molecular beacon should be such that the probe–target hybrid melts at 7–10°C higher than the detection temperature (usually 37°C). A length of 15–25 nt is usually chosen for the probe loop. The molecular beacon should not have any tendency to fold into any other structure besides the designed stem–loop; this can be predicted using the DNA folding program available at http://www.bioinfo.rpi.edu/ applications/mfold/cgi-bin/dna-form1.cgi. The sequence of the molecular beacon used for our example was Cy-3-5′-CUUCGUCCACAAACACAACUCCUGAAG-3′-Black Hole Quencher 2.

3.2.2   Testing the Molecular Beacon

The molecular beacon preparation should be purified using high-pressure liquid chromatography and should not contain any free fluorophores or any oligonucleotide fragments containing the fluorophore but without a quencher. Molecular beacons can be functionally tested in a fluorometer by measuring the change in their fluorescence after adding the complementary oligonucleotide (*see* **Note 3**). Preparations that have at least 30 times more fluorescence in their target-bound state then in their free state are considered good. The thermodynamic characteristics of the molecular beacons are obtained by measuring the denaturing profile in a real-time PCR instrument (*see* **Note 4**). The melting temperature of the probe–target hybrids should be at least 7–10°C higher then the detection temperature (usually 37°C). Finally, the molecular beacons should be tested against full-length target mRNA that is prepared by transcribing a plasmid template in vitro (*see* **Note 5**).

3.3   Cell Culture and Image Acquisition

Despite the amplification of signals afforded by multimeric target sites, the signals from the individual mRNA molecules are rather faint and microscopy conditions are therefore optimized for low-light imaging. Among the steps that we take are: replace normal culture media with phenol-red free media before imaging; use thin glass-bottom culture dishes; use a CCD camera cooled to −30°C; mount the camera on the bottom port of the microscope to avoid light paths with multiple glass elements; and use ×100 oil immersion objectives with high numerical apertures. Since we microinject the probes into cells, we use culture dishes that are open and have appropriate dimensions for access to the microinjection set-up. Furthermore, the temperature of the cells is maintained by controlled heating of the culture dish (T4 culture dishes are coated with conductive material for controlled heating) and the objective.

CHO-GFP-96-mer cells were grown in T4 glass culture dishes in the presence of doxycycline (*see* **Note 6**). For imaging, the culture dishes were placed on the inverted fluorescence microscope with heated stage and ×100 oil-immersion objective. The molecular beacons were dissolved in water at a concentration of 2.5 ng/μL, and a volume of 0.1 to 1 fL was injected into the cells (*see* **Note 7**). Removing the doxycycline from the growth medium started the expression of the GFP-mRNA-96-mer, and usually a bright fluorescent spot appeared in the nuclei between 60 and 90 min after induction. This spot represented the site of transcription, and was actually made up of multiple mRNA molecules and was therefore brighter that the intensity of individual mRNP particles. After this period, individual mRNA molecules were observed as diffraction-limited spots that emanated from the transcription site and distributed uniformly within the volume of the nucleus. Over time, the RNA molecules were found in the cytoplasm where they were functionally translated into GFP.

The exposure time and the time between successive image frames depends on the process that one wishes to follow. To characterize the diffusion of mRNP particles in the nucleus, the exposure time was set to 0.3 sec without any time between successive frames. To achieve higher intensities for the particles in these experiments, the images were recorded at 2× binning.

To study the behavior of the mRNP particles in relation to the chromatin density, heterologous histone H2B fused to GFP was expressed in the CHO cells. The fluorescence intensity of the GFP-H2B directly reflected the density of the chromatin into which it was incorporated *(14)*. This visualization was possible because the simultaneous signal from the destabilized GFP from the reporter RNA was much lower in intensity and was localized in the cytoplasm.

3.4   Tracking Single mRNA Molecules

The mRNP particles were tracked by using custom software developed in MATLAB with its image processing toolbox. The images in a time series were first passed through a median filter, after which they were run through the custom linear filter, which was loosely based on the discrete Laplacian and specifically enhanced diffraction limited spots from their background. We have shown that each spot corresponds to an individual mRNA molecule, thus establishing that the method is a valid way to count the number of mRNA molecules in individual cells *(8)*.

These particles were visually identified in the first frame, after which a local threshold was applied to reveal each particle's outline. The centroid of each outline was used as that particle's position. Its position in subsequent frames was determined with the aid of a nearest maximum algorithm. If the algorithm failed to correctly identify a particle in the subsequent frame, then provision was made for manual identification. The precision of the particle's location was higher than the limit of optical resolution, because the center of a circle with fuzzy boundaries can be located with greater precision than the circle itself. Tracking was stopped if two particles came so close to each other that their identities became confused.

Among the possible types of movements are free diffusion caused by Brownian motion, corralled diffusion within a confined space, directed motion by external agents such as motor proteins, or no movement. All of these motion types are described by the relationship between the magnitude of displacement (MSD) of a particle and the time interval (*see* **Note 8**). We tracked the movement of single mRNAs in the nucleus and calculated their MSD by averaging the squares of all displacements of the particle between frames separated by a given time interval (Fig. 6.3). Furthermore, tracking also allowed us to calculate the diffusion constants for particles showing Brownian motion, the confines for corralled particles, and the speed of potentially directed movement.

The majority of the GFP-mRNA-96-mer particles displayed a linear relationship between the MSD and the time interval, with an average value of their diffusion constant of $0.033\,\mu m^2/s$. However, on average, about half of the mRNP particles were mobile at any moment while the other half were found to be stationary (apparent diffusion constant $0.0006\,\mu m^2/s$). When the tracks of all mRNP particles in one imaging plane of one nucleus were merged with the chromatin density image, it was clear that regions of low chromatin density were frequently visited by mobile

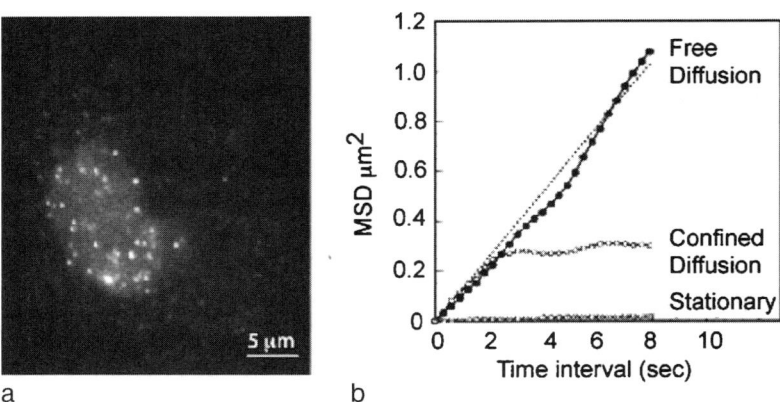

a   b

Fig. 6.3 Relationship the MSD of individual mRNP particles and the time interval of their displacement. **a** Typical frame of a time-lapse movie showing bright mRNP particles in the nucleus. **b** Examples are shown of three different types of movements displayed by the particles. The *dotted line* indicates the behavior of the majority of freely diffusing particles (average diffusion constant of $0.033\,\mu m^2/s$)

Fig. 6.4 Tracks of a few of the mRNP particles in a nucleus overlaid on an image of chromatin density. In *blue*, the fluorescence intensity of GFP-H2B directly reflected the density of the chromatin. The tracks are displayed in *green*, the final positions of mobile particles are shown by *green dots*, and those of stationary particles by *red dots*. To view this figure in color, see COLOR PLATE 4

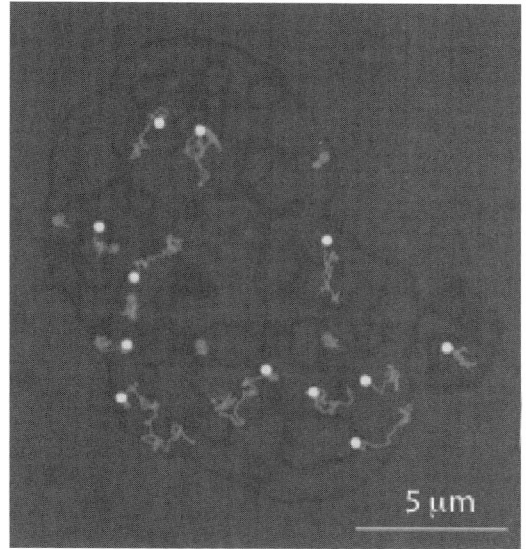

particles while the stationary particles were found in the high-density chromatin (Fig. 6.4). Nucleoli were never visited by the mRNP particles. Interestingly, the diffusion constants of mobile particles were not strongly affected by factors like ATP depletion or lowered temperature, suggesting that the motion of these particles was not controlled by enzymatic processes. These factors did increase the fraction of stalled particles. This, together with the observation of stalled particles resuming a mobile state, suggests that the transition of a particle to assume the mobile state from the stalled state is an ATP-dependent process.

4 Notes

1. In the default settings of *m-fold*, the RNA is folded at 37°C. After folding, the individual secondary structures and their thermodynamic details are presented for viewing. The *ss-count* and *P-nums* for all the structures are presented in separate files and their data can be imported into a spreadsheet program like Microsoft Excel. This program outputs folding results for short sequences (up to 800 nt) immediately; however, for long RNAs you need to submit the job and retrieve the results later from the host server.

2. In this case, the combination of *Sal*I and *Xho*I was chosen; when their compatible cohesive ends are ligated, the resulting site is not recleavable by either enzyme. Other combinations of restriction enzymes are also possible for the same purpose, for example *Bcl*I/*Bam*HI, *Bgl*II/*Bam*HI, *Nsi*I/*Pst*I, or *Xba*I/*Spe*I to suit your own cloning needs.

3. Signal-to-background ratios of molecular beacons are tested in a spectrofluorometer using wavelengths of excitation and emission specific for the fluorophores.
   First measure the signal of 125 μL of the hybridization buffer (F_{buffer}). After 10 μL
   of a 0.5 μM molecular beacon solution is added, a new fluorescence reading is
   taken (F_{closed}). Next, a fivefold molar excess of oligonucleotide target is added and
   the increase in fluorescence is recorded over time. When the maximum fluorescence is achieved (all molecular beacons have bound a target; F_{open}) the signal-to-
   background ratio is calculated as follows: ($F_{open} - F_{buffer}$) / ($F_{closed} - F_{buffer}$).

4. Thermal denaturation profiles of molecular beacons are measured in a spectrofluorometric thermal cycler using wavelengths of excitation and emission
   specific for the fluorophore. Add 50–100 ng of molecular beacon to the hybridization mix in a PCR tube, with and without a fivefold molar excess of oligonucleotide target. Decrease the temperature from 95°C to 25°C in steps of 1°C. To
   allow the equilibrium to form, hold each step for at least 30 sec. The denaturation
   profiles of the probes alone and with their oligonucleotide target should indicate
   correct molecular beacon characteristics (www.molecular-beacons.com).

5. Prepare in vitro-transcribed RNA from a linearized plasmid template using standard procedures. Add 10 μL of a 0.5 μM solution of the molecular beacon to
   125 μL of hybridization solution and measure fluorescence until the signal is stable. Add 10 μL of 100 nM in vitro-transcribed RNA and measure the increase in
   fluorescence intensity of the molecular beacon in a spectrofluorometer using
   wavelengths of excitation and emission specific for the fluorophore at 37°C.

6. The CHO Tet-off cells were regularly cultured at 37°C in the α modification of
   Eagle's minimal essential medium supplemented with glutamine, 10% TET-
   System-Approved FBS (see **Section 2.2.2**), and doxycycline (10 ng/mL). After
   injection with molecular beacons, the doxycycline was removed to allow
   expression. During this time, the cells were placed back in the incubator to keep
   them healthy before imaging commenced. The volume of imaging medium was
   kept large to buffer changes in temperature and CO_2 concentration.

7. Like all small oligonucleotides, molecular beacons are transported to the nucleus
   within minutes of their introduction into living cells *(15)*. Because the free
   probes reside in the nucleus, the other cell compartments are free of background
   fluorescence that may be associated with molecular beacons in living cells.

8. The time intervals ranged from the time elapsed between two successive frames
   to the full duration of the time series. In the averaging, all pairs of time points
   were considered rather than just independent pairs. Only MSD measurements
   determined from time intervals shorter than 25% of the length of time during
   which each particle was tracked were used to calculate the particle's diffusion
   constant, because statistical variations became so large for longer intervals that
   artifactual corrals were predicted to be present where none existed. The diffusion constant of corralled particles (identified by the leveling off of their MSD
   at longer time intervals) was determined in the same manner, but only from data
   obtained during the shorter intervals in which they diffused freely. The average
   diffusion constants determined for each condition were computed by taking the
   weighted average of the diffusion constants for each particle track, where the

weight of each track was proportional to its length because longer tracks yielded statistically more significant data. Whereas this weighting may introduce a bias in the determination of the average diffusion constant in that less mobile particles are more likely to be able to be followed for longer periods, it also minimizes the influence of measurements made on particles whose movements are more difficult to resolve. Although the magnitude of the standard deviations of these measurements was similar to the magnitude of the diffusion constants, wide variations in diffusion constants are natural features of single-particle tracking measurements due to the stochastic nature of diffusion as well as to differences in the microenvironments through which the particles move.

Acknowledgments We thank Diana Y. Vargas and Arjun Raj for their contributions. This work was supported by the National Institutes of Health Grant GM-070357.

References

1. Politz, J. C. and Pederson, T. (2000) Movement of mRNA from transcription site to nuclear pores. *J. Struct. Biol.* **129**, 252–257.
2. Agutter, P. S. (1994) Models for solid-state transport: messenger RNA movement from nucleus to cytoplasm. *Cell Biol. Int.* **18**, 849–858.
3. Lawrence, J. B., Singer, R. H., and Marselle, L. M. 1989 Highly localized tracks of specific transcripts within interphase nuclei visualized by in situ hybridization. *Cell.* **57**, 493–502.
4. Blobel, G. (1985) Gene gating: a hypothesis. *Proc. Natl. Acad. Sci. USA.* **82**, 8527–8529.
5. Colon-Ramos, D. A., Salisbury, J. L., Sanders, M. A., Shenoy, S. M., Singer, R. H., and Garcia-Blanco, M. A. (2003) Asymmetric distribution of nuclear pore complexes and the cytoplasmic localization of beta2-tubulin mRNA in Chlamydomonas reinhardtii. *Dev. Cell.* **4**, 941–952.
6. Casolari, J. M., Brown, C. R., Komili, S., West, J., Hieronymus, H., and Silver, P. A. (2004) Genome-wide localization of the nuclear transport machinery couples transcriptional status and nuclear organization. *Cell* **117**, 427–439.
7. Shav-Tal, Y., Darzacq, X., Shenoy, S. M., Fusco, D., Janicki, S. M., Spector, D. L., and Singer, R. H. (2004) Dynamics of single mRNPs in nuclei of living cells. *Science.* **304**, 1797–1800.
8. Vargas, D. Y., Raj, A., Marras,. S. A., Kramer, F. R., and Tyagi, S. (2005) Mechanism of mRNA transport in the nucleus. *Proc. Natl. Acad. Sci. USA* **102**, 17008–17013.
9. Tyagi S, Bratu, D. P., and Kramer, F. R. (1998) Multicolor molecular beacons for allele discrimination. *Nat. Biotechnol.* **16**, 49–53.
10. Tsourkas, A., Behlke, M. A., and Bao, G. (2003) Hybridization of 2′-O-methyl and 2′-deoxy molecular beacons to RNA and DNA targets. *Nucleic Acids Res.* **31**, 5168–5174.
11. Bratu, D. P. Cha, B. J., Mhlanga, M. M., Kramer, F. R., and Tyagi, S. (2003) Visualizing the distribution and transport of mRNA in living cells. *Proc. Natl. Acad. Sci. USA* **100**, 13308–13313.
12. Zucker, M. (2003) Mfold web server for nucleic acid folding and hybridization prediction. *Nucleic Acids Res.* **31**, 1–10.
13. Robinett, C. C., Straight, A., Li, G., Willhelm, C., Sudlow, G., Murray, A., and Belmont, A. S. (1996) In vivo localization of DNA sequences and visualization of large-scale chromatin organization using lac operator/repressor recognition. *J. Cell Biol.* **135**, 1685–1700.
14. Kanda, T., Sullivan, K. F., and Wahl, G. M. (1998) Histone-GFP fusion protein enables sensitive analysis of chromosome dynamics in living mammalian cells. *Curr. Biol.* **26**, 377–385.
15. Tyagi, S. and Alsmadi, O. (2004) Imaging native β-actin mRNA in motile fibroblasts. *Biophys. J.* **87**, 4153–4162.

Chapter 7
Mapping *Cis*- and *Trans*- Chromatin Interaction Networks Using Chromosome Conformation Capture (3C)

Adriana Miele and Job Dekker

Keywords DNA; Chromatin looping; Long-range gene regulation; *Trans*-regulation; Formaldehyde crosslinking; Spatial organization

Abstract Expression of genes can be controlled by regulatory elements that are located at large genomic distances from their target genes (in *cis*), or even on different chromosomes (in *trans*). Regulatory elements can act at large genomic distances by engaging in direct physical interactions with their target genes resulting in the formation of chromatin loops. Thus, genes and their regulatory elements come in close spatial proximity irrespective of their relative genomic positions. Analysis of interactions between genes and elements will reveal which elements regulate each gene, and will provide fundamental insights into the spatial organization of chromosomes in general.

Long-range *cis*- and *trans*- interactions can be studied at high resolution using chromosome conformation capture (3C) technology. 3C employs formaldehyde crosslinking to trap physical interactions between loci located throughout the genome. Crosslinked cells are solubilized and chromatin is digested with a restriction enzyme. Chromatin is subsequently ligated under conditions that favor intramolecular ligation. After reversal of the crosslinks, the DNA is purified and interaction frequencies between specific chromosomal loci are determined by quantifying the amounts of corresponding ligation products using polymerase chain reaction (PCR). This chapter describes detailed protocols for 3C analysis of chromatin interactions in the yeast *Saccharomyces cerevisiae* and in mammalian cells.

1 Introduction

The spatial organization of a genome plays an important role in genome regulation. For example, a distant regulatory element may come into direct physical contact with target genes or other regulatory elements even when separated by large genomic distances. Such long-range interactions result in the formation of chromatin loops. Long-range interactions between genes and enhancers have been directly implicated in gene regulation, as disruption of these looping interactions abolishes expression *(1, 2)*. The three-dimensional organization of the genome and the mechanisms of long-range gene regulation have been difficult to directly address, mainly due to a lack of suitable assays. The PCR-based technique described here, chromosome conformation capture (3C) has proven to be a powerful method to study physical interactions between genomic elements, e.g., between promoters and enhancers, and to determine the general spatial organization of chromosomes and subchromosomal domains *(3, 4)*.

3C has been successfully utilized in many applications. Most studies have used 3C to detect interactions between genes and regulatory elements located within chromosomal subdomains. For instance, analyses of the human and murine beta-globin loci have revealed long-range interactions between globin genes and several regulatory elements *(5–7)*. Other 3C studies revealed looping interactions involving insulator elements and imprinting centers *(8, 9)*. Thus, it appears that many different types of regulatory elements can act over large genomic distances by engaging in physical interactions with genes and other elements.

In addition to detecting intrachromosomal contacts, 3C has also been used to detect and study functional interactions between elements located on different chromosomes. In yeast, 3C was used to detect such *trans*-interactions between centromeres of different chromosomes, as well as interactions between homologous chromosomes during meiosis *(3)*. More recently, 3C studies on murine chromosomes have reported functional and highly specific *trans*-interactions between loci located on different chromosomes. These interactions appear to play important roles in coordinating expression of multiple genes located throughout the genome *(10, 11)*. Together these studies suggest that *cis*- and *trans*-interactions occur frequently throughout the genome, which has led us to propose that genomes are organized as three-dimensional networks of interacting genes and regulatory elements *(12)*. Analysis of this network of interactions, using 3C technology, promises to reveal new insights into gene expression in particular and genome regulation in general.

The 3C technology uses formaldehyde crosslinking to covalently link interacting chromatin segments in living cells (Fig. 7.1). Crosslinked chromatin is then digested with a restriction enzyme, followed by ligation at dilute DNA concentrations. A low DNA concentration favors intramolecular ligation over intermolecular ligation and thus results in selective ligation of interacting (and crosslinked) genomic elements. The crosslinks are then reversed and the DNA is purified. The resulting template, termed the 3C template, represents a library of ligation products. The abundance of a particular ligation product of two chromosomal fragments reflects how frequently these two sites interact inside of the nucleus.

Upon acquisition of a 3C template, quantitative PCR is performed to determine the relative abundance of various ligation products. A control template must also be generated, which contains equal amounts of every possible ligation product and serves as an excellent control of primer efficiencies. The ratio of the amount of ligation product in the 3C template and the amount of ligation product in the control template is then determined. This ratio is a quantitative measure for the frequency with which the two corresponding DNA fragments interacted inside the nucleus.

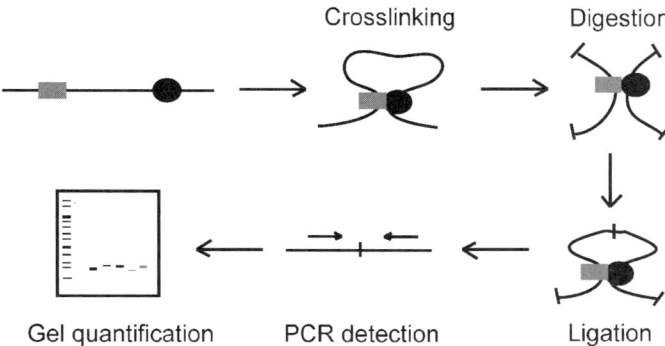

Fig. 7.1 Schematic representation of the 3C technology. Formaldehyde is used to treat either yeast or human cells resulting in covalent crosslinks between interacting DNA fragments via protein interactions (indicated by the *gray square* and *black circle*). Cross-linked chromatin is then digested, followed by ligation under dilute DNA concentrations. The crosslinks are then reversed and the DNA is purified. The PCR products are detected (PCR primers indicated by *arrows*) and can then be quantified by gel electrophoresis

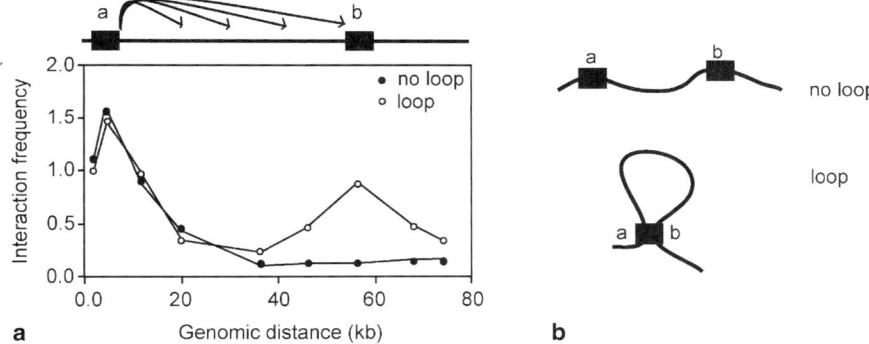

Fig. 7.2 Hypothetical results for a 3C looping experiment. **A** Analysis of a looping interactions between two elements (*a* and *b*) under two conditions: the *open circles* indicate the expected pattern of interactions when (*a*) and (*b*) engage in a specific looping interaction. The *black circles* indicate the expected pattern of interactions when (*a*) and (*b*) do not engage in a specific interactions. For both data sets, the interaction frequency between element (*a*) and element (*b*) was determined as well as a set of interaction frequencies between element (*a*) and loci located in between elements (*a*) and (*b*) and loci located downstream of (*b*). In both cases, (*a*) interacts frequently with sites located very close to it along the length of the chromosome and this interaction frequency decreases as the genomic distance between loci increases. However, in the case of a specific looping interaction, a local peak of high interaction frequencies is observed ~55 kb, which is indicative of a looping interaction at the site of the sequence element (*b*). Thus, the presence of a local peak in interaction frequencies indicates a specific long-range interaction between two elements *(12)*. **B** A hypothetical model inferred from the data sets in (**A**)

Upon determination of a set of interaction frequencies among a number of loci throughout a chromatin domain, the spatial organization of that chromosomal region can be inferred, including the presence of specific looping interactions between genomic elements (Fig. 7.2).

2   Materials

2.1   *Preparation of 3C Template from Mammalian Cells*

1. Mammalian cells growing in appropriate culture medium.
2. 37% formaldehyde (Mallinckrodt, St. Louis, MO, USA) (*see* **Note 1**).
3. 2.5 *M* glycine.
4. Lysing buffer I: 10 m*M* Tris-HCl, pH 8.0, 10 m*M* NaCl, and 0.2% Igepal (Nonidet P-40). Make fresh on the day of the experiment.
5. Protease inhibitor cocktail for use with mammalian cells.
6. Restriction enzyme and corresponding 10× buffer.
7. 1% and 10% sodium dodecyl sulfate (SDS) solutions.
8. 10% (v/v) Triton X-100.
9. Ligation buffer (10×): 500 m*M* Tris-HCl, pH 7.5, 100 m*M* MgCl$_2$, and 100 m*M* dithiothreitol. Store in 1-mL aliquots for up to 1 year at −80°C.
10. 10 mg/mL bovine serum albumin (BSA).
11. 100 m*M* adenosine triphosphate (ATP).
12. T4 DNA ligase (Invitrogen, Carlsbad, CA, USA).
13. 10 mg/mL proteinase K in TE buffer, pH 8.0.
14. Phenol.
15. 1:1 phenol/chloroform.
16. 3 *M* sodium acetate, pH 5.2.
17. 100% and 70% ethanol.
18. TE buffer: 10 m*M* Tris-HCl, 1 m*M* EDTA, pH 8.0.
19. Chloroform.
20. DNase-free RNase A (Sigma-Aldrich, St. Louis, MO, USA), 10 mg/mL.

2.2   *Preparation of Control Template from Mammalian DNA*

1. Bacterial artificial chromosome clones.
2. Ethidium bromide.
3. Agarose.
4. TBE (10×): 108 g Tris base, 55 g boric acid, and 40 mL of 0.5 *M* EDTA, pH 8.0.
5. DNA loading buffer (*see* **Note 2**).
6. TE buffer, pH 8.0.
7. Restriction enzyme and corresponding 10× restriction enzyme buffer.
8. 10 mg/mL bovine serum albumin (BSA).
9. 1:1 phenol/chloroform.
10. 3 *M* sodium acetate, pH 5.2.
11. 70% and 100% ethanol.
12. Ligation buffer (10×): 500 m*M* Tris-HCl, pH 7.5, 100 m*M* MgCl$_2$, and 100 m*M* dithiothreitol. Store in 1-mL aliquots for up to 1 year at −80°C.

13. 100 m*M* adenosine triphosphate (ATP).
14. T4 DNA ligase.
15. Chloroform.

2.3 Analysis of Mammalian 3C and Control Templates by Quantitative PCR

1. 3C and control templates.
2. DNA molecular weight standard of known concentration.
3. Ethidium bromide.
4. Agarose.
5. TBE (10×): 108 g Tris base, 55 g boric acid, and 40 mL of 0.5 *M* EDTA, pH 8.0.
6. DNA loading buffer (*see* **Note 2**).
7. PCR buffer for mammalian templates (10×): 600 m*M* Tris adjusted to pH 8.9 with H_2SO_4 and 180 m*M* $(NH_4)_2SO_4$. Store up to 1 year at −80°C.
8. 100 m*M* dNTPs (Invitrogen).
9. Primers designed for the specific region of interest.
10. *Taq* DNA polymerase (New England Biolabs, Ipswich, MA, USA).
11. 50 m*M* $MgSO_4$.

2.4 Preparation of 3C Template from Yeast Cells

1. *Saccharomyces cerevisiae* cells of interest.
2. Spheroplasting buffer I: 0.4 *M* sorbitol, 0.4 *M* KCl, 40 m*M* sodium phosphate buffer, pH 7.2, and 0.5 m*M* $MgCl_2$. Store up to 6 months at 4°C.
3. 20 mg/mL zymolyase 100-T solution: 20 mg/mL zymolyase 100-T, 2% glucose, and 50 m*M* Tris-HCl, pH 7.5. Make the solution at least 1 day prior to the experiment. It can be stored up to 1 month at 4°C.
4. MES wash buffer: 0.1 *M* MES, 1.2 *M* sorbitol, 1 m*M* EDTA, pH 8.0, and 0.5 *M* $MgCl_2$, adjust to pH 6.4 with NaOH. Store up to 6 months at 4°C.
5. 37% formaldehyde (*see* **Note 1**).
6. 2.5 *M* glycine.
7. Restriction enzyme and corresponding 10× restriction enzyme buffer.
8. 1% and 10% sodium dodecyl sulfate (SDS) solutions.
9. 10% (v/v) Triton X-100.
10. Ligation buffer (10×): 500 m*M* Tris-HCl, pH 7.5, 100 m*M* $MgCl_2$, and 100 m*M* dithiothreitol. Store in 1-mL aliquots for up to 1 year at −80°C.
11. 10 mg/mL bovine serum albumin (BSA).
12. 100 m*M* adenosine triphosphate (ATP).
13. T4 DNA ligase.

14. 10 mg/mL proteinase K in TE, pH 8.0.
15. 1:1 phenol/chloroform.
16. 3 M sodium acetate, pH 5.2.
17. 100% ethanol.
18. 70% ethanol.
19. TE buffer, pH 8.0.
20. DNase-free RNase A, 10 mg/mL.

2.5   Preparation of Control Template from Yeast Genomic DNA

1. *S. cerevisiae* cells of interest.
2. Spheroplasting buffer II: 10 mM sodium phosphate buffer, pH 7.2, 10 mM EDTA, pH 8.0, 1% 2-mercaptoethanol, and 100 µg/mL zymolyase 100-T. Store up to 6 months at 4°C.
3. Lysing buffer II: 0.25 M EDTA, pH 8.0, 0.5 M Tris base, and 2.5% SDS. Make the solution fresh on the day of the experiment.
4. 20 mg/mL proteinase K in TE buffer, pH 8.0.
5. 5 M potassium acetate.
6. 80% ethanol.
7. 100% ethanol.
8. TE buffer, pH 8.0 containing 10 µg/mL DNase-free RNase A.
9. 1:1 phenol/chloroform.
10. 100% isopropanol.
11. 1 mg/mL bovine serum albumin (BSA).
12. 10 mM adenosine triphosphate (ATP).
13. T4 DNA ligase.
14. 0.5 M EDTA, pH 8.0.

2.6   Analysis of Yeast 3C and Control Template by Quantitative PCR

1. Yeast 3C and control templates.
2. DNA molecular weight standard of known concentration.
3. Ethidium bromide.
4. Agarose.
5. TBE (10×): 108 g Tris base, 55 g boric acid, and 40 mL of 0.5 M EDTA, pH 8.0.
6. DNA loading buffer (*see* **Note 2**).
7. PCR buffer for mammalian templates (10×): 600 mM Tris adjusted to pH 8.9 with H_2SO_4 and 180 mM $(NH_4)_2SO_4$. Store up to 1 year at −80°C.
8. 100 mM dNTPs.
9. Primers designed for specific region of interest.
10. *Taq* DNA polymerase.

3 Methods

3.1 Initial Preparation of 3C Experiments for Mammalian and Yeast Cells

1. Determine the genomic region of interest.
2. Select a restriction enzyme (*see* **Note 3**).
3. Design primers throughout the region of interest (*see* **Note 4**).

3.2 Preparation of 3C Template from Mammalian Cells

1. Acquire 1×10^8 (*see* **Note 5**) cells grown in the culture medium appropriate for the cell type (*see* **Note 6**). Centrifuge cells at 450×*g* for 10 min, remove the supernatant, and resuspend the cell pellet in 45 mL of fresh culture medium.
2. Crosslink the cells by adding 1.35 mL of 37% formaldehyde (final concentration 1%) (*see* **Note 1**), mix well by pipetting up and down, and incubate at room temperature for 10 min.
3. To quench the reaction, add 2.5 mL of 2.5 *M* glycine. Mix well by pipetting up and down and incubate at room temperature for 5 min.
4. Store on ice for at least 15 min, centrifuge cells for 10 min at 800×*g*, and resuspend the pellet in 1 mL of ice-cold lysing buffer I containing 0.1 mL protease inhibitor cocktail. Incubate on ice for 15 min.
5. Dounce homogenize the cells on ice with pestle B using 15 strokes, incubating on ice for 1 min, and then stroking an additional 15 times.
6. Transfer the cells to a microcentrifuge tube and centrifuge for 5 min at 2,500×*g*. Wash the pelleted cells with 0.5 mL of the appropriate (1×) buffer for the restriction enzyme chosen. Centrifuge the washed cells under the same conditions and resuspend in 0.5 mL of 1× restriction enzyme buffer.
7. Distribute the cells between 20 individual microcentrifuge tubes (25 μL per tube), centrifuge 5 min at 2,500×*g*, resuspend each pellet in 362 μL of 1× restriction enzyme buffer, add 38 μL of 1% SDS to each tube, and incubate for 10 min at 65°C.
8. Add 44 μL of 10% Triton X-100 to each tube and mix gently by pipetting up and down. Add 400 U of restriction enzyme per tube, mix well, and incubate reactions overnight in the conditions recommended by the manufacturer for the restriction enzyme used.
9. Transfer each reaction to a 15-mL disposable conical tube. Add 86 μL of 10% SDS per tube and incubate for 30 min at 65°C to inactivate the enzyme.
10. Add 745 μL of 10% Triton X-100, 745 of 10× ligation buffer, 80 μL of 10 mg/mL BSA, 5,960 μL of distilled water, and 4,000 cohesive-end units of T4 DNA ligase and incubate for 2 h at 16°C (*see* **Note 7**).
11. Add 50 μL of 10 mg/mL proteinase K in TE buffer, pH 8.0, and incubate overnight at 65°C to reverse the crosslinks.

12. Add 50 μL of 10 mg/mL proteinase K in TE buffer, pH 8.0, and incubate for 2 h at 42°C.

13. Transfer the solutions to 50-mL disposable conical tubes. Extract DNA by adding an equal amount of phenol to each tube, vortex for 30 sec, and centrifuge for 5 min at 2,460×. Collect the upper, aqueous layer and repeat the phenol extraction.

14. Collect the upper, aqueous layer and add an equal volume of 1:1 phenol/chloroform, vortex for 30 sec, and centrifuge for 5 min at 2,460×g.

15. Pool the upper, aqueous layers from the 20 samples into three 250-mL screw-cap centrifuge bottles. Add 1/10 volume of 3 M sodium acetate, pH 5.2, vortex briefly and add 2.5 volumes of ice-cold 100% ethanol. Incubate for at least 30 min at −80°C and centrifuge for 20 min at 12,000×g at 4°C. Decant the supernatant and resuspend the pellets in a total volume of 1 mL of TE buffer, pH 8.0 and transfer to two fresh microcentrifuge tubes (500 μL each).

16. Add an equal volume of phenol, vortex for 30 sec, and centrifuge for 5 min at 2,460×g. Collect the upper, aqueous layer and extract DNA by addition of an equal volume of 1:1 phenol/chloroform, vortex for 30 sec, and centrifuge for 5 min at 2,460×g. Repeat the phenol/chloroform extraction.

17. Add an equal volume of chloroform to each tube containing the upper, aqueous layer, vortex for 30 sec, and centrifuge for 5 min at 1,100×g. Collect the upper, aqueous layer.

18. To the upper layer, add 1/10 volume of 3 M sodium acetate, pH 5.2, and vortex briefly. Precipitate DNA by adding 2.5 volumes of ice-cold 100% ethanol, mix gently, incubate at −20°C and centrifuge for 20 min at 18,000×g at 4°C.

19. Wash the DNA pellet five times with 70% ethanol.

20. Let the DNA pellets dry completely before dissolving in a total volume of 1 mL of TE buffer, pH 8.0. The DNA obtained is termed the 3C template.

21. Remove RNA by adding 2 μL of DNase-free RNase A (10 mg/mL) followed by incubation for 15 min at 37°C.

3.3 Preparation of Control Template from Mammalian DNA

1. Acquire BAC clones (*see* **Note 8**) spanning the entire region of interest (see Note 9), purify the BAC DNA by alkaline lysis with SDS (Qiagen) and estimate the DNA concentration by running 1 μL of DNA on a 0.8% agarose/0.5× TBE gel side by side with a size standard of known concentration (*see* **Note 10**).

2. Digest ~20 μg of BAC DNA (*see* **Notes 11** and **12**) in 10× restriction enzyme buffer, with BSA (10 mg/mL) if recommended by the manufacturer. Add 40 U/ μL of restriction enzyme stock so that the amount of enzyme corresponds to 8.75% of the final digestion volume. Incubate overnight in the conditions recommended by the manufacturer.

3. Extract DNA by adding an equal volume of 1:1 phenol/chloroform, vortex for 30 sec, and centrifuge for 5 min at 18,000×g. Transfer the upper, aqueous layer to a fresh microcentrifuge tube.

4. Precipitate the DNA by adding 1/10 volume of 3 M sodium acetate, pH 5.2, vortex briefly, and add 2.5 volumes of ice-cold 100% ethanol. Mix gently, incubate for at least 15 min at −20°C, and centrifuge for 20 min at 18,000×g at 4°C.

5. Remove the supernatant promptly, wash the DNA pellet with 1 mL of 70% ethanol, and centrifuge for 15 min at 18,000×g at 4°C. Let the pellet air dry briefly (*see* **Note 13**) and resuspend in 161 µL of water. Incubate at 37°C to dissolve the DNA completely.

6. Combine 157 µL of digested BAC DNA, 20 µL of 10× ligation buffer, 2 µL of BSA (10 mg/mL), 2 µL of ATP (100 mM), and 7,600 cohesive end units of T4 DNA ligase, and adjust to 200 µL final volume (*see* **Note 7**). Incubate overnight at 16°C.

7. Inactivate the ligase by incubating the solution for 15 min at 65°C.

8. Purify DNA by adding 200 µL of 1:1 phenol/chloroform, vortex for 30 sec, and centrifuge for 5 min at 18,000×g. Transfer the upper, aqueous layer to a microcentrifuge tube and repeat the phenol/chloroform extraction.

9. To the upper, aqueous layer add 200 µL of chloroform, vortex for 30 sec, and centrifuge for 5 min at 18,000×g. Transfer the upper, aqueous layer to a microcentrifuge tube.

10. Add 1/10 volume of 3 M sodium acetate, pH 5.2, vortex briefly, and add 2.5 volumes of ice-cold 100% ethanol. Mix gently, incubate for 15 min at −20°C, and centrifuge for 20 min at 18,000×g.

11. Carefully decant the supernatant, air-dry the pellet slightly, and resuspend in TE buffer, pH 8.0 to obtain a DNA solution with the final concentration of ~100 ng/µL. The DNA obtained is termed the control template.

12. Remove any RNA by adding 20 µL of DNase-free RNase A (10 mg/mL) followed by incubation for 15 min at 37°C.

3.4   Analysis of Mammalian 3C and Control Templates by Quantitative PCR

1. Determine the DNA concentration of the templates by agarose gel electrophoresis in parallel with DNA length standards (*see* **Note 10**).

2. Make a series of 8–10 twofold dilutions of template DNA starting with a highest amount of 500 ng and ending with a water control.

3. Set up a PCR reaction for each dilution in a 25-µL reaction containing 2 µL of each template dilution (equaling 500 ng of DNA), 1× PCR buffer for mammalian templates, 4 mM MgSO$_4$, 0.2 mM dNTPs, 0.4 µM of each primer, and 1 U of *Taq* DNA polymerase (*see* **Note 14**). For quantitative results use the following hot start parameters to program the thermal cycler: 1 cycle of 1 min at 95°C; 34 cycles of 1 min at 95°C followed by 45 sec at 65°C followed by 2 min at 72°C; and 1 cycle of 1 min at 95°C followed by 45 sec at 65°C followed by 8 min at 72°C.

4. Analyze the PCR products by agarose gel electrophoresis using a 1.5% agarose gel containing 0.5 TBE and 0.5 μg/mL ethidium bromide (*see* **Note 2**). Quantify the PCR products using a gel imaging and documentation system, and determine the amount of product obtained for each concentration of 3C and control template (*see* **Note 15** and Fig. 7.3). Determine the range of template concentrations that are within the linear range of PCR detection, as described in the legend of Fig. 7.3.

5. Using the DNA concentration determined by this PCR titration, set up three reactions for each primer pair of interest for both the 3C and control template. Use the PCR conditions and analyze and quantify the products using the same conditions as described above (*see* **Note 16**).

6. To determine the interaction frequency for each primer pair, divide the amount of PCR product obtained with the 3C template (the average of 3 PCR reactions) by the amount of PCR product obtained with the control template (the average of 3 PCR reactions) (*see* **Note 17**).

7. Determine a set of interaction frequencies from which the spatial organization of the genomic region of interest can be analyzed (*see* **Notes 18** and **19**). A hypothetical 3C experiment is shown in Fig. 7.2.

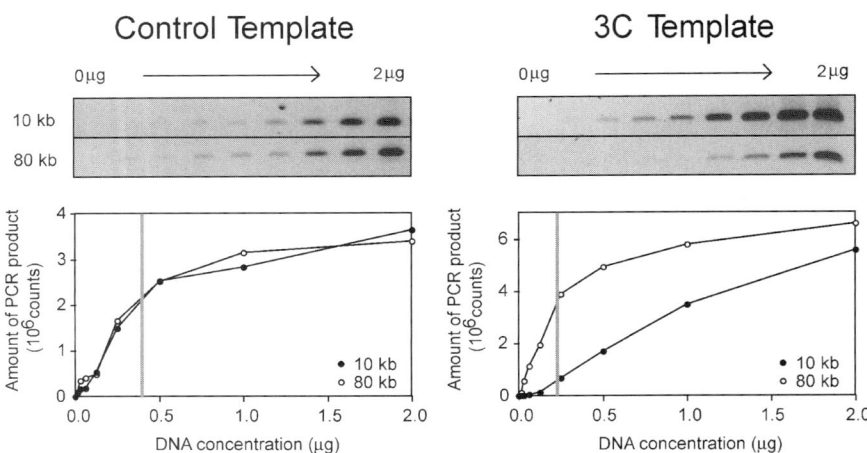

Fig. 7.3 Titration and gel quantification of yeast 3C and control templates. The *left* (control template) and *right panels* (3C template) represent a titration and quantification using two primer pair combinations (detecting interactions between loci separated by 10 kb and 80 kb, respectively). The amount of PCR product is plotted versus the template DNA concentration in micrograms. For both primer pair combinations, the control template yields similar amounts of PCR product, while the 3C template yields significantly more PCR product for the primer pair that detects interactions between loci separated by 10 kb. The linear range for PCR amplification is found to the left of the *gray line*. The template DNA concentration chosen should be high enough to yield sufficient PCR product to allow visualization and quantification on a gel, but should be within the linear range of PCR detection. The same template concentration must be used for all subsequent PCR reactions

3.5 Preparation of 3C Template from Yeast Cells

1. Acquire a 200-mL culture of *S. cerevisiae* cells at $OD_{600} = 1$.
2. Centrifuge the cells for 10 min at 1,250×*g*, remove the supernatant promptly, and resuspend in 10 mL of spheroplasting buffer I.
3. Add 50 μL of zymolyase 100-T solution (20 mg/mL; *see* **Note 20**). Mix the tube gently and incubate in a roller drum for 40 min at 30°C (*see* **Note 21**).
4. Centrifuge the cells (now spheroplasts) for 5 min at 2,460×*g* at room temperature. Wash two times in 10 mL MES wash buffer and resuspend the spheroplasts in 10 mL MES wash buffer.
5. Add 263 μL of 37% formaldehyde (final concentration, 1%) (*see* **Note 1**), mix thoroughly, and incubate for 10 min at room temperature.
6. Quench the reaction by addition of 0.5 mL of 2.5 *M* glycine and incubate for 5 min at room temperature (*see* **Note 22**).
7. Add 50 μL of crosslinked spheroplasts to each of 40 individual microcentrifuge tubes (*see* **Note 23**). Wash spheroplasts three times with 100 μL of 1× restriction enzyme buffer corresponding to the restriction enzyme chosen. Between each wash, mix by pipetting up and down and centrifuge for 3 min at 18,000×*g*. Resuspend the washed pellet in 36.2 μL of 1× restriction enzyme buffer.
8. Add 3.8 μL of 1% SDS per tube, and incubate for 10 min at 65°C (*see* **Note 24**).
9. Add 4.4 μL of 10% Triton X-100 per tube and mix gently by pipetting up and down.
10. Add 60 U of restriction enzyme per tube, mix well, and incubate reactions overnight in the conditions recommended by the manufacturer.
11. Add 8.6 μL of 10% SDS per tube and incubate for 20 min at 65°C to inactivate the enzyme.
12. Add 74.5 μL of 10% Triton X-100, 74.5 μL of 10× ligation buffer, 8 μL of 10 mg/ml BSA, 8 μL of 100 m*M* ATP, 596 μL distilled water, and 800 cohesive end units of T4 DNA ligase per tube (*see* **Note 7**). Incubate for 2 h at 16°C.
13. Add 5 μL of 10 mg/mL proteinase K in TE buffer, pH 8.0, and incubate overnight at 65°C to reverse the crosslinks.
14. Add an additional 5 μL of 10 mg/mL proteinase K in TE buffer and incubate for 2 h at 42°C.
15. Combine ten reactions to end up with four larger pooled reactions in 50-mL centrifuge tubes (*see* **Note 25**).
16. Add an equal volume of 1:1 phenol/chloroform to each of the reactions, vortex for 30 sec, and centrifuge for 5 min at 2,460×*g* in a table-top centrifuge. Remove the upper, aqueous layer promptly, taking care not to include the interface layer. Repeat the phenol/chloroform extraction until the aqueous layer is clear.
17. Transfer the clear aqueous solution to a 30-mL screw-cap centrifuge tube. Add 1/10 volume of 3 *M* sodium acetate, pH 5.2, to the clear aqueous layer and vortex briefly. Precipitate the DNA by adding 2.5 volumes of cold 100% ethanol

and mix gently. Incubate for 15 min at −80°C and centrifuge for 20 min at 10,000×g at 4°C.

18. Remove the supernatant and allow each of the four DNA pellets to dry completely before resuspending in 100 μL of TE buffer, pH 8.0. Pool all the samples to obtain a 400-μL DNA solution.

19. Add an equal volume of phenol/chloroform to the purified DNA. Vortex for 30 sec, and centrifuge for 5 min at 18,000×g. Remove the upper, aqueous layer and transfer to a clean 1.7-mL microcentrifuge tube.

20. To the upper, aqueous layer, add 1/10 volume of 3 M sodium acetate, pH 5.2, and vortex briefly. Precipitate the DNA by adding 2.5 volumes of cold 100% ethanol and mix gently. Centrifuge for 10 min at 18,000×g at 4°C. Remove the supernatant.

21. Wash the DNA pellet by adding 0.5 mL of 70% ethanol. Resuspend the pellet and centrifuge for 5 min at 18,000×g.

22. Remove the supernatant carefully and allow the DNA pellet to dry completely before resuspending in 100 μL of TE buffer, pH 8.0. The DNA sample obtained is termed the 3C template.

23. Remove RNA by adding 2 μL of DNase-free RNase A (10 mg/mL) followed by incubation for 15 min at 37°C.

3.6 Preparation of Control Template from Yeast Genomic DNA

1. Acquire an overnight, saturated culture of the *S. cerevisiae* strain that was used for generation of the 3C template.

2. Centrifuge cells for 10 min at 1,250×g and remove the supernatant promptly.

3. Resuspend the cells in 20 mL of spheroplasting buffer II and distribute between 40 individual microcentrifuge tubes (0.5 mL each; *see* **Note 26**). Incubate at 37°C for 40 min.

4. Add 100 μL of lysing buffer I to each tube.

5. Add 10 μL of 20 mg/mL proteinase K in TE buffer, pH 8.0 to each tube, and incubate at 65°C for 30 min.

6. Add 100 μL of 5 M potassium acetate to each tube and incubate in ice water for 10 min.

7. Centrifuge for 20 min at 18,000×g at 4°C.

8. Transfer the supernatant to a fresh tube containing 0.5 mL of ice-cold 100% ethanol, invert five times, and spin for 10 min at 18,000×g.

9. Remove the supernatant by suction and allow the DNA pellets to dry before dissolving DNA in 500 μL TE buffer, pH 8.0, containing DNase-free RNase A (10 μg/mL). Incubate for 30 min at 37°C or until all DNA is dissolved, occasionally tapping the tubes.

10. Extract DNA once by adding an equal volume of 1:1 phenol/chloroform, vortex for 30 sec, and centrifuge for 5 min at 1,100×g.

11. Transfer the upper, aqueous layer to a clean microcentrifuge tube, add 0.5 mL of isopropanol to precipitate DNA, invert the tubes five times, and centrifuge for 10 min at 18,000×g.

12. Wash the DNA once with 80% ethanol and centrifuge for 10 min at 18,000×g.

13. Allow the DNA pellets to dry and dissolve each pellet in 100 μL of TE buffer. pH 8.0. Pool the samples and determine the DNA concentration by absorption spectroscopy. This is the genomic DNA to be used in generation of the control template.

14. Digest 20 tubes each containing 10 μg of genomic DNA isolated in previous steps in a 400-μL reaction using 60 U of the same restriction enzyme and appropriate buffer as used to prepare the yeast 3C template (*see* **Note 26**). Incubate the tubes for 3 h in the conditions recommended for the enzyme.

15. Extract DNA once by addition of an equal volume of 1:1 phenol/chloroform, vortex for 30 sec, and centrifuge for 5 min at 1,100×g.

16. Transfer the upper, aqueous layer to a fresh tube, add 1/10 volume of 3 M sodium acetate, pH 5.2, and vortex briefly. Precipitate DNA by adding 2.5 volume of ice-cold 100% ethanol, mix gently, and centrifuge for 10 min at 18,000×g. Carefully remove the supernatant.

17. Wash the pellet by addition of 0.5 mL of 80% ethanol, centrifuge for 10 min at 18,000×g, carefully remove the supernatant, and allow the pellet to air-dry completely.

18. Dissolve each DNA pellet in 20 μL of autoclaved water. Add 3 μL of 10× ligation buffer, 3 μL of 1 mg/mL BSA, 3 μL of 10 mM ATP, and 800 cohesive end units of T4 DNA ligase to each tube (*see* **Note 7**) and incubate at 16°C for 1 h.

19. Pool ten reactions together to end up with two larger reactions, and stop the reaction by adding 6 μL of 0.5 M EDTA, pH 8.0 to each pool.

20. Extract DNA once by addition an equal volume of 1:1 phenol/chloroform, vortex for 30 sec, and centrifuge for 5 min at 1,100×g.

21. Transfer the upper, aqueous layer to a fresh tube. Add 1/10 volume of 3 M sodium acetate, pH 5.2, and vortex briefly. Precipitate DNA by adding 2.5 volumes of ice-cold 100% ethanol, mix gently, centrifuge for 10 min at 18,000×g, carefully remove the supernatant, and allow the pellet to air-dry completely.

22. Wash the pellet by addition of 0.5 mL of 80% ethanol. Centrifuge for 5 min at 18,000×g and let the pellet air-dry completely. Dissolve each pellet in 400 μL of TE buffer, pH 8.0, and pool all samples to obtain an 800-μL DNA solution. The DNA sample obtained is termed the control template.

23. Add 2 μL of 10 mg/mL DNase-free RNase and incubate for 15 min at 37°C.

3.7 Analysis of Yeast 3C and Control Template by Quantitative PCR

1. Determine the DNA concentration of the yeast 3C and control templates by agarose gel electrophoresis by running in parallel with a known molecular weight standard (*see* **Note 10**).

2. Make a series of 8–10 twofold dilutions of template DNA starting with the highest amount at 2 μg and ending with a water control.

3. Set up PCR reactions for each dilution in a 50-μL reaction containing 2 μL each template dilution, 1× PCR buffer for yeast templates, 0.5 m*M* dNTPs, 0.4 μM of each primer, and 2.5 U *Taq* DNA polymerase (*see* **Note 14**). For quantitative results use the following hot start PCR parameters to program the thermal cycler: 1 cycle for 1 min at 95°C; 32 cycles for 1 min at 95°C followed by 45 sec at 60°C followed by 2 min at 72°C; 1 cycle for 1 min at 95°C followed by 45 sec at 60°C followed by 8 min at 72°C.

4. Analyze the PCR products by agarose gel electrophoresis using a 1.5% agarose gel containing 0.5 TBE and 0.5 μg/mL ethidium bromide (*see* **Note 2**). Quantify PCR products using a gel documentation setup and determine the amount of PCR product per individual reaction (*see* **Note 15** and Fig. 7.3).

5. Using the DNA concentration determined by the PCR titration, set up three reactions for each primer pair of interest for both the 3C and control template following the same PCR conditions and program as described above. Analyze and quantify the PCR products using the same conditions as described above (*see* **Note 16**).

6. To determine the interaction frequency for each primer pair divide the amount of PCR product obtained with the 3C template (in triplicate) by the amount of PCR product obtained by using the control template (in triplicate) (**Note 17**).

7. Collect a set of interaction frequencies from which the spatial organization of the genomic region of interest can be analyzed (*see* **Notes 18** and **19**). A hypothetical 3C experiment is shown in Fig. 7.2.

4   Notes

1. The formaldehyde used for 3C experiments should not be more than 1 year old, as the effective formaldehyde concentration gradually decreases which will reduce the efficiency of crosslinking.

2. The loading buffer used should not contain bromophenol blue, as this dye will run at the same position as the PCR products and will interfere with quantification. The use of Ficoll is encouraged.

3. Cutting sites for the restriction enzyme should be evenly distributed throughout the genomic region of interest, and if analyzing a potential looping interaction, the putative looping elements should be contained within restriction fragments that are no bigger than 10 kb and no smaller than 1 kb, as this may result in slightly higher or slightly lower interaction frequencies, respectively. Several to many restriction fragments should be located between the two potential looping elements to add confidence to the results and to obtain a high-resolution loop structure (see Fig. 7.2).

4. The primers should be designed with the following criteria: unidirectional, ~80–150 bp 5′ of the restriction cut site to result in a PCR product between 160 and 300 bp in size, around 28 bp in length, and with a GC content of

~50%. Primer combinations should be tested with either the yeast or mammalian control template. Those that behave aberrantly, i.e., do not amplify the correct product, amplify multiple products, or produce high levels of primer dimers, should be discarded and redesigned. Each primer pair should yield roughly the same amount of PCR product when using the control template. If this is not the case, discard the primer. Primers should not be designed on fragments that are greater than 10 kb or less than 1 kb in size.

5. The amount of cells used in this assay can be altered as necessary depending on the amount of interactions one wishes to quantify with the 3C template. The procedure described here is enough to measure ~1,000 interactions.

6. The quality of the starting material is an important factor in the template quality. Therefore, cells should be exponentially growing in the appropriate growth conditions and display healthy growth curves appropriate to the cell type.

7. When setting up ligation reactions, due to the large number of tubes, a master mix of all components is typically made and distributed.

8. Because of the complexity of the mammalian genome, a detectable level of ligation product will not occur in a control template generated from whole genomic DNA. A control template is therefore generated from a BAC clone or set of clones that contains only the genomic region of interest.

9. If more than one BAC clone is needed, the experiment must be designed so that there is minimal overlap and minimal gaps between the clones, as these regions will be over-represented in the case of overlap and absent in the case of gaps.

10. The templates should run as a DNA species of high molecular weight; however, the appearance may differ between samples and may be smear-like. A template titration should be assembled regardless of template appearance.

11. If more than one BAC clone is needed to span the entire region of interest, the clones must be mixed in equimolar ratios before digestion. Therefore, the amount of DNA in the BAC preparations should be quantified. Molar quantities of BAC clones can be determined by real-time PCR using a set of universal primers that amplify a small region within the common vector backbone.

12. The amount of total BAC DNA should be 20 µg if more than one BAC clone is required.

13. Do not let a DNA pellet dry completely, as this will compromise its ability to redissolve.

14. Two primer pairs should be chosen while preparing a template titration. The primers chosen should be of high quality and differ in genomic size separation, i.e., one pair should amplify a ligation product representing an interaction between restriction fragments that are close in location (e.g., separated by ~10 kb along the chromosome) while the other should amplify a ligation product representing an interaction between restriction fragments that are far apart (such as 80 kb).

15. The amount of template DNA chosen for PCR analysis and all subsequent reactions should be in the linear range of amplification. If a linear range of amplifi-

cation cannot be determined, this is most likely due to high concentrations of salt. If this occurs, the template should be repurified by phenol/chloroform extraction, ethanol precipitated, and washed with 70% ethanol as described in the above protocols.

16. PCR analysis of the control template and 3C template for each primer pair should be performed during the same PCR run. Further, the products obtained should be run side by side on the same gel as there is gel-to-gel variation.

17. All PCR reactions are performed in triplicate, and thus for each primer pair three interaction frequencies should be obtained; the average of these three values is the final interaction frequency. The standard error of the mean of the three interaction frequencies found should not be more than 15%, but if this is the case, additional PCR reactions may be necessary to reduce the error.

18. If analyzing a putative looping interaction, it is imperative to test interactions with the elements and sites in between the two putative looping elements. These interactions should be less frequent than the interaction between the two looping elements *(12)*.

19. If comparing interaction frequencies between two different cell types, it is necessary to first normalize the two data sets to each other to correct for differences in DNA concentrations and other experimental differences between the templates. This is done by analyzing interactions (usually 10–20 interactions) throughout a region other than the particular region of interest and that is expected to remain the same between the two cell types at hand. For each interaction, the log ratio between the two data sets is then determined, and the average of all log ratios will serve as the factor to normalize the 3C data sets *(4, 12)*.

20. Twenty milligrams of Zymolyase 100-T per milliliter will not go into solution completely, and the suspension should be mixed well before addition to the cells.

21. The efficiency of cell wall digestion should be tested by cell lysis, which can be done by observation under a microscope; after addition of water to a small amount of cells, ~80% of them should burst open and exhibit hypotonic lysis within 1–2 min.

22. Cells can now be stored for up to a year at −80°C in 1-mL aliquots if desired.

23. Reactions should not be pooled, as this will compromise the quality of the template. Tubes containing 50 µL of cells yield the best results; however, the amount of tubes may vary depending on need. The procedure described here is enough to measure ~350 reactions.

24. This step is essential in template generation and care should be taken to ensure proper incubation temperature.

25. Do not pool more than ten reactions, as this will compromise the quality of the template.

26. Reactions should not be pooled.

Acknowledgements Research in the Dekker laboratory is supported by grants from NIH (HG003143) and the Cystic Fibrosis Foundation.

References

1. Drissen, R., Palstra, R., Gillemano, N., Splinter, E., Grosveld, F., Philipsen, S., and de Laat, W. (2004) The active spatial organization of the ß-globin locus requires the transcription factor EKLF. *Genes Dev.* **18,** 2485–2490.
2. Vakoc, C., Letting, D.L., Gheldof, N., Sawado, T., Bender, M.A., Groudine, M., Weiss, M.J., Dekker, J., and Blobel, G.A. (2005) Proximity among distant regulatory elements at the beta-globin locus requires GATA-1 and FOG-1. *Mol. Cell* **17,** 453–462.
3. Dekker, J. (2002) Capturing chromosome conformation. *Science* **295,**1306–1311.
4. Gheldof, N., Tabuchi, T.M., and Dekker, J. (2006) The active FMR1 promoter is associated with a large domain of altered chromatin conformation with embedded local histone modifications. *Proc. Natl. Acad. Sci. USA* **103,** 12463–12468.
5. Tolhuis, B., Palstra, R.J., Splinter, E., Grosveld, F., and de Laat, W. (2002) Looping interaction between hypersensitive sites in the active beta-globin locus. *Mol. Cell* **10,** 1435–1465.
6. Palstra, R.J., Tolhuis, B., Splinter, E., Nijmeijer, R., Grosveld, F., and de Laat, W. (2003) The β-globin nuclear compartment in development and erythroid differentiation. *Nat. Genet.* **25,** 190–194.
7. Dostie, J., Richmond, R.A., Arnaout, R.A., Selzer, R.R., Lee, W.L., Honan,. A., Rubio, E.D., Krumm, A., Lamb, J., Nusbaum, C., Green, R.D., and Dekker, J. (2006) Chromosome conformation capture carbon copy (5C): A massively parallel solution for mapping interactions between genomic elements. *Genome Res.* **16,** 1299–1309.
8. Liu, Z. and Garrard, W.T. (2005) Long-range interactions between three transcriptional enhancers, active Vkappa gene promoters, and a 3′ boundary sequence spanning 46 kilobases. *Mol. Cell. Biol.* **25,** 3220–3231.
9. Murrell, A., Heeson, S., and Reik, W. (2004) Interaction between differentially methylated regions partitions the imprinted genes *Igf2* and *H19* into parent-specific chromatin loops. *Nat. Genet.* **36,** 889–893.
10. Spilianakis, C.G., Lalioti, M.D., Town, T., Lee, G.R., and Flavell, R.A. (2005) Interchromosomal associations between alternatively expressed loci. *Nature* **435,** 637–645.
11. Lomvardas, S., Barnea, G., Pisapia, D.J., Mendelsohn, M., Kirkland, J., and Axel, R. (2006) Interchromosomal interactions and olfactory receptor choice. *Cell* **126,** 403–413.
12. Dekker, J. (2006) The three C's of chromosome conformation capture: controls, controls, controls. *Nat. Methods* **3,** 17–21.

Chapter 8
Recognition Imaging of Chromatin and Chromatin-Remodeling Complexes in the Atomic Force Microscope

Dennis Lohr, Hongda Wang, Ralph Bash, and Stuart M. Lindsay

Keywords Atomic force microscopy; AFM; SPM; Recognition Imaging; Chromatin; Nucleosomal arrays; BRG1

Abstract Atomic force microscopy (AFM) can directly visualize single molecules in solution, which makes it an extremely powerful technique for carrying out studies of biological complexes and the processes in which they are involved. A recent development, called Recognition Imaging, allows the identification of a specific type of protein in solution AFM images, a capability that greatly enhances the power of the AFM approach for studies of complex biological materials. In this technique, an antibody against the protein of interest is attached to an AFM tip. Scanning a sample with this tip generates a typical topographic image simultaneously and in exact spatial registration with a "recognition image." The latter identifies the locations of antibody–antigen binding events and thus the locations of the protein of interest in the image field. The recognition image can be electronically superimposed on the topographic image, providing a very accurate map of specific protein locations in the topographic image. This technique has been mainly used in in vitro studies of biological complexes and reconstituted chromatin, but has great potential for studying chromatin and protein complexes isolated from nuclei.

1 Introduction

The ability of atomic force microscopy (AFM) to directly visualize single molecules in solution at nanometer resolution makes it uniquely qualified for carrying out studies on the basic properties of biological complexes and the processes in which they are involved *(1–4)*. Molecules are "imaged" by monitoring the flexing of a cantilever attached to a conical tip of end radius 1 to 50 nm as the tip scans over a surface on which the sample has been deposited. The deflection of the cantilever-tip as it responds to the topography of the surface/deposited molecules is measured by reflected laser light (Fig. 8.1) and converted to a digital image map of molecule locations on the surface (x and y coordinates) and their height (z coordinate) profiles.

R. Hancock (ed.) *The Nucleus: Volume 2: Chromatin, Transcription, Envelope, Proteins, Dynamics, and Imaging*,
© Humana Press 2008

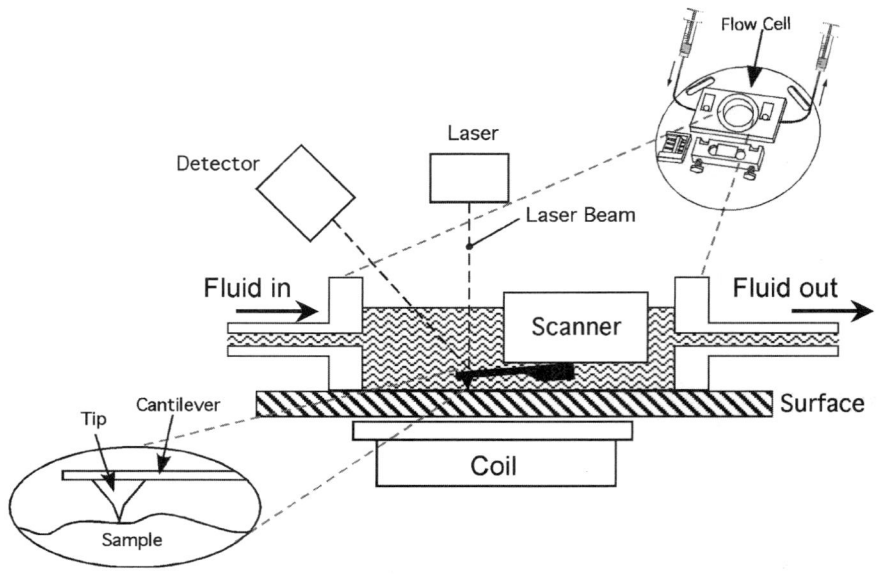

Fig. 8.1 AFM instrumentation. This figure shows a schematic representation of the experimental setup we use for imaging in liquid solutions, an atomic force microscope (*center*) attached to a flow-cell (*upper right*). New solutions (NaCl, ATP, peptides, etc.) are carefully pushed into the flow cell by depressing one of a pair of syringes, for example the left one, while solution in the flow cell is simultaneously withdrawn via the other syringe, for example the right one. The cantilever-tip is shown in *black* at the center and the tip-sample interface is shown enlarged at lower left

AFM approaches have unique advantages for chromatin studies. The technique is well suited to the study of complexes in the megadalton range, like nucleosomal arrays *(5–8)* and chromatin-modifying factors *(9–12)*. The single molecule capability of AFM can provide precise distributions of structural features, such as numbers of nucleosomes and nucleosome locations; analysis of this type of data can yield information on fundamental chromatin properties like histone–DNA binding strength, cooperativity of nucleosome occupation on arrays, and nucleosome positioning *(13)*. The ability to image in solution allows studies to be done under conditions that avoid dehydration or other harsh, potentially structure-altering sample treatments, and also provides the capability to alter the sample environment while continuing to monitor the same individual molecules in repetitive scans. The latter capability permits intrinsic features such as nucleosome stability *(8, 14)* or the actions of chromatin-altering complexes *(9, 10, 15)* to be analyzed in individual nucleosomal array molecules. These studies are carried out in a flow-cell attached to the AFM (Fig. 8.1), which allows one to change solutions without losing scanning position on the set of sample molecules. Solution imaging capabilities are crucial for Recognition Imaging.

Specific types of proteins are in general indistinguishable in traditional topographic AFM images. This is a serious limitation since chromatin, as well as the biological complexes that act on it, are heterogeneous in composition and often undergo compositional changes during biologically relevant processes. Recognition Imaging allows a specific type of protein, and at least in principle other types of biomolecules, to be identified in AFM images of compositionally complex samples *(16)*. In this technique, scanning is carried out with an AFM tip tethered to an antibody against the protein of interest. The scanning process generates a traditional topographic image simultaneously and in exact spatial registration with a recognition image that locates the sites of antigen–antibody binding events, and thus the locations of the protein of interest, within the sample field. The topographic and recognition images can be electronically superimposed to obtain a very accurate map of specific protein locations in the topographic image.

The recognition process between the tip-tethered antibody and a surface-bound antigen can be efficient (>95% of the antigens present can be recognized in the scan of a sample; ref. *(16)*) and specific *(15, 16)*, depending on the biological complex and the antibody (see below). The specificity of the recognition process can be directly tested by rescanning the same sample field (same molecules) in the presence of a peptide (or protein) antigenic to the tip-bound antibody. The presence of this antigen in solution will block the sample antigen–antibody interaction, thus removing those specific recognition signals from the image *(16)*. There are also other ways to test recognition specificity (*see* **below**).

The principle of Recognition Imaging is illustrated in Fig. 8.2. We have used this technique to image specific histones and chromatin complexes *(15, 16)*, to image subunits in the human Swi-Snf ATP-dependent nucleosome remodeling complex *(10)* and to follow changes of specific histones during processes occurring in multicomponent systems *(15, 16)*. Recognition Imaging is a powerful addition to the AFM arsenal and also has great potential for studies of in vivo-derived chromatin and associated proteins.

The basic techniques of AFM are fairly complex *(1, 17–19)*, but they can be mastered by, for example, a dedicated graduate student (R. Bash, Ph.D. dissertation, Arizona State University). Use of the instrument requires substantial training,

Fig. 8.2 The principle of Recognition Imaging. An antibody against the protein of interest (shown as a blob) is attached to the AFM tip via a flexible linker. When the antibody is not interacting with a target antigen on the surface (**a**) the cantilever oscillates with its normal set-point amplitude as illustrated in the leftmost curve in (**c**). On binding to a target antigen on the surface (**b**), the cantilever amplitude is transiently reduced as shown in the middle curve in (**c**). The microscope servo system restores the signal amplitude by pulling the sample away from the probe, as shown in the rightmost (*blue*) curve in (**c**). This causes a small downward shift in the peak signal, shown as ΔA in (**c**). This peak shift provides the signal that is used to make the recognition image; it is acquired simultaneously with the topographic image information. (Reprinted from *(16)* with permission, copyright (2004) National Academy of Sciences, USA)

which the major US manufacturers of AFM instruments (Agilent, http://www. molec.com/; Asylum Research, http://www.asylumresearch.com/; and Veeco, http://www.veeco.com/) are prepared to offer. We will not discuss AFM theory or technical details of instrument use here. For such information, other recent reviews can be consulted *(1, 17–19)*. This chapter focuses on the methodology needed to carry out Recognition Imaging with the atomic force microscope and the potential applications of this powerful technique to the study of chromatin. The theory and broader applications of Recognition Imaging have been recently reviewed *(19)*.

2 Materials

2.1 *Chromatin Substrates*

AFM studies have been carried out with chromatin from several sources:

1. Chromatin fragments purified from nuclei by digestion with micrococcal nuclease *(12, 20)*, a standard way to produce chromatin for in vitro studies.
2. Chromatin fragments derived by restriction enzyme digestion of nuclei prepared by techniques thought to maintain the nuclei in physiological condition *(21)*. Such fragments were kindly provided to us by R. Hancock.
3. Chromatin fragments reconstituted in vitro from purified histones and DNA (*see* **Note 1**).

2.2 *Surface Preparation*

1. A source of ultrapure argon.
2. An ~2.5 L capacity glass desiccator, with a sealable valve in the cover.
3. Mica sheets (Ashville-Schoonmaker, Newport News, VA, USA).
4. 99% distilled *N,N*-diisopropylethylamine (DIPEA) (Sigma-Aldrich, St. Louis, MO, USA).
5. 99% aminopropyltriethoxysilane (APTES) (Sigma-Aldrich) (*see* **Note 2**).
6. Glutaraldehyde (GD), grade I (Sigma-Aldrich) (*see* **Note 3**).

2.3 *Attaching Antibodies to AFM Tips*

1. PD-10 desalting columns, 8.3-mL bed volume (Amersham Biosciences).
2. Buffer A: 100 mM NaCl, 50 mM NaH$_2$PO$_4$, and 1 mM EDTA, pH 7.5.
3. Type-IV Si$_3$N$_4$ cantilevers with a spring constant of 0.1 N/m, which we refer to as "soft" cantilevers (Agilent Technologies AFM, Tempe, AZ, USA.).

4. Antibody stock solution, typically 1 mg/mL (Upstate Cell Signaling Solutions, Lake Placid, NY, USA; or Abcam Ltd., UK).
5. 15 mM stock solution of N-succinimidyl 3-(acetylthio)propionate (SATP) (Sigma-Aldrich) in DMSO (Mallinckrodt Baker Inc., Paris, KY, USA).
6. UV-cleaner (Boekel Industries, Feasterville-Trevose, PA, USA).
7. Bifunctional PEG crosslinker, Pyridyl dithio-PEG succinimidylpropionate (Agilent Technologies; synthesis originally described in ref. *(22)*).
8. Triethylamine (Sigma-Aldrich).
9. Chloroform, CHCl$_3$ (EMD Chemicals Inc, Gibbstown, NJ, USA).
10. Ultrapure Argon.
11. NH$_2$OH reagent: 500 mM NH$_2$OH·HCl and 25 mM EDTA, pH 7.5.
12. Phosphate-buffered saline (PBS) buffer: 150 mM NaCl and 5 mM Na$_2$HPO$_4$, pH 7.5.

2.4 Chromatin Deposition and AFM Imaging

1. Chromatin sample, concentration ≈0.2 nM in DNA.
2. Deposition buffer: 10 mM NaCl and 5 mM NaPO$_4$ buffer, pH 7.5.
3. A solution (30 μg/mL) of a peptide antigenic to the antibody on the AFM tip (*see* **Note 4**).
4. SPM liquid flow cell (Agilent Technologies).
5. PicoPlus AFM with a Picotrec Recognition Imaging attachment (Agilent Technologies).

2.5 Analyzing Recognition Images

1. Adobe Photoshop software.

3 Methods

To carry out Recognition Imaging, an antibody against the protein of interest is attached, via a bifunctional linker, to an AFM tip. Scanning with this modified tip generates a typical AFM topographic image simultaneously and in exact spatial registration with a recognition image, which locates the sites of antibody–antigen binding in the sample field. Electronic superimposition of the two images provides a map of the positions of the protein of interest (the antibody–antigen binding events) in the topographic image. Because the sample molecules are tethered to the surface, the same sample field can be rescanned in the presence (in solution) of an antigen against the tip-tethered antibody, in order to check the specificity of the initial recognition process.

3.1 Chromatin Substrates

1. For procedures to isolate nuclear chromatin by micrococcal nuclease digestion, recent work (e.g., **ref.** *(12)*) should be consulted.
2. Procedures for obtaining chromatin from nuclei maintained in a physiological state have been described (e.g., **ref.** *(21)*).
3. Techniques for in vitro reconstitution of chromatin arrays from purified histones and DNA have been reviewed *(23, 24)*.

3.2 Surface Preparation

See **Note 5.** To make GD-APTES mica, APTES-modified mica (AP-mica) is first prepared, then the GD derivative is added.

1. Clips from which mica strips will be hung are suspended on a glass rod of a length that will fit snugly across the top of a clean, dry glass desiccator (*see* **Note 6**). Two small containers (the tops of 1.5-mL microcentrifuge tubes will do) are placed in the bottom of the desiccator and the desiccator is purged with ultra-pure argon through the cover valve until air and moisture are removed (~2 min). Sheets of mica (3×1/2 inch) attached to Scotch tape are stripped on one side with Scotch tape to expose a smooth surface (the unstripped side is marked before attaching it to the tape) and the strips are immediately placed (in clips) into the desiccator. The desiccator is purged again with argon for ~2 min.
2. The desiccator cover is partially slid open and 10 μL of DIPEA and 30 μL of APTES is carefully pipetted into the two containers on the bottom of the desiccator. It is important to avoid spilling either component in the desiccator. The cover is replaced, the desiccator is purged with argon for ~3 min and then sealed off, leaving the mica exposed to APTES vapor for 1 h (*see* **Note 7**). After this exposure, the APTES container is carefully removed from inside the desiccator, and the desiccator is purged with argon and sealed. The treated mica (AP-mica) can be stored in the sealed desiccator until needed (along with the remaining DIPEA). AP-mica is best when used within 1 week, at least in dry climates.
3. Two hundred microliters of a 1 m*M* to 1 n*M* GD solution (*see* **Note 8**) in water is pipetted onto AP-mica immediately upon removing it from the storage desiccator described above, and allowed to incubate on the mica for ~10 min. The mica is rinsed with ultrapure water and chromatin samples are deposited on the mica immediately (*see* **below**).

3.3 Attaching Antibodies to AFM Tips

The protocol for attaching antibodies to AFM tips is diagrammed in Fig. 8.3. First, the antibody is modified by SATP addition (**step 1 below**) and the AFM tips are

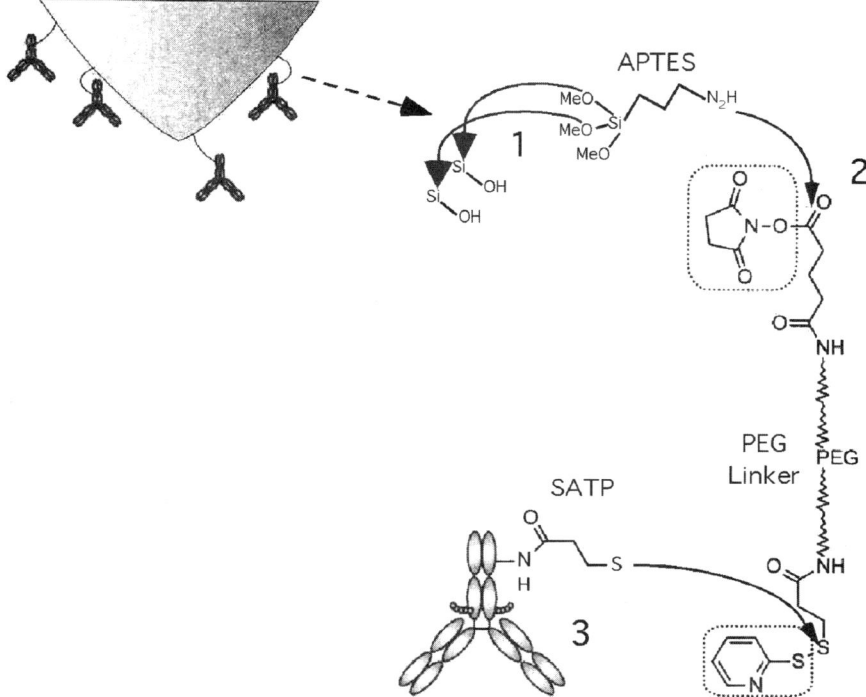

Fig. 8.3 Attaching antibodies to AFM tips. The AFM tip is the downward-facing cone at the *top left* of the illustration. To attach the antibodies (*inverted Ys*), the tip is derivatized with APTES (**step 1**) and the antibody is modified with SATP (reaction not shown), then a bifunctional PEG crosslinker is used to link the modified tip to the SATP-modified antibody (**steps 2** and **3**). See text for details. Note that multiple antibody molecules can become attached to a single tip, whether one or two types of antibodies is (are) used. The number that actually get attached to a given tip during the modification process is unknown and uncontrollable

modified with APTES (**step 2** below). Then the two are linked together via a bifunctional PEG linker (**steps 3** and **4 below**).

1. A PD-10 desalting column is washed with 30 mL buffer A. Then 200 μL of antibody stock solution is diluted to a total volume of 500 μL with buffer A and put over the PD-10 column. The column is washed nine times with 500 μL buffer A per wash, collecting the eluant after each addition. The A_{280} is read to determine in which tube(s) the protein resides. Fractions containing a significant concentration of protein, usually fractions 7 and 8 in our hands, are pooled.

   A tenfold molar excess of SATP stock solution is added to the washed antibody and the solution is incubated for 30–60 min under argon, stirring with a magnetic bar. Two PD-10 columns are washed with 30 mL buffer A, then 500 μL of SATP-modified antibody solution are put on each PD-10 column and the columns are washed as above (nine times with 500 µL buffer A per wash) and the fraction(s) with significant protein collected. The SATP-antibody is stored in 100-µL aliquots at −70°C.

2. Type IV (soft) tips are cleaned in a UV-cleaner for 15 min to get rid of any organic contamination on the tips. Tips are placed in a Petri dish at the bottom of a desiccator and modified with APTES just as in preparing AP-mica (*see* **Section 3.2** above). After the treatment process, the APTES is removed and the treated tips (AP-tip) are stored in the sealed desiccator until needed.

3. To attach the PEG crosslinker to AP-tips, 1 mg of bifunctional PEG crosslinker and 5 µL of triethylamine are mixed in 1 mL CHCl$_3$. AP-modified tips are placed into the solution, and treated for 2–3 h. The tips are washed with CHCl$_3$ and dried with Argon. The succinimide group (Fig. 8.3, upper box) from the bifunctional crosslinker attaches to the amino group on the AP-tip.

4. To attach the PDP (2-(pyridyldithio)propionyl) group on the crosslinker (Fig. 8.3, lower box) to the SATP on the modified antibody, the PEG linker-bound tips (step 3 above) are incubated in 50 µL SATP-antibody, 25 µL of NH$_2$OH-reagent and 50 µL buffer A for 1 h; then the tips are washed once with buffer A and once with PBS buffer. The antibody-tethered tips can be stored in PBS buffer at 4°C and are good for 1 month.

5. To attach two different types of antibodies to AFM tips, the same techniques above are used but with equimolar mixtures of the two types of SATP-modified antibodies (*see* **Note 9**).

3.4 Chromatin Deposition and Imaging

1. A few microliters of a chromatin sample in deposition buffer is pipetted carefully onto a rinsed (as above) GD-APTES surface and the sample allowed to incubate on the surface for 40 min. The surface is then rinsed with 1 mL of the desired buffer (*see* **Note 10**) three times and the sample is mounted into the flowcell and imaged immediately. For nucleosome remodeling studies, hSwi-Snf is added to the chromatin and the two components are incubated in deposition buffer for 25–30 min and then deposited as above.

2. Solution imaging is carried out using the following parameters: 6–8 nm amplitude oscillation at 9 kHz, imaging at 70% set point, and scan speed at <1 Hz. Imaging amplitude is set between 2.0 and 2.5 V. (*See* **Note 11**).

3. We routinely image in a flow cell linked to the AFM (Fig. 8.1), which allows one to alter the sample environment while repetitively imaging the same set of individual molecules. Solutions can be added by gently pushing them in through the "input" syringe and simultaneously pulling the "output" syringe at the same speed. If this is done carefully, reagents can be added while still keeping the same individual molecules within the scanned image.

4. Testing the specificity of recognition is crucial (*see* **Note 12**). One very important advantage of flow-cell imaging is the ability to test that a recognition

reaction can be blocked when a peptide (or protein) antigenic to the antibody on the tip is added to the solution. This important control can be done immediately using the same imaged sample. To carry out this control, an appropriate volume (50 μL in our setup) of the antigenic peptide is flowed into the cell (after the initial scan) and the sample is rescanned. If recognition is specific, the presence of the peptide will block recognition in the deposited sample to a very high degree, often completely, and so remove it from the recognition image *(16)*.

5. To study processes like chromatin remodeling, the sample is scanned twice before in situ activation of remodeling, to assess the frequency of scan-induced changes, then once after activating the remodeling complex (by ATP addition), to assess remodeling-induced changes *(9)*. We typically wait 30 min after adding ATP before scanning again but images are acquired at a rate of ~5 min per image so rescanning can be done at those intervals.

6. Recognition Imaging is an advanced AFM application and not every experiment is successful, for a number of possible reasons (*see* **Note 13**).

3.5 Analyzing Recognition Images

The recognition signal is generally very noisy, but recognition data can be processed to produce analyzable data in the following way:

1. The recognition image is filtered with a median filter (Adobe Photoshop) set to average signal response over ~6 nm, the tether length of the antibody-tip. In a 1-μm-square image containing 512 pixels per line (i.e., approximately 2 nm/pixel), the median filtering is set to a value of three pixels.

2. The background noise level is established by measuring the distribution of pixel heights close to but not coincident with a recognition spot. The distribution of pixel heights across the adjacent recognition spot of interest is then measured by taking perpendicular line traces across the spot and the two distributions are compared in order to establish a cutoff height at which the recognition signal exceeds the noise level. Recognition events thus identified are marked with a colored dot.

3. When each event has been marked, the layer containing the colored dots is overlaid (electronically) on the topographic image; this will place the two types of images to be compared in exact registration.

4   Notes

1. Sample purity is a critical issue in AFM studies because every molecule in a sample shows up; any contaminants in the sample or in the solutions used for dilution, washing, etc., will appear in the image. Furthermore, one component (or contaminant) in a complex sample may bind the surface better than other molecules, severely compromising the analysis. The selectivity of Recognition

Imaging is helpful, but even this can be compromised by highly contaminated samples. DNA fragments electroeluted from agarose gels *(17)* and histone octamers isolated from nuclei by standard procedures *(5)* or recombinant histones (renatured as described in ref. *(23)*) have proven satisfactory for AFM studies of in vitro reconstituted chromatin *(17)*.

2. APTES can be used as received. Redistilling APTES seems to be necessary only if it is older than ~2 months or has been exposed to air for periods of hours. The integrity of APTES is the most likely source for problems in making a surface. Do not hesitate to buy new APTES if difficulties are encountered.

3. Open a fresh vial of GD each time a surface is prepared.

4. Usually, the peptide that was used to elicit the antibody response is used but we have also used proteins for blocking. The blocking protein has to be antigenic; nonspecific proteins will not block the recognition response *(16)*. Peptides can often be purchased from the company supplying the antibodies or can be synthesized inhouse.

5. The surface is very important in AFM studies. Surface chemistry and surface effects impact both the success of the technique and the results themselves. For example, surface attachment constrains the freedom of attached molecules, and the ionic environment of imaged molecules (at the surface) is likely to differ from the bulk solution environment. Mica is typically the basic component of AFM surfaces because it has atomic level flatness, but it is usually treated chemically to enhance the attachment of the desired molecules, e.g., chromatin, to the surface. Many kinds of derivative agents have been used *(17)*. Since 2002, we have used glutaraldehyde-modified APTES (GD-APTES) for our studies. GD-APTES mica is extremely reliable and is suitable for deposition/imaging of DNA, proteins, or combinations of the two *(10)*. The mode of molecular attachment to the surface permits repetitive scanning of the same individual molecules. This feature allows one to add reagents, for example, NaCl to assess nucleosome stability *(8, 14)*, ATP to follow the actions of nucleosome remodeling complexes on chromatin substrates *(9, 10, 15)*, or antibody-blocking peptides to test the specificity of Recognition Imaging *(15, 16)*, while continuing to monitor the same individual molecules before and after reagent addition. The importance of being able to track the same molecules through a process cannot be overstated. The ability to repetitively scan the same molecules also provides a way to assess the contribution of instrument-induced changes to the biological process under study *(9)*, an important control.

6. It is very important that the desiccator and especially the seal area be kept clean of spilled materials. Alcohol should be used to wipe out the desiccator thoroughly before each new run. Do not put any desiccant in the desiccator.

7. Although we usually use 1 h exposures, exposure times can be varied between 30 min and 2 h with no apparent effect on the process.

8. The degree of GD modification of the surface can be varied *(14)* by varying the GD concentration. Higher concentrations make a surface with more sites of attachment; lower GD concentrations allow more freedom for deposited sample molecules. We have typically used ~1 μM GD concentrations. At

these levels, DNA can be progressively loosened and ultimately completely removed from nucleosomal arrays by the in situ addition of NaCl *(8, 14)*, and in situ activation of ATP-dependent nucleosome remodeling complexes can produce striking changes, including complete DNA release from nucleosomes *(9, 10)*, and major histone release *(15)*. Thus, although attached to the surface strongly enough to allow repetitive scans, the DNA and even the histones in chromatin molecules maintain a good deal of freedom.

9. The antibody attachment process is not quantifiable. Thus, it is not possible to know how many of a given type of antibody molecules attach to a tip during the modification process (it is very possible that multiple antibodies get attached to a single tip), where on the tip they are attached, or, when attaching two types of antibodies, whether both types get attached. Modifying tips with a single type of antibody is usually successful, producing tips that recognize their antigen with high efficiency. When modifying tips with two types of antibodies, it is necessary to test the tips to find tips that are capable of recognizing both antigens.

10. We usually wash with H_2O to lessen the possibility that proteins are washed off the chromatin, but buffer or salt solutions can be used.

11. "Soft" cantilevers with low frequency (6–9 KHz) should be used for Recognition Imaging. Stiffer cantilevers (higher frequency) do not produce good recognition images. Also, scan speeds >1 Hz should be avoided as they increase the leakage of topography (white spots) into the recognition image.

12. In addition to testing the ability of antigenic peptides to block the recognition reaction, control experiments to test the strength of the specific recognition reaction as well as the strengths of cross-reactions with appropriate non-antigens should be done. For example, in the Recognition Imaging studies of H2A-H2B release from nucleosomal arrays *(15)*, the strengths of the reactions between anti-H2A or anti-H2B antibodies and H2A-H2B dimers and H3-H4 tetramer/DNA particles were checked by quantifying the Recognition Imaging efficiency and by ELISA assays. This turned out to be very important. The anti-H2B and anti-H2A antibodies both recognized their specific antigens (H2A-H2B) well but the anti-H2B antibodies showed significant nonspecific reaction (with H3-H4 tetramers) and would not have been suitable for use in that analysis. On the other hand, the anti-H2A antibodies from the same commercial source did not recognize H3-H4 and were very suitable. Note that our studies indicate that Recognition Imaging is more sensitive to nonspecific interactions than ELISA assays.

For in vivo studies, it is also important to carry out in vitro control experiments. For example, in studies of Recognition Imaging of Ku protein in chromatin fragments derived from nuclei using anti-Ku antibodies, an unexpectedly large number of recognition events were observed (H. Wang et al., unpublished results). Using in vitro reconstituted chromatin, we were able to show that the Ku antibodies strongly recognized nucleosomes, thus accounting for the high level of recognition in the nuclear chromatin fragments (and precluding the use of those antibodies). This nonspecific recognition

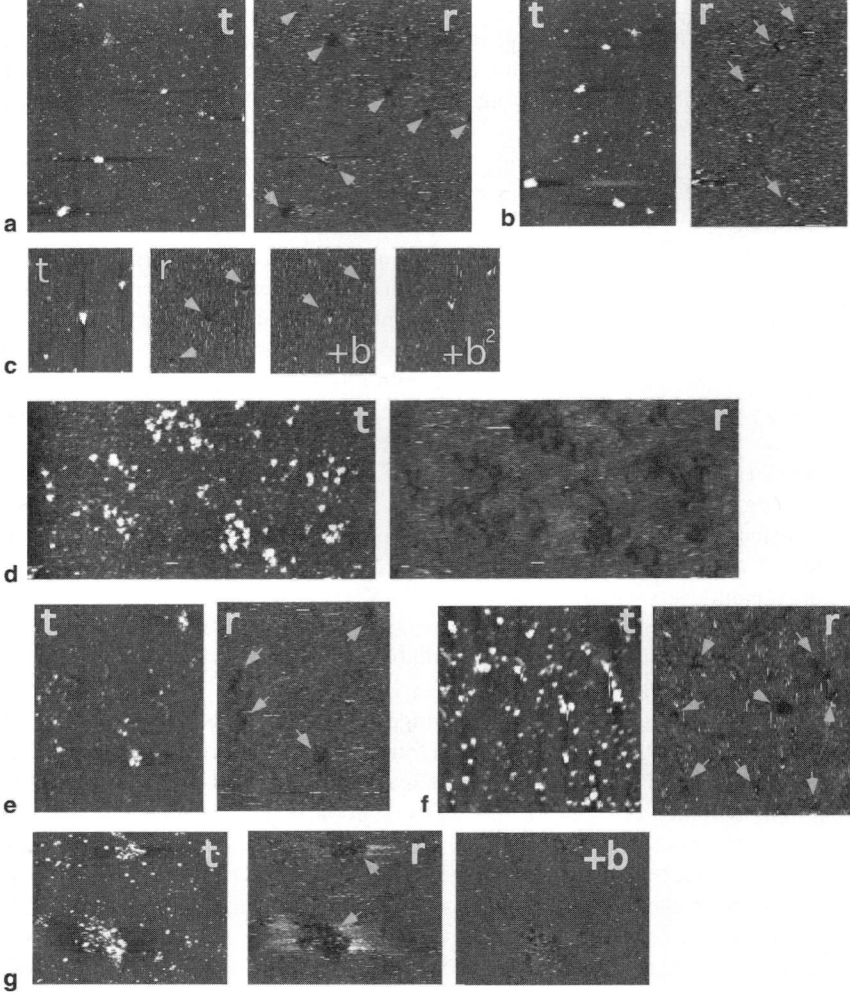

Fig. 8.4 Examples of Recognition Imaging. **a–c** Topographic (*t*) and corresponding recognition (*r*) images of human Swi-Snf samples scanned with an AFM tip containing either anti-BRG1 antibodies (**a**), anti-β-actin antibodies (**b**), or both anti-β-actin and anti-BRG1 antibodies (**c**). As in all these examples, *yellow arrows* point to recognition "events." The +*b* image in (**c**) was obtained from the scan carried out in the presence (in solution) of a peptide antigenic to β-actin antibodies and the +b^2 image was obtained from the scan carried out in the presence of peptides antigenic to both β-actin and BRG1 antibodies. The *small white dots* (+b^2) are sometimes seen in recognition images at sites where recognition has been blocked. **d** Topographic (*t*) and corresponding recognition (*r*) images of reconstituted nucleosomal arrays (mouse mammary tumor virus [MMTV] DNA/HeLa cell histones) obtained from scans made with an AFM tip containing anti-H3 histone antibodies. **e–g** Several examples of Recognition Imaging of fragments derived from restriction digestion of nuclei. Samples were scanned with tips containing an antibody to histone H3. Again, topographic images are identified by "*t*" and the corresponding recognition images by "*r*." The "+*b*" image in (**g**) was obtained by rescanning the sample shown in image "*r*" in the presence of a peptide that blocks H3 recognition. To view this figure in color, see COLOR PLATE 5

of chromatin is not unique to Ku antibodies; both commercial anti-BRG1 and anti-β-actin antibodies recognize nucleosomal arrays and thus cannot be used to identify hSwi-Snf complexes in the presence of chromatin *(10)*. Recently, we used phage display technology to produce a more specific antibody *(25)*, and the use of nucleic acid aptamers may provide a solution to the specificity problem *(26)*. Unfortunately, antibody specificity currently remains a major concern and limitation for the AFM Recognition Imaging technique.

13. Experiments may fail because of surface problems, difficulties in antibody attachment to the tip, etc. Perseverance and patience are essential. Success rates also appear to vary with the nature of the imaged complex and perhaps the antibody used; for example, recognition efficiencies of histones in nucleosomal arrays are >90% but are 60–70% for BRG1 in human Swi-Snf complexes and even lower for another protein in the complex, β-actin (H. Wang et al., unpublished results). Note that experiments with two antibodies on a single AFM tip are much more difficult than those using a single type of antibody, and the pitfalls are not completely worked out. For example, tips do not always get modified with both antibodies (only recognizing one of the two antigens) and there may be interference effects among the tip-bound antibodies, thus decreasing the recognition efficiency of surface-bound antigens, which can already be fairly low *(10)*. Nevertheless, some success has been achieved (Fig. 8.4). Reliable application of this double-probe approach awaits the development of tethering reagents that will reliably place two different antibodies (or other recognition ligands) in equimolar amounts on the end of AFM tips.

5   Prospects

Figure 8.4 presents some examples of Recognition Imaging. In all cases, "t" identifies the topographic image, "r" identifies the corresponding recognition image, and yellow arrows identify recognition "events." Images of human Swi-Snf (hSwi-Snf) complexes are shown in Fig. 8.4a–c. Samples were scanned with a tip containing an antibody against BRG1 (Fig. 8.4a), the major ATPase subunit in this nucleosome remodeling complex *(27)* or an antibody against β-actin (Fig. 8.4b), another subunit in the complex *(28)*. These results demonstrate that it is possible to identify each of these proteins in images of the hSwi-Snf complex. Recognition is typically weaker and less frequent in scans made with the actin than with the BRG1 antibody, which may reflect a lower exposure of the actin subunit in the complex.

Images in Fig. 8.4c are from scans carried out with tips containing both anti-BRG1 and anti-β-actin antibodies. Recognition images obtained with such dual-modified tips ("r") can contain recognition events from both protein antigens. To distinguish the two types of events, the same molecules (same image field) are rescanned first in the presence (in solution) of a peptide antigenic to one of the tip-tethered antibodies, β-actin in this case (+b), blocking that type of recognition. Note that the recognition of the lowest arrow in the initial "r" image has now disappeared. The same field was then rescanned in the presence of peptides anti-genic to both β-actin and BRG1 antibodies (+b²), to block BRG1 and β-actin recognition in the deposited sample. Note that recognition from the top two arrows in the initial "r" image has now disappeared. These results show that it is possible to identify two proteins in the same image, but we have so far been unable to identify both within the same individual complex.

Figure 8.4d shows images of in vitro reconstituted mouse mammary tumor virus (MMTV) nucleosomal arrays scanned with a tip containing anti-histone H3 antibodies. Note the high recognition efficiency; virtually every array molecule is recognized and the recognition pattern in array molecules mimics the pattern of nucleosomes in individual arrays.

Another major goal is to apply this technique to chromatin derived from in vivo (nuclear) sources. Figure 8.4e–g show images of chromatin fragments obtained by restriction digestion of nuclei prepared by physiological approaches and imaged with an AFM tip containing an antibody against histone H3. Several points are clear. Firstly, fragments of various sizes and compaction states are recognized by the tip-tethered antibody (Fig. 8.4e) and the recognition pattern suggests recognition of multiple nucleosomes in the fragments. Secondly, not all particles are recognized (Fig. 8.4f). Given the high efficiency of recognition by the H3 antibody with chromatin (>90% *(16)*, also Fig. 8.4d), it is likely that this discrimination reflects the selectivity of Recognition Imaging. Thirdly, Fig. 8.4g shows that the recognition is specific because it largely disappears when images are taken in the presence of H3 blocking peptide (image r vs +b), even though the chromatin fragments are often highly compacted and thus not well exposed.

The results obtained with Recognition Imaging thus far *(10, 15, 16)* suggest several possible improvements in the technique. The efficient recognition of histones in nucleosomal arrays is probably due to their multicopy presence and relatively significant exposure to the environment (and the AFM tip), due to the extended nature of the subsaturated nucleosomal arrays we study *(15, 16)*. In large complexes like hSwi-Snf, components may be more or less buried and/or hidden, permanently by their location within the complex and/or sporadically depending on how a given complex deposits on the surface. Antibodies with higher recognition efficiencies for proteins in such large complexes might be obtained by raising antibodies against protein regions known (or thought) to be exposed in the complex. For example, the H3 and H2A antibodies were made against the tail and "acidic patch" *(29)* regions respectively. Both of these regions are exposed in nucleosomes. Recognition studies of proteins that are released during processes, either one or both proteins of interest, could also be very successful due to the enhanced exposure of the released component(s). For example, released H2A histone was detected in a recent Recognition Imaging study *(15)*.

Acknowledgments Support from the National Institutes of Health is gratefully acknowledged. Also, Ralph Bash died in October 2007; this chapter is dedicated to our longtime colleague and coworker whose skills in AFM analysis of chromatin contributed immeasurably to the development of these approaches.

References

1. Lindsay, S.M. (2000) The scanning probe microscope in biology. In: *Scanning Probe Microscopy, Techniques and Applications*, 2nd edn (Bonnell, D., ed), John Wiley, New York, NY, pp. 289–336.
2. Yip, C. M. (2001) Atomic force microscopy of intermolecular interactions. *Curr. Opin. Struct. Biol.* **11**, 567–572.
3. Frederix P., Akiyama, T., Staufer, U., Gerber, C., Fotiadis, D., Muller, D., and Engel, A. (2003) Atomic force bio-analytics. *Curr. Opin. Chem. Biol.* **7**, 641–647.
4. Hansma, H., Kasuya, K., and Oroudjev, E. (2004) Atomic force microscopy imaging and pulling of nucleic acids. *Curr. Opin. Struc. Biol.* **14**, 380–385.
5. Yodh, J.G., Lyubchenko, Y.L., Shlyakhtenko, L.S., Woodbury, N., and Lohr, D. (1999) Evidence for nonrandom behavior in 208-12 subsaturated nucleosomal array populations analyzed by AFM. *Biochemistry* **38**, 15756–15763.
6. Yodh, J., Woodbury, N., Shlyakhtenko, L., Lyubchenko, Y., and Lohr, D. (2002) Mapping nucleosome locations on the 208-12 by AFM provides clear evidence for cooperativity in array occupation. *Biochemistry* **41**, 3565–3574.
7. Bash, R., Yodh, J. Lyubchenko, Y., Woodbury, N., and Lohr, D. (2001) Population analysis of subsaturated 172-12 nucleosomal arrays by Atomic Force Microscopy detects nonrandom behavior that is favored by histone acetylation and short repeat length. *J. Biol. Chem.,* **276**, 48362–48370.
8. Bash, R., Wang, H., Yodh, J., Hager, G., Lindsay, S.M., and Lohr, D. (2003) Nucleosomal arrays can be salt-reconstituted on a single-copy MMTV promoter DNA template: their properties differ in several ways from those of comparable 5 S concatameric arrays. *Biochemistry* **42**, 4681–4690.
9. Wang, H., Bash, R., Yodh, J.G., Hager, G., Lohr, D., and Lindsay, S.M. (2004) Using AFM to track nucleosome remodeling on individual nucleosomal arrays in situ. *Biophys. J.* **87**, 1964–1971.
10. Wang, H., Bash, R., Lindsay, S.M., and Lohr, D. (2005) AFM studies of human Swi-Snf and its interactions with MMTV DNA and chromatin. *Biophys. J.* **89,** 3386–3398.
11. Schnitzler, G., Cheung, C., Hafner, J., Saurin, A., Kingston, R., and Lieber, C. (2001) Direct imaging of human Swi-Snf remodeled mono- and polynucleosomes by atomic force microscopy employing carbon nanotube tips. *Mol. Cell. Biol.* **21**, 8504–8511.
12. Kepert, J., Mazurkiewicz, J., Heuvelman, G., Tóth, K., and Rippe, K. (2005) NAP1 modulates binding of linker histone H1 to chromatin and induces an extended chromatin fiber conformation. *J. Biol. Chem.* **280**, 34063–34072.
13. Solis, F. J., Bash, R., Yodh, J., Lindsay, S., and Lohr, D. (2004) A statistical thermodynamic model applied to experimental AFM population and location data is able to quantify DNA-histone binding strength and internucleosomal interaction differences between acetylated and unacetylated nucleosomal arrays. *Biophys. J.* **87**, 1–16.
14. Wang, H. Bash, R., Yodh, J. Hager G., Lohr, D., and Lindsay, S. (2002) Glutaraldehyde-modified mica: a new surface for atomic force microscopy of chromatin *Biophys. J.* **83**, 3619–3625.
15. Bash, R. Wang, H., Anderson, C., Yodh, J., Hager, G., Lindsay, S. M., and Lohr, D. (2006) AFM imaging of protein movements: histones H2A-H2B release during nucleosome remodeling. *FEBS Lett.* **580**, 4757–4761.
16. Stroh, C., Wang, H., Bash, R., Ashcroft, B., Nelson, J., Gruber, H., Lohr, D., Lindsay, S.M., and Hinterdorfer, P. (2004) Single molecule recognition imaging microscopy. *Proc. Natl. Acad. Sci. USA* **101**, 12503–12507.
17. Lohr, D., Bash R., Wang, H., and Lindsay, S.M. (2007) Using atomic force microscopy to study chromatin and nucleosome remodeling. *Methods* **41**, 333–341.
18. Leuba, S., Bennink, M., and Zlatanova, J. (2004) Single molecule analysis of chromatin. *Methods Enzymol.* **376**, 73–105.

19. Kienberger, F., Ebner, A., Gruber, H., and Hinterdorfer, P. (2006) Molecular recognition imaging and force spectroscopy of single biomolecules. *Acc. Chem. Res.* **39**, 29–36.

20. Leuba, S., Yang, G., Robert, C., Samori, B., van Holde, K., Zlatanova, J., and Bustamante, C. (1994). 3-dimensional structure of extended chromatin fibers as revealed by tapping-mode scanning force microscopy. *Proc. Natl. Acad. Sci. USA.* **91**, 11621–11625.

21. Jackson, D, Dickinson, P., and Cook, P.R. (1990) Attachment of DNA to the nucleoskeleton of HeLa cells examined using physiological conditions. *Nucleic Acids Res.* **18**, 4385–4393.

22. Haselgrubler, T., Amerstorfer, A., Schindler, H., and Gruber, H. (1995) Synthesis and applications of a new poly(ethyleneglycol) derivative for the crosslinking of amines with thiols. *Biocojugate Chem.* **6**, 242–248.

23. Luger, K., Rechstiner, T., and Richmond, T.J. (1999) Preparation of nucleosome core particles from recombinant histones. *Meth. Enzymol.* **304**, 3–18.

24. Carruthers, L., Tse, C., Walker, K., and Hansen, J. (1999) Assembly of defined nucleosomal and chromatin arrays using pure components. *Meth. Enzmol.* **304**, 19–34.

25. Marcus, W.D., Wang, H., Lohr, D. Sierks, M.R., and Lindsay, S. M. (2006) Isolation of an scFv targeting BRG1 using phage display. *Biochem. Biophys. Res. Commun.* **342**, 1123–1129.

26. Lin, L., Wang, H. Liu, Y., Yan, H., and Lindsay, S.M. (2006) Recognition imaging with a DNA aptamer. *Biophys. J.* **90**, 4236–4238.

27. Kingston, R. E. and Narlikar, G. J. (1999) ATP-dependent remodeling and acetylation as regulators of chromatin fluidity. *Genes Dev.* **13**, 2339–2352.

28. Olave, I. A., Reck-Peterson, S. L., and Crabtree, G. R. (2002) Nuclear actin and actin-related proteins in chromatin remodeling. *Annu. Rev. Biochem.* **71**, 755–781.

29. Luger, K., A. Mader, R. Richmond, D. Sargent, and T. Richmond. (1997) Crystal structure of the nucleosome core particle at 2.8 A resolution. *Nature* **389**, 251–260.

Chapter 9
Using Cells Encapsulated in Agarose Microbeads to Analyse Nuclear Structure and Functions

Dean Jackson

Keywords Agarose-encapsulated cells; Cell extraction; Chromatin; DNA loops; RNA transcription; DNA replication; Nucleoskeleton

Abstract It is now generally agreed that the nuclei of higher eukaryotes, and particularly of mammalian cells, are highly structured and that different aspects of this structure contribute to the regulation of function *(1, 2)*. Despite the general consensus, the key mechanisms that link nuclear structure and function have proved elusive. A major reason for this is a lack of techniques that allow nuclei to be manipulated in a way that preserves the complex architectural features that are present in vivo. Historically, significant progress in understanding the makeup of nuclei from mammalian cells has been made using cells that are permeabilised in a physiological buffer after being encapsulated in agarose microbeads. By using such beads, cells are protected from shear forces that otherwise can degrade crucial elements of the architecture that it is essential to preserve.

1 Introduction

Nuclei of higher eukaryotes are so complex that it is inevitable that many different techniques, each with their own scope and limitations, will be used in their analysis. It was inevitable, for example, that biochemical tools would be used to purify key components to understand the architecture of their constituent parts and key functional criteria. The reductionist approach does, however, come with limitations and specifically falls short when attention is focused on the significance of three- and four-dimensional organisation. Historically, purified nuclei have proved an amenable experimental system to analyse and understand how nuclear components might function inside the cell. Of course, their use has both advantages and disadvantages. A major disadvantage is that nuclei and the chromatin they contain are prone to damage and aggregation if manipulated under physiological conditions. This means that experiments have generally been performed in the presence of unphysiological environments, which might themselves impact—in unpredictable ways—on the behaviour of the nuclear components under study.

Cells permeabilised in agarose microbeads *(3)* present an accessible nuclear compartment that is protected from damage so that experiments can be performed in an environment that mimics that found in vivo *(4)*. The environment can be manipulated at will simply by changing the buffer environment. Hence, in an optimised environment, the system presents nuclei that are freely accessible to molecules added to the medium in which they are suspended, but which maintain the organisation of chromosomal and sub-chromosomal compartments that are present inside the cell prior to permeabilisation. This experimental system has been used to advance our understanding of nuclear organisation in a number of important ways. Following the lead of experiments performed on cells extracted with hypertonic salt *(5)*, typically buffers containing $2\,M$ NaCl, encapsulated permeabilised cells were used to confirm that RNA and DNA synthesis are performed in association with a nucleoskeleton *(6, 7)*. Key features of this analysis were the preservation of endogenous chromatin structure and the use of restriction endonucleases and electroelution to separate the nucleoskeleton-associated and detached chromatin fragments. Further experiments, using the same procedure, were instrumental in defining the architecture of putative core filaments of the mammalian nucleoskeleton *(8)* and the organisation of centres where groups of active complexes are clustered *(9, 10)*. These dedicated active compartments were termed "factories" because they contain accumulations of all the activities needed to perform complex tasks such as DNA and RNA synthesis.

Encapsulated cells provide a wide range of opportunities for understanding the architecture of nuclei in mammalian cells. Throughout the following detailed description, I endeavour to emphasise the power of the system and where complementation with other techniques can be achieved. Specifically, it is crucial to recognise the limitation of any experimental approach. And here it is worth emphasising how information developed using various in vitro systems can be used to supplement in vivo experiments.

The following protocols describe some applications of a simple procedure for encapsulating living cells in agarose microbeads (Fig. 9.1). Cells in phosphate-buffered saline (PBS) supplemented with molten agarose are homogenised with an immiscible phase of liquid paraffin; on cooling, suspended agarose droplets containing cells gel into microbeads *(3, 11)*. The pores in the beads allow exchange of very large molecules but not of chromosomal DNA. Therefore encapsulated cells can be grown to give microcolonies. Alternatively, cells can be permeabilised in solutions containing a "physiological" concentration of salts; such cells are permeable to triphosphates and continue to replicate and transcribe their intact chromatin template at rates close to those in vivo.

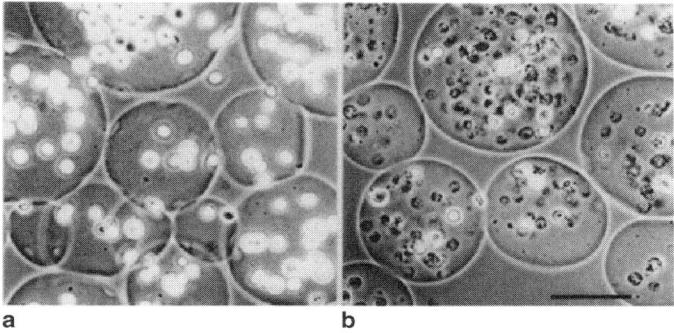

Fig. 9.1 Cells encapsulated in agarose microbeads. HeLa cells were encapsulated in agarose microbeads at 5×10^6 cells/mL of agarose. The beads were visualised by phase contrast microscopy either before (**a**) or after (**b**) cell permeabilisation with 0.25% Triton X-100 in an isotonic buffer

2 Materials

2.1 Encapsulating Cells in Agarose Microbeads

1. HeLa cells or other cells of interest, preferably growing in suspension culture (*see* **Note 1**).
2. Phosphate-buffered saline (PBS): for 10× PBS, dissolve 80 g of NaCl, 2 g of KCl, 14.4 g of Na_2HPO_4, and 2.4 g of KH_2PO_4 in 800 mL of dH_2O. Add dH_2O to 1 L and adjust the pH to 7.4 with HCl. Store at room temperature.
3. Low melting temperature agarose, 2.5% w/v in PBS (*see* **Note 2**). Some enzymes used in molecular biology are particularly sensitive to inhibition by compounds that are found naturally in agarose; lower temperature-gelling types have fewer sulphate groups, and for work with DNA tend to have fewer inhibitors. If reactions such as restriction digestion and ligation prove to be inefficient, the agarose should be purified using DEAE-cellulose as follows:

   Dissolve 5 g of low-gelling temperature agarose (Type VII; Sigma-Aldrich, Gillingham, Dorset, UK) in 200 mL of PBS and mix with 100 mL of fine DEAE-cellulose (Whatman, Maidstone, Kent, UK) pre-equilibrated with PBS at 50°C for 15 min. Pour the slurry into 50-mL Falcon tubes, pellet the DEAE-cellulose using a bench centrifuge at 1,000×g for 5 min at 37°C, and decant the molten agarose. Store the purified agarose in 10-mL aliquots at −20°C.

4. Physiological buffer (PB): 100 mM potassium acetate, 30 mM KCl, 10 mM Na_2HPO_4, 1 mM $MgCl_2$, 1 mM Na_2ATP (Type I; Sigma-Aldrich, Gillingham, Dorset, UK), 1 mM dithiothreitol, and 0.2 mM phenylmethylsulphonyl fluoride (PMSF) in HPLC-pure water. Bring the pH to 7.4 by adding <1% (v/v) of 200 mM KH_2PO_4; the amount required varies slightly as the acidity of ATP

varies from batch to batch. The free Ca^{2+} levels of this buffer are below $0.3\,\mu M$ and can be clamped at $0.1\,\mu M$ using $40\,\mu M$ EGTA. Optionally, addition of 1 mg/mL of BSA gives optimal preservation of nuclear structure.

5. PB containing 0.5% v/v Triton X-100 (Pierce, Chester, UK).
6. Liquid paraffin (BDH; VWR, Lutterworth, Leicestershire, UK; cat. 29436 or Merck, West Drayton, UK; cat. 7162).
7. Flask shaker (Gallenkamp, Loughborough, UK).
8. (Optional) monofilament nylon filters (R. Cadisch, Finchley, London, UK) in Swinnex filters.
9. Centrifuge tubes, 50 mL (Falcon; BD Biosciences, Oxford, UK).

2.2 Chromatin Loops Under "Physiological" Conditions

1. (^{3}H]Thymidine (~50 Ci/mmol; Amersham; GE Healthcare, Little Chalfont, Buckinghamshire, UK).
2. Clear polycarbonate 10-mL centrifuge tubes.
3. PB containing 0.5% v/v Triton X-100 (Pierce).
4. Restriction endonucleases appropriate to the particular experiment (*see* **Section 3.2**).
5. Glass-fibre filters (GF/C; Whatman, Maidstone, UK).
6. 5% w/v Trichloroacetic acid, in H_2O.
7. Vertical electrophoresis gel box with buffer recirculation.
8. TEA buffer: 40 mM Tris, pH 8.3, 20 mM sodium acetate, and 2 mM EDTA.
9. *N*-laurylsarcosine (sarkosyl) (Sigma-Aldrich).
10. Sodium dodecyl sulphate (SDS).
11. RNase A (Sigma-Aldrich).
12. Proteinase K (Sigma-Aldrich).
13. Phenol, chloroform, diethylether, and ethanol.

2.3 Resin-less Electron Microscopy to Visualise the Nucleoskeleton

1. Glutaraldehyde (SPI Supplies, West Chester, PA, USA).
2. 0.1 M Sorensen buffer, pH 7.4 (SB): 81 mL of 0.1 M Na_2HPO_4 and 19 mL of 0.1 M NaH_2PO_4 in 100 mL.
3. SB containing 0.02 M glycine.
4. Osmium tetroxide solution: 0.5% w/v OsO_4 (SPI) in SB.
5. Ethanol solutions: 30%, 50%, 70%, 96%, and 100% v/v in water.
6. Uranyl acetate (UAc) solution: 2% UAc (SPI) in 70% ethanol.
7. n-Butanol, acetone.
8. Diethylene glycol distearate (DGD) embedding medium (Electron Microscopy Sciences, Hatfield, PA, USA).

9. Microtome to cut 500-nm-thick sections.
10. Pioloform-coated electron microscope grids (Agar Scientific, Stansted, UK).

2.4 Using Nuclear Derivatives to Analyse Links Between Nuclear Structure and Function

1. Triton X-100, 0.02–0.05% in PB.
2. Digitonin (Sigma-Aldrich), 0.01–0.02% in PB.
3. Lysolecithin (Sigma-Aldrich), 0.02–0.05% in PB.
4. Saponin (Sigma-Aldrich), 0.01–0.02% in PB.

2.5 Labelling Sites of DNA Replication In Vitro

1. Microscope slides, cover glasses.
2. Saponin (Sigma-Aldrich).
3. BrdU or a DNA precursor conjugated with biotin, digoxygenin, or a fluorochrome (e.g., Sigma-Aldrich, Boehringer, Amersham, Du Pont, or other suppliers) (see **Note 3**).
4. 10× concentrated replication initiation mix (10× RIM): 2.5 mM dATP, dCTP, and dGTP, 1 mM CTP, GTP, and UTP, and 0.1–1 mM of the selected labelled DNA precursor and MgCl$_2$ at a molarity equal to the triphosphates.
5. Paraformaldehyde solution: 4% w/v in PBS with 0.05% w/v bovine serum albumin (BSA) and 0.1% v/v Tween 20 (PBS/BSA/Tween).
6. Primary antibody directed against the DNA precursor (e.g., monoclonal anti-BrdU; Boehringer; or goat anti-biotin; Sigma-Aldrich).
7. If required, appropriate fluorochrome-coupled secondary antibody: e.g., FITC-donkey anti-goat (Jackson Laboratories).
8. Mounting medium (Vectashield; Vector Laboratories, Peterborough, UK).
9. Nail varnish.

2.6 Labelling Sites of RNA Transcription In Vitro

1. Microscope slides, cover glasses.
2. A biotin- or digoxygenin-coupled RNA precursor (e.g., Br-UTP, biotin-11-UTP, biotin-14-CTP [Gibco-BRL], or digoxigenin-11-UTP [Roche], or a fluorescent NTP analogue [e.g., fluorescein-12-UTP; Invitrogen]).
3. PB supplemented with 0.5 U/mL of human placental ribonuclease inhibitor (HPRI) (Amersham).

4. 10× concentrated transcription initiation mix (10× TIM): PB containing 1 mM CTP and GTP, 500 μM UTP, 0.02–1 mM labelled RNA precursor, MgCl$_2$ at a molarity equal to the triphosphates, and 100 U/mL of HPRI.
5. Appropriate fluorochrome-coupled secondary antibody.
6. Mounting medium (Vectashield).
7. Nail varnish.

2.7 Analysing the Organisation of Nuclear Proteins: the Microarchitecture of Sites of Replication and Transcription

1. Microscope slides, cover glasses.
2. PB with 100 μg/mL saponin.
3. PB with 4% paraformaldehyde.
4. PB with 0.5% w/v BSA and 0.1% v/v Tween 20.
5. Primary antibodies directed against the selected DNA or RNA precursor and protein (*see* **Section 3.7**).
6. Appropriate fluorochrome-coupled secondary and tertiary antibodies (e.g., FITC-donkey anti-goat IgG or Cy3-donkey anti-mouse IgG (Jackson Immunoresearch; Stratech Scientific, Newmarket, Suffolk, UK).
7. 4′,6-diamindino-2-phenylindole dihydrochloride (DAPI) or TOTO-3 iodide (Molecular Probes; Invitrogen, Paisley, UK), 0.02 μg/mL in PBS.
8. Mounting medium (Vectashield).
9. Nail varnish.

2.8 Analysing Nuclear Architecture by Step-Wise Deconstruction

1. Triton X-100 (or a related mild detergent), 0.05–0.5% v/v.
2. 0.2–2 M NaCl.
3. 0.1–1 M (NH$_4$)$_2$SO$_4$.
4. 0.02–0.5% w/v sarkosyl.
5. 0.5–2.5 M urea.
6. 0.02–0.5% w/v SDS, in PB with Na$^+$ in place of K$^+$, incubate at 20°C.
7. Standard equipment and buffers for SDS-PAGE.

3 Methods

3.1 Encapsulating Cells in Agarose Microbeads

1. If necessary, prepare purified agarose as described in **Section 2.1.3.**

2. Encapsulate cells at a density ranging from 10^6 to 10^7 cells/mL of beads for most purposes. Harvest cells and resuspend enough to give the desired cell density in 4 mL of PBS. Warm the cells to 37°C in a 100-mL round-bottomed flask.

3. Melt 2.5% agarose in PBS, cool to 37°C, add 1 mL to the cell suspension, and mix thoroughly.

4. Pour 10 mL of liquid paraffin at 37°C into the flask, cover with Parafilm, and shake at room temperature using about 800 cycles/min on a flask shaker, until a creamy emulsion forms (~10 sec). If a flask shaker is not available, shake by hand as fast as possible for 10 sec (*see* **Note 4**). Plunge the flask into ice-cold water and leave for ~10 min until the agarose gels.

5. To recover the beads, mix the cooled slurry with 35 mL of ice-cold PBS, transfer to a 50-mL Falcon centrifuge tube, and pellet the beads using a bench centrifuge at 500×*g* for 2.5 min. If some beads remain at the PBS/paraffin interface, remove most of the paraffin, mix, and centrifuge again.

6. Aspirate the PBS and any remaining paraffin and resuspend the encapsulated cells in the desired medium.

7. If the cells are to be grown in agarose beads, add growth medium to the pelleted beads. Growth of encapsulated HeLa continues normally for one or two cycles, but then the tightly confined cells eventually slow and then stop growing.

3.2   The Organisation of Chromatin Domains In Vitro

3.2.1   Chromatin Loops Under "Physiological" Conditions

Ideally, any analysis of chromatin domain structure should be performed in situ. Though this is not practical, it is sensible to perform experiments under conditions as similar as possible to those existing inside the cell. However, nuclei and derivatives prepared under isotonic condition tend to lyse and aggregate upon manipulation, making analysis difficult.

The precise conditions inside a mammalian nucleus are not known, and it is therefore unlikely that we can establish an environment that preserves all the interactions that exist in vivo. Nevertheless, it is clear that a buffer system that is as close as possible to the environment in vivo will offer advantages. In vivo the intranuclear salt concentration is likely to be roughly 150 mM, predominantly K^+ and Na^+. Mg^{2+} and Ca^{2+} ions play important roles, but at much lower concentrations; the combination of equimolar amounts of Mg^{2+} and ATP gives sufficient free Mg^{2+} for enzymes to work but is likely to leave chromatin in its natural state. Acetate and chloride provide convenient counterions in vitro; in the nucleus, much of the positive charge is balanced by negative charges on chromatin and proteins. Dithiothreitol is added to maintain an appropriate redox environment; glutathione would be more physiological, but its cost is prohibitive. Finally, protease and nuclease inhibitors are not found in vivo but should be added as required.

1. To allow quantitation of DNA, grow cells for 24 h in medium supplemented with [³H]thymidine (0.1 μCi/mL).
2. Prepare cells encapsulated at ~2×10⁶ cells/mL of microbeads as described in **Section 3.1**.
3. Collect the beads by centrifugation for 1–2 min in a bench centrifuge at 500×*g*. Aspirate the medium, add ice-cold PBS to the pelleted beads to give 1 mL of bead slurry/10 mL, invert the tube to resuspend and collect the beads by centrifugation, as before. To reduce bead losses, use clear polycarbonate 10-mL tubes for this and subsequent steps.
4. Resuspend the bead pellets in ice-cold PB, collect the beads by centrifugation, resuspend the washed pellet in ice-cold PB containing 0.5% Triton X-100 and incubate on ice for 10 min, mixing intermittently (*see* **Note 5**).
5. Collect the beads by centrifugation and repeat the incubation in PB containing 0.5% Triton X-100 as in **step 4.**
6. After lysis, wash the beads 3× in ice-cold PB to remove Triton.
7. Add the appropriate restriction endonucleases and incubate at 33°C for 15 min (see **Note 6**). Incubate a control sample without added enzymes in parallel.
8. Prepare a 6-mm-thick 1% agarose gel in TEA buffer in a vertical gel box, using a comb to produce deep wells (each well should hold ~0.5 mL of bead slurry). Pre-run the gel for 1 h at 1–1.5 V/cm using 2 L of PB (without BSA) (*see* **Section 2.1.5**) at 4°C. To prevent pH drift, recirculate the buffer at ~50 mL/min (*see* **Note 7**).
9. Wash the beads once in ice-cold PB, pellet, and load the bead slurry into the gel wells with ~250 μL bead slurry/well. Perform electrophoresis at 4°C at 1–1.5 V/cm for 15 h or at 5 V/cm for 4 h to remove detached chromatin fragments. Recirculate the buffer from the bottom to the top reservoir at ~50 mL/min throughout to maintain the pH at 7.4±0.1 (*see* **Note 8**).
10. Recover the beads from the gel wells and measure ³H in equivalent samples of digested and control beads: add 100 μL of beads to 300 μL of 1% SDS for 1 h, pipette 3×100 μL aliquots onto glass-fibre filters, extract with 5% trichloroacetic acid, dry the filters with ethanol and then diethylether, and measure ³H in acid-insoluble material by scintillation counting.
11. To recover DNA from the beads, extract them with RNase (2 μg/mL) with sarkosyl (0.5% w/v) for 1 h at 37°C followed by proteinase K (50 μg/mL) with 0.2% w/v SDS for 5 h at 37°C. Extract twice with phenol (required to remove agarose), twice with phenol/chloroform, once with chloroform, once with diethylether, and precipitate DNA with ethanol using standard methods. Resuspend the precipitate and analyse DNA fragments by gel electrophoresis and ethidium bromide staining (Fig. 9.2), and Southern blotting if required, using standard procedures.

3.2.2 Chromatin Loops in Different Nuclear Derivatives

While encapsulated cells allow chromatin loops to be analysed under isotonic conditions, they also provide a system for comparing the structure of loops in different

Fig. 9.2 Chromatin loops in human cells. Encapsulated HeLa cells were permeabilised with Triton X-100 and chromatin was digested in PB containing *Eco*RI (2,000 U/mL) and *Hae*III (200 u/mL) at 33°C for 30 min. After digestion, samples were subjected to electroelution to remove detached chromatin fragments (5 V/cm for 5 h in PB). Beads were recovered and digested for a second time with a mixture of eight further restriction enzymes *(22)* and the detached and attached chromatin fragments were recovered. Purified DNA from these samples was separated in an agarose gel to define the average fragment size: note the preservation of nucleosomal structure under the conditions used. The amounts of DNA in the different fractions were determined from ^{3}H-thymidine labelling, and the loop sizes were calculated after first calculating the number-average molecular weight as described in *(22)*. Lane 2, 4.3% of DNA remaining, average length 6.6 kbp; average loop size is therefore 89.3 kbp; Lane 4, 1.4% remaining, average length 2.1 kbp; average loop size is therefore 84.6 kbp

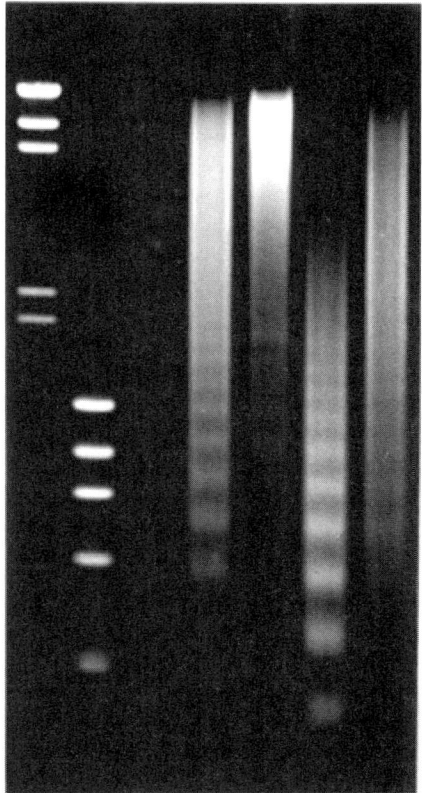

preparations. Thus encapsulated cells can be treated with different low or high salt buffers and the chromatin/DNA loops analysed. The beads prevent damage to the fragile preparations. If necessary, nuclei can be prepared and encapsulated as described above using a reduced shaking speed to prevent breakage.

Different procedures used to analyse chromatin organisation give different impressions of chromatin domain structure. This probably reflects the complexity in vivo and the behaviour of different interactions when treated in different ways. For example, DNA undergoing replication is associated with the remnant nuclear structures. However, this DNA is likely to account for a small proportion of the interaction existing in vivo. Under physiological conditions, active genes are associated with the nucleoskeleton *(12)*. However, as individual sequences are rarely bound in all cells of a population, the interactions responsible must be dynamic. This is believed to reflect the transient association of transcribed parts of the genome with the nucleoskeleton in vivo.

Inappropriate Mg^{2+} and Ca^{2+} concentrations alter chromatin structure and loop size. High Mg^{2+} concentrations (e.g., >2.5 mM) irreversibly fix chromatin into smaller loops. It is very important to balance the Mg^{2+} concentration with that of the nucleoside or deoxynucleoside triphosphates; a 1:1 molar ratio of Mg^{2+}:triphosphate is safe as far as fixing the chromatin is concerned, suppresses the effects of nucleases, and allows sufficient free triphosphates and Mg^{2+} for both cutting with restriction enzymes and efficient DNA or RNA synthesis. Ca^{2+} and Cu^{2+} ions and spermine and spermidine will similarly "fix" chromatin and should be avoided or used only at appropriate concentrations *(13)*.

Given the complexity of eukaryotic nuclei, it is not that surprising that different preparations of nuclei give a different picture of their organisation. It seems probable, however, that the majority of interactions described in vitro will reflect those existing in vivo, though some preparations may drive in vivo interactions away from their natural equilibrium so that their overall contribution is exaggerated. In view of this possibility, preparations that best preserve the endogenous organisation will be most likely to provide reliable information about nuclear structure and function inside the mammalian cell.

3.3   Resin-less Electron Microscopy to Visualise the Nucleoskeleton

Cell encapsulated in agarose beads and permeabilised under isotonic conditions have nuclei that retain the major features seen in vivo; as expected, nucleoli and chromatin are the dominant morphological features. However, if chromatin is removed (*see* **Section 3.2**) the immense complexity of the underlying structure becomes apparent *(8–10)* (Fig. 9.3).

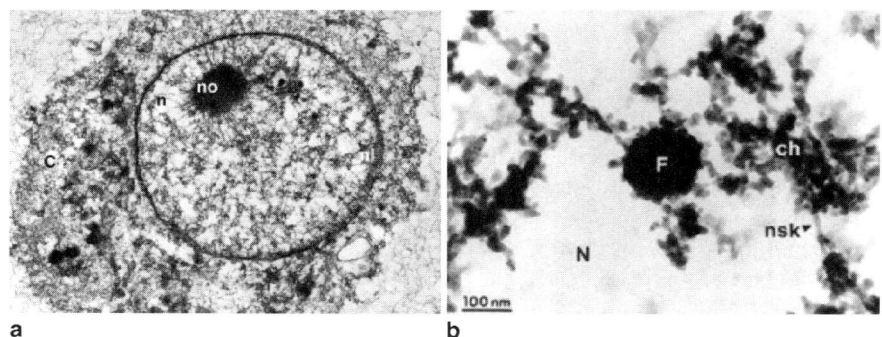

a   b

Fig. 9.3 Visualising the nucleoskeleton. Encapsulated HeLa cells with biotin-dUTP incorporated into replication factories were treated with restriction endonucleases and ~90% of the chromatin was removed. After immunolabelling, 500-nm-thick resinless sections were prepared. A network of agarose filaments is seen to surround the cell. The nuclear lamina (*nl*) separates the cytoplasmic and nuclear (*n*) compartments. Densely staining nucleoli (*nu*) are clearly seen. The image on the right shows a single replication factory (*F*) deep inside the nucleus (*N*). Both the factory and residual chromatin (*ch*) are attached to the residual nucleoskeleton (*nsk*). Main image reproduced from *(23)* with permission; replication factory image courtesy of Pavel Hozak

The resin-less electron micrograph shown in Fig. 9.3 gives the best available impression of the organisation of non-chromatin structures inside mammalian nuclei. Remarkably, even when almost all chromatin is removed, a complex residual structure of coated intermediate filaments *(8)* and associated complexes, such as replication and transcription factories and interchromatin granule clusters, can be seen. As the spatial organisation of these structures is maintained in the absence of most chromatin, it is likely that an underlying "nucleoskeleton" will play a pivotal role in coordinating these vital nuclear functions.

1. Follow **steps 1–9** of **Section 3.2.1** and after electroelution recover the beads from the gel. Pellet the beads using a brief spin on a bench centrifuge (e.g., 20 sec at 1,000×g) to preserve the morphology of the underlying nucleoskeleton.
2. Fix the cells from which chromatin was eluted with 0.5% (if performing step 3) or with 2.5% glutaraldehyde in PB for 10 min at 0°C, wash 2× in Sorensen buffer, incubate for 10 min at 0°C with SB containing $0.02 M$ glycine, and rewash in SB.
3. If the samples were pre-labelled using nucleotide analogues, perform immunolabelling using a standard procedure (e.g., ref. *(10)*), fix in 2.5% glutaraldehyde in SB, and wash 3× in SB.
4. Postfix the samples with 0.5% OsO_4 in SB, dehydrate through a series of increasing ethanol concentrations (*see* **Section 2.3.5**), with a 30 min incubation in 70% ethanol containing 2% uranyl acetate. Replace the ethanol with three changes of n-butanol and embed the samples in the removable compound diethylene glycol distearate (DGD): melt DGD at 70°C, transfer the samples to a mixture of 1:1 molten DGD and n-butyl alcohol, and incubate for at least 45 min at 60°C. Then impregnate with pure DGD for 2–3 h, allow to harden, and cut 500-nm-thick sections onto Pioloform-coated electron microscope grids. Extract the DGD with n-butanol 3× for 1 h at 20°C, then with acetone to replace the butanol, and finally critical-point dry.

3.4 Using Nuclear Derivatives To Analyse Links Between Nuclear Structure and Function

Early experiments suggested close links between nuclear structure and function (reviewed in refs. *(5, 12)*). However, nuclear matrix preparations made by high-salt extraction are not of general utility for studying nuclear function, as almost all interesting activities are extracted. While active RNA polymerases remain in nuclear matrices extracted with $0.2 M$ $(NH_4)_2SO_4$, cells encapsulated in agarose beads and permeabilised under "physiological" conditions provide by far the most versatile system to analyse nuclear structure and function.

The preservation of nuclear structure requires that cells are permeabilised in the gentlest possible way and maintained in a structure-friendly environment. A variety of reagents can be used to permeabilise the cell membrane while leaving the nuclear membrane essentially intact. This is achieved either by the use of selective reagents, or combinations of detergent concentration and time that are insufficient to cause significant internal damage.

Biological detergents such as streptolysin O, α-toxin and complement can also be used to selectively permeabilise the external membrane of mammalian cells. These small proteins assemble multimeric complexes in cell membranes, generating pores of characteristic structure. Streptolysin generates ~15-nm pores in cholesterol-containing membranes, sufficient to allow the passage of large proteins like antibodies. As excess protein can be removed prior to lysis, internal membranes should remain intact. α-Toxin pores are only ~2 nm, allowing passage of small molecules such as nucleic acid precursors, but few proteins.

Different lysis protocols have advantages and disadvantages. Saponin is suitable for most purpose, and gives a reproducibly uniform lysis with good nuclear morphology. The extent of lysis is empirical and critical to the success of most experiments. To define conditions, use a twofold dilution series of detergent in PB and assess the level of permeabilisation using trypan blue exclusion: add 50 μL of 1% trypan blue in PB to 50 μL of packed microbeads, after 10 min inspect by light microscopy and score the per cent of permeabilised, dark-blue, cells. Choose the detergent concentration that permeabilises >95% of the cells. Note that absolute cell numbers critically affect the extent of cell lysis. The following (and related) detergents can be used; guideline concentrations (v/v) for use with 5×10^6 cells in 1 mL of beads and 10 mL of PB are indicated:

1. 0.02–0.05% Triton X-100.
2. 0.01–0.02% digitonin.
3. 0.02–0.05% lysolecithin.
4. 0.01–0.02% saponin.

3.5 Labelling Sites of DNA Replication In Vitro

Encapsulated cells permeabilised under isotonic conditions retain all the replication complexes that were active in the cell at the moment of lysis, allowing the sites of DNA synthesis to be labelled under controlled conditions. The combination of labelling in vitro, chromatin extraction, and immunostaining in combination with resin-less EM allowed the first morphological description of replication "factories" in mammalian cells *(10)* (Fig. 9.3).

Although it is a routine matter to label sites of DNA synthesis in vivo using bromodeoxyuridine (BrdU) as a precursor *(14)*, the use of dNTP analogues in vitro does have several advantages: 1) after permeabilisation, precursor pools can be depleted by washing; 2) elongation rates can be modified at will by adjusting the precursor concentration; and 3) a range of different precursors is available, allowing double immunostaining of replication sites together with any protein of interest for which antibodies are available (*see* **Section 3.7**). In addition, the use of fluorescent analogues allow sites of DNA synthesis to be visualised directly, so avoiding fixation *(15)*, and does not require DNA denaturation so that the sample has better-preserved nuclear morphology *(16)* (note that incorporated BrdU can only be detected after denaturing DNA).

The concentration of modified precursor and the duration of labelling can be adjusted to suit individual requirements; the following provides some guidelines. A 15-min incubation with a $20\,\mu M$ biotin- or digoxygenin-coupled precursor gives good indirect immunofluorescence signals, and longer incubations give correspondingly stronger signals. Five and $2\,min$ incubations with $100\,\mu M$ biotin-16-dUTP allow detection by light and electron microscopy, respectively. Incorporated label can be detected after 30–60-min incubations with $20\,\mu M$ fluorescent precursors.

1. Encapsulate the cells at $\sim$2–10$\times$10^6 cells/mL beads as described in **Section 3.1** (*see* **Note 9**).
2. Recover the beads by centrifugation at $\sim$100$\times g$ for $2\,min$ and wash 1$\times$ in PBS and 1$\times$ in PB.
3. Permeabilise the cells by incubating the beads in PB with 100–250$\,\mu$g/mL of saponin for $3\,min$ at $0°C$, pellet the beads, and wash them 2$\times$ in PB. If inhibitors of replication are to be used, add them to the beads for $10\,min$ at $0°C$ before transferring to $33°C$.
4. Warm the beads in PB to $33°C$ and initiate replication by adding 1/10th volume of prewarmed 10$\times$ replication mix (*see* **Note 3**).
5. Incubate at $33°C$ for 5–$30\,min$.
6. Wash the beads 3$\times$ in >10 volumes of ice-cold PB; pellet the beads, aspirate the supernatant, add fresh buffer, mix by inverting the tube, and stand the sample on ice for $5\,min$.
7. Fix in fresh 4% w/v paraformaldehyde in PB for $15\,min$ at $4°C$, and wash 4$\times$ in PBS.
8. If using indirect detection, incubate with the primary antibody directed against the modified DNA precursor (2.5$\,\mu$g/mL in PBS/BSA/Tween) for $4\,h$ at $20°C$, wash 4$\times$ in PBS/BSA/Tween, incubate with secondary antibody (2.5$\,\mu$g/mL in PBS/BSA/Tween) for $16\,h$ at $4°C$, and wash 4$\times$ in PBS/BSA/Tween.
9. Mount the beads in Vectashield under a coverslip.

3.6 Labelling Sites of RNA Transcription In Vitro

Permeabilised cells supplemented with a mixture of NTPs and incubated under optimal conditions perform transcription at rates approaching those seen in vivo. This is only possible, however, using salt conditions optimal for RNA polymerase elongation, i.e., $\sim$350 mM $(NH_4)_2SO_4$. At physiological salt concentrations, the polymerase complexes transcribe through nucleosomes with a lower efficiency, presumably implying that lysis or the buffer environment are not perfect. Nevertheless, the use of encapsulated cells has provided interesting details about the arrangement of sites of transcription in mammalian cells. Most notably, this approach has demonstrated that transcription, like replication, takes place at specialised nuclear sites, transcription "factories", where many active units operate together *(9, 17)*.

Though many modified NTPs are available, only BrUTP and biotin-14-CTP have been shown to be incorporated by mammalian cell RNA polymerases. As BrUTP is inexpensive, it can be used at the same concentration as the unmodified precursors; incorporation is ~75% of the level seen with UTP and immunolabelling can be performed after 5–30 min of synthesis. Biotin-14-CTP should be used as described for the replication analogues (*see* **Section 3.5**).

The different RNA polymerase activities can be distinguished by using inhibitors. To suppress polymerase I activity, grow encapsulated cells in medium supplemented with 0.1–0.2 µg/mL of actinomycin D for 15 min before permeabilisation. Polymerases II and III activities are suppressed by 2 and 250 µg/mL of α-amanitin, respectively, added to the beads for 15 min at 0°C before transferring to 33°C.

1. Prepare PB with HPRI (0.5 U/mL).
2. Prepare 10× concentrated transcription initiation mix (10× TIM) as described in **Section 2.6**.
3. Follow the protocol in **Section 3.5** but instead of RIM for replication, use 10× TIM for transcription (*see* **Note 10**).

3.7 Analysing the Organisation of Nuclear Proteins: The Micro-Architecture of Sites of Replication and Transcription

If is often useful to know how particular nuclear proteins are distributed with respect to different nuclear compartments. For example, if a protein activates genes at a particular time in the cell cycle it should be possible to establish if it associates with transcription factories throughout the cell cycle or only while the gene is active. This type of analysis is conveniently performed on permeabilised cells in which the replication or transcription sites are pre-labelled with an appropriate analogue. Once these sites are labelled, preparations are fixed and the active sites and protein of interest immunolabelled.

Performing immunolabelling in encapsulated cells provides a versatile alternative to the more usual labelling of permeabilised cells that have been grown on glass coverslips (*see* **Note 11**).

1. Encapsulate cells in microbeads (for routine experiments use 2–5×10⁶ cells/mL of agarose) as described in **Section 3.1** and regrow them in growth medium for 1 h.
2. If required, first perform chromatin extraction (*see* **Section 2.2**) or in situ replication or transcription assays (*see* **Sections 3.5** and **3.6**).
3. Permeabilise the cells by immersing the beads in 5 volumes of ice-cold PB with 100 µg/mL of saponin for 2 min, and wash the beads 2× in 10 volumes of PB.

4. For routine immunolabelling experiments, use 50–100 μL of beads per sample in 1.5-mL microcentrifuge tubes.

5. Rinse the beads 3× in ice-cold PB, fix in ice-cold PB with 4% paraformalde-hyde for 15 min, and rinse 3× in ice-cold PB containing 0.5% BSA and 0.1% Tween 20.

6. Disperse the beads in 5 volumes of PB/BSA/Tween containing a 1:500–1:1,000 dilution of the primary antibody and incubate at 0°C for 1 h with intermittent mixing. Wash the beads 3× in ice-cold PB/BSA/Tween.

7. Disperse the beads in PB/BSA/Tween containing an appropriate dilution (gen-erally 1:100–1:2,000, determined empirically) of the chosen secondary anti-body (typically a mouse monoclonal) and incubate at 0°C for 1 h. Wash the beads 3× in ice-cold PB/BSA/Tween.

8. Disperse the beads in PB/BSA/Tween containing 1:1,000 dilutions of appropri-ate fluorochrome-coupled tertiary antibodies, together (e.g., FITC-donkey anti-goat IgG, Cy3-donkey anti-mouse IgG) and incubate at 0°C for 1 h. Wash the beads 3× in ice-cold PB/BSA/Tween.

9. To stain DNA, disperse the beads and incubate for 10 min at room temperature in ice-cold PBS/BSA/Tween with DAPI (excited by UV light) or TOTO-3 (excited by far red light), according to the capabilities of your microscope.

10. Wash the beads 2× in PBS, mount ~5 μL of bead slurry with 2.5 μL of Vectashield, seal with nail varnish, and inspect using a suitable fluorescent microscope.

3.8 Analysing Nuclear Architecture by Step-Wise Deconstruction

The types of analysis described above are usually complicated by an over-abun-dance of protein molecules, the majority of which are non-functional at any time (*18*). The splicing proteins are a good example, as most of these (perhaps as much as 90%) are concentrated at sites (SC35 speckles) that appear to be sites of storage and not of active splicing, most of which takes place at the site of tran-scription (*19*).

Cells encapsulated in agarose beads afford excellent opportunities for the analy-sis of proteins that interact with different nuclear compartments. Even when all protein has been extracted from encapsulated cells, the remaining DNA is protected from damage and remains supercoiled (*20*). Consequently, it is a simple matter to extract encapsulated cell using a wide range of reagents, either separately or sequentially, and to analyse the structure of the active compartments and how these relate to the distribution of different proteins under the different conditions used (Fig. 9.4).

1. Encapsulate cells in agarose beads at $1–10×10^6$ cells/mL as described in **Section 3.1**. The cell density should be appropriate for the analysis to be used; for exam-ple, if performing detergent extraction after labelling transcription sites, low cell

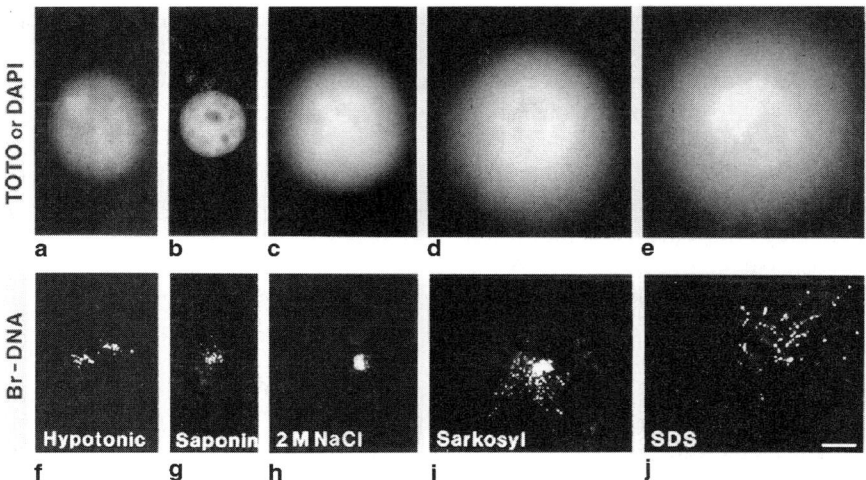

Fig. 9.4 Deconstructing chromosome territories. HeLa cells were labelled with BrdU for 20 min at the onset of the S phase. Labelled territories were visualised by immunofluorescence 10 days later after extraction under the conditions shown. Note the structure and clear boundaries of individual territories and their sensitivity to extraction under different conditions. After 10 days, most cells have a single labelled chromosome territory (**g–j**) though about 5% of the cells have two (**f**) or more. Note that remnant territories are seen even after extractions that destroy nuclear structure and allow DNA (**a–e**) to diffuse into the surrounding agarose (see **c–e**); such extensive extraction is only possible with samples protected in agarose microbeads. Bar, 5 μm

densities should be used. Higher cell densities are necessary if proteins are to be analysed using polyacrylamide gels. Supplement PB with nuclease and/or protease inhibitors appropriate for individual experiments and cell types.

2. If required, label active sites of replication or transcription (*see* **Sections 3.5** and **3.6**).

3. If required, it may be informative to include a restriction endonuclease or ribonuclease digestion. Simply add an appropriate concentration of enzyme and incubate at 30°C for 15 min.

4. Wash the beads 3× in PB and incubate for 10 min in ice-cold PB containing the reagent to be tested, such as Triton X-100 or a related mild detergent (0.05–0.5% v/v); NaCl (0.2–2 M); $(NH_4)_2SO_4$ (0.1–1 M); sarkosyl (0.02–0.5% w/v); urea (0.5–2.5 M); or SDS (0.02–0.5% w/v in PB with Na^+ in place of K^+ and incubated at 20°C).

5. Fix and immunolabel as described in **Section 3.7**. For a typical labelling reaction, use 50–100 μL of beads, add 200–500 μL of ice-cold PB/BSA/Tween containing antibody for 1 h, and wash 3× with 1 mL of ice-cold PB/BSA/Tween.

6. The same samples can be processed for electron microscopy using the procedure described in **Section 3.3**.

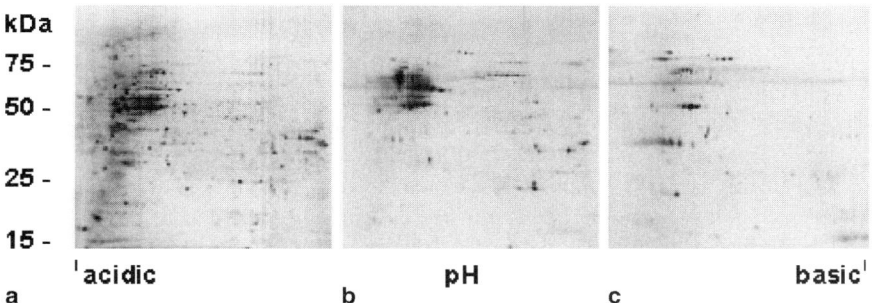

Fig. 9.5 2D gel electrophoresis of proteins involved in nuclear architecture. Samples were extracted as described in Fig. 9.4 and the extracted proteins were analysed using 2D gel electrophoresis. Electrofocusing was performed on Immobiline DryStrips (pH 3–10 NL; Amersham) and electrophoresis in 10% acrylamide. Proteins were detected by silver staining. The 13×13-cm regions show all resolved proteins from samples (**a**–**c**) in Fig. 9.4. After extraction with saponin, $2M$ NaCl, or sarkosyl, the samples were washed extensively in buffer and extracted with 0.25% SDS. The gels contained 200, 100, and 50 µg of protein respectively, with cell equivalents in the ratio ~1:5:25

7. To analyse proteins that resist extraction, take 100 µL of bead slurry from each sample, wash briefly in PB diluted 1:5 with water and containing 1 mM MgCl$_2$, and incubate with 50 U of DNase I at 33°C for 15 min. Add SDS-PAGE sample buffer and perform SDS-PAGE using a standard procedure. If quantitative gels are required (Fig. 9.5), take an appropriate volume of bead slurry and dissolve residual proteins in 1% SDS for 15 min at 20°C, centrifuge the beads, and concentrate the extracted proteins after dialysis to remove SDS.

3.9   Conclusions

It is now generally accepted that the nuclei of mammalian cells are compartmentalised, so that different nuclear functions are performed at specialised, dedicated sites. This was first demonstrated for sites of replication and later for sites of transcription. In both cases, the density of active sites in mammalian cells turned out to be far lower than expected from estimates of the number of active replicons or transcribing genes, implying that multiple units were active in each site. The different approaches described here have been instrumental in the analysis of these active sites and have confirmed that synthesis takes place within specialised compartments, synthetic "factories". Such spatial coordination of related events presumably increases the efficiency of these complicated processes.

4   Notes

1. Although almost any cell type can be encapsulated in this way, some are more sensitive than others to the encapsulation process. For example, HeLa cells grown in suspension are robust and will survive much more vigorous shaking than freshly isolated lymphocytes, which are rather fragile. Optimal conditions for each cell type should be established by trial and error.

2. The choice of agarose is determined by its gelling/melting temperature and purity. When encapsulating mammalian cells, use an agarose that remains molten at 37°C (the Sigma-Aldrich catalogue lists the gelling temperature of various agaroses).

3. A range of DNA precursors conjugated with biotin, digoxygenin, or various fluorochromes is available. Most will support replication by endogenous eukaryotic DNA polymerases when added in place of the equivalent unmodified precursor, although the rate of elongation is typically 10–25% of the level seen with the unmodified precursor. Incorporation can be monitored using ^{32}[P]TTP; incubate samples at 33°C for 2–60 min and sample the incorporation of ^{32}P into acid-insoluble material at appropriate intervals. Two-20 μM TTP and 50–100 μCi/mL of ^{32}[P]TTP give a good level of incorporation and a high efficiency.

4. Bead size is controlled by the shaking speed, temperature, viscosity, and interfacial tension between water and the immiscible liquid used. The protocol detailed in Section 3.1 gives spherical beads 50–100 μm in diameter; beads prepared by hand are often non-spherical and more varied in size. Occasional large beads can be removed by filtering dilute solutions through monofilament nylon filters in a Swinex filter. Filtration through a 150-μm mesh filter should remove few beads and leave a filtrate that passes freely through yellow pipette tips.

5. Note that beads containing permeabilised cells tend to stick to pipette tips, so it is advisable to siliconise tips to prevent losses.

6. For most cells, restriction with *Hae*III plus electroelution should leave 6–10% of the total DNA associated with the residual nuclei in the beads. Because chromatin (not naked DNA) is cut, the limit digest for HeLa nuclei treated with *Hae*III gives 6.3% DNA remaining; to obtain a limit digest use <2×10^6 cells/mL of beads and 1,000 U/mL of *Hae*III in a 10-mL incubation. Enzymes that cut less frequently give larger fragments and more attached DNA; e.g., following digestion with *Eco*RI, 15–25% of the DNA remains. To minimise the general deterioration of nuclear structure, digestions should not be performed at higher temperatures or for longer than 15 min. Use lower temperatures and shorter times, if possible.

7. If it is necessary to recover both attached and detached chromatin fragments, prepare a gel with a single deep well and perform the electroelution with the beads inside wide dialysis tubing, with the direction of electrophoresis reversed. Pellet the beads and prepare DNA from beads and supernatant as described.

8. The pores in 0.5% agarose permit egress of nucleoprotein complexes of up to 150×10^6 daltons and allow added enzymes to equilibrate throughout beads within seconds.

9. If synchronised cells are required, prepare them using a standard procedure (e.g., ref. *(21)*).

10. Levels of incorporation can be monitored using ^{32}[P]UTP; 5 μM UTP, and 50–100 μCi/mL of ^{32}[P]UTP, which give good incorporation with reasonable efficiency. In addition, RNA labelled with ^{32}P can be purified from beads and used to assess the relative density of RNA polymerase molecules on different regions of the genome.

11. For double immunolabelling, as with all immunodetection systems the quality of reagents will determine the success of multiple-labelling experiments. It is usually necessary to titrate reagents to establish the concentration that gives the best signal with little or no background of non-specific binding. It is also important to use controls to confirm that different reagents are not cross-reacting. In some cases, it is also crucial to confirm that the order in which antibodies are added does not influence the extent of binding; this is particularly important when attempting to establish if particular proteins are found at or close to sites of transcription or replication. If the first antibody appears to reduce binding of the second, add them both together.

References

1. Lanctot, C., Cheutin, T., Cremer, M., Cavalli, G., and Cremer, T. (2007). Dynamic genome architecture in the nuclear space: regulation of gene expression in three dimensions. *Nat. Rev. Genet.* **8**, 104–115.

2. Misteli, T. (2005). Concept in nuclear architecture. *BioEssays* **27**, 477–487.

3. Jackson, D.A. and Cook, P.R. (1985). A general method for preparing chromatin containing intact DNA. *EMBO J.* **4**, 913–918.

4. Pombo, A., Jackson, D.A., Hollinshead, M., Wang, Z., Roeder, R.G., and Cook, P.R. (1999). Regional specialization in human nuclei: visualization of discrete sites of transcription by RNA polymerase III. *EMBO J.* **18**, 2241–2253.

5. Berezney, R., Mortillaro, M.J., Ma, H., Wei, X., and Samarabandu, J. (1996). The nuclear matrix: a structural milieu for genome function. *Int. Rev. Cytol.*, **162A**, 1–65.

6. Jackson, D.A. and Cook, P.R. (1985). Transcription occurs at a nucleoskeleton. *EMBO J.* **4**, 919–925.

7. Jackson, D.A. and Cook, P.R. (1986). Replication occurs at a nucleoskeleton. *EMBO J.* **5**, 1403–1410.

8. Jackson, D.A. and Cook, P.R. (1988). Visualization of a nucleoskeleton. *EMBO J.* **7**, 3667–3677.

9. Jackson, D.A., Hassan, A.B., Errington, R.J., and Cook, P.R. (1993). Visualization of focal sites of transcription within human nuclei. *EMBO J.* **12**, 1059–1065.

10. Hozak, P., Hassan, A.B., Jackson, D.A., and Cook, P.R. (1993). Visualization of replication factories attached to a nucleoskeleton. *Cell* **73**, 361–373.

11. Nilsson, K., Scheirer, W., Merten, O.W., Ostberg, L., Liehl, E., Katinger, H.W.D., and Mosbach, K. (1983). Entrapment of animal cells for production of monoclonal antibodies and other biomolecules. *Nature* **302**, 629–631.

12. Jackson, D.A. and Cook, P.R. (1995). The functional basis of nuclear structure. *Int. Rev. Cytol.* **162A**, 125–149.

13. Guo, X.-W. and Cole, R.D. (1989). Chromatin aggregation depends on the anion species of the salt. *J. Biol. Chem.* **264**, 16873–16879.

14. Dolbeare, F. (1996). Bromodeoxyuridine: A diagnostic tool in biology and medicine. *Histochem. J.* **28**, 531–575.

15. Hassan, A.B. and Cook, P.R. (1993).Visualization of replication sites in unfixed human cells. *J. Cell Sci.* **105**, 541–550.

16. Nakayasu, H. and Berezney, R. (1989). Mapping replication sites in the eukaryotic cell nucleus. *J. Cell Biol.* **108**, 1–11.

17. Iborra, F.J., Pombo, A., Jackson, D.A., and Cook, P.R. (1996). Active RNA polymerases are localized within discrete transcription 'factories' in human nuclei. *J. Cell Sci.* **109**, 1427–1436.

18. Grande, M.A., van der Kraan, I., de Jong, L., and van Driel, R. (1997). The distribution of transcription factors in relation to sites of transcription and RNA polymerase II. *J. Cell Sci.* **110**, 1781–1791.

19. Pombo, A. and Cook, P.R. (1996). The localisation of sites containing nascent RNA and splicing factors. *Exp. Cell Res.* **229**, 201–203.

20. Cook, P.R. (1984). A general method for preparing intact nuclear DNA. *EMBO J.* **3**, 1837–1842.

21. Jackson, D.A. (1995). S-phase progression in synchronized human cells. *Exp. Cell Res.* **220**, 62–70.

22. Jackson, D.A., Dickinson, P., and Cook, P.R. (1990) The size of chromatin loops in HeLa cells. *EMBO J.* **9**, 567–571.

23. Hozak, P., Jackson, D.A., and Cook, P.R. (1994). Replication factories and nuclear bodies: the ultrastructural characterization of replication sites during the cell cycle. *J. Cell Sci.* **107**, 2191–2202.

Part III
The Nuclear Envelope

Chapter 10
Investigation of Nuclear Envelope Structure and Passive Permeability

Victor Shahin, Yvonne Ludwig, and Hans Oberleithner

Keywords Atomic force microscopy; Confocal microscopy; Nuclear envelope structure and permeability; Nuclear hourglass technique; Nuclear pore complex; Xenopus laevis oocyte

Abstract We present an experimental approach by the help of which structure and passive permeability of the nuclear envelope (NE) can be investigated thoroughly, by combining imaging, fluorescent, and electrophysiological techniques. A mature *Xenopus laevis* oocyte features a large nucleus offering an excellent system for these investigations. Using the emerging technique of atomic force microscopy, NE structure and the conformational state of nuclear pore complexes (NPCs), known to rule NPC and inevitably NE permeability, can be visualised at high resolution and under near physiological conditions. Passive NE permeability to macromolecules can be determined by the long-established confocal laser scanning microscopy, applying fluorescent macromolecules (dextran). Passive NE permeability to small molecules, which has long remained confounded by the lack of an appropriate technique, can finally be investigated following development of a proper technique designated the "nuclear hourglass" technique. The experimental approach presented here thus opens unique perspectives towards understanding the correlation between NE structure and passive permeability. This chapter describes in detail the protocols for performing such investigations.

1 Introduction

Nuclear pore complexes (NPCs) *(1,2)* perforate the nuclear envelope at regular distances *(3,4)*. They mediate bidirectional transport of molecules into and out of the nucleus in a highly selective manner *(5–11)*. As shown in Fig. 10.1, the current assumption is that the NPC features two distinct transport pathways, a large one for macromolecules, the NPC central channel, and a smaller one for small molecules, the NPC peripheral channels *(2, 12–14)*. The NPC central channel is long since accepted

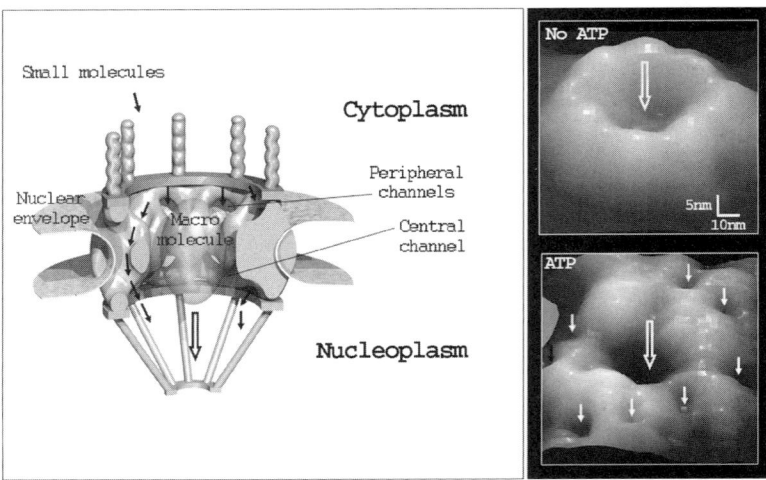

Fig. 10.1 Three-dimensional architecture (*left*) and AFM images (*right*) of the NPC showing that it harbours two distinct transport pathways, a central channel (*large arrow*) for macromolecule transport and eight peripheral channels (*small arrows*) through which small molecule transport proceeds. The peripheral channels seem to open up, thus becoming more visible, following exposure to ATP. *Left* and *right* images are modified from Pante and Aebi (*2*) and Shahin et al. (*3*), respectively

as being the gate through which at least macromolecules enter or leave the nucleus. NPC peripheral channels, however, have just started to gain general acceptance as being further transport pathways. These pathways are more and more believed to be the preferential gates through which small molecules (*12*), at least inorganic ions (*15–17*), enter or leave the nucleus. This is supported by several lines of structural and functional evidence (*2,12,15–17*). Recently, we could show that a transient change in NPC conformational state (e.g., contraction and dilation) was paralleled by a transient change in NPC permeability to both macromolecules and small molecules (*3,16–18*). This in turn not only inevitably implemented changes in whole nuclear envelope permeability, but also implied altered permeability of either NPC transport pathway (*3,15–17*). Such changes in NPC conformational state and permeability underlined the structural and functional plasticity of the NPC, and were concluded to be of key importance for steroid-induced gene expression to be mediated (*3,19, 20*). Exposure of responsive cells to steroid hormones/analogues proved to change the NPC conformational state (*3,4,18*) transiently, inevitably implementing transient changes in passive NPC permeability to macromolecules (fluorescent dextrans) (*3,18*) and small molecules (inorganic ions) (*3,16,18,21,22*). In the light of the current assumption of the NPC possessing two distinct transport pathways, we set out to investigate each NPC transport pathway separately. Indeed, using fluorescence microscopy, the macromolecule transport pathway has been able to be readily investigated since a long time ago. A separate investigation of the small molecule transport pathway, however, has long remained a serious problem confounded by the lack of an appropriate technical approach *(13)*. Our current knowledge about passive nuclear envelope permeability to small molecules (mainly inorganic ions) is, to a large extent, based on studies using large amphibian oocyte nuclei *(15)*. Unfortunately, these nuclei cannot be investigated with the microelectrode technique for technical reasons caused by their extraordinarily large size *(15)*. Another established technique, patch clamp, is restricted to preparations with electrically closed NPCs *(15)*. Induced by these restrictions, we developed an appropriate technique designated the nuclear hourglass technique *(15)*. Having developed this technique, determination of passive nuclear envelope permeability to small molecules has been rendered possible *(3,15–18,21–23)*. Moreover, a separate investigation of the NPC transport pathway for small molecules has become practicable for the first time *(23)*. After a separate investigation of either NPC transport pathway was rendered possible, we next aimed to select one of several high-resolution imaging techniques available that enable visualisation of, and any change in, NPC conformational state. Moreover, this technique had to be applicable when the nuclear envelope was intended to be imaged under fluid (near physiologic conditions) without the need of any elaborate preparation steps. This unique requirement eliminated all high-resolution imaging techniques one after another, but singled out atomic force microscopy (AFM) *(24)*. AFM was developed in 1985 to overcome the limitations of its ancestor, the scanning tunnelling microscope (STM) *(25)*, in imaging non-conducting samples. AFM immediately attracted the attention of the biophysical community, and step-by-step, AFM was established in diverse research fields, among them, biomedical and life sciences *(3,4,17,18,26–37)*. At the beginning, the emphasis was mainly on the improved imaging resolution compared with that of optical microscopy, but it soon became clear that AFM was much more than just a high-resolution imaging technique. The possibilities of spectroscopic analysis, surface modification, and molecular manipulation *(3,28,34,35,37–39)* gave rise to real breakthroughs in the realm of AFM use. Now that appropriate techniques have been found to visualise the actual NPC conformational state (and any change in this), and to investigate the resulting permeability changes of either NPC transport pathway separately, a suitable test object (system) remained to be selected out of many possibilities. This selection, however, turned out to be dead easy: mature *Xenopus laevis* oocytes, widely used, suggested themselves automatically. Firstly, and as mentioned above, our current knowledge about passive nuclear envelope permeability to small molecules is, to a large extent, based on studies using these nuclei. Secondly, these oocytes proved to be an excellent system, widely applied in studies investigating nuclear envelope permeability and structure. For such studies, they reveal many advantages over other systems. The large sizes of these oocytes (1,000–1,200 µm) and of their nuclei (400–500 µm) make their injection relatively easy. Oocytes could, thus, be microinjected with physiologic stimuli and the resulting changes in nuclear envelope structure and permeability could be traced readily *(3,4,16,18,22)*. Another major advantage is the extraordinary high density of NPCs, 40–50 per µm^2 *(3,40)*. Having found appropriate techniques and a suitable system, we developed an experimental approach by which NPC permeability and structure can be investigated and linked to one another. This chapter describes in detail the protocols for performing such investigations.

2 Materials

2.1 Oocyte Preparation and Microinjection

1. *Xenopus laevis*: female, size up to 110 mm.
2. Ethyl m-aminobenzoate methanesulfonate (Serva, Heidelberg, Germany): 0.1% solution buffered with 600 mg/L sodium bicarbonate.
3. Stereomicroscope.
5. Dumont #5 and #55 tweezers (INOX 5/55; World Precision Instruments, Sarasota, FL, USA).
6. Modified Ringer solution (HEPES Ringer solution [HRF]): 87 mM NaCl, 3 mM KCl, 1.5 mM CaCl$_2$, 1 mM MgCl$_2$, 10 mM HEPES, 100 U penicillin, and 100 μg/mL streptomycin (Sigma-Aldrich, Munich, Germany), pH 7.4.
7. Microinjection equipment (optional).

2.2 Isolation of the Cell Nucleus and Preparation of the Nuclear Envelope

1. Bovine serum albumin (BSA) (Sigma-Aldrich): 1% w/v solution in distilled H$_2$O.
2. Superglue.
3. Pasteur pipets (150 mm).
4. Microscope slides and 15-mm-diameter glass coverslips.
5. Cell-tak: 0.2–0.25 mg/mL solution in distilled H$_2$O (BD Biosciences, Heidelberg, Germany).
6. Minuten needles (Entomologie-Bedarf Meier, Munich, Germany; order number 300/10).
7. Double-faced adhesive tape.
8. Polyvinylpyrrolidone (PVP): M_r 40 kDa (Sigma-Aldrich).
9. Nuclear isolation medium (NIM): 90 mM KCl, 10 mM NaCl, 2 mM MgCl$_2$, 1.1 mM EGTA, 10 mM HEPES, and 1.5% PVP, pH 7.4 (*see* **Note 1**).

2.3 Determination of Passive Nuclear Envelope Permeability

2.3.1 Passive Permeability to Small Molecules Using the Nuclear Hourglass Technique

Equipment for the nuclear hourglass technique is not commercially available. The technical aspects of the method and its application to isolated nuclei are described in detail in ref. *(15)* and the electrical circuitry of the setup is shown in Fig. 10.5.

1. Tapered glass tube that narrows in its middle part to two thirds of the diameter of the nucleus, with a massive Ag/AgCl electrode at each end. Glass capillaries with 2-mm outer and 1.7-mm inner diameters (Servoprax, Wesel, Germany) are narrowed using heat and pulling force from a vertical patch pipette puller (LM-3P-A; List Electronics, Darmstadt, Germany) to introduce an initial narrowing to ~50% of the initial outer diameter, and in a second step only heat and no pulling force is applied. The walls of the narrowed part of the glass capillary are thickened so that the inner diameter is reduced but the outer remains essentially unchanged. Capillaries are incubated for 1 h in 1% BSA to prevent adhesion of nuclei to the glass.

2. Current generator: up to 1 mA.

3. Two conventional microelectrodes to measure the voltage drops across the cell nucleus.

2.3.2 Passive Permeability to Fluorescent Macromolecules Using Confocal Laser Scanning Microscopy

1. Perfusion chamber for nuclei: we use a chamber produced in our workshop that features a "nuclear chamber" of ~98.2 mm^3 and an outflow of ~1,820 mm^3 (*see* Fig. 10.7).

2. Perfusor with silicone perfusion hose (Secura Perfusor; Braun, Melsungen, Germany).

3. Fluorescent-labeled dextrans for perfusion: tetramethylrhodamine isothiocyanate (TRITC)-dextran, 64-kDa and fluorescein isothiocyanate (FITC)- or (Ca^{2+} Green)-10-kDa dextran (Sigma-Aldrich), 0.01 mM in NIM.

4. Confocal laser scanning microscope mounted on an inverted microscope: e.g., Olympus IX70, IOlympus, Hamburg, Germany).

5. Image measurement/analysis software: e.g., Fluoview version 2.1.39 (Olympus).

2.4 Investigation of Nuclear Envelope Structure

1. AFM: we use a BioScope (EMSL, Richland, WA, USA) or a Multimode AFM (NanoScope IIIa controller; Digital Instruments, Santa Barbara, CA, USA).

2. AFM cantilevers: aluminum-coated, V-shaped 290-μm-long silicon cantilevers with pyramidal tips of radius typically 5–10 nm (Ultrasharp; MikroMasch, Anfatec, Germany).

3. For recording, a computer with a Digidata 2000 ADC board and PCLAMP7 software (Axon Instruments, Foster City, CA, USA).

3   Methods

In general, the methods for studying nuclear envelope structure as well as passive nuclear envelope permeability to macromolecules and small molecules include preparation of microinjected (microinjection is optional) *Xenopus* oocytes, isolation of cell nuclei, and preparation of the nuclear envelope. Figure 10.2 displays the sequential steps starting at surgical removal of a *Xenopus* ovary cluster and ending at isolation of the oocyte nucleus and preparation of the nuclear envelope. Figure 10.3 shows a scheme of the experimental approach. As seen in Fig. 10.3, AFM is used to visualise the surface topography of the nuclear envelope and to view the NPC conformational state. Confocal laser scanning microscopy and the nuclear hourglass technique are applied to investigate the passive permeability of either NPC transport pathway, respectively.

3.1   Oocyte Preparation and Microinjection

3.1.1   Preparation

1. Anaesthetise female *Xenopus laevis* by placing the frog in 0.1% ethyl m-aminobenzoate methanesulfonate solution for 60 min. Remove a portion of the frog's ovaries (ovary cluster) surgically. Wash the ovary cluster with HRF repeatedly to remove traces of blood and debris from oocytes.

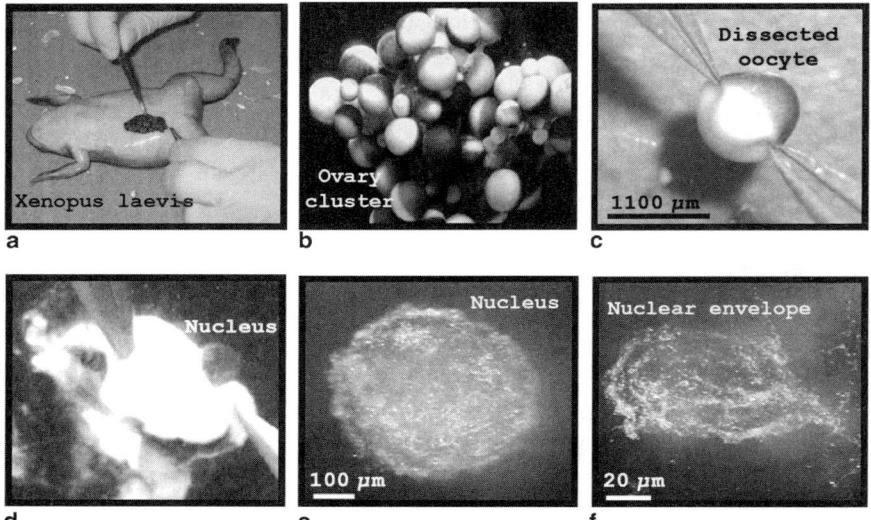

Fig. 10.2 Surgical removal of a portion of a female *Xenopus laevis* ovary (**a**, **b**), isolation of dissected oocytes (**c**), the nucleus (**d**, **e**), and preparation of its nuclear envelope (**f**)

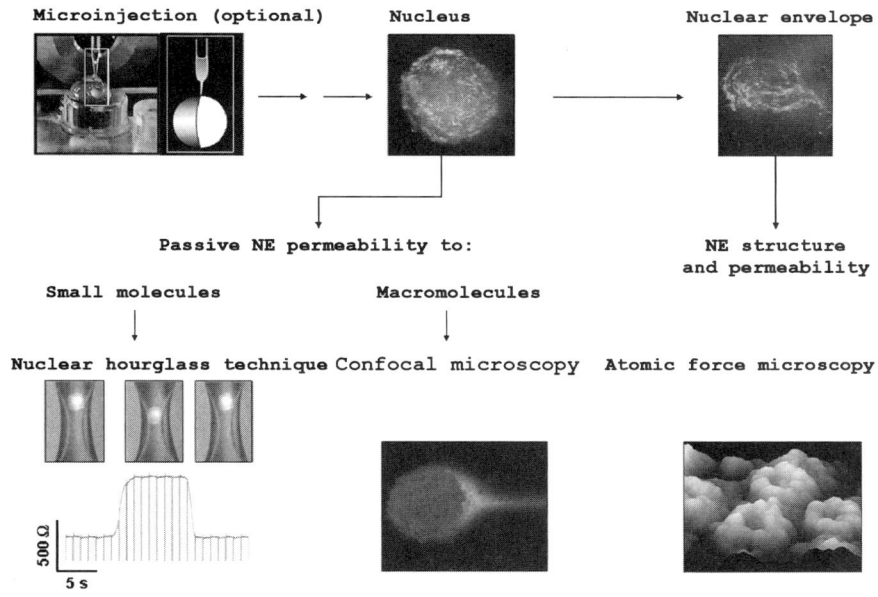

Fig. 10.3 A scheme of the experimental approach of the present chapter describing how to study structure and passive permeability of the *Xenopus* oocyte nuclear envelope (*NE*). *NPC*, nuclear pore complex

2. Keeping the ovary cluster in HRF, use two Dumont #5 tweezers to dissect stage VI oocytes (*see* **Note 2**) in a HRF-filled Petri dish under a stereomicroscope, and wash them three times with HRF. Transfer them one by one to a sterile Petri dish and then keep them at 4°C, a condition under which the oocytes remain generally in good condition for the next 5 days (*see* **Notes 3** and **4**).

3.1.2 Microinjection (Optional)

Oocytes can be microinjected with diverse substances. This step, however, is optional. Should you intend to microinject oocytes, please follow the next steps:

1. If the oocytes were kept at 4°C, let them warm to room temperature in a HRF-filled Petri dish before microinjecting them.
2. For cytoplasmic injection, place the oocyte as shown in Fig. 10.4. Prior to nucleoplasmic injection, oocytes may need to be centrifuged at low speed ($500 \times g$ for 15 min). This renders the nucleus visible through the animal pole, thus increasing the accuracy of nucleoplasmic injection. The injected volume is up to you, but should not exceed 50 and 100 nL in the nucleus and the cytosol, respectively.

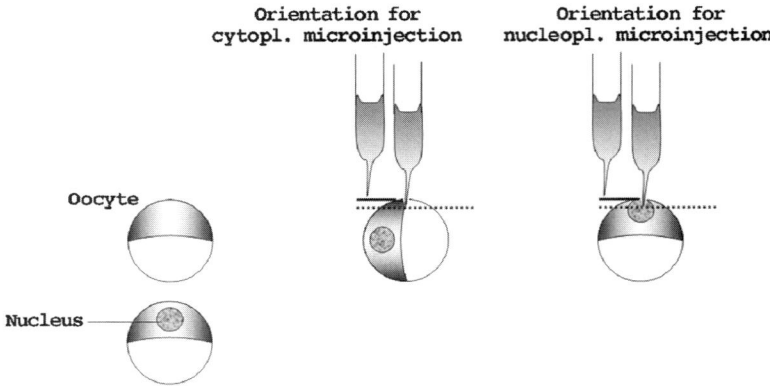

Fig. 10.4 A scheme displaying the orientation of the oocyte for cytoplasmic (*cytopl*) or nucleo-lasmic (*nucleopl*) injection

3.2   Isolation of the Cell Nucleus and Preparation of the Nuclear Envelope

Each individual step in this section is dealt with in detail on our homepage at http://physiologieii.klinikum.uni-muenster.de/forschung/frogs-text.html and is described briefly in the following.

3.2.1   Preliminary Setup

1. Make needles for the preparation: fill a droplet of superglue into the stretched end of a Pasteur pipet (150 mm) until 5–7 mm of the tip is filled with glue. Use a pair of tweezers to stick a so-called minuten needle (**Section 2.11**) into the tip in such a way that 5 mm remain outside. Leave the prepared needles for 1–2 days in order to let the glue dry.
2. Prepare "nuclear tips": use a scalpel to cut off 5–7 mm of a 200-μL Eppendorf micropipet tip. Melt the truncated end by approaching each tip to a flame, taking care that the resulting "bead" is not too thick.
3. Prepare coverslips:

   (a) Clean glass coverslips (15 mm) with 70% ethanol using a Kimwipe, wash twice by rinsing with distilled H_2O, and dry with Kimwipes.
   (b) Pipet 3 μL of Cell-tak solution onto the coverslips and dry the coverslips for 2 h or more in Petri dishes.

(c) On the day of the experiment, wash the coverslips by rinsing 1× with 70% ethanol and 2× with distilled H_2O. After drying, use a marker to mark the outline of the Cell-tak droplet from the backside.

4. Coat all vials and tools that will come in contact with nuclei with BSA in order to prevent the nuclei from sticking to their surface:

(a) Petri dishes (35 mm): fill with 1–2 mL of 1% w/v BSA in distilled H_2O and leave overnight at 4°C or at least 2 h at room temperature.

(b) "Nuclear tips": inset in 1-mL Eppendorf vials filled with at least 300 μL of 1% BSA. Keep the tips in the vials during preparation while you are not using them.

(c) Needles: insert horizontally in 1-mL Eppendorf vials filled with at least 300 μL of 1% BSA. Keep the tips in the vials while you are not using them.

3.2.2 Isolation of the Cell Nucleus

1. Use two pairs of INOX 55 tweezers to open the oocyte (Fig. 10.2) kept in a NIM-filled Petri dish. Detach the yolk from the nucleus using the same tweezers (*see* **Note 5**).

3.2.3 Preparation of the Nuclear Envelope

1. Transfer the nucleus with a "nuclear tip" onto a Cell-tak-coated coverslip kept in another NIM-filled Petri dish (*see* **Note 6**). Note that the nucleus must be positioned at the border of the Cell-tak, not on it.

2. Carefully open the nucleus with two needles prepared as described above. Widen the gap so that you can remove the chromatin.

3. Attach two small pieces of double-faced adhesive tape to microscope slides in such a way that later on, the coverslips can be easily attached to the slides.

4. Fill a Petri dish with NIM, submerge a Cell-tak-coated glass coverslip and keep it down with a small weight at its edge. Prepare an additional Petri dish with distilled H_2O to rinse the preparations.

5. Raise the nuclear envelope with two needles, pull it over the Cell-tak coating and spread it as flat as possible (*see* **Note 7**). Remove the chromatin with a pipet. Take the coverslip out of the Petri dish using tweezers, and put it carefully into another dish filled with NIM.

6. Should you intend imaging under fluid, do not let the coverslip dry! For imaging, take the coverslip out and add 200 μL of NIM.

7. Should you image in air, however, transfer the sample into a Petri dish filled with distilled H_2O. Leave it there for 5 min to wash away the bulk of salt (electrolytes originating from the NIM in which the nuclear envelope was prepared). Take out the coverslip, dry the backside with Kimwipe, and attach the coverslip to the prepared microscope slide.

3.3 Determination of Passive Nuclear Envelope Permeability

3.3.1 Passive Permeability to Small Molecules Using the Nuclear Hourglass Technique

3.3.1.1 Principle of Measurement

The technical aspects of the nuclear hourglass technique and its application to isolated cell nuclei have been described in detail previously *(15)*. The electrical circuitry of the setup is shown in Fig. 10.5 and the way the nuclear envelope conductance is determined is shown in Fig. 10.6. In short, the method uses an NIM-filled tapered glass tube that narrows in its middle part to two thirds of the diameter of the nucleus. A current of up to 1 mA is injected via two massive Ag/AgCl electrodes through either end of the glass tube. The voltage drops across the cell nucleus are measured with two conventional microelectrodes. Since current and voltage are measured simultaneously, the resistance can be calculated continuously. To start the experiment, the nucleus is sucked into the tapered part of the capillary by gentle fluid movement. Thus, the whole current now flows through the accessible parts of the nuclear envelope. The resulting rise in total electrical resistance is caused predominantly by the nuclear envelope and thus indicates the envelope's electrical resistance. The reciprocal of the measured electrical resistance gives the electrical conductance of the nuclear envelope. This in turn is an indicator of the passive nuclear envelope permeability for small molecules (inorganic ions in NIM).

3.3.1.2 Performing Measurements

1. Isolate the nucleus as described previously, and move it into the tapered part of the glass capillary by gravity and gentle downward flow of the fluid.
2. Measure the nuclear envelope's electrical resistance. Figure 10.6 shows a representative original tracing of an individual nuclear envelope electrical resistance measurement using the nuclear hourglass technique. When the nucleus is sucked gently into the tapered part of the capillary, nuclear envelope electrical resistance is measured as a sudden increase in resistance followed by a new steady state value. As long as the nucleus remains in place, nuclear envelope electrical resistance remains constant. The raw data obtained from the experiment represents the increase in resistance of the NIM-filled capillary caused by the nucleus that occludes the narrowing. This resistance of the whole nucleus, R_{Nuc}, is composed of three components: the electrical resistance of the upper and lower nuclear envelope (NE) surface, R_{NE1} and R_{NE2}, plus that of the nuclear interior, R_{Inter}.
3. From these values, the electrical resistance of one single nuclear envelope surface can be readily calculated: $R_{NE1} = R_{NE2} = (R_{Nuc} - R_{Inter})/2$. However, it is convenient to use the reciprocal of the electrical resistance, the conductance G_{NE}: $G_{NE1} = R_{NE2} = 2/(R_{Nuc} - R_{Inter})$. G_{NE} in turn reflects the passive nuclear envelope permeability to inorganic ions (small molecules) of which the NIM is composed.

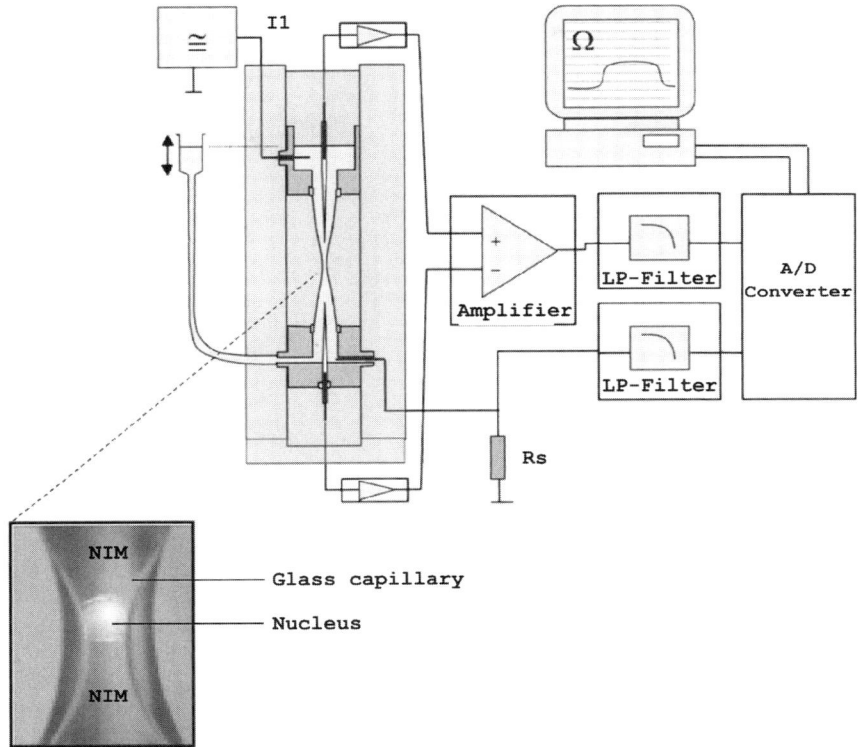

Fig. 10.5 *Top*: electrical circuitry of the setup for the nuclear hourglass technique. Two Ag/AgCl electrodes and a pulse generator are used to drive an alternating current of up to 1 mA through the capillary. Current is measured as a voltage drop over a series resistor of 100 Ω. This signal is recorded after filtering through a Bessel-type low-pass (*LP*) filter. Two conventional KCl-filled microelectrodes are connected to a differential amplifier to measure the voltage drop across the center part of the capillary. The output of this amplifier is also filtered through a Bessel-type low-pass filter before recording (*see* **Section 2.17**). From the current and voltage, the resistance is calculated online and displayed on the monitor. *Bottom*: An image of the glass capillary used in the nuclear hourglass technique. The isolated nucleus is brought into the tapered part of the nuclear isolation medium (NIM)-filled glass capillary in order to measure the nuclear envelope's electrical resistance. Modified from *(15)*

3.3.2 Passive Permeability to Fluorescent Macromolecules Using Confocal Laser Scanning Microscopy

3.3.2.1 Principle of Measurement

Passive nuclear envelope permeability to macromolecules can be measured with fluorescence-labeled macromolecules of defined sizes using a confocal laser-scanning microscope. The underlying principle is that a change in the functional opening

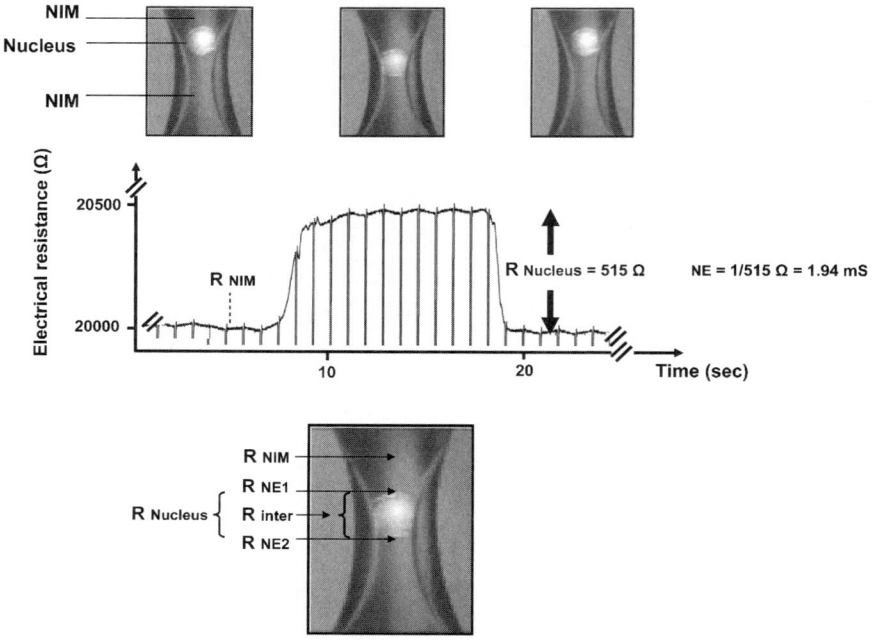

Fig. 10.6 Principle of determination of the electrical conductance (*G*) of the nuclear envelope or of its permeability to inorganic ions, respectively, using the nuclear hourglass technique. *Top*: locating the nucleus into the tapered part of the glass capillary increases the electrical resistance, caused by the NIM (base line ~20,000 Ω). *Middle*: an original tracing displaying the electrical resistance of the nucleus (500 Ω) and describing how to derive nuclear envelope conductance from this. *Bottom*: the electrical resistance is composed of the resistance (*R*) of the two nuclear envelope (*NE*) segments R_{NE1} and R_{NE2}, and the resistance of the interior of the nucleus R_{Inter}, in series

state of the NPC central channel is followed by a change in the passive diffusion rate of fluorescence-labeled macromolecules (of defined sizes) across the nuclear envelope. Comparison of diffusion rates under different experimental conditions allows conclusions to be drawn about NPC permeability to such macromolecules. Note that these macromolecules lack a nuclear localisation signal (NLS), which is known to be essential for active transport through the NPC central channel. Due to the lack of a NLS, the applied fluorescent macromolecules fail to induce any enlargement of the 9-nm-wide NPC central channel. Using isolated *Xenopus laevis* oocyte nuclei, you can apply fluorescent dextrans of defined molecular weights and diameters *(41)* to measure passive nuclear envelope permeability to macromolecules. If so, you need a permeable dextran to determine the rate of macromolecule diffusion across the nuclear envelope and an impermeable dextran to check for the integrity of the isolated nucleus. Dextran molecules lack a NLS and thus diffuse passively in and out of the nucleus through the NPC. This in turn is decreasingly permeable to dextran sizes up to ~40 kDa *(42)* and is virtually impermeable to 64-kDa dextran *(43)*. Typically, 10-kDa (fairly permeable) and 64-kDa dextrans are used to test for the passive diffusion rate across the NPC and for overall integrity of the isolated nucleus, respectively *(43)*. Ten-kilodalton ("small") and 64-kDa ("control") dextrans should be labeled with different fluorescent markers such as Ca^{2+} Green (alternatively, FITC) and TRITC, respectively. Depending on the fluorescent dye, use excitation wavelengths of 488 nm (Ca^{2+} Green) or 568 nm (TRITC), respectively. This strategy allows testing of the integrity of each nucleus after a successful measurement with 10-kDa dextran.

3.3.2.2 Performing Measurements

1. Coat the perfusion chamber ("nuclear chamber") (Fig. 10.7) with Cell-tak to make the nucleus stick to the bottom of the chamber and to prevent it from moving.
2. Bring an isolated nucleus into the chamber along with 100 µL of NIM.
3. Place the chamber on the stage of the confocal microscope, which in turn is mounted on an inverted microscope and equipped with measurement/analysis software.
4. The method used for nucleus perfusion is up to you; you can perfuse manually, but you would be well advised to use a commercially available perfusor (*see* **Section 2.3.6**).
5. Adjust the perfusion rate to 20–30 mL/min, allowing permanent perfusion and fast changes in perfusion solutions.
6. A representative experiment is shown in Fig. 10.8 (*43*). The measurement starts with perfusion of 10-kDa dextran labeled with either Ca^{2+} Green or FITC.

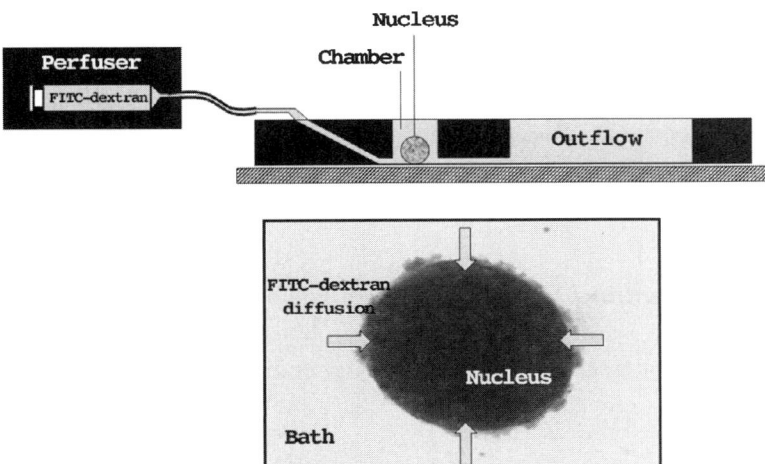

Fig. 10.7 A scheme of the perfusion chamber used to determine passive nuclear envelope permeability to fluorescent macromolecules (for example) applying confocal laser scanning microscopy. Shown is a perfusion chamber with a nucleus attached to the bottom of the chamber. A perfusor, connected to the perfusion chamber, allows complete exchange of the perfusate

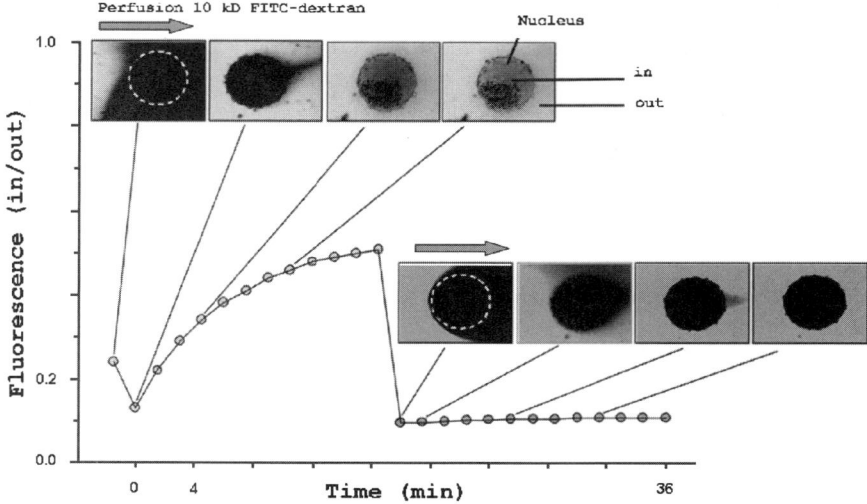

Fig. 10.8 A representative experiment to measure passive nuclear envelope permeability to fluorescent macromolecules (dextran). The graph represents the time course of an experiment. The fluorescence ratio (internal/external) is given on the y-axis. Images show the nucleus exposed to 10-kDa Ca-Green-dextran, and subsequently to 64-kDa TRITC-dextran to test for overall nuclear envelope integrity. Modified from *(43)*

   After the start of perfusion, the nucleus is scanned 12 times every 90 sec. Subsequently, the membrane integrity of the nucleus is tested by using the 64-kDa TRITC–dextran.
7. To evaluate the fluorescence quantitatively, determine the fluorescence intensity in the nucleus and in the bath in each scanned image. Each fluorescence intensity value of the nucleus (in arbitrary units) should be related to bath fluorescence intensity at the same time. The initial slope can be derived from the straight line set through the first three test points (first three ratio values). The initial slope correlates directly with macromolecule nuclear envelope permeability.

3.4   Investigation of Nuclear Envelope Structure

3.4.1   Visualisation of NPC Conformational State Using AFM

3.4.1.1   Principle of Measurement

The AFM works by scanning a fine tip over a surface much in the same way as a phonograph needle scans a record. The tip is positioned at the end of a cantilever beam shaped much like a diving board and as it is repelled by or attracted to the surface, the cantilever beam deflects. The magnitude of the deflection is captured by a laser that reflects at an oblique angle from the very end of the cantilever. A plot of the laser deflection versus the tip position on the sample surface provides the resolution of the hills and valleys that constitute the topography of the surface. The AFM can work with the tip touching the sample (contact mode), or the tip can tap across the surface (tapping mode) much like the cane of a blind person.

3.4.1.2 Performing Experiments

1. Prepare the nuclear envelope as described previously. To image its cytoplasmic side, spread out the nuclear envelope on a glass coverslip with the nucleoplasmic side facing downwards. To image its nucleoplasmic side, however, spread out the nuclear envelope with the cytoplasmic side facing downwards. Figure 10.9 displays two representative AFM images of either nuclear envelope side (*see* **Note 8**).

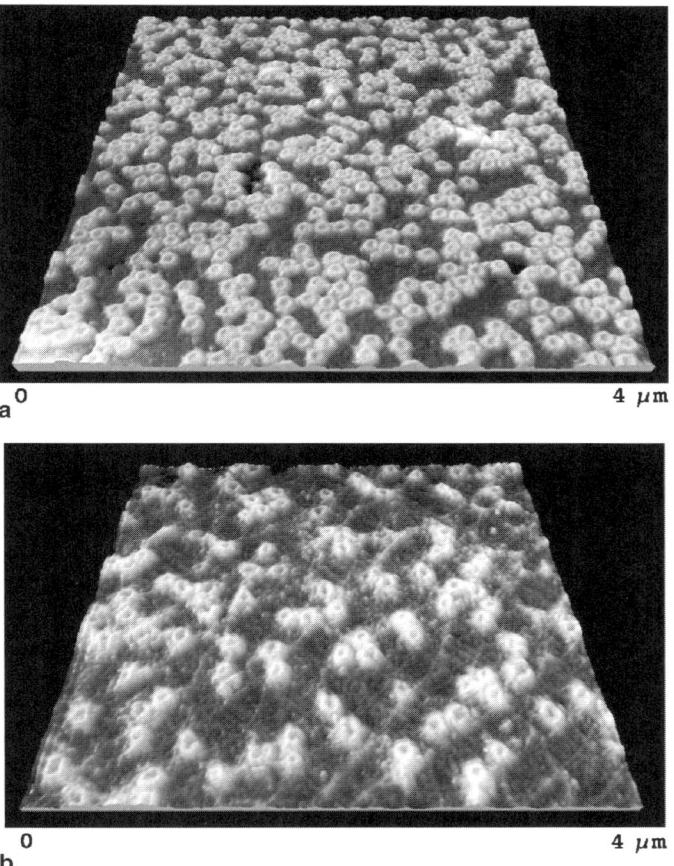

Fig. 10.9 Representative AFM images of the cytoplasmic (*top*) and nucleoplasmic (*bottom*) faces of a *Xenopus* oocyte nuclear envelope

2. What type of AFM and tips you use for imaging is up to you (*see* **Section 2.16**). The following initial control settings are suggested before using *Engage* (driving the AFM tip to the surface of the nuclear envelope):

Scan controls: before making changes to the *Scan controls* panel screen parameters, go to the *Other controls* panel (*Panels/Other*) and verify that the AFM Mode is set to *Contact*. Click on *Panels/Scan*. Set the *Scan size* as large as desired (up to 100×100 μm, dependent on the scanner). Set the *Scan Angle* to 0.0 degrees, the *Scan rate* to 3 Hz, *Aspect ratio* 1:1, *X* and *Y offsets* 0, *Scan Angle* 0 degree, *Scan rate* 3 Hz, *Samples/line* 512, and *Slow scan axis* enabled. Set the feedback controls to: *Integral gain* ≥ 5, *Proportional gain* ≥ 7.5, and *Deflection setpoint* 1. Set the other controls to: *Z limit* 440 volts, *Units* metric, and *Color Table* 12. Go to *Interleave Controls* by clicking with the mouse on *Panels/Interleave*. Verify that the *Interleave Mode* field is set to disabled. Channel 1: on the *Channel 1* panel, set *Data type* to height and *Data scale* to 40–60 nm. Set *Z range* to a reasonable value for the sample (between 200 and 1,500 nm for nuclear envelopes; lower yields better image clarity). *Line direction* can be set to either Trace or Retrace. Set *Scan line*

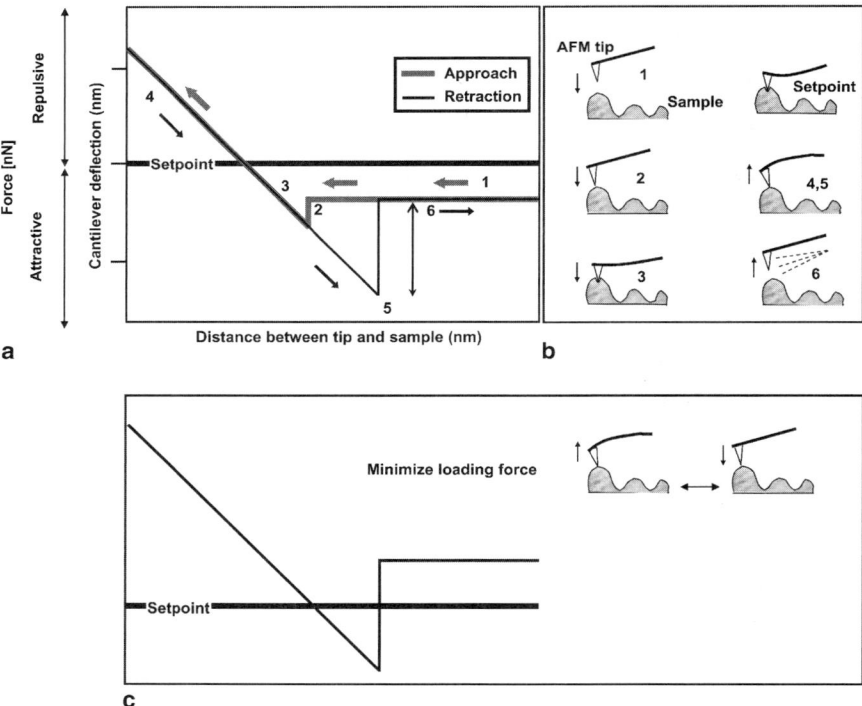

Fig. 10.10 a Schematic force-distance curve describing a single approach-retract cycle of the AFM tip, which is continuously repeated during surface scanning. **b** Before and after contact, several attractive and repulsive forces are experienced between the tip and the sample. **c** How to minimise the loading force after having approached the sample surface

to main, *Realtime planefit* to line, *Offline planefit* to none, *Highpass filter* to off, and *Lowpass filter* to off.

3. Having set the control parameters, engage the AFM. After good engagement is obtained, set the *Samples/line* to 256 to expedite setup; later the value must be increased to 512 for better image resolution. Gradually reduce the *Deflection setpoint* offset to give smaller contact force. The most accurate way to minimise the setpoint value is by using *Force calibration* (*Real time/View/Force mode/Plot*). This consists of obtaining a force plot when adjusting the setpoint value until the tip almost pulls free of the surface, as displayed schematically in Fig. 10.10. In this way, it is possible to minimise tip-sample forces and reduce wear on both the tip and the surface. Having minimised the *Deflection setpoint* value, return to *Realtime image mode*, select an area of interest, *zoom in* gradually (start with 20×20 μm) and finally record (*Capture*). Having captured (*Capture done*) you can start analysing the saved raw data in the *Offline mode*. For analysis and better image quality, you may have to flatten the files after opening (*Image→Select* left/right image→*Modify* →*Flatten→Flatten order* 1).

4 Notes

1. Addition of PVP compensates for the lack of macromolecules in NIM, mimicking the intact cytosol. The presence of PVP is of crucial importance to prevent swelling of total nuclear volume, which occurs instantaneously during isolation in pure electrolyte solution.
2. To distinguish stage VI oocytes from others at different developmental stages, make use of the following information: stage VI oocytes are the largest among all, with diameters ranging between 1,000 and 1,200 μm. In addition, they possess two distinctive hemispheres with good contrast, a black animal and a creamy-coloured vegetal hemisphere (Fig. 10.2).
3. When kept at 4°C, oocytes generally remain in good condition for the next 5 days. However, should you intend to microinject the oocytes, is advisable to revert to oocytes dissected within the first 2 days after surgical removal of the ovaries.
4. Transfer of dissected oocytes should be done carefully and you would be well advised to do it as follows to avoid disrupting the oocytes by an improper suction method: prepare a micropipet tip cut at the end and use a 20-μL pipet to suck and transfer the oocytes.
5. The way the tweezers should be used to clip and then open the oocyte is shown in Fig. 10.2. Pulling the tweezers apart opens the oocytes and renders the distinctive nucleus clearly visible.
6. Transfer the nucleus to a NIM-filled Petri dish and keep it there for 3–5 min to let the nucleus swell. Swelling causes dissociation of the nuclear envelope from the chromatin, an important step that makes separation/preparation of the envelope remarkably easier.

7. This step is very tricky and is very likely not to succeed at the first try.
8. Visualisation of NPC peripheral channels by AFM is elaborate, tricky, dependent on the sample quality, and very likely to require a pretreatment of the prepared nuclear envelope with an ATP-regenerating system (3 mM ATP, 3 mM creatine phosphokinase, and 10 mM phosphocreatine) and 0.1 µM Ca^{2+}. According to our findings by AFM, NPC peripheral channels seem to open up following nuclear envelope exposure to ATP and Ca^{2+}. This, however, does not necessarily mean that visualisation of NPC peripheral channels is only possible when ATP and Ca^{2+} are present. No matter under which condition you choose to image NPC peripheral channels, you would be well advised to use the sharpest tip (radius of curvature ≤5 nm) available. Such tips are commercially available.

Acknowledgments I thank Yvonne Ludwig, Lilian Kastrup, and Claudia Schäfer for their enthusiastic cooperation. The excellent technical assistance of Barbara Windoffer is gratefully acknowledged. This study was supported by grants from the Interdisciplinary Center of Clinical Research Muenster (IZKF project no. Küh3/064/04), the Deutsche Forschungsgemeinschaft (DFG project no. KU 1314/2-1), the "Innovative Medizinische Forschung" (no. SH-110315 and SH-520404), SFB 629, and EU grant "Tips4cells."

References

1. Fahrenkrog, B., Stoffler, D., and Aebi, U. (2001) Nuclear pore complex architecture and functional dynamics. *Curr. Top. Microbiol. Immunol.* **259**, 95–117.
2. Pante, N. and Aebi, U (1993) The nuclear pore complex. *J. Cell Biol.* **122**, 977–984.
3. Shahin, V., Ludwig, Y., Schafer, C., Nikova, D., and Oberleithner, H. (2005) Glucocorticoids remodel nuclear envelope structure and permeability. *J. Cell Sci.* **118**, 2881–2889.
4. Shahin, V., Albermann, L., Schillers, H., Kastrup. L., Schafer, C., Ludwig, Y., Stock, C., and Oberleithner, H. (2005) Steroids dilate nuclear pores imaged with atomic force microscopy. *J. Cell Physiol.* **202**, 591–601.
5. Allen, T. D., Cronshaw, J. M., Bagley, S., Kiseleva, E., and Goldberg, M. W. (2000) The nuclear pore complex: mediator of translocation between nucleus and cytoplasm. *J. Cell Sci.* **113**, 1651–1659.
6. Bayliss, R., Corbett, A. H., and Stewart, M. (2000) The molecular mechanism of transport of macromolecules through nuclear pore complexes. *Traffic* **1**, 448–456.
7. Corbett, A.H. and Silver, P. A. (1997) Nucleocytoplasmic transport of macromolecules. *Microbiol. Mol. Biol. Rev.* **61**, 193–211.
8. Gorlich, D. and Kutay, U. (1999) Transport between the cell nucleus and the cytoplasm. *Annu. Rev. Cell. Dev. Biol.* **15**, 607–660.
9. Mattaj, I. W. and Englmeier, L. (1998) Nucleocytoplasmic transport: the soluble phase. *Annu. Rev. Biochem.* **67**, 265–306.
10. Rout, M. P., Aitchison, J. D., Magnasco, M. O., and Chait, B. T. (2003) Virtual gating and nuclear transport: the hole picture. *Trends Cell Biol.* **13**, 622–628.
11. Wente, S. R. (2000) Gatekeepers of the nucleus. *Science* **288**, 1374–1377.
12. Hinshaw, J. E., Carragher, B. O., and Milligan, R. A. (1992) Architecture and design of the nuclear pore complex. *Cell* **69**, 1133–1141.
13. Mazzanti, M., Bustamante, J. O., and Oberleithner, H. (2001) Electrical dimension of the nuclear envelope. *Physiol. Rev.* **81**, 1–19.

14. Oberleithner, H. (2004) Nuclear envelope: nanoarray responsive to aldosterone. *J. Membr. Biol.* **199**, 127–134.

15. Danker, T., Schillers, H., Storck, J., Shahin, V., Krämer, B., Wilhelmi, M., and Oberleithner, H. (1999) Nuclear hourglass technique: novel approach detects electrically open pores in xenopus laevis oocyte. *Proc. Natl. Acad. Sci. USA* **96**, 13531–13535.

16. Schafer, C., Shahin, V., Albermann, L., Hug, M. J., Reinhardt, J., Schillers, H., Schneider, S. W., and Oberleithner, H. (2002) Aldosterone signaling pathway across the nuclear envelope. *Proc. Natl. Acad. Sci. USA* **99**, 7154–7159.

17. Shahin, V., Danker, T., Enss, K., Ossig, R., and Oberleithner, H. (2001) Evidence for Ca2+- and ATP-sensitive peripheral channels in nuclear pore complexes. *FASEB. J.* **15**, 1895–1901.

18. Kastrup, L., Oberleithner, H., Ludwig, Y., Schafer, C., and Shahin, V. (2006) Nuclear envelope barrier leak induced by dexamethasone. *J. Cell Physiol.* **206**, 428–434.

19. Kroeger, H. (1966) Elektrophysiologische und cytologische Untersuchungen an den Speicheldrüsen von Chironomus Thummi. *Exp. Cell Res.* **41**, 64–80.

20. Wünsch, S., Schneider, S., Schwab, A., and Oberleithner, H. (1993) 20-OH-ecdysone swells nulcear volume by alkalinization in salivary glands of Drosophila melanogaster. *Cell Tissue Res.* **274**, 145–151.

21. Oberleithner, H., Schafer, C., Shahin, V., and Albermann, L. (2003) Route of steroid-activated macromolecules through nuclear pores imaged with atomic force microscopy. *Biochem. Soc. Trans.* **31**, 71–75.

22. Schafer, C., Shahin, V., Albermann, L., Schillers, H., Hug, M. J., and Oberleithner, H. (2003) Intracellular calcium: a prerequisite for aldosterone action. *J. Membr. Biol.* **196**, 157–162.

23. Danker, T., Shahin, V., Schlune, A., Schafer, C., and Oberleithner, H. (2001) Electrophoretic plugging of nuclear pores by using the nuclear hourglass technique. *J. Membr. Biol.* **184**, 91–99.

24. Binnig, G., Quate, C. F., and Gerber, C. (1986) Atomic force microscope. *Phys. Rev. Lett.* **56**, 930–933.

25. Binnig, G., Rohrer, H., Gerber, C., and Weibel, E. (1982) Surface studies by scanning tunneling microscopy. *Phys. Rev. Lett.* **49**, 57–61.

26. Henderson, R. M. and Oberleithner, H. (2000) Pushing, pulling, dragging, and vibrating renal epithelia by using atomic force microscopy. *Am. J. Physiol. Renal Physiol.* **278**, F689–F701.

27. Hillebrand, U., Hausberg, M., Stock, C., Shahin, V., Nikova, D., Riethmuller, C., Kliche, K., Ludwig, T., Schillers, H., Schneider, S. W., and Oberleithner, H. (2006) 17beta-estradiol increases volume, apical surface and elasticity of human endothelium mediated by Na+/H+ exchange. *Cardiovasc. Res.* **69**, 916–924.

28. Hinterdorfer, P., Baumgartner, W., Gruber, H. J., Schilcher, K., and Schindler, H. (1996) Detection and localization of individual antibody-antigen recognition events by atomic force microscopy. *Proc. Natl. Acad. Sci. USA* **93**, 3477–3481.

29. Hinterdorfer, P., Schilcher, K., Baumgartner, W., Gruber, H. J., and Schindler, H. A. (1998) Mechanistic study of the dissociation of individual antibody-antigen pairs by atomic force microscopy. *Nanobiology* **4**, 177–188.

30. Oberleithner, H., Brinckmann, E., Schwab, A., and Krohne, G. (1994) Imaging nuclear pores of aldosterone sensitive kidney cells by atomic force microscopy. *Proc. Natl. Acad. Sci. USA* **91**, 9784–9788.

31. Oberleithner, H., Schneider, S., Lärmer, J., and Henderson, R. M. (1996) Viewing the renal epithelium with the atomic force microscope. *Kidney Blood Press. Res.* **19**, 142–147.

32. Oberleithner, H., Schneider, S. W., and Henderson, R. M. (1997) Structural activity of a cloned potassium channel (ROMK1) monitored with the atomic force microscope: the "molecular-sandwich" technique. *Proc. Natl. Acad. Sci. USA* **94**, 14144–14149.

33. Oberleithner, H., Ludwig, T., Riethmuller, C., Hillebrand, U., Albermann, L., Schafer, C., Shahin, V., and Schillers, H. (2004) Human endothelium: target for aldosterone. *Hypertension* **43**, 952–956.

34. Oberleithner, H. (2005) Aldosterone makes human endothelium stiff and vulnerable. *Kidney Int.* **67**, 1680–1682.

35. Oberleithner, H., Riethmuller, C., Ludwig, T., Shahin, V., Stock, C., Schwab, A., Hausberg, M., Kusche, K., and Schillers, H. (2006) Differential action of steroid hormones on human endothelium. *J. Cell Sci.* **119**, 1926–1932.

36. Perez-Terzic, C., Behfar, A., Mery, A., van Deursen, J. M., Terzic, A., and Puceat, M. (2003) Structural adaptation of the nuclear pore complex in stem cell-derived cardiomyocytes. *Circ. Res.* **92**, 444–452.

37. Shahin, V., Hafezi, W., Oberleithner, H., Ludwig, Y., Windoffer, B., Schillers, H., and Kuhn, J. E. (2006) The genome of HSV-1 translocates through the nuclear pore as a condensed rod-like structure. *J. Cell Sci.* **119**, 23–30.

38. Hansma, H. G., Kim, K. J., Laney, D. E., Garcia, R. A., Argaman, M., Allen, M. J., and Parsons, S. M. (1997) Properties of biomolecules measured from atomic force microscope images: a review. *J. Struct. Biol.* **119**, 99–108.

39. Rotsch, C. and Radmacher, M. (1997) Mapping local electrostatic forces with the atomic force microscope. Langmuir **13**, 2825–2832.

40. Maul, G. G. (1977) The nuclear and the cytoplasmic pore complex: structure, dynamics, distribution, and evolution. *Int. Rev. Cytol. Suppl.* **6**, 75–186.

41. Lenart, P., Rabut, G., Daigle, N., Hand, A. R., Terasaki, M., and Ellenberg, J. (2003) Nuclear envelope breakdown in starfish oocytes proceeds by partial NPC disassembly followed by a rapidly spreading fenestration of nuclear membranes. *J. Cell Biol.* **160**, 1055–1068.

42. Peters, R. (1986) Fluorescence microphotolysis to measure nucleocytoplasmic transport and intracellular mobility. *Biochim. Biophys. Acta.* **864**, 305–359.

43. Enss, K., Danker, T., Schlune, A., Buchholz, I., and Oberleithner, H. (2003) Passive transport of macromolecules through Xenopus laevis nuclear envelope. *J. Membr. Biol.* **196**, 147–155.

Chapter 11
Reconstitution of Nuclear Import in Permeabilized Cells

Aurelia Cassany and Larry Gerace

Keywords Nuclear protein import; Digitonin-permeabilized cells; Recombinant protein expression; Recombinant protein purification; Shuttling nuclear transport factors; Karyopherins; Importins; Ran

Abstract The trafficking of protein and RNA cargoes between the cytoplasm and the nucleus of eukaryotic cells, which is a major pathway involved in cell regulation, is mediated by nuclear transport sequences in the cargoes and by shuttling transport factors. The latter include receptors (karyopherins) that recognize the cargoes and carry them across the nuclear pore complex (NPC), and the small GTPase Ran, which modulates karyopherin–cargo binding. Nuclear import can be studied in vitro using digitonin-permeabilized cells, which are depleted of shuttling transport factors. Nuclear import can be reconstituted in the permeabilized cells with exogenous cytosol or with purified recombinant transport factors, and can be quantified by light microscopy of fluorescently labeled cargoes or by immunofluorescence staining. Here we describe procedures for in vitro nuclear import in permeabilized mammalian cells, and for the preparation of recombinant transport factors (importin α, importin β, importin 7, transportin, Ran, NTF2) and other reagents commonly used in the assay. This assay provides means to characterize the molecular mechanisms of nuclear import and to study the import requirements of specific cargoes.

1 Introduction

In eukaryotic cells, molecules traffic between the cytoplasm and the nucleus through nuclear pore complexes (NPCs), large proteinaceous channels that perforate the nuclear envelope. Whereas small molecules are able to diffuse passively through the NPC, proteins and nucleoprotein complexes larger than ~20–40 kDa require signals and receptors for their nuclear import and export (*1*). The facilitated transport of receptor–cargo complexes through NPC is saturable and energy dependent, and commonly involves nucleocytoplasmic shuttling receptors of the importin β/karyopherin β family and the small GTPase Ran (*2,3*).

Nuclear import and export is mediated by nuclear localization sequences (NLSs) and nuclear export sequences (NESs), respectively, in the cargoes. NLSs typically are short amino acid stretches enriched in basic residues, which bind to their cognate karyopherins in the cytoplasm either directly or via adaptors. The resulting import complexes then pass through the NPC by means of multiple interactions between the karyopherins and NPC proteins (nucleoporins) containing Phe-Gly (FG)-repeats. The directionality of nuclear import is regulated by Ran, which exists predominantly in its GTP-bound form in the nucleus, and in its GDP-bound form in the cytoplasm. The nucleocytoplasmic Ran-GTP gradient results from the nuclear localization of Ran guanine nucleotide exchange factor (RanGEF, RCC1), which generates Ran-GTP from Ran-GDP, and from the cytoplasmic localization of RanGTPase-activating protein (RanGAP), which promotes GTP hydrolysis on Ran. When import complexes reach the nucleoplasm, Ran-GTP binds directly to the karyopherins and thereby induces the release of cargoes *(4)*. The Ran-GTP-karyopherin complexes then are translocated back to the cytoplasm, where the Ran-GTP is hydrolyzed by RanGAP and the karyopherins are released for further import. Ran is transported back into the nucleus by NTF2, a receptor for Ran-GDP. Ran and its regulators also play a key role in the directionality of export, since Ran-GTP promotes the binding of NESs to export karyopherins, and the hydrolysis of Ran-GTP by RanGAP in the cytosol promotes the disassembly of export complexes.

An in vitro nuclear import assay involving digitonin-permeabilized cells, developed in our laboratory over 15 years ago, has been widely used to characterize the shuttling factors involved in nuclear import, the import requirements of specific cargoes, and the mechanisms of nuclear transport *(5)*. This method is simple and can be used with virtually any cultured mammalian cells (adherent or non-adherent). Methods for in vitro nuclear assembly and import using extracts from Xenopus eggs also have been developed and extensively used, especially to study nuclear assembly *(6)*. Digitonin is a nonionic detergent that binds cholesterol selectively, and at low concentrations permeabilizes the plasma membrane while leaving the nuclear envelope and endoplasmic reticulum (ER) intact and functional for transport *(7)*. During the process of permeabilization and subsequent washing, most of the shuttling nuclear transport factors are released from the cells, rendering them unable to support nuclear import of exogenous cargoes. However, efficient nuclear import can be restored if cells are reconstituted with exogenous cytosol to provide shuttling nuclear transport factors. Moreover, transport can be reconstituted in permeabilized cells with purified recombinant transport factors instead of cytosol, i.e., karyopherins, Ran, and NTF2. This assay can be used for analyzing endpoints of transport assays (i.e., the level of cargo accumulated in the nucleus at a fixed time) or for real-time imaging of cargo accumulation.

Fluorescently labeled cargoes, which are easily detected by light microscopy, are commonly used in nuclear import assays. In addition, cargo import can be monitored by immunofluorescent staining of cells after the import reaction. Import can be quantified by digital light microscopy to determine the nuclear-localized fluorescence (for adherent cells or for non-adherent cells attached to a slide after the transport reaction) or by flow cytometry to determine the cell-associated fluorescence (for non-adherent cells). The quantification of nuclear import by flow cytometry allows facile analysis of a large number of cells (~10–50,000) and is rapid, although the number of individual samples that can be analyzed is limited. Flow cytometry measures the total fluorescence of cells (both nuclear and cytoplasmic), and is not recommended for nuclear import assays reconstituted with recombinant factors or involving the import inhibitor wheat germ agglutinin (WGA), because nonspecific cytoplasmic binding of cargoes can occur in these cases. Quantification of import by digital light microscopy requires the analysis of at least 50–300 cells (typically three microscope fields with a ×40 objective), since there often is considerable cell-to-cell variability in nuclear transport levels (8). While quantitative analysis of import by light microscopy is laborious if done manually, this method can be automated for high throughput analysis using a computer-controlled microscope with a motorized *x-y-z* stage together with image analysis software.

Here we describe protocols for analyzing the nuclear import of proteins in digitonin-permeabilized adherent cells reconstituted with cytosol or recombinant transport factors. We provide methods for analyzing several of the karyopherin import pathways that have been characterized, but in principle the protocol can be extended to analysis of any shuttling receptor that can be expressed recombinantly. Moreover, nuclear export can be similarly analyzed in this system, using either exogenous or endogenous cargoes *(9, 10)*. We describe cargoes and recombinant factors that can be used to analyze the import pathways involving importin β (karyopherin β1), importin α/β, importin 7, the importin 7/β heterodimer, and transportin (karyopherin β2).

2 Materials

2.1 Cells

1. HeLa S3 cells, a clonal derivative of the parent HeLa cell line adapted for continuous growth in suspension (ATCC, Manassas, VA, USA; #CCL-2.2).
2. HeLa medium: Joklik's modified minimal essential medium (Sigma-Aldrich, St. Louis, MO, USA; #M0518-10L), 10% v/v fetal bovine serum, and 1% w/v penicillin and streptomycin.
3. NRK (normal rat kidney) cells (ATCC; #CRL-6509).
4. NRK medium: DMEM medium (4.5 g/L D-glucose, L-glutamine, 110 mg/L sodium pyruvate without sodium carbonate; Gibco-BRL, Invitrogen, Carlsbad, CA, USA; #12800-082), 10% fetal bovine serum, and 1% penicillin/streptomycin.

2.2 Preparation of Cytosol

1. Phosphate-buffered saline (PBS) (1 L): 10 mM sodium phosphate, pH 7.4, 140 mM NaCl.

2. Dithiothreitol (DTT): 1 M stock solution in distilled water, store at −20°C.

3. Wash buffer (500 mL): 10 mM N-2-hydroxyethylpiperazine-N'-2-ethanesulfonic acid (HEPES), pH 7.3, 110 mM potassium acetate, 2 mM magnesium acetate, and 2 mM DTT.

4. Hypotonic buffer (100 mL): 5 mM HEPES, pH 7.3, 10 mM potassium acetate, 2 mM magnesium acetate, 2 mM DTT, 1 mM phenylmethylsulfonyl fluoride (PMSF), and 1 μg/mL each of aprotinin, pepstatin, leupeptin.

5. Digitonin (Calbiochem, EMD Chemicals, Gibbstown, NJ, USA): 10% w/v stock solution in dimethyl sulfoxide (DMSO), store at −20°C.

6. Trypan Blue solution (0.4% w/v) (Sigma; #T-8154).

7. Transport buffer (TB): 20 mM HEPES, pH 7.3, 110 mM potassium acetate, 2 mM magnesium acetate, 1 mM ethylene glycol-bis (β-aminoethyl ether)-N,N,N',N;-tetraacetic acid (EGTA), 2 mM DTT, 1 mM PMSF, and 1 μg/mL each of aprotinin, pepstatin, and leupeptin.

8. Spectra/Por dialysis membranes (6,000–8,000-kDa cut-off) (Spectrum Laboratories, Rancho Dominguez, CA, USA; #132645) and clamps.

2.3 Preparation of FITC-labeled Transport Cargoes

1. NLS peptide: synthetic peptide containing the SV40 large T antigen wild-type nuclear localization signal (11, 12), preceded by a tri-glycine spacer and an N-terminal cysteine for coupling (CGGGPKKKRKVED).

2. 1% v/v acetic acid.

3. Sephadex G10 (Sigma; #G10120), chromatography column (10-mL bed volume, 1-cm diameter).

4. 50 mM HEPES, pH 7.0.

5. Ellmann's buffer: 0.1 M sodium phosphate buffer, pH 7.4, and 5 mM EDTA.

6. Ellmann's reagent: 1 mM dithiobisnitrobenzoic acid in methanol.

7. FITC isomer I (Molecular Probes, Invitrogen; #F1906) in dimethylformamide (DMF). Prepare fresh. DMF is toxic and all procedures using this chemical should involve the use of gloves and be done under a chemical fume hood.

8. Coupling buffer 1: 0.1 M sodium bicarbonate, pH 9.

9. PBS: 10 mM sodium phosphate, pH 7.4, 140 mM NaCl.

10. Sulfosuccidimidyl [N-maleimidomethyl]cyclohexane carboxylate (SMCC) (Pierce; Rockford, IL, USA; 22322): 20 mM in DMSO, prepared immediately before use.

11. PD-10 column (prepacked G25 column, 10-mL bed volume, Amersham Biosciences, Piscataway, NJ, USA; #17-0851-01).

12. Fatty acid-free bovine serum albumin (BSA) (Roche, Boehringer Mannheim, Indianapolis, IN, USA; #100062).

13. Histone H1: Solution of 20 mg of histone H1 in 1 mL of water (Upstate, St. Charles, MO, USA; #14-155).

14. Coupling buffer 2: 130 mM sodium carbonate, pH 7.0.

2.4   Plasmids, Expression, and Purification Systems

2.4.1   Cargoes: GST-M9 and GST-IBB

1. Isopropyl-β-Δ-thiogalactopyranoside (IPTG):1 M stock solution in water, aliquot, store at −20°C.
2. pGEX2T-IBB: this construct contains GST fused to residues 1–65 of human importin α1 (i.e., importin beta binding [IBB] domain) inserted into *Bam*HI-*Eco*RI sites of pGEX2T (Amersham Biosciences; #27-4801-01) *(13)*.
3. pGEX5X-M9: this construct contains GST fused to residues 263–306 of human hnRNP A1 (termed the "M9 domain") inserted into *Eco*RI-*Xho*I sites of pGEX5X (Amersham Biosciences) *(13)*.
4. GST lysis buffer: 5% v/v glycerol, 50 mM Tris-HCl, pH 8.0, 0.5 M NaCl, 1 mM magnesium acetate, 2 mM DTT, 1 mg/mL lysozyme, 10 µg/mL DNase I, 0.1% v/v Triton X-100, and 1 µg/mL each of leupeptin, pepstatin, and aprotinin.
5. Glutathione Sepharose 4B (Amersham Biosciences; #17-0756-01).
6. GST washing buffer: 50 mM Tris-HCl, pH 8.0, 0.5 M NaCl, and 1 µg/mL each of leupeptin, pepstatin, and aprotinin.
7. Reduced glutathione (Sigma; #6529).
8. GST elution buffer: 50 mM Tris-HCl, pH 8.0, and 15 mM reduced glutathione.

2.4.2   Importin α1

1. pRSETB-importin α1: this construct contains a 6×-His tag at its N-terminus and an Xpress epitope for detection (pRSETB vector; Invitrogen; #V351-20) and human importin α1 (hSRP1 α, karyopherin α2) sequence inserted into *Bam*H1-*Xho*1 sites *(14)*.
2. PMSF: 17.4 mg/mL PMSF in isopropanol (100 mM). Divide the solution in aliquots and store at −20°C. *PMSF is toxic and all procedures using this chemical should involve the use of gloves and be done under a chemical fume hood.*
3. Importin α lysis buffer: 5% v/v glycerol, 50 mM Tris-HCl, pH 8.0, 0.5 M NaCl, 1 mM β-mercaptoethanol, 1 mM PMSF, and 1 µg/mL each of leupeptin, pepstatin, and aprotinin.
4. Talon metal affinity resin (Clontech, Mountain View, CA, USA; #635503).

2.4.3   Importin β

1. pTYB4-importin β: this construct contains intein/chitin-binding domain at the C-terminus and human importin β sequence inserted into the *Nco*I-*Not*I sites *(13)*.
2. ER2566 cells (New England Biolabs, Ipswich, MA, USA).
3. Tris (2-carboxyethyl)-phosphine hydrochloride (TCEP) (Pierce, #20490): 0.2 M stock solution in water, store at −20°C.

4. Importin β lysis buffer: 50 mM Tris-HCl, pH 8.0, 0.5 M NaCl, 2 mM magnesium chloride, 10 mM CHAPS, 0.2 mM TCEP (add fresh), and 1 μg/mL each of leupeptin, pepstatin, and aprotinin.
5. Chitin beads (New England Biolabs; #S6651L).
6. Importin β elution buffer: 50 mM Tris-HCl, pH 8.0, 0.5 M NaCl, 2 mM MgCl₂, 1 mM CHAPS, 30 mM DTT, and 1 μg/mL each of leupeptin, pepstatin, and aprotinin.

2.4.4   Importin 7

1. pQE9-importin 7 (RanBP7): this construct contains a 6×-His tag at the N-terminus (pQ9 vector, Qiagen, Valencia, CA, USA; #32915) and the human importin 7 sequence *(15)*.
2. TG-1 (Zymo Research, Orange, CA, USA; #T3017) and M15 (Qiagen; #34210) cells.
3. Importin 7 phosphate buffer: 50 mM sodium phosphate, pH 8.0, 300 mM NaCl, and 1 μg/mL each of leupeptin, pepstatin, and aprotinin.

2.4.5   Transportin

1. pET28c-transportin: this construct contains a 6×-His-tag and a T7 tag at its N-terminus, with a thrombin cleavage site between the two tags (pET28c vector, Novagen, EMD Chemicals, Gibbstown, NJ, USA; #69866) and the human transportin sequence inserted into the *Bam*H1-*Sal*I sites *(16)*.
2. BL21 (DE3) (Stratagene, Agilent, Santa Clara, CA, USA; #200131).
3. Transportin lysis buffer: 50 mM Tris-HCl, pH 8.0, 0.5 M NaCl, 1 mM magnesium acetate, 5% v/v glycerol, 2 mM CHAPS, 1 mg/mL lysozyme, 10 μg/mL DNase I, and 1 μg/mL each of leupeptin, pepstatin, and aprotinin.
4. Transportin washing buffer: 50 mM Tris-HCl, pH 8.0, 0.5 M NaCl, 1 mM magnesium acetate, 1 mM CHAPS, and 1 μg/mL each of leupeptin, pepstatin, and aprotinin.
5. Transportin elution buffer: 50 mM Tris-HCl, pH 6.8, 0.5 M NaCl, 1 mM magnesium acetate, and 1 μg/mL each of leupeptin, pepstatin, and aprotinin.

2.4.6   Ran

1. Ran vector: Human Ran is in modified pET11d (Novagen; #69439-3). The pET11d was digested by *Nco1* and *Bam*H1 such that there is no purification tag. The *Nco*I site was introduced at the start codon of Ran by PCR and the fragment was inserted into *Nco*I-*Bam*HI sites of the modified pET11d (17).
2. Ran lysis buffer: 50 mM Tris pH 8.0, 75 mM NaCl, 1 mM MgCl₂, 1 mM DTT, 1 mM PMSF, 1 mg/mL lysozyme, 10 μg/mL Dnase I, and 1 μg/mL each of leupeptin, pepstatin, aprotinin.

3. DE52: Pre-swollen microgranular DEAE cellulose (Whatman, Maidstone, Kent, UK).
4. Transport buffer (TB): 20 mM HEPES pH 7.3, 110 mM potassium acetate, 2 mM magnesium acetate, 1 mM EGTA, 2 mM DTT, 1 mM PMSF, and 1 µg/mL each of aprotinin, pepstatin, leupeptin.
5. Hiload 26/60 Superdex 200 prep grade (Amersham Biosciences).
6. SP buffer: 50 mM Tris pH 6.8, 10 mM KCl, 1 mM MgCl$_2$, 1 mM DTT, and 1 µg/mL each of aprotinin, pepstatin, and leupeptin.
7. PD-10 column (prepacked G25 column, 10-mL bed volume, Amersham Biosciences).
8. SP-Sepharose^T Fast Flow (Amersham Biosciences).

2.4.7 NTF2

1. BL21 (DE3) pLysS (Stratagene; #200132).
2. pET23b-NTF2: This plasmid contains human NTF2. The pET23b vector (Novagen; #69746-3) was modified to include a stop codon at the HincII site, thereby eliminating translation of the histidine tag (18).
3. TNE buffer: 50 mM Tris-HCl pH 8.0, 100 mM NaCl, and 10 mM EDTA.
4. S200 HR: Superdex 200 10/300 GL (Amersham Biosciences).
5. Q Sepharose Fast Flow (Amersham Biosciences).
6. Q buffer: 50 mM HEPES pH 7.4, 1 mM MgCl$_2$, 1 mM DTT, and 1 µg/mL each of aprotinin, pepstatin, and leupeptin.

2.5 Nuclear Import Assay

1. Ten-well microscope slides (ICN Biomedicals, Irvine, CA, USA; #6041805).
2. Hydrophobic slide marker for staining procedure (Research Products International, Mt. Prospect, IL, USA; #195505).
3. Transfer pipet (Fisher Scientific, Pittsburgh, PA, USA; #13-711-5A).
4. Square cell culture plate (Fisher Scientific; #0875711A) that can accommodate three ten-well slides.
5. Two metal blocks, one prechilled on ice and the other prewarmed to 30°C.
6. Humidified chamber at 30°C. We use a plastic box covered by aluminum foil, which contains the prewarmed metal block on moistened absorbent paper. Temperature is maintained at 30°C using an incubator or water bath.
7. Cytosol from HeLa S3 cells: 10 mg/mL cytosol in TB (see Section 3.2).
8. Fluorescent transport cargoes (see Section 3.3) or unlabeled transport cargoes (see Section 3.4). The cargoes are in TB and stored as aliquots at −80°C.
9. Transport buffer (TB): 20 mM HEPES pH 7.4, 110 mM potassium acetate, 2 mM magnesium acetate, 1 mM EGTA, 2 mM DTT, and 1 µg/mL of each leupeptin,

pepstatin, and aprotinin. The TB solution is prepared freshly using a 20-fold concentrated stock solution containing HEPES, potassium acetate, and magnesium acetate (stored at 4°C), a solution stock of $0.2\,M$ EGTA, pH 7.4 (stored at 4°C), and $1\,M$ stock solutions of DTT and protease inhibitors (stored at −20°C).

10. Digitonin (Calbiochem; #300410): Make a 10% w/v stock in DMSO and store at −20°C.

11. Adenosine triphosphate (ATP) (Sigma): $200\,\text{m}M$ stock in $20\,\text{m}M$ HEPES, pH 7.4, and $100\,\text{m}M$ magnesium acetate. Dissolve ATP in $100\,\text{m}M$ magnesium acetate, adjust the pH to ~7.4 with $10\,M$ NaOH, and add HEPES to $20\,\text{m}M$ from a $1\,M$ HEPES, pH 7.4 stock. Aliquot and store at −20°C.

12. Creatine phosphate (CP) (Calbiochem): $160\,\text{mg/mL}$ stock solution in water. Aliquot and store at −20°C.

13. Creatine phosphate kinase (CPK) (Calbiochem): $2,000\,\text{U/mL}$ stock solution in $20\,\text{m}M$ HEPES, pH 7.4 and 50% glycerol. Aliquot and store at −20°C.

14. ATP regenerating system: freshly prepared from the stock solutions of ATP, CP, and CPK, mixed at a ratio of 1:1:1.

15. Guanosine triphosphate (GTP) (Calbiochem): $5\,\text{m}M$ stock solution in modified TB (lacking EGTA, DTT, and protease inhibitors), filter with a 0.22-μm filter, aliquot and store at −20°C.

16. Wheat germ agglutinin (WGA) (Sigma): Make a $2\,\text{mg/mL}$ stock in modified TB (lacking EGTA and DTT), aliquot and store at −20°C.

17. Formaldehyde, 37% (Fisher Scientific).

2.6 Immunostaining

1. Gelatin type B from bovine skin (Sigma): 2% stock solution in PBS, aliquot and store at −20°C.

2. Purified goat anti-GST antibody (Amersham).

3. FITC-labeled mouse anti-goat antibody (Pierce).

4. Solution of Topro-3 (Molecular Probes) or DAPI (Sigma). Aliquot and store at −20°C. Stock solution of DAPI at $1\,\text{mg/mL}$ in PBS, aliquot, and store at −20°C.

5. SlowFade Antifade kit (Molecular Probes).

6. Coverglasses ($22 \times 60\,\text{mm}$; VWR Scientific).

3   Methods

3.1 Cells

1. For preparation of HeLa cytosol, grow $8\,\text{L}$ of HeLa S3 cells at 37°C in spinner flasks, always maintaining a density of $4\text{–}10 \times 10^5$ cells/mL. During expansion of cultures, the cell density is checked daily using a hematocytometer and cultures

are diluted to a density of 4×10^5 cells/mL. Starting with 100 mL of culture at 4×10^5 cells/mL, growth of 8 L of cells should take ~6–8 days. When harvested at ~6–8×10^5 cells/mL, 8 L of culture yields ~20 mL of packed cells.

2. NRK cells are used for the nuclear import assay. The cells are grown in NRK medium at 37°C with 5% CO_2 in culture flasks. NRK cells are plated on 10-well slides the day before the experiment at a density yielding 12,000 cells/well. Before seeding, the slides are rinsed briefly with a solution of 70% ethanol, dried using Kimwipes, sterilized in a Bunsen burner, and put in the square culture plate using sterile tweezers. The cells should be at ~80% confluence the next day for nuclear import assays.

3.2   Preparation of HeLa Cell Cytosol

1. Harvest the cells from 8 L of HeLa S3 culture by centrifugation at $1,000 \times g$ for 15 min at 4°C in 500-mL conical bottles using a Beckman JS5.2 rotor.
2. Wash the cells twice with 100 mL of ice-cold PBS per 2 L of original culture, pool the bottles, and centrifuge again.
3. Resuspend the pellet in ice-cold wash buffer so the final volume reaches 50 mL, transfer the cell suspension to a 50-mL Falcon tube, and centrifuge as before.
4. Resuspend the cell pellet in an equal volume of cold hypotonic lysis buffer and let the cells swell on ice for 10 min (total volume ~40 mL).
5. Lyse the cells on ice by adding digitonin to the point where ~90–95% of the cells are permeable to Trypan Blue. To accomplish this, start by adding 120 µL of 10% digitonin solution (25 µL per 10^9 cells), then add smaller (10 µL) aliquots in a stepwise manner until adequate permeabilization is obtained. After the addition of each aliquot, wait 2–5 min and examine a sample of cells in the microscope to assess Trypan Blue permeability. Stop adding digitonin when 90–95% permeabilization is reached. Avoid adding excess digitonin, as this could interfere with the import assay (see below). Approximately 160 µL of 10% digitonin solution are needed for a 40-mL cell suspension to get suitable permeabilization.
6. Centrifuge the cells at $1,500 \times g$ for 15 min at 4°C to remove permeabilized cells and debris.
7. Centrifuge the supernatant from the previous step at $15,000 \times g$ for 20 min at 4°C in a Beckman JA-20 rotor, collect the supernatant, and centrifuge the latter at $100,000 \times g$ for 1 h at 4°C using a Beckman 70Ti rotor.
8. Dialyze the final supernatant (~20 mL) at 4°C in preboiled dialysis tubing (6,000–8,000 kDa cut-off) in 2 L of transport buffer with three changes of buffer.
9. After dialysis, the concentration of protein in the cytosol is determined by the Bradford protein assay, and adjusted to 10 mg/mL by dilution with TB or by concentration with a Centricon concentrator (Amersham). Aliquot, freeze in liquid nitrogen, and store at −80°C.

3.3   Preparation of FITC-labeled Transport Cargoes

The FITC-BSA-NLS is prepared in two steps. First, BSA is conjugated with FITC
(**Section 3.3.1.2**), and then a peptide containing the wild-type SV40 NLS is coupled
to the FITC-BSA via the N-terminal cysteine on the peptide (**Section 3.3.1.3**).
Since the sulfhydryl group in the peptide tends to be oxidized, it is necessary to
reduce the peptide with DTT prior to coupling (**Section 3.3.1.1**). The FITC-histone
H1 is prepared in a single step, following the same procedure as conjugation of
BSA with FITC, although the coupling buffer is modified to maintain the solubility
of histone H1 (**Section 3.3.2**).

3.3.1   FITC-BSA-NLS

3.3.1.1   Reduction of Synthetic Peptide

1. Wash a Sephadex G10 column with 25 mL of 1% v/v acetic acid.
2. In the meantime, dissolve 10 mg of NLS peptide in 1 mL of 50 mM Hepes, pH
   7.4. Add 10 mg of DTT, vortex, and incubate for 1 h at room temperature.
3. Separate the peptide from free DTT by gel-filtration on the acetic acid-equili-
   brated G10 column. Load the peptide and elute the column with 1% acetic acid,
   collecting 0.5-mL fractions; 30 fractions should be enough to elute the NLS
   peptide and the DTT.
4. Analyze the fractions using Ellman's reagent. For this, add 10 µL of each fraction to
   900 µL of Ellman's buffer and 100 µL of Ellman's reagent in an Eppendorf tube and
   vortex. The sulfhydryl residues yield a bright yellow color. The first yellow peak
   contains the peptide and the second (brighter) yellow peak contains the DTT.
5. Pool the fractions containing the peptide, aliquot into eight preweighed
   Eppendorf tubes, cover the tubes with perforated Parafilm, and dry the peptide
   using a Speed Vac. Weigh the tubes again (there should be ~1 mg peptide per
   tube) and store at −80°C.

3.3.1.2   Preparation of FITC-BSA

1. Dissolve 20 mg of fatty acid-free BSA in 1 mL of Coupling Buffer 1.
2. Add 200 µL of freshly prepared FITC solution (10 mg/mL FITC isomer I in
   DMF) to BSA and incubate at room temperature for 1 h in the dark.
3. In the meantime, wash a PD-10 column with 25 mL of PBS.
4. Adjust the volume of the FITC-BSA conjugate to 2.5 mL with PBS and load on
   the column, let the sample run into the column, and elute with 3.5 mL of PBS.
   Collect the peak of FITC-BSA that elutes prior to the peak of uncoupled FITC.
5. Dilute the peak of eluted FITC-BSA twofold with PBS to a final concentration
   of ~2.5 mg/mL.
6. Freeze 1-mL aliquots in liquid nitrogen and store at −80°C.

3.3.1.3   Conjugation of the NLS Peptide to FITC-BSA

1. Thaw a 1-mL aliquot of FITC-BSA (**Section 3.3.1.2**), add 50 µL of freshly prepared 20 mM SMCC and incubate for 30 min at room temperature in the dark.
2. In the meantime, wash a PD-10 column with 25 mL of PBS.
3. To remove the unconjugated crosslinker, apply the mixture to the PD-10 column
   and let the sample run into the column.
4. Apply 1.5 mL of PBS to the column and allow it to enter.
5. Elute the column with a further 3.5 mL of PBS. Collect the bright yellow peak
   in a separate tube (in ~2 mL).
6. Dissolve 1 mg of lyophilized NLS peptide (from **Section 3.3.1.1**) with the column eluate and incubate for 2 h at room temperature in the dark.
7. Meanwhile, equilibrate the PD-10 column with 25 mL of TB.
8. To remove uncoupled peptide from the FITC-BSA-NLS, apply the mixture to
   the PD-10 column, let the sample run in and elute with TB. Collect only the
   bright yellow peak (~2 mL, now at ~1 mg/mL). Freeze 20-µL aliquots in liquid
   nitrogen and store at −80°C.

3.3.2   FITC-Histone H1

1. Dilute 3 mg of histone H1 in 1 mL of Coupling Buffer 2 (130 mM sodium
   carbonate pH 7.0) and mix with 30 µL of FITC dissolved in DMF (2 mg FITC
   in 50 µL DMF). The concentration of DMF should be <2% to avoid precipitation of histone H1. Incubate for 1 h at room temperature in the dark.
2. In the meantime, wash the PD-10 column with 25 mL PBS.
3. To remove free FITC, load the mixture on the PD-10 column, let it run in, and
   elute with 5 mL PBS. Collect the first bright yellow peak in a single tube
   (~4 mL).
4. Freeze in 20-µL aliquots in liquid nitrogen and store at −80°C.

3.4   Expression and Purification of Cargoes: GST-M9 and GST-IBB

1. Transform samples of BL21 (DE3) with pGEX2T-IBB and pGEX5X-M9
   plasmids.
2. Grow 2 L of transformed cells in LB medium with ampicillin (50 µg/mL) at
   37°C until the cultures reach an OD$_{600\,nm}$ of 0.6.
3. Induce protein expression with 50 µM IPTG for 2 h at 37°C.
4. Harvest the cells by centrifugation at 1,500×g for 15 min at 4°C. The pellets can
   be frozen and stored at −80°C before recombinant protein purification (below).
5. Resuspend the pellets in 30 mL of GST lysis buffer per liter of original culture.
6. Freeze the suspension in liquid nitrogen and thaw in an ice-water bath.

7. Sonicate on ice, three times for 30 sec with 30-sec intervals.
8. Clear the lysate by centrifugation at 100,000×g for 20 min at 4°C in a Beckman Ti 45 rotor and collect the supernatant (cleared lysate).
9. Incubate the cleared lysate with glutathione–Sepharose beads (1 mL of glutathione beads slurry in PBS (1:1) per liter of original culture) on a rotating wheel for 2 h at 4°C.
10. Centrifuge at 500×g for 10 min at 4°C. Discard the supernatant.
11. Wash the beads three times with 50 mL of GST washing buffer.
12. Elute the recombinant proteins with 2 mL of GST elution buffer.
13. Check the purity of the GST fusion proteins on an SDS gel. GST-M9 and GST-IBB migrate at ~29 kDa and ~32 kDa, respectively. If the proteins are not pure enough you can add another step of purification by chromatography on a S200 Superdex column.
14. Dialyze into TB. Determine the concentration of protein using the Bradford protein assay. Freeze in 20-μL aliquots in liquid nitrogen and store at −80°C.

3.5 Expression and Purification of Transport Factors

3.5.1 Importin α

1. Transform BL21 (DE3) with the plasmid pRSETB-importin α1.
2. Grow 4 L of transformed bacteria in LB medium with ampicillin (50 μg/mL) at 37°C until cultures reach an $OD_{600\,nm}$ of 0.6.
3. Induce protein expression with 200 μM IPTG for 3 h at 37°C.
4. Collect the bacteria by centrifugation at 1,500×g for 15 min at 4°C. The pellets can be frozen and stored at −80°C before recombinant protein purification (below).
5. Resuspend the pellets by pipetting or vortexing in 50 mL of importin α lysis buffer per liter of original culture.
6. Add lysozyme to 0.5 mg/mL and keep on ice for 20 min.
7. Add Triton X-100 to 0.2 % v/v final concentration and add fresh PMSF to 1 mM.
8. Sonicate on ice, three times for 30 sec with 30-sec intervals.
9. Clear the lysate by centrifugation at 100,000×g for 20 min at 4°C in a Beckman Ti 45 rotor and collect the supernatant (cleared lysate).
10. Add imidazole, pH 7.0 from a 1 M stock solution to the cleared lysate to a final concentration of 10 mM, then add to the Talon beads previously equilibrated with importin α lysis buffer (0.5 mL resin per liter of original culture).
11. Incubate for 1–2 h on a rotating wheel at 4°C.
12. Load the beads into a column.
13. Wash the column with importin α lysis buffer containing 10 mM imidazole until no more protein is released. Check the protein concentration of the eluate using the Bradford protein assay. You will need ~20 mL to wash the column fully.

14. Elute the importin α with five column volumes of importin α lysis buffer containing 100 mM imidazole.
15. Dialyze into TB containing an additional 390 mM potassium acetate (500 mM total) and 5% v/v glycerol (*see* **Note 1**).
16. Check the purity of the importin α on an SDS gel. Importin α migrates at ~60 kDa. Determine the final concentration using the Bradford protein assay (*see* **Note 2**), freeze in 20-μL aliquots in liquid nitrogen and store at −80°C.

3.5.2 Importin β

1. Transform ER2566 cells with the pTYB4-importin β1 construct. Grow the cells in LB medium with ampicillin (100 μg/mL).
2. Grow 8 L of transformed bacteria in LB medium with ampicillin (100 μg/mL) at 37°C until cultures reach an $OD_{600\,nm}$ of 0.4. Cool cultures to room temperature.
3. Induce with 1 mM IPTG for 4 h at 20–25°C.
4. Harvest the cells by centrifugation at 3,000×g for 15 min at 4°C.
5. Wash the pellets with 100 mL of PBS containing 0.2 mM TCEP. The pellets can be frozen and stored at −80°C before recombinant protein purification (below).
6. Resuspend the pellets by pipetting or vortexing in 40 mL of importin β lysis buffer per liter of original culture (*See* **Note 3**).
7. Sonicate on ice, three times for 30 sec with 30-sec intervals.
8. Clear the lysate by centrifugation at 70,000×g for 20 min at 4°C in a Beckman Ti 45 rotor and collect the supernatant (cleared lysate).
9. In the meantime, wash 9 mL of packed chitin beads. Use 1 mL of beads per liter of original culture. Wash two times with 40 mL of importin β lysis buffer. Spin down the beads at 500×g for 5 min at 4°C.
10. Add the cleared lysate to the beads and incubate for 2–3 h on a rotating wheel at 4°C.
11. Pour the cleared lysate/bead mixture into a column at 4°C and allow to pack by gravity flow.
12. Wash the resin with 10 column volumes of importin β lysis buffer.
13. Pass three column volumes of importin β elution buffer over the column. Stop the column flow and incubate overnight at 4°C.
14. Elute importin β from the column with the addition of three columns volume of importin β elution buffer.
15. Add 50 μL of chitin beads (equilibrated in importin β elution buffer) to the total eluate and incubate for 15 min on a rotating wheel to remove the excess tag. Remove the beads by centrifugation at 500×g for 5 min at 4°C.
16. Add 50 μL of Q Sepharose Fast Flow beads (equilibrated in importin β elution buffer) to the supernatant to remove nucleic acids, which selectively bind to Q Sepharose Fast Flow beads in importin β elution buffer. Incubate for 15 min on a rotating wheel and remove the beads by centrifugation at 500×g for 5 min at 4°C.

17. Dialyze the supernatant against TB with 0.01 mM CHAPS (*see* **Note 4**).
18. Check the purity of the importin β on an SDS gel. Importin β migrates at ~97 kDa. Determine the protein concentration using the Bradford protein assay. Freeze in 50-μL aliquots in liquid nitrogen and store at −80°C.

3.5.3 Importin 7

1. For the expression of importin 7, we use TG-1 or M15 *Escherichia coli* cells. Both contain the Lac repressor for tightly induced expression, since low expression of importin 7 is toxic. After transformation, grow TG-1 cells in LB medium with ampicillin (50 μg/mL) or M15 cells in LB with kanamycin (50 μg/mL) and ampicillin (50 μg/mL). Expand cultures to 8 L at 37°C until they reach an OD$_{600\,nm}$ of 0.6.
2. Induce with 500 μM IPTG for 5 h at 20°C. Bacteria are chilled in the cold room before induction to decrease the temperature from 37°C to 20°C.
3. Harvest the cells at 1,500×g for 15 min at 4°C.
4. Wash the cells with importin 7 phosphate buffer.
5. Resuspend the pellet by pipetting or vortexing in 30 mL of importin 7 phosphate buffer per liter of original culture.
6. Sonicate on ice, three times for 30 sec with 30-sec intervals.
7. Clear the lysate by centrifugation at 17,000×g for 20 min at 4°C in a Beckman Ti 45 rotor and collect the supernatant (cleared lysate)
8. Incubate the cleared lysate with Talon beads (equilibrated with importin 7 phosphate buffer) for 2 h at 4°C on a rotating wheel. Use 0.5 mL of beads per liter of original culture.
9. Pour the supernatant and bead mixture into a column.
10. Wash the beads with five column volumes of importin 7 phosphate buffer.
11. Elute with three column volumes of importin 7 phosphate buffer containing <20 mM imidazole (*see* **Note 5**). Collect 0.5-mL fractions.
12. Check the purity of the importin 7-containing fractions on an SDS gel. Importin 7 migrates at ~120 kDa on an SDS gel. The eluate contains most of the importin 7 but is substantially contaminated with proteins with a M_r lower than 50 kDa. To eliminate the latter, pool fractions containing importin 7 and use an Amicon Ultra centrifugal filter unit (Millipore) with a cut-off of 50 kDa to remove the small proteins and to exchange the phosphate buffer with TB. If a further purification step is needed, chromatography on an S200 column can be employed. Measure the protein concentration using the Bradford assay. Aliquot, freeze in liquid nitrogen, and store at −80°C.

3.5.4 Transportin

1. Transportin is expressed using BL21 (DE3) cells. Grow the cells in LB medium with kanamycin (50 μg/mL).

2. Grow 2 L of transformed bacteria in LB at 37°C until cultures reach an $OD_{600\,nm}$ of 0.7.
3. Cool cultures to room temperature.
4. Induce with 1 mM IPTG at 17°C for 4 h. Bacteria are chilled in the cold room before induction to decrease the temperature to 17°C.
5. Collect the bacteria by centrifugation at 3,000×g for 15 min at 4°C. The pellet can be frozen in liquid nitrogen and stored at −80°C before recombinant protein purification (below).
6. Resuspend the pellet by pipetting or vortexing in transportin lysis buffer. Add 50 mL of transportin lysis buffer per liter of culture.
7. Freeze at −80°C for 1 h and thaw the lysate in an ice-water bath. This step of freeze–thaw increases the lysis yield.
8. Sonicate on ice, three times for 30 sec with 30-sec intervals.
9. Clear the lysate by centrifugation at 17,000×g for 20 min at 4°C in a Beckman Ti 45 rotor and collect the supernatant (cleared lysate).
10. Incubate the cleared lysate with Talon beads (preequilibrated with transportin washing buffer) for 2 h at 4°C on a rotating wheel. Add 0.75 mL of beads per liter of original culture.
11. Pour the supernatant and beads into a column.
12. Wash the beads with five column volumes of transportin washing buffer and 10 mM imidazole.
13. Elute with three column volumes of transportin elution buffer containing 100 mM imidazole. Collect 1-mL fractions in Eppendorf tubes.
14. Check the transportin-containing fractions and the purity of transportin on a SDS gel. The transportin migrates at ~101 kDa. Pool the transportin-containing fractions and dialyze into TB containing 0.05 mM CHAPS. The small amount of detergent from transportin does not interfere with nuclear import. Freeze in 20-μL aliquots in liquid nitrogen and store at −80°C. If a higher degree of purity is needed, transportin can be chromatographed on a Superdex S200 column.

3.5.5 Ran

1. BL21 (DE3) cells are used for the expression of Ran. After transformation with the Ran expression vector, the bacteria are grown in LB medium with ampicillin (50 μg/mL). Typically 2 L of cells are used for protein purification.
2. Grow at 37°C until cultures reach an $OD_{600\,nm}$ of 0.6.
3. Induce the rest of the culture with 0.5 mM IPTG at 37°C for 2 h (*see* **Note 6**).
4. Harvest the cells by centrifugation at 3,000×g for 15 min at 4°C. The pellets can be frozen in liquid nitrogen and stored at −80°C before recombinant protein purification (below).
5. Thaw the pellet from 1 L of culture with 50 mL of Ran lysis buffer in an ice-water bath.

6. Resuspend the pellet by vortexing. Incubate for 30 min on ice during cell lysis and vortex again.

7. Clear the lysate by centrifugation at 70,000×g for 20 min at 4°C in a Beckman Ti 45 rotor. Collect the supernatant (cleared lysate).

8. Keep an aliquot of the cleared lysate.

9. For the first purification step, pass the cleared lysate over a DEAE-cellulose column. Prepare the column during the lysis or the day before (store at 4°C), containing 10 mL bed volume per liter of original culture. The column is prewashed with Ran lysis buffer.

10. Collect and save the flow-through and wash with 10 mL of Ran lysis buffer.

11. Run a 12.5% SDS gel to compare 10 μL of the cleared lysate and 10 μL of the flow-through from the DEAE column. Ran should migrate as a double band close to a 25-kDa marker and should be present in large amounts in the cleared lysate and the flow-through (*see* **Note 7**). Most contaminants adsorb to the DEAE column, so Ran in the flow-through should be significantly more enriched than in the cleared lysate.

12. Take the flow-through and while stirring at 4°C, slowly add solid ammonium sulfate to 65% saturation to precipitate the protein (40.4 g/100 mL). Let stir for at least 1 h after dissolving (or overnight, for convenience).

13. Centrifuge at 70,000×g for 20 min at 4°C and resuspend the pellet in 6 mL of TB (filtered on a 0.22-μm filter) per liter of original culture.

14. Clarify the resuspended pellet by centrifuging at 100,000×g for 10 min at 4°C.

15. Pass the supernatant through a 0.22-μm syringe filter (Nalgene; #190-2520) to remove any precipitate.

16. Load the filtered supernatant on a S200 gel filtration column (26/60). The column and pumps are first washed with TB. Use a flow rate of 2 mL/min and collect 3-mL fractions.

17. Check the content and purity of fractions that have absorbance at 280 nm on an SDS gel. Pool the Ran-containing fractions that lack most contaminants.

18. Determine the concentration of Ran using the Bradford protein assay. You should get around 15 mg Ran per liter of original culture. Freeze in 50-μL aliquots in liquid nitrogen and store at −80°C.

19. If the S200 fractions are not sufficiently pure (*see* **Note 8**), add another purification step involving an SP-Sepharose cation-exchange column. Exchange the buffer in the Ran pool to SP buffer using a PD-10 column. Load 5 mL onto the column pre-equilibrated with buffer SP, wash the column until the eluate has a baseline absorbance at 280 nm, and then load the sample. Elute with a 50–500 mM gradient of KCl in SP buffer. Check the content of the fractions on an SDS gel, pool the fractions containing pure Ran, and precipitate with ammonium sulfate as described previously (*see* **Note 9**). Resuspend in TB to ~2 mg/mL, determine the concentration of Ran using Bradford protein assay, aliquot, and freeze in liquid nitrogen and store at −80°C.

3.5.6 NTF2

1. The plasmid coding for NTF2 is transformed into BL21 (DE3) pLysS cells. Transformants are grown in LB containing ampicillin (50 μg/mL) and chloramphenicol (10 μg/mL) overnight at 37°C.
2. Grow cultures at 37°C to an OD_{600nm} of 0.6. Typically, we prepare 4 L of culture.
3. Induce with 0.4 mM IPTG for 2 h at 37°C (*see* **Note 10**).
4. Collect the bacteria by centrifugation at 6,000×g for 15 min at 4°C. The pellets can be frozen in liquid nitrogen and stored at −80°C before recombinant protein purification (below).
5. The cell pellets from each 1 L of culture are resuspended in 50 mL of TNE containing 0.2 % v/v Triton X-100, 1 mM PMSF, 5 mM DTT, and 1 μg/mL each of aprotinin, leupeptin, and pepstatin (*see* **Note 11**).
6. Freeze by putting the resuspended cell suspensions at −20°C or in liquid nitrogen, and thaw in an ice-water bath.
7. Sonicate on ice, three times for 30 sec with 30-sec intervals.
8. Clear the lysate by centrifugation of the sample, which will be viscous, at 70,000×g for 20 min at 4°C in a Beckman Ti 45 rotor.
9. Precipitate the proteins in the cleared lysate by slowly adding solid ammonium sulfate, while stirring, to 50% saturation (29.5 g/100 mL) and stir for 1 h at 4°C after total dissolution.
10. Collect the precipitate by centrifugation at 70,000×g for 15 min at 4°C in a Beckman Ti 45 rotor, and resuspend the pellet in TB.
11. Clarify the sample by centrifugation at 140,000×g for 30 min in a Beckman TLA 100.3 rotor.
12. Chromatograph the clarified sample on a S200 HR column that has been equilibrated with two column volumes of TB. The composition of fractions is checked by analysis on an SDS gel and the fractions containing the peak of NTF2 are pooled. NTF2 migrates at ~28 kDa.
13. NTF2 is further purified using Q Sepharose anion-exchange column. The column is pre-equilibrated with two columns volumes of Q buffer. Dilute the pooled fractions from the S200 HR column 200-fold in buffer Q and incubate batch-wise with the resin on a rotating wheel for 2 h at 4°C.
14. Pour the sample into a column and wash with one column volume of Q buffer.
15. Elute NTF2 with Q buffer containing 150 mM KCl. Collect 1-mL fractions and check their composition on an SDS gel. Pool the fractions containing NTF2 and dialyze against TB. Determine the protein concentration using the Bradford assay. You should get ~2 mg/mL (~8 mg per liter of original culture). Freeze in 10-μL aliquots in liquid nitrogen and store at −80°C.

3.6 Nuclear Import Assay

We describe implementation of the nuclear import assay with permeabilized NRK cells, but the assay can easily be adapted to other adherent cell lines. Whatever cells are used, in our experience the most homogeneous and reproducible nuclear import is obtained when cells are at 50–80% confluence (*see* **Note 12**). The assay can be carried out on cover slips, but 10-well slides are much easier to manipulate. Furthermore, their use provides the capacity for simple implementation of 10 or more nuclear import assays at the same time (*see* **Note 13**). The transport buffer and all reaction mixtures are prepared before the beginning of the assay. The reaction mixture containing fluorescently labeled cargo should be kept in the dark.

3.6.1 Import Reaction Mixtures

1. The simplest nuclear import assay involves the incubation of permeabilized cells with a reaction mix containing cargo (e.g., FITC-BSA-NLS), cytosol, and "energy," i.e., an ATP-regenerating system and GTP (+Cytosol +E). Several negative controls should be performed to demonstrate that accumulation of cargo in the nucleus is physiologically relevant, i.e., that it depends on energy, exogenous nuclear transport factors (which are provided by cytosol), and passage through the NPC (which is inhibited by WGA). These negative controls are: cargo alone in TB (−Cytosol −E); cargo with an ATP-regenerating system and GTP (−Cytosol +E); and cytosol with an ATP-regenerating system, GTP and WGA (+Cytosol +E +WGA) (see Fig. 11.1). The concentration of cytosol can be adjusted (*see* **Note 14**). The standard transport mix for wells of a 10-well

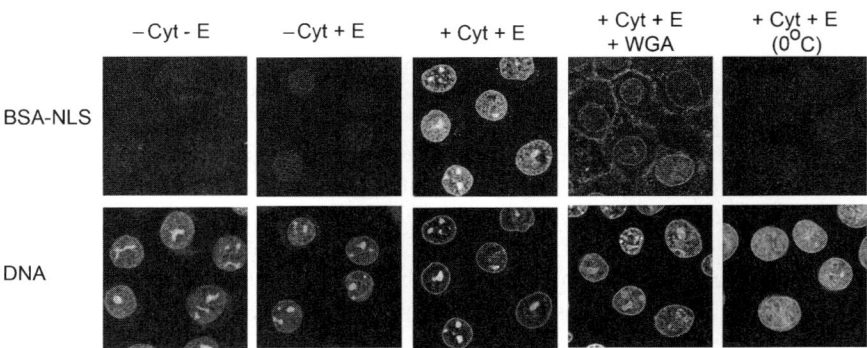

Fig. 11.1 Nuclear import of FITC-labeled BSA-NLS in digitonin-permeabilized NRK cells supplemented with cytosol. The assays contained $2\,\mu M$ FITC-labeled BSA-NLS and were incubated at 30°C for 30 min, except as noted, with the following conditions: in the absence of an ATP-regenerating system and GTP ("energy" [E]) and without cytosol (−Cyt −E); in presence of energy but without cytosol (−Cyt +E); with energy and $4\,\mu g/\mu L$ of cytosol (+Cyt +E); with energy and cytosol except the reaction was incubated at 0°C (+Cyt +E [0°C]), or with energy, $4\,\mu g/\mu L$ of cytosol, and WGA (+Cyt +E +WGA). The cells were examined with a Bio-Rad 1024 laser scanning confocal microscope using a ×63 objective, and the images were collected with Bio-Rad Lasersharp 2000 software. Fluorescent cargo is imaged in the *top row*, and DNA staining with Topro-3 is shown in the *bottom row*

slide is a 50-μL solution containing 1.5 μM cargo (e.g., FITC-BSA-NLS), cytosol to a final protein concentration of 4 mg/mL , an ATP-regenerating system (1 mM ATP, 1 mg/mL CP, and 15 U/mL CPK), and 0.1 mM GTP. When included, WGA is used at 0.8 mg/mL.

2. To study the nuclear import of cargo with recombinant import factors, the reaction mix contains cargo together with the import receptor, an ATP regenerating system, GTP, Ran, and NTF2 (+Receptor +E). Recommended negative controls are cargo alone in TB (−Receptor −E); cargo with an ATP-regenerating system, GTP, Ran, and NTF2 (−Receptor +E); and cargo with receptor, an ATP-regenerating system, GTP, Ran, NTF2, and WGA (+Receptor +E +WGA) (see Fig. 11.2). The reaction mix is a 50-μL solution containing appropriate concentrations of cargo, receptor, and Ran (see below) together with an ATP-regenerating system (1 mM ATP, 1 mg/mL CP, and 15 U/mL CPK), 0.1 mM GTP, and 1 μM NTF2. When used, WGA is present at 0.8 mg/mL. The concentration of each cargo, receptor, and Ran should be titrated to determine concentrations that yield optimal import (see **Note 15**) (See Fig. 11.3a, b). To analyze the nuclear import of FITC-BSA-NLS, we use 2 μM FITC-BSA-NLS, 1.5 μM importin α, and 1 μM importin β; to analyze GST-IBB, we use 100 nM GST-IBB and 120 nM importin β; to analyze GST-M9, we use 100 nM GST-M9 and 500 nM transportin; and to analyze FITC-Histone H1, we use 0.4 μM FITC-Histone H1 and 0.3 μM importin 7, or 0.3 μM importin β, or a combination of 0.3 μM importin 7 and 0.3 μM importin β.

3.6.2   Nuclear Import Assay in Permeabilized NRK Cells

1. Plate NRK cells on 10-well slides the day before the experiment.
2. To begin the assay, move the slides to a prechilled metal block on ice.
3. Add some cold TB to dilute the NRK medium and prevent the cells from drying out.
4. Draw boundaries around each well with a hydrophobic pen to keep the assay solution in each well separate.
5. Wash three times with cold TB to remove NRK medium and anything that remains from the hydrophobic pen. A wash step involves applying a ~100 μL drop of TB to each well using a plastic transfer pipet, and removing the liquid by vacuum aspiration (see **Note 16**).
6. Permeabilize the cells with 50 μL of digitonin solution (0.005% or 50 μg/mL)/well for 5 min at room temperature (see **Note 17**). The 10% w/v digitonin stock is diluted to 0.005% in TB. The slide is removed from the prechilled block to the bench for 5 min and put back on the block for the washes.
7. Wash three times with ice-cold TB to remove the digitonin.
8. Incubate the cells with TB alone or with solutions containing WGA for 15 min at 30°C on a prewarmed metal block in a humidified chamber or on ice.
9. Wash five times with TB to remove the cytosol (see **Note 18**).

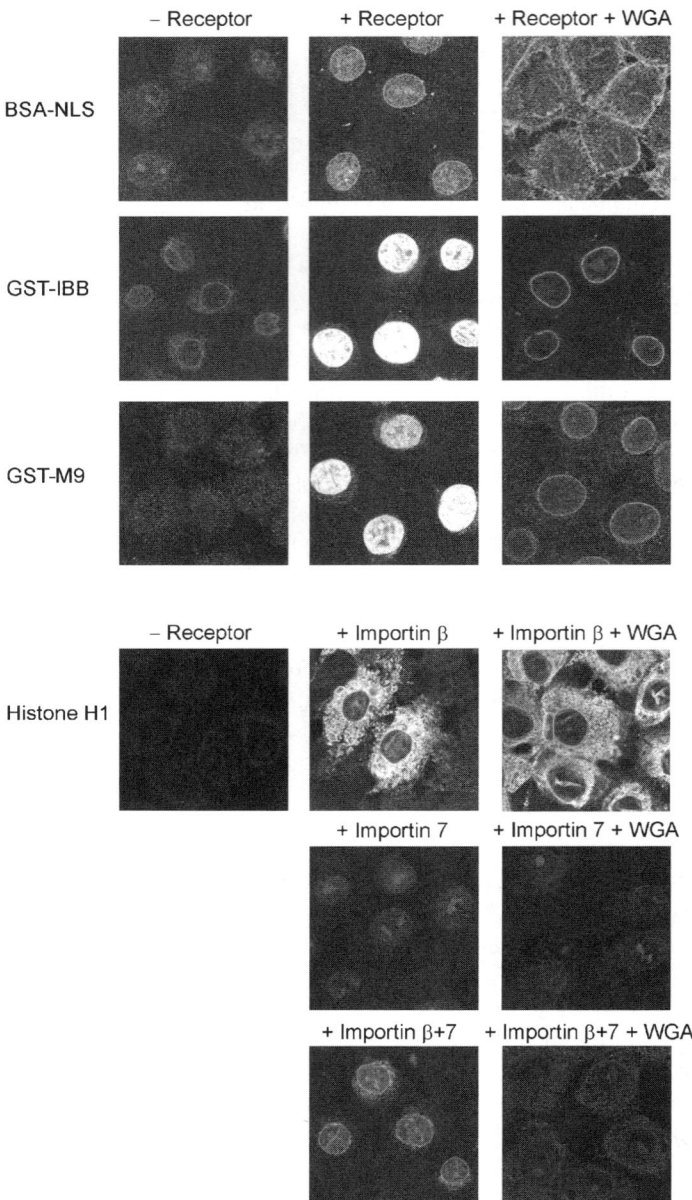

Fig. 11.2 Nuclear import assays of various cargoes in digitonin-permeabilized NRK cells reconstituted with recombinant import factors. The localization of each cargo was visualized in the absence of receptor (–Receptor), in the presence of receptor (+Receptor) and in the presence of receptor and WGA (+Receptor +WGA). The assays involved the following concentrations of cargoes and receptors: for BSA-NLS: $2\,\mu M$ FITC labeled-BSA-NLS, $1.5\,\mu M$ importin α, and $1\,\mu M$ importin β; for GST-IBB: $100\,nM$ GST-IBB and $120\,nM$ importin β; for GST-M9: $100\,nM$ GST-M9 and $500\,nM$ transportin; and for histone H1: $0.4\,\mu M$ FITC-labeled histone H1 and either $0.3\,\mu M$ importin 7, $0.3\,\mu M$ importin β, or both $0.3\,\mu M$ importin 7 and $0.3\,\mu M$ importin β. GST-IBB and GST-M9 were detected with an anti-GST antibody. Images were obtained as in Fig. 11.1

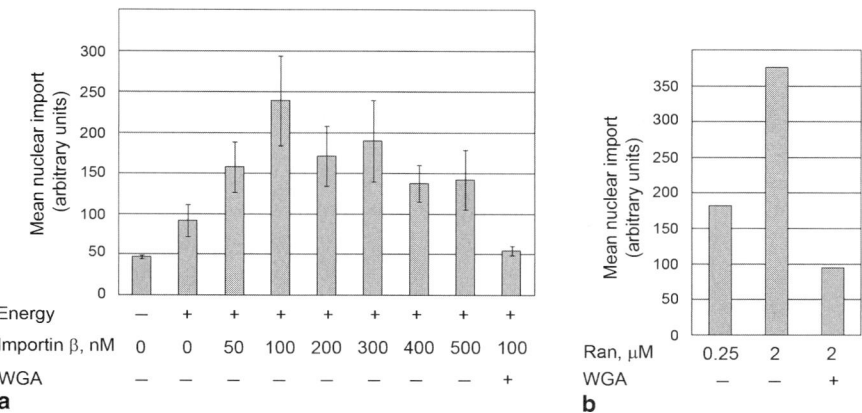

Fig. 11.3 Titration of recombinant factors to optimize nuclear import in digitonin-permeabilized NRK cells. The nuclear import assays were performed with reactions reconstituted with recombinant transport factors and cargoes. GST-IBB and GST-M9 were detected with an anti-GST antibody. The cells were examined using a LEICA DM IRE2 microscope and a ×40 objective. The pictures were collected and the mean nuclear intensity was quantified using SimplePCI software (Compix, Sewickley, PA, USA). Error bar represents the standard deviation of the mean nuclear fluorescence found in three different fields, each containing 30–50 cells. **a** Titration of importin β for the nuclear import of GST-IBB. The nuclear import assay was performed in presence of 1 μM NTF2, 2 μM Ran, an ATP-regenerating system and GTP (Energy), WGA, 100 nM GST-IBB, and different concentrations of importin β as indicated. **b** Titration of Ran for the nuclear import of GST-M9 by transportin. The reactions contained 1 μM NTF2, 0.25 or 2 μM Ran, an ATP-regenerating system, and GTP, WGA (in some cases), 100 nM GST-M9, and 500 nM transportin

10. Incubate the cells with 50 μL of import reaction mix for 30 min at 30°C on a prewarmed metal block in a humidified chamber, or for 30 min on ice (*see* **Note 19**).
11. Wash three times with TB.
12. Fix the cells for 10 min at room temperature with 3.7% formaldehyde in PBS.
13. Wash three times with PBS.
14. Proceed directly to immunostaining if this is needed for cargo detection. Alternatively, slides may be stored overnight at 4°C in PBS with 0.05% w/v NaN$_3$ before immunostaining.

3.6.3 Immunostaining to Detect GST-Cargoes

All steps are performed at room temperature.

1. Wash the wells three times with PBS.
2. Permeabilize for 10 min with 50 μL of PBS solution containing 0.2% v/v Triton X-100.
3. Wash three times with PBS.
4. Pre-block wells with 50 μL of PBS solution containing 0.3% w/v gelatin for 5 min.

5. Aspirate the gelatin/PBS and incubate for 1 h with 25 μL of purified goat anti-GST antibody diluted 1:100 in PBS with 0.3% gelatin.
6. Wash three times with PBS to remove unbound antibody.
7. Wash once with PBS containing 0.3% gelatin.
8. Aspirate and incubate in the dark for 1 h with 25 μL of FITC-labeled mouse anti-goat antibody diluted 1:100 in PBS with 0.3% gelatin.
9. Wash three times with PBS to remove unbound antibody.
10. Stain the DNA with DAPI diluted 1:5,000 in PBS from a stock solution of 1 mg/mL, or with Topro-3 diluted 1:500 in PBS, for 15 min in the dark (*see* **Note 20**).
11. Wash three times with PBS to remove the unbound dye.
12. Apply SlowFade Antifade or an equivalent antifade mounting reagent to the slide.
13. Put a cover glass (22×60 mm) over the entire slide and aspirate away excess mounting reagent from the edges of the cover glass.
14. Seal with nail varnish.

4 Notes

1. Importin α precipitates if its concentration is too high; make sure its concentration is ≤0.5 mg/mL before dialysis.
2. In the purification of importin α, the protein concentration should be determined by the Bradford assay since imidazole absorbs at 280 nm.
3. For the importin β lysis buffer, it is important to use TCEP instead of DTT or β-mercaptoethanol, since these latter reducing agents are not compatible with the binding of the intein tag. Moreover, lysozyme cannot be used in the initial lysis of the bacteria since it inactivates chitin beads.
4. The addition of a small amount of CHAPS to the dialysis buffer for importin β helps to increase protein yield and does not interfere in the nuclear import assay (the final concentration of CHAPS in the assay will be ~1 μM).
5. His-tagged importin 7 does not bind tightly to the Talon beads, and is eluted by the addition of 1 mM imidazole.
6. Most of the recombinant Ran is not soluble, but the level of expression is so high that even with this low solubility, ~30% of Ran is in the soluble pool.
7. Sometimes β lactamase from the bacteria, which is about 30–32 kDa, is strongly induced during the induction of Ran. Expanding cultures using colonies from freshly transformed plates without letting the bacteria reach the stationary phase helps to reduce the levels of β lactamase.
8. The final column is necessary if β lactamase is present at significant levels in the Ran peak from the S200 column.
9. To exchange the buffer to TB after the cation exchange column we recommend precipitating Ran with ammonium sulfate added to 65% saturation, since Ran precipitates during dialysis to TB if the pH before dialysis is below 6.8.
10. Though only 10–20% of NTF2 remains soluble under these preparative conditions, our attempts to recover active NTF2 from the insoluble fraction have been unsuccessful.

11. For the NTF2 lysis buffer, lysozyme should be avoided because its size (~20 kDa) is too close to that of the NTF2 dimer (28 kDa) and it could contaminate the NTF2 from the S200 column.

12. The optimal seeding density for the nuclear import assay can vary between different cell lines. The most important point is that the cells should not be confluent. The efficiency and specificity of plasma membrane permeabilization with digitonin is essential for the success of import assays, and have to be checked for each cell line and for each seeding. We recommend testing a range of digitonin concentrations from 0.002 to 0.008%. The integrity of the nuclear envelope can be tested using antibodies against histones or lamins, as well as with large FITC–dextrans (150 kDa). Antibodies and large dextrans are excluded from intact nuclei, while small FITC–dextrans (e.g., 9 kDa) have access *(5, 19)*.

13. We suggest starting with one or two slides to become familiar with the procedure. We recommend analyzing the nuclear import of FITC-BSA-NLS in the absence or presence of cytosol to check the procedure, before starting experiments with different cargoes or recombinant receptors.

14. The concentration of cytosol that gives optimal nuclear import can vary with different cargoes, with different concentrations of the same cargo, and with different batches of cargo. An excess of cytosol can have an inhibitory effect on transport. A high concentration of WGA is typically used, but its concentration can be adjusted according to the specific cells used in the assay. The concentrations of cytosol and WGA that we commonly use in the nuclear import assay are 4 mg/mL and 0.8 mg/mL, respectively.

15. Each cargo and recombinant receptor has to be titrated to obtain optimal levels of nuclear import (see Fig. 11.3). Cargo is titrated to determine the concentration that yields the optimal ratio between nuclear import and background (0°C control, no receptor, or no cytosol controls). The fluorescently labeled cargoes typically are used at a concentration of 0.5–2 μM to be readily detected, whereas unlabeled cargoes detected by immunofluorescence staining with anti-GST antibody can be analyzed from 100 nM. Nuclear import involving receptors is saturable. In addition, if the concentration of receptor is too high compared with the Ran concentration, nuclear import can be a limiting factor for nuclear import with recombinant receptors, so its concentration should also be optimized.

16. The washes are performed well by well and one well at a time. It is very important to keep a thin layer of liquid on top of the slides to prevent the cells from drying out. Therefore, the addition and aspiration of solutions on one well should be closely coordinated. If the cells become dry during the nuclear import assay, import is nonspecifically inhibited and strong cytoplasmic staining can be obtained.

17. The concentration of digitonin for cell permeabilization should be tested for each new supply of digitonin and for each new cell line examined. The appropriate concentration permeabilizes the plasma membrane but not the nuclear envelope, and yields cytosol-dependent nuclear import of FITC-BSA-NLS.

18. This step is essential to remove the endogenous cytosolic factors. We recommend increasing the number of washes from three to five. Alternatively, the cells can be incubated in presence of TB with energy or in presence of TB and RanQ69L for 15 min at 30°C. The presence of residual receptors in the permeabilized cells can induce nuclear import of cargo in the presence of energy alone.

19. Nuclear import is already detectable after 2 min at 30°C, and low levels of transport can be observed even at 10°C. Optimal import conditions usually involve 30 min incubation at 30°C. The permeabilized cells tend to become detached from cover glasses after ~60 min, and ATP supplies can also be depleted by this time. Nuclear import involving receptors is temperature dependent and is totally inhibited at 0°C. To perform the negative control of nuclear import at 0°C, we keep the cells on ice during the preincubation and incubation periods.

20. DAPI and Topro-3 are commonly used to stain DNA for fluorescence microscopy. The choice between these two dyes is dictated by the filters available on your microscope. Marking the nuclear space by fluorescence staining is a useful alternative to other techniques for visualizing the nucleus (e.g., phase contrast microscopy).

Acknowledgments The writing of this chapter was supported by the National Institutes of Health (NIH) grant AI55729 to LG. AC was supported by a fellowship from the French Foundation for Medical Research (FRM), SPE20041102385. We are grateful to Geza Ambrus-Aikelin for comments on the manuscript. We thank the postdocs of the Gerace lab for their protocols.

References

1. Fahrenkrog, B., Koser, J., and Aebi U. (2004) The nuclear pore complex: a jack of all trades? *Trends Biochem. Sci.* **29**, 175–182.
2. Mosammaparast, N. and Pemberton, L.F. (2004) Karyopherins: from nuclear-transport mediators to nuclear-function regulators. *Trends Cell Biol.* **14**, 547–556.
3. Macara, I.G. (2001) Transport into and out of the nucleus. *Microbiol. Mol. Biol. Rev.* **65**, 570–594.
4. Pemberton, P.L. and Paschal B. (2005) Mechanisms of receptor-mediated nuclear import and nuclear export. *Traffic* **6**, 187–198.
5. Adam, S.A., Marr, R.S., and Gerace L. (1990) Nuclear protein import in permeabilized mammalian cells requires soluble cytoplasmic factors. *J. Cell. Biol.* **111**, 807–816.
6. Chan, R.C. and Forbes, D.J. (2005) In vitro study of nuclear assembly and nuclear import using Xenopus egg extracts. In: *Xenopus protocols* (Liu, J.X., ed.), Humana Press, Totowa, NJ, pp. 289–300.
7. Colbeau, A., Nachbaur, J., and Vignais, P.M. (1971) Enzymatic characterization and lipid composition of rat liver subcellular membrane. *Biochim. Biophys. Acta* **249**, 462–492.
8. Melchior F. (1998) Nuclear protein import in a permeabilized cell assay, in *Protein targeting protocols* (Clegg, R.A., ed.), Humana Press, Totowa, NJ, pp. 265–273.
9. Kehlenbach, R.H. and Gerace L. (2002) Analysis of nuclear protein import and export in vitro using fluorescent cargoes, in *GTPase protocols* (Manser, E. and Leung, T., ed), Humana Press, Totowa, NJ, pp. 231–245.

10. Paraskeva, E., Izaurralde, E., Bischoff, F.R., Huber, J., Kutay, U., Hartmenn, E., Luhrmann, R., and Gorlich, D. (1999) CRM1-mediated recycling of snurportin 1 to the cytoplasm. *J. Cell Biol.* **145**, 255–264.
11. Kalderon, D., Roberts, B. L., Richardson, W. D., and Smith, A.E. (1984) A short amino acid sequence able to specify nuclear location. *Cell* **39**, 499–509.
12. Dingwall, C. and Laskey, R.A. (1991) Nuclear targeting sequences-a consensus? *Trends Biochem. Sci.* **16**, 478–481.
13. Lyman, S.K., Guan, T., Bednenko, J., Wodrich, H., and Gerace, L. (2002) Influence of cargo size on Ran and energy requirements for nuclear protein import. *J. Cell Biol.* **159**, 55–67.
14. Weis, K., Mattaj, I.W., and Lamond, A.I. (1995) Identification of hSRP1 alpha as a functional receptor for nuclear localization sequences. *Science* **268**, 1049–1053.
15. Jakel, S., Albig, W., Kutay, U., Bischoff, F.R., Schwamborn, K., Doenecke, D., and Gorlich, D. (1999) The importin beta/importin 7 heterodimer is a functional nuclear import receptor for histone H1. *EMBO J.* **18**, 2411–2423.
16. Pollard, V.W., Michael, W.M., Nakielny, S., Siomi, M.C., Wang, F., and Dreyfuss G. (1996) A novel receptor-mediated nuclear import pathway. *Cell* **86**, 985–994.
17. Melchior, F., Paschal, B., Evans, J., and Gerace L. (1993) Inhibition of nuclear protein import by nonhydrolyzable analogues of GTP and identification of the small GTPase Ran/TC4 as an essential transport factor. *J. Cell Biol.* **123**, 1649–1659.
18. Paschal, B. and Gerace, L. (1995) Identification of NTF2, a cytosolic factor for nuclear import that interacts with nuclear pore complex protein p62. *J. Cell Biol.* **129**, 925–937.
19. Paine, P.L., Moore, L.C., and Horowitz S.B. (1975) Nuclear envelope permeability. *Nature* **254**, 109–114.

Chapter 12
Nuclear Envelope Formation In Vitro: A Sea Urchin Egg Cell-Free System

Richard D. Byrne, Vanessa Zhendre, Banafshé Larijani, and Dominic L. Poccia

Keywords Nuclear envelope; Sea urchin; Phosphatidylinositol; Phosphatidyl-choline; Cholesterol; Phospholipase C; PtdIns(4;5)P2; Diacylglycerol; Liquid NMR; Liquid nuclear magnetic resonance

Abstract The formation of the nuclear envelope (NE) typically occurs once during every mitotic cycle in somatic cells, and also around the sperm nucleus following fertilization. Much of our understanding of NE assembly has been derived from systems modeling the latter event in vitro. In these systems, demembranated sperm nuclei are combined with fertilized egg cytoplasmic extracts and an ATP-regenerating system and in a multistep process they form the functional double bilayer of the NE. Using a system that we developed from sea urchin gametes, we have demonstrated that NE assembly is regulated by membrane vesicles in a spatial and temporal fashion, emphasizing the roles of phosphoinositides, particularly phosphatidylinositol 4,5-bisphosphate (PtdIns(4,5)P$_2$), diacylglycerols (DAG), and lipid-modifying enzymes in NE assembly.

1 Introduction

Cell-free systems derived from the gametes of a number of species have been used to elucidate the mechanisms of nuclear envelope assembly. The sea urchin has a number of advantageous features *(1–3)*. First, sea urchins produce millions of gametes when fertile, allowing the purification of large amounts of protein and lipids. Second, the in vitro system is an open one, allowing easy manipulation with pharmacological inhibitors and recombinant proteins *(4–6)*. Third, sea urchin eggs have completed meiosis, and after fertilization they progress through the first mitotic cycles with a high degree of synchrony, allowing extracts of various cell cycle stages to be prepared. Finally, the components of the cell-free system, the egg cytoplasmic extract (S10) and demembranated nuclei, can be further fractionated *(7, 8)*, the former to membrane vesicle fractions (MVs) and cytosol (S150) and the latter to chromatin and detergent-resistant membranous structures (corresponding to sperm nuclear envelope remnants also seen in vivo *(1)* and previously referred to as lipophilic structures [LSs] *(2)*).

R. Hancock (ed.) *The Nucleus: Volume 2: Chromatin, Transcription, Envelope, Proteins, Dynamics, and Imaging*,
© Humana Press 2008

The sea urchin cell-free system has proven to be suited to nuclear envelope studies due to the stepwise assembly of the envelope (3). Following addition of nuclei to egg cytoplasmic extracts and an ATP-regenerating system, MVs initially bind to the surface of decondensing chromatin. In the presence of GTP, in a step requiring an endogenous phospholipase C (PLC) activity and the hydrolysis of PtdIns $(4,5)P_2$ to form the fusogenic lipid DAG, these vesicles fuse to form a double bilayer (4, 5). In a final step, the pronucleus expands approximately twofold in diameter in the presence of additional ATP (4), incorporating more membrane and functional nuclear pores. Nuclei at these different steps of assembly can be isolated as required, and their lipid contents determined and quantified (4, 5).

We have previously described membrane domains within both the demembranated nuclei and the egg cytoplasmic extracts that contribute in an obligatory manner to the formation of the nuclear envelope in vitro (7, 8). For example, MV1 is a polyphosphoinositide-rich membrane fraction (5) containing very high levels of phosphatidylinositol (PI)–PLC and PtdIns $(4,5)P_2$ (6), which can be separated from other egg cytoplasmic membranes on sucrose gradients by virtue of its low buoyant density (7). Also, detergent resistant membrane structures (sperm nuclear envelope remnants) on demembranated sperm can be isolated (2). Analysis of these fractions has revealed novel features that indicate how they participate in nuclear envelope assembly (8).

2   Materials

2.1   Gamete Collection, Preparation of Egg Cytoplasmic Extracts, and Sperm Nuclei Isolation

1. $0.5\,M$ KCl. Stored at room temperature.
2. 22-gauge, 1.5-inch needles (Sigma-Aldrich, Dorset, UK or St. Louis, MO, USA).
3. Medium-size weighing boats.
4. Millipore-filtered artificial seawater (MPSW). Reef salt (Instant Ocean; Aqua-Medic Commercial, Telford, UK) is dissolved at a concentration of $33\,g/L$ and stirred overnight. This is filtered through a 0.22-μm membrane filter (500 mL GP Express PLUS; Millipore, Watford, UK or Billerica, MA, USA). The filtered seawater is hereafter referred to as MPSW (*see* **Note 1**).
5. 3-amino-1,2,4-triazole (ATA), $30\,mM$: prepared fresh as a 10× stock in distilled water no more than 30 min before use and stored on ice.
6. Nylon mesh of 64, 100, 120, and 210 μm (Sefar, Bury, UK or Kansas City, MO, USA).
7. Lysis buffer (LB): $10\,mM$ Hepes, pH 8.0, $250\,mM$ NaCl, $5\,mM$ MgCl$_2$, $110\,mM$ glycine, $250\,mM$ glycerol, and $1\,mM$ DTT, supplemented with $1\,mM$ PMSF immediately before use.

8. Nuclei extraction buffer (SXN): 50 m*M* Hepes, pH 7.2, 250 m*M* sucrose, 150 m*M* NaCl, 0.5 m*M* spermidine, 0.15 m*M* spermine, and 300 m*M* glucose, aliquot in 50-mL volumes and store at −80°C.

9. Bath sonicator: Ultrawave U50 230 V 50 Hz, 500 mL (Ultrawave, Cardiff, UK or Fischer Scientific, Fair Lawn, NJ, USA).

10. Nuclei freezing buffer: 10 mL of SXN, 2 mL of 3% (w/v) BSA in SXN, 6 mL of glycerol. Prepare fresh on day of use.

2.2 Nuclear Envelope Assembly Assay

1. TN buffer: 10 m*M* Tris-HCl, pH 7.2, 150 m*M* NaCl in distilled water, aliquot in 50-mL volumes and store at −80°C.

2. 100 m*M* ATP, 5 m*M* GTP (Sigma), 1 *M* creatine phosphate, and 2.5 mg/mL creatine phosphokinase. All are made up in LB, aliquoted in 10-μL amounts and stored at −80°C.

3. CaCl$_2$ stock solutions: 500 n*M* for *L. pictus* and 1.25 μ*M* for *P. lividus*, and 250 m*M* EGTA stock solution, all in LB.

4. TN supplemented with 5 m*M* MgCl$_2$ and 0.5 *M* sucrose, frozen at −20°C in 1-mL aliquots.

5. Microscope slides, circular coverslips (13-mm diameter), and nail varnish.

6. 3′3-dihexyloxacarbocyanine iodide (DiOC$_6$) and 1,1′-didodecyl-3,3,3′,3′-tetramethylindocarbocyanine perchlorate (DiIC$_{12}$; Invitrogen, Paisley, UK or Carlsbad, CA, USA) are dissolved in methanol at 10 mg/mL. These dye solutions are stored in the dark at room temperature. Immediately before use they are diluted 10-fold in LB (for DiOC$_6$) or methanol (for DiIC$_{12}$; *see* **Note 2**).

2.3 Preparation of Fertilized Egg Soluble Protein (S150) and Subfractionation of Egg MV Fractions

1. Membrane wash buffer (MWB): 50 m*M* Hepes, pH 7.5, 250 m*M* sucrose, 50 m*M* KCl, 1 m*M* DTT, 1 m*M* ATP, and 1 m*M* PMSF in distilled water, frozen in 50-mL aliquots at −80°C.

2. Soniprep 150 probe sonicator (Sanyo-Gallenkamp, Leicester, UK or Palisades Park, NJ, USA).

3. 2 *M* sucrose in TN, frozen in 25-mL aliquots at −20°C.

4. Gradient maker (Hoefer, San Francisco, CA, USA) and gradient fractionator (Auto Densi-flow; GRI, Essex, UK or Buchler Instruments, Fort Lee, NJ, USA).

5. Mineral oil.

6. Ultracentrifuge tubes (Beckman, Buckinghamshire, UK or Fullerton, CA, USA; *see* **Note 3**).

7. Ultracentrifuge and SW40 rotor (Beckman).
8. 22-gauge, 1.5-inch needles.

2.4 Isolation of Detergent-Resistant Membranes from 0.1% Triton X-100-Treated Nuclei

1. Triton X-100 solution: 10% v/v in distilled water, stored at room temperature.
2. Bath sonicator: Ultrawave U50 230 V, 50 Hz, 500 mL (Ultrawave).

2.5 Detection of Sea Urchin Proteins by Western Analysis

1. 10 mL LB supplemented with 0.1% Triton X-100 and a Complete Mini protease inhibitor cocktail tablet (Roche, Mannheim, Germany or Indianapolis, IN, USA; cat. 1836153).
2. 4× sodium dodecyl sulfate (SDS) sample buffer: 250 mM Tris, pH 6.8, 20% (v/v) glycerol, 4% SDS, 0.01% bromophenol blue, 50 mM β-mercaptoethanol.
3. Heating block.
4. NuPAGE Novex 4–12% bis-tris pre-cast gels and NuPAGE MOPS SDS running buffer (Invitrogen).
5. Mini-electrophoresis cell (XCell Surelock; Invitrogen).
6. Blotting paper and Immobilon-P transfer membrane (Millipore, Watford, UK or Billerica, MA, USA).
7. Methanol.
8. Transfer buffer: 25 mM Tris base, 192 mM glycine, and 20% (v/v) methanol.
9. Semi-dry transfer cell (Trans-Blot SD; Bio-Rad, Hertfordshire, UK or Hercules, CA, USA).
10. Phosphate-buffered saline (PBS): 137 mM NaCl, 2.7 mM KCl, 10 mM Na$_2$HPO$_4$, and 1.76 mM KH$_2$PO$_4$, pH 7.2), supplemented with 0.2% (v/v) Tween-20 to make PBST.
11. Milk powder and ultrapure 30% w/v BSA solution (Sigma). These are added to PBST to make 5% (w/v) and 3% (v/v) solutions respectively.
12. Anti-goat HRP (Perbio, Northumberland, UK or Rockford, IL, USA), anti-mouse HRP serum and ECL detection reagents (both from GE Healthcare, Amersham, UK or Piscataway, NJ, USA).
13. X-ray film (Cronex-5; AGFA, Mortsel, Belgium or Greenville, SC, USA).

2.6 Lipid Extraction Procedure—Modified Folch Method

1. 3% (v/v) solution of dimethyldichlorosilane in HPLC-grade toluene (Fischer Scientific, Fair Lawn, NJ). This solution is made fresh on the day of silanation.

2. HPLC-grade methanol and HPLC-grade water (Fischer Scientific).

3. Chloroform/methanol 2.5:1 (Fischer Scientific). This is made by first acidifying the methanol with a drop of 10 N HCl (1 drop per 100–200 mL methanol) and subsequently adding the required volume of chloroform. The solution is stored at room temperature.

4. Probe sonicator (Soniprep 150; Sanyo-Gallenkamp, UK or Palisades Park, NJ, USA).

5. Filter apparatus: 0.22-µm GV Durapore membrane filters, conical flask with side arm, funnel, and clamp (Millipore).

6. K_4EDTA, pH 6.0 (5): 5.84 g of EDTA (purified grade, Sigma) is dissolved in 100 mL distilled water. Thirty pellets of KOH (Sigma) are added periodically with stirring.

7. Nitrogen source, sample incubator (Techne Dri-block DB3), and sample concentrator (Barloworld Scientific, Staffordshire, UK).

2.7   Solution Nuclear Magnetic Resonance (NMR) Lipid Analysis

1. Deuterated solvents (chloroform [$CDCl_3$], methanol [CD_3OD], and 0.2 M EDTA in D_2O (Sigma)) made up as $CDCl_3/CD_3OD/K_4$ EDTA (100:40:20, v/v).

2. 5-mm NMR tubes (Wilmad NMR tubes, Royal Imperial grade, Sigma-Aldrich).

3. Bruker AMX-2 500 MHz NMR spectrometer with a Bruker 5-mm broadband indirect probe tuned to 202 MHz and 500.13 Hz, stabilized at 298 K.

4. Dipalmitoyl phosphatidylcholine (DPPC), dipalmitoyl phosphatidylserine (DPPS), dipalmitoyl phosphatidylethanolamine (DPPE), and phosphatidylinositol (PI) sodium salt (Sigma; or Avanti Polar Lipids, Alabaster, AL, USA).

5. XWIN-NMR software (Bruker, Billerica, MA, USA).

3   Methods

This section details the methods we have developed to isolate fractions from sea urchin egg and sperm that contribute to the assembly of the nuclear envelope. We emphasize additions and corrections to detailed methods previously published. We present a typical Western blotting procedure useful for localization of protein antigens to individual membrane fractions. We have used 2-D NMR and mass spectrometry methods to quantify phospholipid species in sea urchin extracts. A procedure for 2-D solution NMR is presented here. Lipid mass spectrometry of phospholipids is described elsewhere in this series (9).

3.1 Collection of Gametes

1. *Lytechinus pictus*, *Strongylocentrotus purpuratus*, and *Paracentrotus lividus* sea urchins are injected twice intracoelomically (100 μL each injection for *L. pictus*, approximately 300 μL for the latter two species) with 0.5 *M* KCl followed by swirling the animals to distribute the KCl (*see* **Note 4**).
2. Eggs (orange) are collected by placing the inverted sea urchin over a full beaker of MPSW. Sperm (white/pale brown) are collected "dry" by placing an inverted sea urchin into a plastic weighing boat. Female sea urchins are kept moist during this procedure by the regular application of MPSW; males are covered with MPSW-soaked tissue to avoid dilution of the sperm.

3.2 Egg Lysis and Preparation of Fertilized Egg Cytoplasmic Extracts

1. All the following steps take place at 16°C unless otherwise stated.
2. Eggs from individual animals are first checked to ensure they fertilize correctly. Sperm is diluted approximately 500-fold in MPSW and a few drops are added to eggs before viewing under a light microscope (preferably inverted, ×10 objective). Fertilized eggs (90% of eggs raising a fertilization envelope within 30 sec of sperm addition) are pooled and concentrated by centrifugation at 100×*g* for 1 min.
3. Eggs are resuspended in 5 volumes of MPSW, pH 5.0 and collected by centrifugation as in step 2. Residual acidic MPSW is removed by two washes in MPSW as described in step 2 (*see* **Note 5**).
4. Approximately 10 mL of eggs are resuspended in 100 mL MPSW supplemented with 3 m*M* ATA (*see* **Note 6**).
5. Sperm is diluted approximately 500-fold and added to the eggs to give an approximate final sperm/egg ratio of 10–100:1. Eggs are viewed as above to ensure correct fertilization.
6. Two minutes after fertilization, eggs are filtered through a nylon filter (*L. pictus*, 100 μm filter; *S. purpuratus*, 64-μm filter) and washed twice in 10 volumes of MPSW as described in step 2 (*see* **Note 7**). These steps take place 10–15 min after fertilization to produce a G1-phase extract.
7. Eggs are washed three times in 10 volumes of ice-cold LB as described in step 2 (one more wash than previously described in **ref.** *(10)*) to thoroughly remove contamination of the extracts by MPSW.
8. Eggs are finally resuspended in 1 volume of LB and homogenized by twice vigorously drawing the suspension into a 10-mL syringe fitted with a 22-gauge, 1.5-inch needle, followed by vigorous expulsion. *See* **Notes 8** and **9**, and Figs. 12.1 and 12.2.

Fig. 12.1 Rate of nuclear decondensation with varying pH of the LB buffer added. The pH of the resultant extract is lower

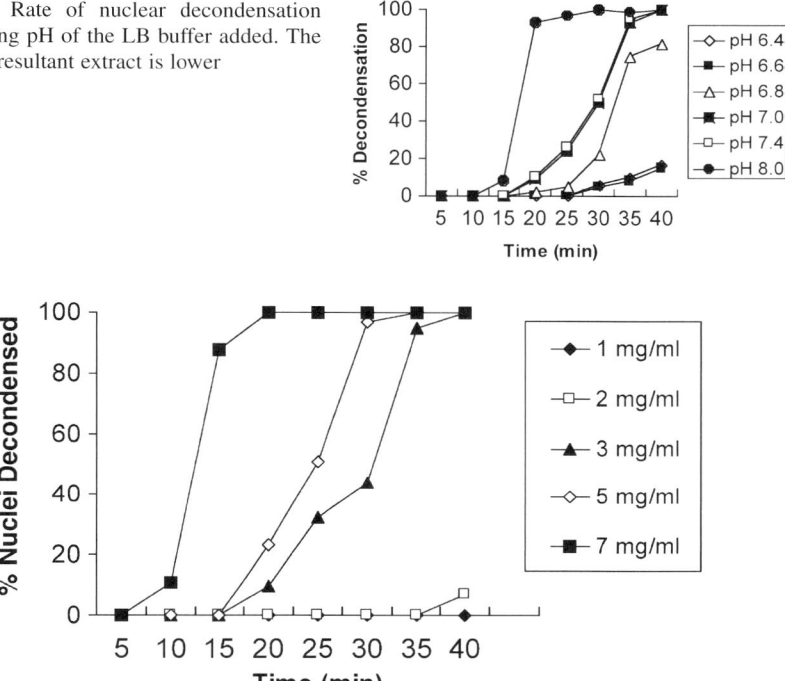

Fig. 12.2 Variation of the rate of nuclear decondensation with the concentration of extract in milligrams protein per milliliter. Dilute extracts may not support decondensation or nuclear envelope formation

9. The homogenate is centrifuged at 10,000×g for 10 min at 4°C to produce three layers: an upper yolky layer, a lower pigmented pellet, and in the middle the egg cytoplasmic extract (referred to hereafter as S10; *see* **Note 10**).

10. The S10 is removed with a P200 pipette and snap-frozen in liquid nitrogen in 0.2-, 0.5-, or 1.0-mL aliquots.

3.3 *Sperm Nuclei Isolation and Membrane Removal*

1. Before preparation, samples of sperm from each male are collected and observed under ×40 magnification to ensure that the sperm are active and fertilize eggs (as described in **Section 3.2**).

2. Sperm are concentrated in a microcentrifuge tube for 10 min at 500×g, 4°C (*see* **Note 11**).

3. 250 µL of concentrated sperm are resuspended in 10 mL of ice-cold SXN buffer in a 15-mL centrifuge tube, using a plastic transfer pipette.

4. Sperm are centrifuged at 2,600×*g* for 5 min at 4°C.
5. The supernatant is aspirated off and the pellet resuspended in 1.5 mL of SXN.
6. The sample is bath sonicated for 6 min with continual mixing with a plastic transfer pipette (*see* **Note 12**).
7. Sperm nuclei are centrifuged at 2,600×*g*, 4°C for 90 sec. The supernatant is removed and each pellet is resuspended in 990 μL SXN and 10 μL of 10% (v/v) Triton X-100 (final concentration 0.1%).
8. Samples are initially shaken vigorously to allow the detergent to react, and further mixing with plastic pipettes takes place every 5 min for 15 min at room temperature (*see* **Note 13**).
9. The Triton-treated nuclei are pelleted at 2,600×*g*, 4°C for 1 min and washed twice with 1 mL of SXN as above.
10. The final pellet is resuspended in 500 μL of SXN and 500 μL of freezing solution. This suspension is thoroughly mixed with progressively smaller pipette tips (P1000–P200 size), and aliquoted in 250-μL volumes. Samples are cryogenically frozen in liquid nitrogen and stored at −80°C. Previously, nuclei were only stored for short periods at 4°C *(10)*.
11. We no longer use the lysolecithin method of sperm nuclei permeabilization *(10)*.

3.4   Nuclear Envelope Assembly Assay

1. 0.1% Triton-treated nuclei, S10, and ATP-GS components (ATP, creatine kinase, and creatine phosphate) are thawed on ice.
2. Nuclei are pelleted at 1,500×*g*, 4°C for 2 min in a microcentrifuge and gently resuspended in 50 μL of TN buffer (*see* **Note 14**).
3. In a 1.5-mL microcentrifuge tube, 3 μL of nuclear suspension are added to 20 μL of S10 and supplemented with 1.2 μL of ATP-GS (final concentration approximately 6×10⁵ nuclei/μL).
4. This mixture is incubated at room temperature for 1 h with periodic agitation. Decondensation of nuclei from a conical to spherical shape is confirmed under a ×100 oil-immersion objective. It is of note that we often see full decondensation after 1 h and not 40 min as we previously reported (Fig. 12.2) *(10)*. This is probably due to variations in the ratio of nuclei to cytoplasm concentration.
5. At this point, samples may be additionally treated with 5 μL of GTP stock (1 m*M* final concentration) for 2 h to induce fusion of the bound membrane vesicles. The fusion step has been lengthened from 40 min *(10)* to 120 min to maximize the number of nuclei displaying nuclear envelopes (Fig. 12.3). The rate of completion of nuclear envelope formation varies with the extract, but the reaction is usually complete by 90 min *(6)*.
6. To control variations of free [Ca²⁺], samples are supplemented to a final concentration of 12.5 m*M* EGTA and either 20 n*M* (for *L. pictus*) or 50 n*M* (for *P. lividus*) CaCl₂ at the same time as GTP is added. We have optimized these

calcium conditions, as well as calculating the free calcium in the reaction mix using the program WinmaxC32 2.50 (http://www.stanford.edu/ cpatton/downloads.htm) as illustrated in Fig. 12.4.

7. To remove unbound membranes, samples from the reactions are underlain with 1 volume of TN containing 5 mM MgCl$_2$ and 0.5 M sucrose. The nuclei are pelleted for 15 min at 750×g, 4°C and the pellet is gently resuspended in 20 μL of LB using a 200-μL pipette.

8. To visualize chromatin-bound membrane vesicles, 3 μL of either diOC$_6$ or diIC$_{12}$ is added to each sample yielding either green or red emission.

9. 3 μL of dye-stained sample is spotted on a slide and covered with a coverslip, which is attached to the slide with nail varnish at 2 points. Visualization of decondensed nuclei takes place by phase contrast microscopy as in Step 4.

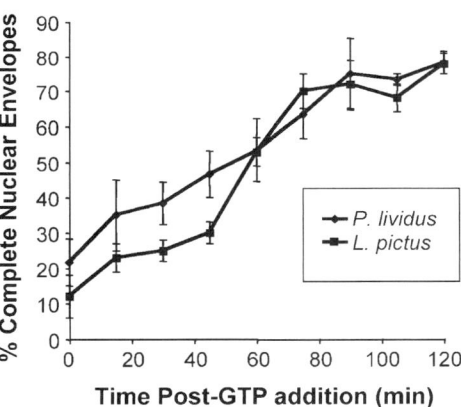

Fig. 12.3 Progress of complete nuclear envelope formation. The maximal percentage of nuclei exhibiting complete nuclear envelope formation is usually achieved by 90 min after addition of GTP to the system

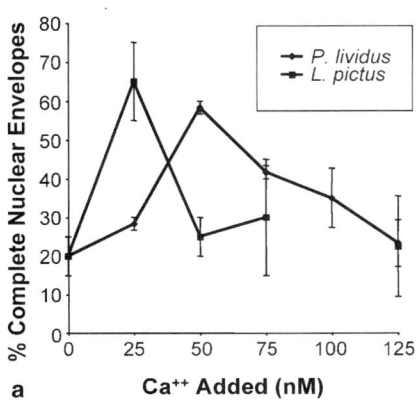

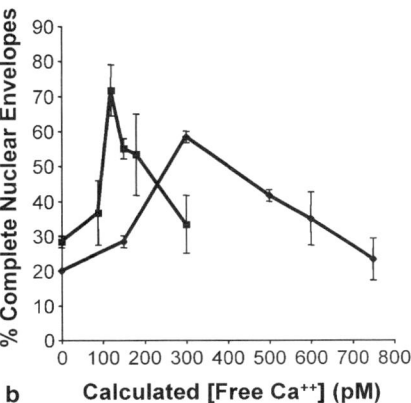

Fig. 12.4 Buffering of free Ca^{2+} in the cell-free system. The optimum free Ca^{2+} can vary between species. **a** Various concentrations of Ca^{2+} were added to LB containing 12.5 mM EGTA. **b** Concentrations of free Ca^{2+} were calculated using the WinmaxC32 program

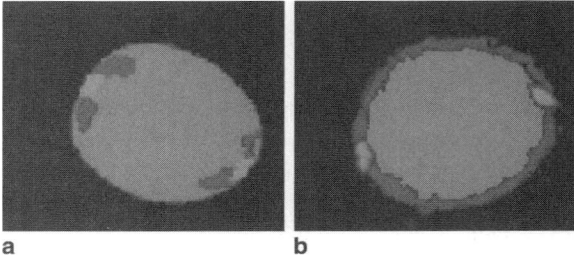

a   b

Fig. 12.5 Binding of membrane vesicle fractions MV1 and MV2beta to decondensed sea urchin sperm chromatin. **a** MV1 is labeled with DiIC$_{18}$ (*red*), sperm remnant membranes with DiOC$_6$ (*green*), and nuclear DNA with Hoechst 33342 (*blue*). **b** MV2 labeled with DiIC$_{18}$ (*red*), sperm remnant membranes with DiOC$_6$ (*green*), and nuclear DNA with Hoechst 33342 (*blue*). (Adapted from **ref.** *(7)*). To view this figure in color, see COLOR PLATE 6

Alternatively, dyes are visualized under a ×100 oil immersion objective with excitation with an argon/krypton laser and the resulting fluorescence is separated using a combination of a dichroic beamsplitter (Q495LP; Chroma) and a HQ510/20 nm emission filter. Images are captured with a Hamamatsu Orca camera and processed in OpenLab. Binding of membrane vesicle fractions to decondensed chromatin is shown in Fig. 12.5.

10. The integrity of the envelope can be additionally confirmed by the addition of 2 μL of ATP-GS to the reaction mix for 30 min after the fusion step. Nuclei are subsequently visualized as above. Under these conditions, nuclei will swell from a diameter of 4 to 8 μm only if the nuclear envelope is fully formed *(11)*.

11. We have reported several manipulations of the steps described above using a variety of tools. In particular, we have demonstrated that the MV fusion step is dependent upon the hydrolysis of PtdIns(4,5)P$_2$ to DAG by an endogenous PLC activity *(4, 5)*. For example, the addition of PtdIns(4,5)P$_2$-consuming PI 5-phosphatases prior to GTP inhibits MV fusion, as does the specific PLC inhibitor U73122, but not its inactive analogue U73343. We can bypass the need for GTP in fusion by using recombinant human PLCγ2 protein, or the functional analog of its DAG product, phorbol 12-myristate 13-acetate (PMA).

3.5 Preparation of Egg Cytosolic Extracts (S150) and Purified MV Subfractionation

1. S10 is centrifuged at 150,000×*g* for 2 h at 4°C, after which the supernatant is removed and centrifuged again for a further 1 h as above. The cleared supernatant (S150) contains soluble egg proteins. S150 is snap-frozen in liquid nitrogen in 100- and/or 500-μL aliquots and stored at −80°C.

2. The resulting pellets from the above centrifugations (MV0) are gently resuspended in MWB (200 µL per 2 mL starting material) using a P1000 pipette and probe sonicated for 3 sec at power 22 to produce a uniform suspension.

3. MV0 in MWB is diluted to 1 mL with 800 µL of TN (*see* **Note 15**).

4. 1.5 mL of 2 M sucrose in TN ("heavy" solution) is diluted 20-fold in TN to produce a 0.1 M sucrose in TN ("light" solution).

5. The heavy and light sucrose solutions are loaded into the gradient maker (5.5 mL of each) with a stirring bar added to the heavy chamber (*see* **Note 16**). With stirring, the plug between the columns is opened, allowing the two solutions to mix. The 0.1–2 M sucrose gradient is layered using the Densi-flow into an ultracentrifuge tube at a speed (~2.5) such that it takes approximately 12 min to pour a gradient.

6. The resuspended MV0 is slowly layered onto the top of the gradient, and topped off with mineral oil.

7. The gradient is centrifuged at 150,000×g for 20 h at 4°C.

8. MV subfractions appear in the gradient as pale bands that are removed from the centrifuge tubes by side puncture with a 22-gauge, 1.5-inch needle on a 1- or 2-mL syringe.

9. Collected MV subfractions are washed once in 4 volumes of MWB at 150,000×g for 30 min at 4°C, and finally resuspended in 250 µL of MWB and washed as above.

10. The final pellets are resuspended in an appropriate volume of MWB (100 µL per 5 mL S10) and aliquoted in 10-µL amounts. These are snap-frozen in liquid nitrogen and stored at −80°C.

11. The four fractions (MV1, MV2, MV3, and MV4) formed have densities of 1.02, 1.04–1.08, 1.13, and 1.18, respectively.

12. The binding profile of the isolated MVs can be assessed in the nuclear assembly assay (*see* **Section 3.4**) by combining 20 µL of S150, 1.2 µL of ATP-GS, and 2 µL of the relevant MV fraction prestained with a lipophilic dye (*see* **Note 17**).

3.6   Isolation of Detergent-Resistant Membranes from 0.1% Triton X-100-Treated Nuclei

1. 0.1% Triton X-100-extracted nuclei are thawed on ice, pelleted at 1,500×g for 2 min at 4°C and resuspended in 45 µL of TN (*see* **Note 18**).

2. 5 µL of 10% Triton X-100 is added and the sample is bath sonicated for 30 min.

3. The nuclei are collected by centrifugation at 1,500×g for 10 min at 4°C. The pellet containing these 1% Triton X-100-treated nuclei is washed twice in TN (as in **Section 3.6.1**) and resuspended in 50 µL of TN.

4. The supernatant containing the sperm nuclear membrane material is diluted 70-fold in TN and stirred for 1.5 h at 4°C to allow reassembly. This dilution

factor is larger than we have previously described *(10)* in order to bring Triton X-100 below its critical micelle concentration (CMC), preventing it from perturbing the reassembly reaction.

5. The suspension is pelleted at 150,000×g for 1 h at 4°C and the pellet is finally resuspended in 50 µL of TN buffer.

3.7   Detection of Sea Urchin Proteins by Western Blot Analysis

1. S10 is separated into its constituent fractions as described in **Section 3.5**.
2. The MV0 pellet is resuspended in LB/0.1% Triton X-100/protease inhibitors to the same volume as the S150 from which it was obtained.
3. Protein concentration of samples is determined by the Bradford assay (Biorad), then 4× SDS buffer is added and samples are heated for 10 min at 95°C.
4. Equal volumes of samples are loaded onto a NuPAGE 4–12% gel. Gels are run at 75 V to allow samples to concentrate in the stacking gel followed by 125 V until the dye front leaves the bottom of the gel.
5. Gels are removed from their plastic casts and assembled on the semi-dry transfer cell as follows: four pieces of transfer buffer-soaked blotting paper, methanol-activated Immobilon-P transfer membrane (Millipore), the gel, and four pieces of transfer buffer-soaked blotting paper. Care is taken to ensure air pockets are not trapped between the sandwich layers.
6. Transfer takes place for 90 min at 12 V.
7. The membrane is removed from the sandwich and air-dried, protein side up, until white in color.
8. The membrane is incubated with agitation at room temperature for 1 h in PBST supplemented with 3% BSA (w/v).
9. The buffer is removed and replaced by the same solution supplemented with the desired primary antibody. We have used antibodies against phosphoinositide monophosphate kinase I (PIPK) and PLCγ at dilutions of 1:1,000 and 1:2,500, respectively. The membrane and antibody in buffer are agitated overnight at 4°C.
10. The membrane is washed 5× ~10 min at room temperature in PBST (*see* **Note 19**).
11. The membrane is incubated with agitation in PBST supplemented with 5% (w/v) milk powder and the appropriate secondary antibody (anti-goat for PIPK, 1:10,000; anti-mouse for PLCγ, 1:5,000) for 1–2 h at room temperature.
12. Wash the membrane as in step 10.
13. The membrane is finally incubated with ECL detection reagent, wrapped in Saran wrap, and exposed to X-ray film for an appropriate time. Films are subsequently developed in an automated developer.
14. We have used this protocol to detect a PIPK in sea urchin extracts, and have been able to show that it is localized both in the soluble S150 and MV0 membrane fractions. By contrast, PLCγ is virtually entirely membrane localized *(6)* (Fig. 12.6). This methodology could equally well be applied to probing subfractionated MV1–4 for proteins, or detecting proteins in 0.1% Triton-treated nuclei and extracted sperm nuclear membrane material.

Fig. 12.6 Detection of phosphatidylinositol 5-monophosphate kinase (*PIPK*) and PLCγ in sea urchin cytoplasmic (*S10*), cytosolic (*S150*), and total membrane (*MV0*) fractions. PIPK was detected in *S. purpuratus* and PLCγ in *L. pictus*

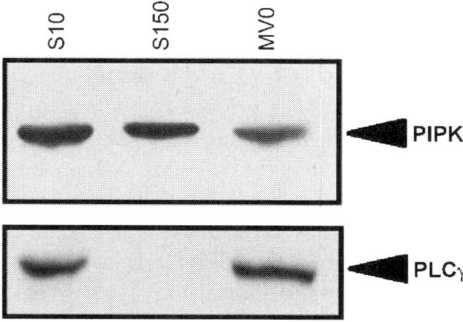

3.8 Lipid Extraction Procedure—Modified Folch Method (12)

1. Prior to the extraction all glassware is silanated to enhance the recovery of phospholipids, particularly the phosphoinositides. Glassware is silanated by immersion into a 3% (v/v) solution of dimethyldichlorosilane (*toxic! work under a fume hood*) in HPLC-grade toluene for 1 h. The glassware is subsequently rinsed twice in HPLC-grade methanol, twice in HPLC-grade water, and dried in a heat cabinet or under a fume hood before use.
2. Biological samples are added to ice-cold acidified chloroform/methanol in glass screw-cap universal tubes (*see* **Note 20**). Four mL of solvent is used per 200 μL sample (if the sample size is smaller than 200 μL, a minimum of 4 mL solvent is used). If necessary, lipid standards for further applications are added at this point.
3. The samples are probe sonicated for 10 sec at power 22, and left for 1 h at room temperature. Samples can be stored at −80°C after sonication if necessary.
4. The samples are removed from the sample tubes with a glass Pasteur pipette, and filtered through a 0.22-μm GV Durapore membrane filter into another glass sample tube.
5. Sample phase separation is induced by adding 0.2 volumes of K_4EDTA in distilled water followed by centrifugation at 800×g for 15 min at 4°C.
6. The lower phase is removed with a glass Pasteur pipette, transferred to a new glass tube, and dried under nitrogen at 30°C. Dried lipid extracts can be stored at −80°C for a number of months.

3.9 Solution NMR Lipid Analysis

1. Extracted lipids (*see* **Section 3.7**) are dissolved in the $CDCl_3/CD_3OD/K_4$ EDTA solvent mix and left to equilibrate in a glass tube for 1 h.
2. The samples are transferred to 5-mm NMR tubes (*see* **Note 21**).
3. Samples are scanned using an indirectly detected 2D $^{31}P–^1H$ method (*see* **Note 22**) established via multiple quantum coherence (HMQC) giving a single

phosphorus resonance for each phospholipid in the projection along the F_1 axis and proton resonances along the F_2 axis.

4. Scanning is for 2 h (48 scans) on concentrated samples or 20 h (480 scans) where lipid is present at a less than micromolar concentration. During scanning, the field is locked to the $CDCl_3$ signal to aid reproducibility.

5. The 2D spectra scanning parameters are as follows: ^{1}H and ^{31}P 90-degree pulses of 8.0 and 9.5 μsec, respectively, relay delay 2 sec, acquisition time 0.68 sec, sweep width F_2 2,008 Hz in 2,750 real points, sweep width F_1 607 Hz in 64 slices, Malcolm Levitt decoupling scheme (MLEV) 17 mixing at 7.1 kHz power for 101 msec (45 cycles), globally optimized alternating rectangular pulse (GARP) ^{31}P decoupling during acquisition at 2.4 k Hz power, refocus delay 0.071 sec (J_{P-H} = 7 Hz).

6. Data sets are processed with zero-shifted sine-bell window functions in both directions and displayed in magnitude mode for 2-D integration.

7. To allow the quantification of phospholipids, three sets of standards are made (*see* **Note 23**). The first is comprised of DPPC, DPPE, DPPS, and PI each at 3.9 μM; the second set of 0.40 μM DPPC, 0.40 μM DPPE, and 0.39 μM PI; and the third set of 0.04 μM DPPC, 0.04 μM DPPE, and 0.039 μM PI.

8. The standards are dissolved in 700 μL of the $CDCl_3/CD_3OD/K_4$ EDTA solvent mix and analyzed as described in Step 5, with XWIN-NMR used to integrate the peaks.

9. Using the known phospholipid molar concentrations [PL] and the integration values of peaks (A), the efficiency coefficient of each phospholipid (K) relative to DPPC can be calculated from each of the standard sets using Eq. 12.1:

$$K_{2/1} = \frac{[PL_1] \times A_2}{[PL_2] \times A_1}.$$ (Eq. 12.1)

10. The molarities of individual phospholipids are calculated as follows. The ratio of A_1/A_2 is multiplied by its K value. Thus the molarity of PI relative to DPPC is calculated by Eq. 12.2 (*see* **Note 24**):

$$\left| \frac{PI}{DPPC} \right| = K_{PC/PI} \frac{A_{PI}}{A_{PC}}.$$ (Eq. 12.2)

4   Notes

1. Overnight stirring ensures complete solvation of the salt, especially when prepared in large batches. Approximately 2 L of MPSW can be prepared using one filter. Other salt sources work well, and large amounts of unfiltered sea water can be prepared in a cylindrical container with a circulating pump at the bottom that

mixes the solution overnight (Rolf C Hagen Ltd, Castlefield, UK or Mansfield, MA, USA). The pH and density are checked after final preparation.

2. $DiIC_{12}$ is more sensitive than $DiOC_6$ at equal concentrations. However, $DiIC_{12}$ will form crystals in aqueous solution, interfering with visualization of nuclei under the microscope and is best diluted in methanol.

3. We have found 14×95-mm centrifuge tubes (Beckmann #344060) to be the most suited to the MV0 subfractionation.

4. Injection of animals with an excessive volume of KCl can lead to inhibition of gamete shedding.

5. These washes should be performed as quickly as possible to ensure that eggs are exposed to an acidic environment for a short period of time only. We have also dejellied eggs by passing them through a nylon filter (210 μm for *L. pictus*, and 120 μm for *S. purpuratus*), as previously described *(13)*. We find that this is just as efficient as the pH 5 wash.

6. ATA prevents hardening of the fertilization envelope, and thus aids the homogenization step to produce extracts. We add ATA to the eggs a few minutes prior to fertilization.

7. The nylon filter strips the eggs of the fertilization envelope, and is size adjusted to the egg diameter of the species.

8. It is useful to occasionally monitor the pH of the homogenate with pH paper. The final pH should be above 7.3, since the eggs become alkaline after fertilization. The homogenate pH is always lower than the pH of the LB used. The effect of altering the initial pH of LB on the rate of decondensation is shown in Fig. 12.1.

9. It is important to keep extracts concentrated. Usually the protein concentration of the extract is around 6 mg/mL. The effect of diluting extracts to lower protein concentrations on the rate of decondensation is shown in Fig. 12.2.

10. The S10 contains membrane vesicles (MVs), soluble proteins, ribosomes, and some mitochondria. It is free of female pronuclei, yolk granules, the majority of the mitochondria, and large cytoplasmic structures.

11. After concentration, three layers are visible, an upper aqueous phase, the middle "concentrated viable sperm" phase, and a small black pellet. Only the sperm are kept.

12. This step removes most of the sperm tails. The anchor of the tail in the deepest section of the centriolar fossa may remain in some sea urchin species. For more thorough removal of tails, sonication has been extended to 2 min from the previously described method *(10)*.

13. This process is monitored with lipophilic dyes to confirm that 0.1% Triton removes the sperm head lateral membranes. We perform this step at room temperature instead of 4°C as described previously, to allow more stringent removal of lateral nuclear membranes *(10)*.

14. After resuspension, the concentration of nuclei is around 5×10^9 nuclei/mL (confirm by counting in a hemocytometer).

15. If not used immediately, this MV0 can be snap-frozen in liquid nitrogen and stored at −80°C.

16. A small amount of heavy solution is allowed to run into the light solution column to displace the air in the tube that connects them, before the addition of the light solution.

17. Under these conditions, only MVs and not nuclei are stained to allow the distinction to be made between nuclei-bound MVs and sperm nuclear envelope remnants.

18. We start the extraction with at least four of the frozen aliquots described in **Section 3.3** (250 μL of concentrated sperm equivalent). This provides enough material for lipid analysis by liquid chromatography electrospray ionization tandem mass spectrometry. For less sensitive applications, more material will be required.

19. If antibodies display a high background, the washes can be supplemented with 1% (w/v) milk to increase the stringency.

20. Plastics can contaminate lipid extracts, so glassware should always be used for handling and storage of solvents and lipid extracts.

21. An aqueous phase will form above the sample, preventing solvent evaporation during analysis. It is important that the interface is clear of emulsified water, as this degrades the homogeneity of the magnetic field.

22. This method has a 9.5-fold increase in signal sensitivity over ^{31}P detection alone.

23. Saturated acyl fatty-acid chains are used in the standards, since their chemical shifts are the most stable.

24. The mole percent of phospholipid is calculated relative to the most abundant lipid present, which is assigned the arbitrary value of 1.

Acknowledgments This work was supported by an Amherst College Faculty Research Award of the H. Axel Schupf '57 Fund for Intellectual Life (DLP) and core funding from Cancer Research UK (BL).

References

1. Longo, F.J. and Anderson, E. (1968) The fine structure of pronuclear development and fusion in the sea urchin, Arbacia punctulata. *J. Cell Biol.* **39**, 339–368.
2. Collas, P. and Poccia, D. (1995) Lipophilic organizing structures of sperm nuclei target membrane vesicle binding and are incorporated into the nuclear envelope. *Dev. Biol.* **169**, 123–135.
3 Cothren, C.C. and Poccia, D.L. (1993) Two steps required for male pronucleus formation in the sea urchin egg. *Exp. Cell Res.* **205**, 126–133.
4. Collas, P. and Poccia, D.L. (1995) Formation of the sea urchin male pronucleus in vitro: membrane-independent chromatin decondensation and nuclear envelope-dependent nuclear swelling. *Mol. Reprod. Dev.* **42**, 106–113.
5. Larijani, B., Poccia, D.L., and Dickinson, L.C. (2000) Phospholipid identification and quantification of membrane vesicle subfractions by 31P-1H two-dimensional nuclear magnetic resonance. *Lipids* **35**, 1289–1297.
6. Byrne, R., Garnier, M., Han, K., Dowicki, M., Totty, N., Cho, A., Pettit, R., Wakelam, M.J.O., Poccia, D., and Larijani, B. (2006) PLCg and phosphoinositides are highly enriched in membrane vesicles required for nuclear envelope assembly. *Cell Signalling* **19**, 913–922.

7. Collas, P. and Poccia, D. (1996) Distinct egg membrane vesicles differing in binding and fusion properties contribute to sea urchin male pronuclear envelopes formed in vitro. *J. Cell Sci.* **109**, 1275–1283.

8. Larijani, B. and Poccia, D. (2007) Protein and lipid signaling in membrane fusion: nuclear envelope assembly. *Signal Transduction* **7,** 142–153.

9. Garnier-Lhomme, M., Dufourc, E.J., Larijani, B., and Poccia, D. (2007) Lipid Quantification and structure determination of nuclear envelope precursor membranes in the sea urchin. *Methods Mol. Biol.* in press.

10. Collas, P. and Poccia, D. (1998) Methods for studying in vitro assembly of male pronuclei using oocyte extracts from marine invertebrates: sea urchins and surf clams. *Methods Cell Biol.* **53**, 417–452.

11. Byrne, R.D., Barona, T.M., Garnier, M., Koster, G., Katan, M., Poccia, D.L., and Larijani, B. (2005) Nuclear envelope assembly is promoted by phosphoinositide-specific phospholipase C with selective recruitment of phosphatidylinositol-enriched membranes. *Biochem. J.* **387**, 393–400.

12. Meneses, P. and Glonek, T. (1988) High resolution 31P NMR of extracted phospholipids. *J Lipid Res.* **29**, 679–689.

13. Foltz, K.R., Adams, N.L., and Runft, L.L. (2004) Echinoderm eggs and embryos: procurement and culture. *Methods Cell Biol.* **74**, 39–74.

Part IV
Modifications of Nuclear Proteins

Chapter 13
Detection and Analysis of (O-linked β-*N*-Acetylglucosamine)-Modified Proteins

Natasha E. Zachara

Keywords Protein glycosylation; O-GlcNAc; Signal transduction; Posttranslational modification; Affinity purification; Glycomics; Site-mapping; Mass spectrometry

Abstract Glycosylation is one of the most common and complex forms of posttranslational modifications of proteins in eukaryotes. Seven different protein–carbohydrate linkages have been characterized on nuclear and cytoplasmic glycoproteins, the most widespread of which is the modification of Ser/Thr residues with monosaccharides of O-linked β-*N*-acetylglucosamine (O-GlcNAc). O-GlcNAc modification is concentrated in nuclear proteins. O-GlcNAc is thought to regulate protein function in a manner analogous to phosphorylation; and is implicated in the regulation of transcription, the proteasome, insulin and MAP kinase signaling, the cell cycle, and the cellular stress response. In this chapter we focus on methods for the detection of O-GlcNAc-modified proteins and discuss general techniques for the detection and subsequent analysis of other protein–carbohydrate conjugates.

1 Introduction

Biochemical data, lectin-binding studies (Table 13.1), and the existence of nuclear and cytoplasmic lectins and glycosyltransferases suggest that there are nuclear and cytoplasmic glycoconjugates (**ref. (*1*)** and references therein).

In fact, at least 500 nuclear and cytoplasmic glycoproteins have now been identified, and these fall into seven groups (Table 13.2). These are 1), mannose (man)-containing proteoglycans in rat brain; 2) heparin sulfate in rat liver nuclei; 3) glucosyl residues attached to hydroxy-tyrosine, a primer for glycogen synthesis (*2*); 4) O-linked man on cytoplasmic proteins, which is subsequently modified by a novel glucose (Glc) phosphotransferase (*3–5*); 5) the modification, by *Clostridium* toxin, of small G-proteins by ADP-ribose or β-*N*-acetylglucosamine (GlcNAc) (*6*); 6) the modification of Skp1 (a ubiquitin ligase subunit) of *Dictyostelium discoideum* by a GlcNAc-terminating oligosaccharide (*7*); and 7) the addition of a monosaccharide of βGlcNAc through an O-glycosidic linkage to Ser/Thr residues

R. Hancock (ed.) *The Nucleus: Volume 2: Chromatin, Transcription, Envelope, Proteins, Dynamics, and Imaging,*
© Humana Press 2008

Table 13.1 Lectin binding in the nucleus

Lectin[a]	Competing sugar	Nuclear binding
WGA	D-GlcNAc, sialic acid	Yes
GS-1, GS-2	D-GlcNAc	—
Con A	α-D-glucoside, α-D-mannosides	Yes
RCA 1, RCA 2	D-GalNAc, D-galactose	Yes
LCA	α-D-mannosides	Yes
UEA-I	α-L-fucose	Yes
APA	α-L-fucose	Yes
TTA	D-GlcNAc	Yes
SBA	D-GalNAc	Yes
PHA	Oligosaccharides	Yes
DSA	D-GlcNAc	Yes
SJA	D-galactose	Yes[b]
PNA	—	Yes[b]
MMA	Sialyllactose	Yes[b]
SNA	Sialyllactose	Yes[b]

See **ref.** (1) and references therein

[a] Lectins: agglutinins from APA, *Abrus precatorius*; Con-A, *Canavalia ensiformis*; DSA, *Datura stramonium*; GS-1 and GS-2, *Griffonia simplicifolia* 1 and 2; LCA, *Lens culinaris*; MAA, *Maackia amurensis*; PHA, *Phaseolus vulgaris*; PNA, Peanut; RCA 1 and RCA 2, *Ricinus communis* 1 and 2; SBA, *Glycine max*; SJA, *Sophora japonica*; SNA, *Sambucus nigra*; sWGA, succinyl-wheat germ (*Triticum vulgare*); TTA, *Tachypleus tridentatus*; UEA-I, *Ulex europaeus*; WGA, Wheat Germ (*Triticum vulgare*)

[b] See **ref.** (55)

Table 13.2 Intracellular glycoproteins

Protein(s)	Carbohydrate	Nuclear
>400 nuclear and cytoplasmic proteins[a]: nuclear pore proteins, chromatin-associated proteins, Sp1, AP-1, CTF, HNF-1, V-*erb*A, SRF, C-Myc, p53, ER-α,β, β-catenin, NF-κB, ELF-1, PAX-6, YY1, PDX-1, CREB, RB, RNA pol II, Tau, nuclear tyrosine phosphatase p65, heat shock proteins, OGT	β-*N*-acetylglucosamine-1-O-Ser/Thr	Yes
Small G proteins[b]: Rho, Rac, cdc42, R-Ras, Ral, Rap, Ha-Ras, Ran	ADP-Ribose, Glc-, βGlcNAc-Ser/Thr	Some
Skp1	αGal1-?αGal1-3αFuc1-2βGal1-3αGlcNAc1-O-HydroxyPro	Yes
Glycogenin	$(αGluc1-4)_n$-tyrosine	No
Phosphoglucomutase	$αGlc-PO_4-Man-X_n$-Ser/Thr	No
O-Linked mannose-containing proteoglycans	Mannose	No
Proteoglycans	Chondroitin sulfate, heparin sulfate	Yes
α-synuclein[c]	Sialyl-βGal1-3αGalNAc1-O-Ser/Thr	Yes
High mobility group glycoproteins	GlcNAc, Man, Gal, Glc, Fuc	Yes
Cytokeratins[c]	αGalNAc1-3βX-?	No
VP24, 35[c]	$Sialyl-(X/GlcNAc)_n$-?	Yes

[a] For a complete list, refer to **ref.** (1) and references therein

[b] While this modification occurs on proteins of eukaryotes, it is transferred to the proteins by *Clostridium* toxins

[c] Supporting structural work for these proteins has not yet been published

(O-GlcNAc) *(1, 8, 9)*. Some studies have indicated the presence of other glycoproteins/ glycoconjugates (Tables 13.1 and 13.2), but supporting structural data is yet to be published.

The most widespread, and perhaps the most intensively studied, form of protein glycosylation is addition of O-GlcNAc to Ser and Thr residues (Table 13.1) *(1, 8, 9)*, a modification that is concentrated in nuclear proteins. This modification is dynamic and responds to numerous signals including hormones, phorbol esters, ionomycin, neutrophil agonists, development, extracellular Glc concentrations, the cell cycle, and cellular stress. These data, and those showing that O-GlcNAc and phosphate can compete for the same Ser/Thr residue, has led to the conclusion that O-GlcNAc is a modulator of protein function analogous to phosphorylation. Certain key regulatory proteins are modified by O-GlcNAc, and changes in their modification state are implicated in regulation of the proteasome, insulin and MAP kinase signaling, the cell cycle, and the cellular stress response. Deletion of O-GlcNAc transferase, which adds O-GlcNAc, is lethal in mammalian and plant systems, highlighting the importance of O-GlcNAc in regulating cellular function(s) *(10–12)*.

This chapter focuses on methods to study O-GlcNAc modification of proteins. However, where possible, we will indicate methods that can be adapted to analyze other carbohydrate modifications (Table 13.3, Adaptable). Any novel structure identified on a nuclear or cytoplasmic protein should be characterized in the most rigorous way possible; the existence of nuclear and cytoplasmic glycoproteins remained under debate for some time due, in part, to lack of structural data, and it was suggested that much of the initial data resulted from contamination of nuclear preparations by plasma membrane or endoplasmic reticulum and/or false-positive lectin staining reactions *(1)*.

The characterization of a glycoprotein can be broken down into three questions (Fig. 13.1): first, is my protein glycosylated?; second, what is the structure of the carbohydrate(s)?; and third, which residue(s) is/are modified? Numerous methodologies are available (Fig. 13.1) to answer each of these questions for O-GlcNAc modification, ranging from simple immunoblotting to high-throughput mass spectrometry-based techniques. Each offers advantages and disadvantages, which are discussed below and summarized in Table 13.3, and ultimately the choice will be based on the equipment available and the focus of the study.

Perhaps the fastest and easiest way to determine if a protein is modified by O-GlcNAc is by immunoblotting. Numerous O-GlcNAc-specific antibodies have been described including HGAC85 (*see* **Note 1**) *(13)*; RL2 *(14, 15)*; CTD110.6 *(16)*; and MY95 *(17)*. Of these, CTD110.6 appears to recognize the widest range of O-GlcNAc-modified proteins and is the most rigorously characterized. Site-specific glycosylation-dependent antibodies have been raised and characterized against the c-Myc proto-oncogene *(18)* and neurofilament proteins *(19)*. The effect of raising and lowering the levels of O-GlcNAc (*see* **Section 2.1**) or of treating samples with hexosaminidase (*see* **Section 2.9**) increases the confidence in positive signals. Several GlcNAc-specific lectins are also available (*see* **Section 2.4**) but most of these also react with numerous N-linked carbohydrate structures and it is important to pretreat samples with peptide:*N*-glycosidase F (PNGase F) (*see* **Section 2.10**).

Table 13.3 Methods for the detection of O-GlcNAc

Method	Adaptable	Modified	Components	Structure	Site mapping	Glycomics	Comments
CTD110.6 RL2	No	Indicative	Indicative	Indicative	No	No	False positives possible
Lectin binding	Yes	Indicative	Suggestive	No	No	No	False positives possible; N-linked sugars react
ITT	No	Indicative	Suggestive	Indicative	No	No	False positives possible; Useful for screening
Gal-T	No	Definitive	Definitive	Definitive	Yes	No	False positives possible; ^{3}H label not sensitive, useful for later analyses; Best with purified proteins or extracts treated with PNGase F
GlcNAz/Staudinger ligation	Yes	Indicative	Suggestive		Yes	Yes	Low risk of incorporation into cell surface proteins; GlcNAz not available commercially
Gal-T (Y289l)	No	Indicative	Suggestive	No	Yes	Yes	False positives possible; reagents not available commercially
DIONEX	Yes	Indicative	Definitive	Definitive	No	No	Quantitative; Best combined with Gal-T labeling
MS	Yes	Indicative	Suggestive	No	Yes	Yes	O-GlcNAc is labile; MS compatible with site-mapping; Neutral loss possible; Sugars are isobaric; False positives possible
BEMAD	Yes	Indicative	Suggestive	No	Yes	Yes	False positives possible; Density labeling possible; Proline-rich peptides problematic

Confidence in data: Definitive > indicative > suggestive; *modified*: My protein is modified by a glycan; *components*: What sugars make up the glycan; *structure*: What is the structure of the glycan?; *site-mapping*: Where is my protein modified by a glycan?; *glycomics*: high-throughput method

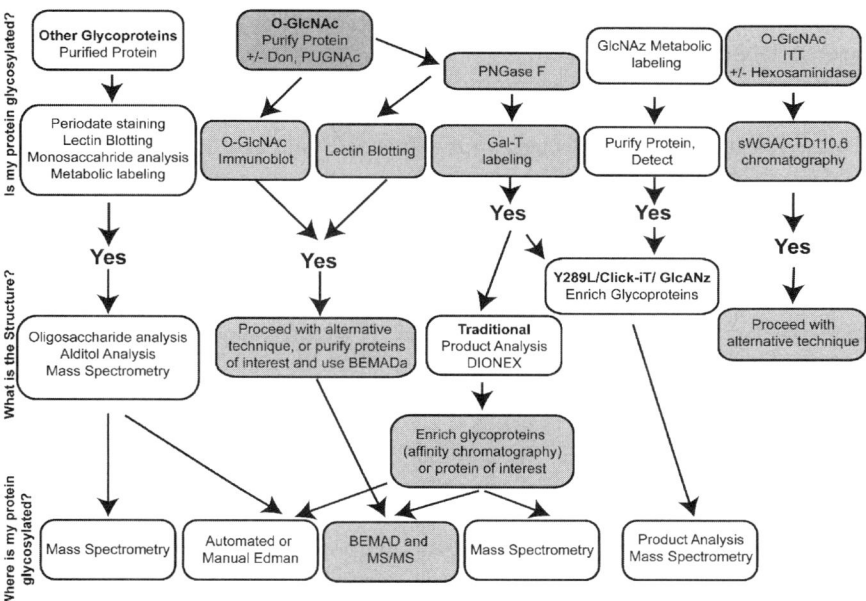

Fig. 13.1 Flow chart describing the analysis of glycoproteins. Methods described in the text are highlighted in *grey*

The gold standard for the detection and analysis of O-GlcNAc is labeling with UDP-[³H]Gal mediated by bovine milk UDP-galactose:D-glucose1-4-β-galactosyl-transferase (Gal-T) *(20–21)* (*see* **Section 2.5**). Gal-T will label any terminal GlcNAc/Glc residue with a Gal residue; thus when labeling O-GlcNAc, the resulting product is βGal1-4βGlcNAc1-O-Ser/Thr. One advantage of this technique is the incorporation of the [³H]Gal label, which is very useful for subsequent analyses such as site-mapping and purification of proteins/peptides of interest. Recently, a mutant Gal-T (Y289L) has been used to incorporate a Gal-ketone biotin tag, rather than a [³H]Gal residue, into O-GlcNAc-modified proteins; the resulting biotin tag allows for rapid detection and purification of the proteins *(22)*. However, product analysis with this tag is more challenging, and many of the reagents are not commercially available. The Click-iT O-GlcNAc Enzymatic Labeling System (Invitrogen, Carlsbad, CA, USA), which combines the Y289L Gal-T labeling method with the GlcNAz reagent developed by Bertozzi and colleagues *(23)*, is very promising, but as yet we have not tested it thoroughly.

Determining the site of O-GlcNAc attachment is challenging. Traditionally, proteins have been labeled by Gal-T and [³H]Gal and the radioactive label has been used to track proteins/peptides through subsequent rounds of purification *(20)*. In some cases the resulting disaccharide βGal1-4βGlcNAc has been used in combination with *Ricinus communis* agglutinin (RCA) affinity chromatography to enrich proteins/peptides of interest *(24–27)*. Site-assignment has been determined using manual Edman degradation, which, while labor intensive, can be performed with standard laboratory equipment and has been used to assign sites on as little as 10 pmol of starting material. Recently, it has been shown that carbohydrate-modified amino acids can be recovered on automated Edman sequencers, and without labeling the sugar can be detected on as little as 10 pmol of starting material (28).

Mass spectrometry has also been used to assign sites on peptides that have been Gal-T labeled (22–27, 30–37), but until recently has offered few advantages over manual Edman degradation. The addition of a single GlcNAc residue to the peptide backbone means that the glycopeptide is not ionized as well as unmodified peptides, reducing the signal in the mass spectrometer fivefold (38). In addition to being both acid- and base-labile, the O-linked β-GlcNAc bond is highly labile in conventional electrospray mass spectrometers (MS) and O-GlcNAc is often released at lower collision energies than those required to sequence peptides. Using an ion trap MS, even though the sugar is lost it is still possible to sequence the now unmodified peptide, but the question of the location of the GlcNAc on a particular Ser/Thr remains. To overcome the loss of sensitivity due to release of O-GlcNAc in MS, Greis and co-workers analyzed peptides that had been Gal-T-labeled and enriched by RCA1 affinity chromatography after β-elimination of the carbohydrate. Beta-elimination of carbohydrates from Ser (89 atomic mass units [amu]) and Thr (101 amu) residues results in 2-aminopropenoic acid (69 amu) and α-aminobutyric acid (83 amu), respectively (24–26). As these have different masses to their parent amino acids, they can be used to map the site of attachment. One elegant adaptation of this technique is the addition of dithiothreitol (DTT) back to the β-eliminated peptide via a Michael addition (BEMAD, see **Section 2.11**), which offers three advantages over β-elimination alone: first, there is a greater mass difference from the starting peptide(s); second, the DTT can be used to enrich peptides of interest by thiol-affinity chromatography; and third, the DTT can be density-labeled, facilitating quantitative MS (29, 30). The obvious disadvantage of these techniques is the possible detection of phosphorylation sites or other O-linked carbohydrate modification sites. Recently two new forms of MS, electron capture dissociation (ECD) and electron transfer dissociation (ETD) have been reported to retain labile phosphate-amino acid bonds, and these hold promise for the analysis of O-GlcNAc modified proteins/peptides (39).

2   Materials

2.1   Modulating O-GlcNAc Protein Modification in Cell Culture

1. O-(2-acetamido-2-deoxy-D-glucopyranosylidene)amino-N-phenylcarbamate (PUGNAc; Toronto Research Chemicals, Ontario, Canada; cat. A157250). Dissolve in water at 20 mM, filter sterilize, aliquot, and store at −20°C for 3–6 months.

2. 6-Diazo-5-oxo-L-norleucine (DON; Sigma-Aldrich, St Louis, MO, USA; cat. D2141). Dissolve at 2–5 mM in water, filter sterilize, and store at −20°C for up to 6 months.
3. Mammalian cells of interest grown in the medium of choice, preferably with 1 g/L of Glc (*see* **Note 2**).

2.2 Detecting O-GlcNAc-Modified Proteins by Western Blotting with Antibody CTD110.6

1. Negative and positive control proteins (*see* Table 13.4).
2. Proteins of interest (*see* **Notes 3** and **4**) separated by sodium dodecyl sulfate (SDS)-polyacrylamide gel electrophoresis (PAGE) or two-dimensional gel electrophoresis (2D-PAGE) and transferred to nitrocellulose or polyvinylidene fluoride (PVDF) membranes. Note that duplicate blots are required (*see* **Note 5**).
3. Tris-buffered saline (TBS): 10 mM Tris-HCl, 150 mM NaCl, pH 7.5.
4. TBST: TBS with 0.05% v/v Tween-20.
5. Blocking solution: 2% w/v skimmed-milk powder in TBST.
6. Antibody incubation buffer: 2% w/v BSA in TBST.
7. Competition buffer: 100 mM GlcNAc in antibody incubation buffer, made fresh (*see* **Notes 5** and **6**).

Table 13.4 Positive and negative control proteins.

	Protein				
	Ovalbumin[a]	Ovalbumin +PNGase F	BSA[a]	BSA-GlcNAc	BSA-Gal, -GalNAc
Positive controls					
CTD110.6 (*see* **Section 2.2**)	No	No	No	5–20 ng	No
SWGA (*see* **Section 2.4**)	Yes (20–100 ng)	No	No	Yes (5–20 ng)	No
Gal-T (*see* **Section 2.5**)	Yes (1 μg)	No	No	Yes (1 μg)	No
PNGase F (*see* **Section 2.10**)	Yes	No	No	No	No
Hexosaminidase (*see* **Section 2.9**)	Yes	No	No	No	No
Negative controls					
CTD110.6 (*see* **Section 2.2**)	Yes (20–100 ng)	No	Yes (5–20 μg)	No	Yes (20–100 ng)
sWGA (*see* **Section 2.4**)	No	Yes (20–100 ng)	Yes (1 μg)	No	Yes
Gal-T (*see* **Section 2.5**)	No	Yes (1 μg)	Yes (1 μg)	No	Yes
PNGase F (*see* **Section 2.10**)	No	Yes	No	No	No
Hexosaminidase (*see* Section 2.9)	No	Yes	No	No	No

[a] Sigma-Aldrich

8. Nonspecific competition buffer: 100 mM sugar, for example, Glc or galactose (Gal), in antibody incubation buffer (*see* **Note 7**).

9. Primary antibody: CTD110.6 (antigen purified; *see* **Section 2.3**), 1 mg/mL in PBS:glycerol (1:1 v/v) (*see* **Notes 8** and **9**).

10. Secondary antibody: anti-mouse IgM-horse radish peroxidase (Sigma-Aldrich; cat. A8786).

11. Detection reagents and autoradiography film for horse radish peroxidase (e.g., ECL; Amersham Biosciences, Piscataway, NJ, USA, or an equivalent detection system).

2.3   Purification of Antibody CTD110.6

1. GlcNAc-agarose (Sigma-Aldrich; cat. 01143) or GlcNAc-Sepharose (EY Labs, San Mateo, CA, USA; cat. CG-003-5) (*see* **Note 10**).

2. Protease inhibitors: typically, we include a cocktail that inhibits mammalian proteases (e.g., Sigma-Aldrich; cat. P8340) (*see* **Note 11**).

3. Chromatography column housings (e.g., Econo-Pac columns; Bio-Rad, Hercules, CA, USA).

4. Phosphate-buffered saline (PBS): 10 mM phosphate buffer, 136 mM sodium chloride, and 2.6 mM potassium chloride, pH 7.4.

5. High-speed (>10,000×g) centrifuge tubes (~10 mL).

6. 1 μm pore size filters (nonsterile).

7. 1 M Tris-HCl, pH 8.0.

8. CTD110.6 ascites fluid (Covance Research Products, Princeton, NJ, USA; cat. MMS-248R).

9. Nonspecific competition buffer: 100 mM Gal in PBS (*see* **Note 12**).

10. Elution buffer: 1 M GlcNAc in PBS.

11. 0.05% w/v sodium azide in PBS.

12. Dialysis equipment.

13. Glycerol 80% v/v in water, autoclaved.

2.4   Detection of O-GlcNAc-Modified Proteins with Succinylated Wheat Germ Agglutinin

1. Positive and negative control proteins (*see* **Section 1.12**).

2. Proteins of interest (*see* **Note 4**) separated by SDS-PAGE or 2D-PAGE and transferred to a nitrocellulose or PVDF membrane. Duplicate blots are required (*see* **Note 5**).

3. TBS and TBST.

4. Blocking solution: 2% w/v BSA in TBST (*see* **Note 13**).

5. Succinyl wheat germ (*Triticum vulgare*) agglutinin (sWGA)-HRP conjugate (EY Labs): 1 μg/mL in blocking solution (*see* **Notes 14** and **15**).
6. Competition buffer: 100 mM GlcNAc in TBST.
7. High-salt TBST (HS-TBST): 10 mM Tris-HCl, 1 M NaCl, 0.05% v/v Tween-20, pH 7.5.
8. Detection reagents and autoradiography film for horse radish peroxidase (e.g., ECL, Amersham Biosciences, or equivalent detection system).

2.5 Labeling O-GlcNAc-Modified Proteins with UDP-(^{3}H)Gal and Gal-T

1. Protein sample(s), positive and negative control proteins (*see* **Section 1.12**).
2. UDP-[^{3}H]Gal (Amersham Biosciences): 1.0 mCi/mL (17.6 Ci/mmol) in 70% v/v ethanol.
3. 5′-adenosine monophosphate (5′-AMP) (Sigma-Aldrich): 25 mM in Milli-Q water, pH 7.0, store aliquots at −20°C for up to 6 months.
4. Buffer H: 50 mM HEPES, 50 mM NaCl, and 2% v/v Triton X-100, pH 6.8.
5. 10× labeling buffer: 100 mM HEPES, pH 7.5, 100 mM Gal, and 50 mM MnCl$_2$.
6. Autogalactosylated Gal-T, 30–50 U/mL (*see* **Section 1.6**).
7. UDP-Gal (unlabeled): 100 mM in water (store aliquots at −20°C for up to 6 months).
8. Stop solution: 10% w/v SDS and 100 mM EDTA.
9. Desalting column: Sephadex G-50 (30×1 cm, 20–80-μm mesh; Sigma-Aldrich) equilibrated in 50 mM ammonium formate, 0.1% w/v SDS (*see* **Note 16**).
10. Acetone.

2.6 Autogalactosylation of Gal-T (See **Note 17**)

1. 10× Gal-T buffer: 100 mM HEPES, pH 7.4, 100 mM Gal, and 50 mM MnCl$_2$.
2. Centrifuge tubes (30–50 mL).
3. Aprotinin (Sigma-Aldrich).
4. β-mercaptoethanol (Sigma-Aldrich).
5. UDP-Gal (unlabeled) (Sigma-Aldrich).
6. Gal-T: 25 U (Sigma-Aldrich; cat. G-5507).
7. Saturated ammonium sulfate: ~17.4 g (NH$_4$)$_2$SO$_4$ in 25 mL of Milli-Q water.
8. 85% ammonium sulfate: ~14 g (NH$_4$)$_2$SO$_4$ in 25 mL of Milli-Q water.
9. Gal-T storage buffer: 2.5 mM HEPES, pH 7.4, 2.5 mM MnCl$_2$, and 50% v/v glycerol.

2.7 Purifying O-GlcNAc-Modified Proteins Using the Antibody CTD110.6

1. Protein extracts (*see* **Note 18**).
2. 20 mM PUGNAc in water (use at 200 nM in extracts, *see* **Section 1.1**) and a mammalian protease inhibitor cocktail (Sigma-Aldrich).
3. Sepharose 4B (or agarose; *see* **Note 19**).
4. CTD110.6 antibody covalently coupled to Sepharose (or agarose; *see* **Note 20**).
5. Sepharose (or agarose) coupled to a nonspecific mouse IgM (*see* **Note 21**).
6. 15- and 50-mL centrifuge tubes.
7. Chromatography column housings (e.g., Econo-Pac or Poly-Prep columns, BioRad).
8. PBS-Tx (*see* **Note 22**).
9. Anti-mouse IgM-agarose (Sigma-Aldrich; cat. A4540) or anti-mouse IgM-Sepharose, Invitrogen; cat. 046841).
10. Nonspecific competition buffer: 100 mM Gal in PBS-Tx.
11. Elution buffer: 1 M GlcNAc in PBS-Tx.
12. 1 M GlcNAc in 100 mM sodium acetate buffer, 500 mM NaCl, pH 4.5 (*see* **Note 23**).
13. 0.05% w/v sodium azide in PBS.
14. Methanol.

2.8 In Vitro Transcription–Translation of O-GlcNAc-Modified Proteins

1. cDNA coding for a positive control protein, for example, nuclear pore protein p62, and for your protein(s) of interest, subcloned into an expression vector with an SP6 or T7 promoter (~0.5–1 μg/μL; *see* **Note 24**).
2. Rabbit reticulocyte lysate in vitro transcription–translation kit (Promega, Madison, WI, USA).
3. Labeled [^{35}S]Met, [^{35}S]Cys, or [^{14}C]Leu (Amersham Biosciences) (*see* **Note 25**).
4. 1-mL tuberculin syringe with a glass-wool frit.
5. Sephadex G-50, 20–80-μm mesh (Sigma-Aldrich).
6. sWGA-agarose (Vector Labs, Burlingame, CA, USA; cat. AL-1023 S; or EY labs; cat. A-2102-5).
7. PBS and PBS-Tx (*see* **Note 22**).
8. Nonspecific competition buffer: 1 M Gal in PBS-TX
9. Elution buffer: 1 M GlcNAc in PBS-TX.
10. Liquid scintillation counter.
11. Equipment and buffers for SDS-PAGE.
12. Gel dryer.

2.9 Removing O-GlcNAc from a Protein with Hexosaminidase

1. Positive control protein such as ovalbumin (*see* Table 13.4).
2. 2% w/v SDS.
3. *N*-acetyl-β-D-glucosaminidase (New England Biolabs, Ipswich, MA, USA; cat. P0721)
4. α_2-Macroglobulin (Sigma-Aldrich).
5. 2× reaction mixture: 80 m*M* citrate-phosphate buffer, pH 4.0, 1–10 U *N*-acetyl-β-D-glucosaminidase, 8% v/v Triton X-100 (*see* **Note 22**), 0.01 U/mL of aprotinin, 1 μg/mL of leupeptin, and 1 μg/mL of α_2-macroglobulin.

2.10 Removing N-Linked Glycosylation from a Protein with PNGase F

1. Purified or crude protein samples and a control protein such as ovalbumin (2–5 μg; *see* **Section 1.12**).
2. 10× PNGase F denaturing buffer: 50 m*M* sodium phosphate buffer, pH 7.5, 5% w/v SDS, and 10% v/v β-mercaptoethanol.
3. 10% v/v Nonidet P-40.
4. 10× PNGase F reaction buffer: 500 m*M* sodium phosphate buffer, pH 7.5.
5. Peptide N: glycosidase F (PNGase F) (New England Biolabs; cat. P0704).

2.11 Mapping of Sites of O-GlcNAc-Modification Using the BEMAD Strategy

1. Trypsin, sequencing grade, modified (Promega).
2. 40 m*M* ammonium bicarbonate, pH 8.0.
3. Trifluoroacetic acid (TFA), sequencing grade.
4. Performic acid oxidation solution (make fresh): 45% v/v formic acid, 5% v/v hydrogen peroxide.
5. $MgCl_2$.
6. Alkaline phosphatase (Promega).
7. DTT, high purity (GE Healthcare, Piscataway, NJ, USA).
8. BEMAD solution: 1% v/v triethylamine, 0.1% v/v NaOH, and 10 m*M* DTT, make fresh.
9. C_{18} reversed-phase macro-spin columns (The Nest Group, Southborough, MA, USA).
10. Buffer A: 1% v/v TFA in Milli-Q water.
11. Buffer B: 75% v/v acetonitrile, 1% v/v TFA in Milli-Q water.
12. Thiol column buffer: PBS and 1 m*M* EDTA, make fresh and degas.
13. Thiol column elution buffer: PBS, 1 m*M* EDTA, and 20 m*M* DTT, make fresh and degas.

14. Thiopropyl-Sepharose 6B (GE Healthcare).
15. 1% v/v acetic acid.
16. Speed Vac concentrator.
17. Finnigan LCQ with nanospray source.
18. Control peptides (*see* **Note 26**).
19. ~1–100 pmol of protein sample(s) in 40 mM ammonium bicarbonate, pH 8.0 (*see* **Note 27**).
20. Seal-Rite Natural microcentrifuge tubes (USA Scientific, Ocala, FL, USA; *see* **Note 28**).

3 Methods

3.1 Modulating O-GlcNAc Protein Modification in Cell Cultures

Increasing or decreasing the levels of O-GlcNAc on your protein(s) of interest is a useful specificity control for immunoblotting with both O-GlcNAc-specific antibodies and lectins. For site-mapping studies, it is favorable to increase the stoichiometry as much as possible. Several inhibitors of O-GlcNAcase *(40–46)*, which removes O-GlcNAc, have been reported and of these two are commercially available. We prefer PUGNAc to streptozotocin (STZ), as PUGNAc is a more potent inhibitor in vitro with a Ki of 0.22 μM compared with 2.5 mM for STZ *(41, 42)*. Moreover, in some cell types STZ can initiate apoptosis via an O-GlcNAc-independent mechanism *(47, 48)*. Several specific inhibitors of O-GlcNAc transferase (OGT) have been reported, but unfortunately they are not commercially available *(49)*. O-GlcNAc levels have also been reduced using two nonspecific OGT inhibitors *(50–52)*. The first, Alloxan, has been reported to inhibit O-GlcNAcase as well as OGT, while benzyl-α-GalNAc has never been shown to inhibit OGT in vitro and is a known inhibitor of endoplasmic reticulum (ER)/Golgi glycosylation; as such, we strongly advise against the use of both these compounds. To lower O-GlcNAc levels pharmacologically, we reduce the levels of UDP-GlcNAc by blocking the hexosamine biosynthetic pathway using DON *(53)*.

Cells grown in dishes are exposed to PUGNAc (40–100 μM) or to DON (1–20 μM) for 4–18 h (*see* **Notes 30** and **31**), harvested, and extracted as desired.

3.2 Detecting O-GlcNAc-Modified Proteins by Immunoblotting with Antibody CTD110.6

A general protocol for immunoblotting with antibody CTD110.6 is described here. However, the controls used are appropriate for any of the O-GlcNAc-specific antibodies discussed above. Controls include treating proteins with

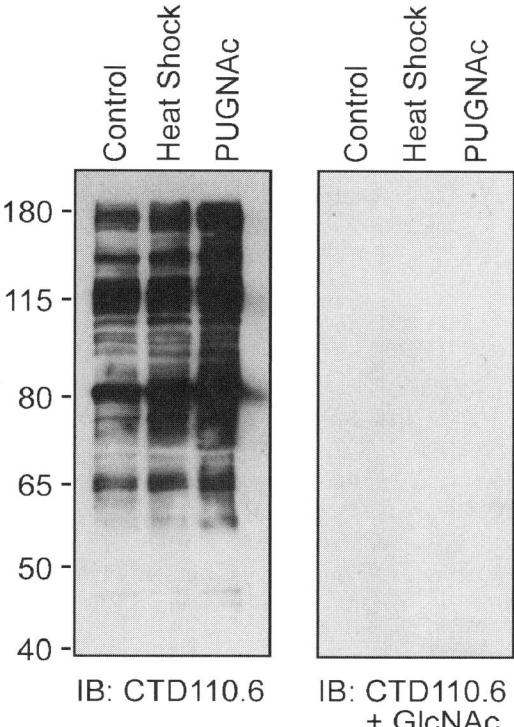

Fig. 13.2 Immunoblotting with antibody CTD110.6. Thirty micrograms of Cos-7 cell lysate from control, heat shocked, or PUGNAc-treated (100 μ*M*, 18 h) cells were separated by SDS-PAGE (7.5% gel) and transferred to nitrocellulose membranes. The membranes were probed with CTD110.6, or CTD110.6 + 100 m*M* GlcNAc

hexosaminidase (*see* **Section 1.9**); raising (PUGNAc) or lowering (DON) the levels of O-GlcNAc in cell culture (*see* **Section 1.1**); running positive and negative control proteins (*see* **Section 1.12**) such as ovalbumin (negative) and BSA-GlcNAc (positive); and competing the antibody with a specific sugar (GlcNAc) and a nonspecific sugar (*see* **Note 7**; *see* Fig. 13.2). This latter control is *very* important when immunoblotting immunoprecipitates. Note that as many of these antibodies have some peptide dependence, a negative result does not mean that the protein of is *not* modified by O-GlcNAc.

1. Incubate the membranes in blocking solution for 1 h at room temperature. This and all following incubations are with shaking or rocking.
2. Wash 3× 10 min in washing buffer at room temperature.
3. Incubate overnight at 4°C with:

Blot 1: CTD110.6 1 μg/mL in antibody incubation buffer.
Blot 2: CTD110.6 1 μg/mL in GlcNAc competition buffer.

4. Wash 3× 10 min in washing buffer at room temperature.
5. Incubate with secondary antibody (1:5,000) in antibody incubation buffer for 1 h at room temperature.
6. Wash 4× 10 min in wash buffer, then 1× 10 min in TBS at room temperature (*see* **Note 32**).
7. Develop the HRP reaction with ECL reaction and detect (*see* **Note 33**).

3.3   Purification of Antibody CTD110.6 (see **Note 8**)

1. Defrost CTD110.6 ascites fluid on ice, dilute 10-fold with PBS, add protease inhibitors, and centrifuge at >10,000×*g* for 20 min at 4°C.
2. Decant the supernatant and filter through a 1-μm pore-size filter. Adjust the pH to 7.0–8.0 with 1 *M* Tris-HCl, pH 8.0 if necessary.
3. Pack 4 mL of GlcNAc-Sepharose slurry (50% v/v) in a column housing (*see* **Note 34**), resulting in a column with a 2-mL bed volume (*see* **Note 35**).
4. Wash the column with a minimum of 10 column volumes (~20 mL) of PBS at a flow rate of 0.5–1 mL/min (*see* **Note 36**).
5. Apply the antibody solution to the column at a flow rate of 0.2–0.5 mL/min (*see* **Note 37**). Collect the material that flows through and designate as the "flowthrough."
6. Reapply the flowthrough fraction to the column and save the material that flows through as Flowthrough 2.
7. Wash the column with at least 10 column volumes of PBS (~20 mL), then 5 column volumes (~10 mL) of nonspecific competition buffer, then at least 5 column volumes (~10 mL) of PBS.
8. Elute the bound antibody from the column by adding 10× 2 mL of GlcNAc elution buffer, collecting 2-mL fractions and designating them "eluant 1–10."
9. Wash the column with at least 5 column volumes (~10 mL) of PBS.
10. Assay the eluted fractions for protein to determine which contain the CTD110.6. The antibody can be assayed by protein estimation, A_{280}, SDS-PAGE, or enzyme-linked immunoabsorbent assay (ELISA).
11. Combine the fractions containing antibody, and remove the GlcNAc by dialysis against PBS at 4°C (*see* **Note 38**) or spin-filtration.
12. Measure the protein concentration of the dialyzed antibody. Concentrate if necessary. Dilute to 2 mg/mL and add an equal volume of glycerol, giving a final antibody concentration of 1 mg/mL in 50% v/v glycerol. The antibody can be stored at −20°C for at least 1 year.
13. Test the flowthrough fractions to confirm that all CTD110.6 was recovered (*see* **Note 39**).
14. The column should be stored in PBS with 0.05% w/v sodium azide.

3.4 Detecting O-GlcNAc-Modified Proteins with sWGA

The lectins sWGA, *Griffonia simplicifolia* 1 (GS-1), and GS-2 recognize O-GlcNAc (*see* **Note 40**). This protocol is adaptable to other lectins, but TBST should be substituted for HS-TBST, and GlcNAc in the competition buffer should be substituted by the appropriate sugar (Table 13.1).

1. Wash duplicate blots for 10 min in TBST (*see* **Note 5**).
2. Incubate in blocking buffer for 60 min at room temperature (*see* **Note 13**).
3. Wash 3× 10 min in TBST.
4. Incubate in 1 µg/mL sWGA-HRP in blocking solution with and without 1 *M* GlcNAc overnight at 4°C.
5. Wash in HS-TBST 6× 10 min and in TBS 1× 10 min.
6. Develop the HRP reaction.

3.5 Detecting O-GlcNAc-Modified Proteins by Labeling with UDP-(³H)Gal and Gal-T

Gal-T requires Mn^{2+} ions but is inhibited by concentrations >5 mM and by low concentrations of Mg^{2+}. For stoichiometric labeling, it is best to first denature the protein samples by boiling in 10 mM DTT and 0.5% w/v SDS for 5 min. The SDS, which inhibits Gal-T, should be titrated out by adding NP-40 to a 10-fold higher concentration. Digitonin should be avoided as it can react with Gal-T.

1. Place the UDP-[³H]Gal (~1–2 µCi/reaction) in a microcentrifuge tube and remove the solvent in a Speed Vac or under a stream of nitrogen (*see* **Note 41**).
2. Resuspend the UDP-[³H]Gal in 25 mM 5′-AMP (*see* **Note 42**). Use 50 µL per reaction.
3. Set reactions up as shown in Table 13.5 (*see* **Note 43**).
4. Incubate the samples for 2 h at 37°C or overnight at 4°C.
5. Add unlabeled UDP-Gal to a final concentration of 0.5–1.0 mM and 2–5 µL of fresh Gal-T (*see* **Note 45**). Incubate for a further 2–4 h.

Table 13.5 Reactions setup

Reagent	Volume
Sample (0.5–5 mg)	Up to 50 µL
Buffer H	350 µL
10× labeling buffer	50 µL
Calf intestinal alkaline phosphatase[a]	1–4 U
UDP-[³H]Gal/5′-AMP	50 µL
Gal-T 30–50 U/mL[b]	2–5 µL
Milli Q water to a final volume of	500 µL

[a] *See* **Note 44**

[b] It is best to use autogalactosylated Gal-T (*see* **Section 1.6** and **Note 17**)

6. Add 50 μL of stop solution to each reaction and heat at 100°C for 5 min.
7. Separate protein(s) from unincorporated label on a Sephadex G-50 column (30×1 cm equilibrated in 50 m*M* ammonium formate, 0.1% w/v SDS. Collect 1-mL fractions (*see* **Note 16**).
8. Measure radioactivity in an aliquot (50 μL) of each fraction using a liquid scintillation counter (*see* **Note 46**).
9. Combine fractions in the void volume and lyophilize to dryness.
10. Resuspend the sample(s) in 100–1,000 μL of Milli-Q water, add at least 3 volumes of ice-cold acetone, and incubate for 2–18 h at −20°C.
11. Pellet protein(s) at a minimum of 3,000×*g* for 10 min at 4°C (*see* **Note 47**).
12. Decant the acetone, air-dry the pellets (*see* **Note 48**), and resuspend in your buffer of choice.
13. To confirm that Gal was added to a monosaccharide of O-linked GlcNAc and not to other O-linked or N-linked carbohydrates, the reaction product must be characterized (*see* **Note 49**).

3.6   Autogalactosylation of Gal-T

1. Resuspend 25U of Gal-T (Sigma-Aldrich; cat. G5507) in 1 mL of 1× Gal-T buffer.
2. Transfer the Gal-T to a 30–50-mL centrifuge tube.
3. Remove a 5-μL aliquot for an activity assay.
4. Add 10 μL of aprotinin, 3.5 μL of β-mercaptoethanol, and 1.5–3.0 mg of UDP-Gal.
5. Incubate the sample on ice for 30–60 min.
6. Add 5.66 mL of prechilled saturated ammonium sulfate drop-wise to the reaction, mixing well, and incubate on ice for 30 min.
7. Centrifuge at > 10,000×*g* for 15 min at 4°C.
8. Resuspend the pellet in 5 mL of cold 85% ammonium sulfate, incubate on ice for 30 min.
9. Centrifuge at > 10,000×*g* for 15 min at 4°C.
10. Resuspend the pellet in 1 mL of Gal-T storage buffer.
11. Aliquot the autogalactosylated Gal-T (50 μL), and store at −20°C for up to 1 year.
12. Assay 5 μL of the autogalactosylated and nongalactosylated enzyme to determine the activity (above).

3.7   Purifying O-GlcNAc-Modified Proteins

O-GlcNAc-modified proteins can be affinity-purified using sWGA-Sepharose or CTD110.6-Sepharose. Galactosylated proteins (*see* **Section 2.5**) can also be enriched using RCA1 affinity chromatography with minor changes to the protocol (*see* **Note 50**). While the method works well using typical immunoaffinity/immunoprecipitation procedures, a column format gives cleaner data and is more appropriate for large-scale purifications. The protocol described below is appropriate for ~10–50 mg of cell extract and is also compatible with sWGA affinity chromatography (*see* **Note 51**). CTD110.6 is covalently coupled to cyanogen bromide-activated Sepharose and lysates are precleared using Sepharose 4B or a nonspecific IgM coupled to cyanogen bromide-activated Sepharose.

1. Equilibrate Sepharose 4B in PBS-TX (at least 10 column volumes, 20 mL).
2. Preclear cell lysate by adding 1 mL of IgM-Sepharose or Sepharose 4B (*see* **Note 52**) to 10–50 mg of cell extract. Incubate with rocking for 2 h at 4°C.
3. Separate the IgM-Sepharose or Sepharose 4B from the cell lysate either by centrifuging the mixture at 1,500×*g* for 5 min and recovering the supernatant, or by placing the mixture in a column housing and collecting the flowthrough.
4. Equilibrate 0.5 mL of CTD110.6-Sepharose (2 mg/mL) and IgM-Sepharose (2 mg/mL (*see* **Note 52**) in PBS-TX (at least 10 column volumes, ~5 mL).
5. Add 5 mg of cell extract to the CTD110.6 affinity column and 5 mg to the IgM affinity column. Incubate with mixing overnight at 4°C.
6. Pack 200 μL of anti-IgM Sepharose into a column housing and equilibrate with at least 10 volumes of PBS-TX (~2 mL).
7. Place an additional frit on the top of the anti-IgM agarose (*see* **Note 53** and Fig. 13.3).

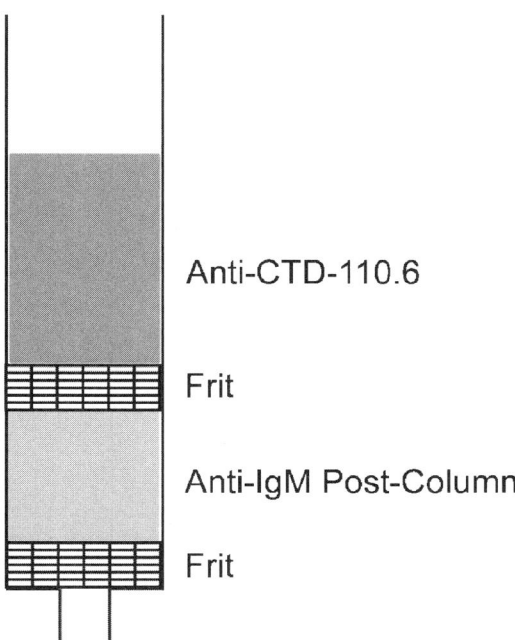

Fig. 13.3 Schematic of CTD110.6 affinity columns. A 200-μL anti-IgM column is poured and capped with a frit. CTD110.6/IgM-Sepharose that has been incubated with the sample overnight is loaded on top of the anti-IgM column, as shown. The anti-IgM column traps any antibody that leaches from the CTD110.6/IgM-Sepharose column

8. Add the cell lysate/affinity-matrix from **Step 5** to the anti-IgM columns and collect the flowthrough.

9. Wash the columns with 10 volumes (7 mL) of PBS-TX (*see* **Note 54**) at ~0.2–0.5 mL/min.

10. Wash the columns with 10 column volumes (7 mL) of nonspecific competition buffer (*see* **Note 12**) followed by 10 column volumes (7 mL) of PBS-TX.

11. Elute O-GlcNAc-modified proteins in 2 column volumes (~1.4 mL) of GlcNAc elution buffer and repeat four times. Collect the eluted fractions separately.

12. Wash the columns with 10 column volumes (~7 mL) of 100 mM sodium acetate, pH 4, 0.5 M NaCl, and 0.5 M GlcNAc (*see* **Note 23**) followed by 10 column volumes (~7 mL) of PBS.

13. Determine the elution pattern of O-GlcNAc-modified proteins by SDS-PAGE and immunoblotting.

14. Samples are either desalted (below) or acidified and injected directly onto a reversed-phase column. Desalting methods include precipitation with methanol (*see* **Note 55**), spin-filtration, or dialysis.

15. Store the columns in PBS with 0.05% w/v sodium azide.

3.8   *In Vitro Transcription–Translation to Screen for O-GlcNAc Modification*

It is possible to screen for the O-GlcNAc modification on low-abundance proteins by synthesizing the proteins in an in vitro rabbit reticulocyte lysate transcription–translation (RRL ITT) system and determining if they bind sWGA or CTD110.6. Typically, [^{14}C]leucine, [^{35}S]cysteine, or [^{35}S]methionine are incorporated during translation, and elution of the proteins from either sWGA or CTD110.6 is followed by liquid scintillation counting. Due to the high concentration of unincorporated label, these experiments are best performed in column format.

1. Synthesize the protein of choice using a RRL ITT system according to the manufacturer's instructions. Include a negative control (for example, luciferase supplied with the kit), a positive control (for example, the nuclear pore protein p62), and a no-DNA control.

2. Treat half of each sample with hexosaminidase (*see* **Note 56**).

3. Desalt the samples as follows:

   (a) Pour exactly 1 mL of Sephadex G-50 into a tuberculin syringe that has been packed with a small glass wool frit (~100 µL) and wash the column with 5 mL of PBS-TX.

   (b) Load the RRL reaction mix onto the column; the sample volume can be up to 200 µL.

   (c) Wash the column with PBS-TX so that the total volume of this wash and the protein sample is 350 µL (e.g., if the sample volume is 150 µL, add 200 µL of PBS-TX).

    (d) Transfer the column to a clean prechilled tube. Elute proteins with 200 μL of PBS-TX. This is the desalted sample.

4. Equilibrate ~150 μL of sWGA-agarose in ~3 mL of PBS-TX; perform this and the following steps at 4°C or with ice cold buffers.
5. Cap the outlet of the column.
6. Apply the desalted sample to the column and stand at 4°C for 30 min.
7. Uncap the column and allow unbound material to run though the column.
8. Wash the column with 15 mL of PBS-TX at ~0.5 mL/min, collecting 0.5-mL fractions.
9. Cap the column and load with 500 μL of 100 m*M* Gal in PBS-TX, stand for 20 min.
10. Wash the column with 5 mL of 100 m*M* Gal in PBS-TX, collecting 0.5 mL fractions.
11. Repeat steps 9 and 10 using 1 *M* GlcNAc in PBS-TX.
12. Count 25 μL of each fraction using a liquid scintillation counter.
13. Pool the radioactive fractions that are eluted by GlcNAc and precipitate using TCA or methanol (*see* **Note 55**).
14. Analyze the pellet by SDS-PAGE and autoradiography to confirm that the label has been incorporated into a protein of an appropriate size.

3.9 Removing O-GlcNAc Using Hexosaminidase

Terminal GlcNAc residues can be removed using commercial hexosaminidases. While this will show that the reactivity is GlcNAc specific, it will not confirm that the signal was O-GlcNAc dependent unless used in conjunction with an O-GlcNAc-specific antibody. There are numerous other hexosaminidases that will provide useful specificity controls for other carbohydrates, and the manufacturer's instructions should be followed. Note that these are often contaminated with other activities, and care should be taken when interpreting the results.

1. Protein samples (include a positive control such as ovalbumin; *see* Table 13.4).
2. Mix the sample 1:1 with 2% w/v SDS and boil for 5 min.
3. Mix the sample 1:1 with reaction mixture and incubate at 37°C for 4–24 h.

3.10 Removing N-linked Glycosylation Using PNGase F

1. Add 1/10 of the sample volume of 10× PNGase F denaturing buffer to each sample and heat at 100°C for 10 min.
2. Add 1/10 of the sample volume of 10× PNGase F reaction buffer and 1/10 of the sample volume of 10% v/v NP-40, mix (*see* **Note 57**).
3. Add 1 μL of PNGase F and incubate at 37°C for 1 h to overnight.
4. Analyze deglycosylation by SDS-PAGE and sWGA immunoblot (*see* **Note 58**).

3.11   Mapping of Sites of O-GlcNAc-Modification Using the BEMAD Strategy

Performic acid oxidation (*see* **Note 59**):

1. Suspend the protein sample(s) in 300 μL of performic acid oxidation buffer.
2. Spike with 1–10 pmol of control peptides.
3. Incubate on ice for 1 h.
4. Dry down in a Speed Vac concentrator.

Trypsin digestion:

5. Resuspend the protein sample(s) in ~100 μL of 40 m*M* ammonium bicarbonate.
6. Digest by the addition of 1:10 (w/w) of sequencing-grade trypsin, and incubate overnight at 37°C.9
7. Acidify the digest by addition of TFA (~1% v/v).
8. Desalt peptides using a C$_{18}$ reversed-phase column according to the manufacturer's instructions.
9. Dry the peptides using a Speed Vac concentrator.

Phosphatase treatment:

10. Resuspend the samples in 100 μL of 40 m*M* ammonium bicarbonate, 1 m*M* MgCl$_2$.
11. Add alkaline phosphatase (1 U/10 μL) and incubate at 37°C for 4 h.
12. Dry down in a Speed Vac.

BEMAD treatment (*see* **Note 59**):

13. Resuspend the samples in 500 μL of BEMAD solution and adjust the pH to 12.0–12.5 with triethylamine if necessary.
14. Incubate the reactions at 50°C for 2.5 h.
15. Stop the reaction by adding TFA to 1% (v/v).
16. Desalt the samples on C$_{18}$ reversed-phase columns according to the manufacturer's instructions.
17. Dry down the peptides in a Speed Vac.

Thiol affinity column (*see* **Note 60**):

18. Swell and wash thiopropyl-Sepharose resin in degassed thiol column buffer.
19. Resuspend the peptides in thiol column buffer.
20. Bind the peptides to the thiol column at room temperature for 1 h (*see* **Note 61**).
21. Wash the column with 20 mL of thiol column buffer.
22. Elute peptides three times sequentially with 150 μL of Thiol elution buffer.
23. Acidify the peptides by adding TFA to 1% v/v.
24. Desalt the peptides using C$_{18}$ reversed-phase chromatography according to the manufacturer's instructions.

25. Dry down the peptides in a Speed Vac for liquid chromatography (LC)-MS/MS analysis (*see* **Note 62**):

26. Resuspend the peptides in 1% v/v acetic acid.

27. Load the sample(s) onto a C$_{18}$ reversed-phase nanobore column (10 cm×0.075 mm, 5-μm beads) equilibrated with 1% v/v acetic acid.

28. Peptides are separated using a 75-min linear gradient of acetonitrile, 1% v/v acetic acid at a flow rate of ~200 nL/min into the MS source (Finnigan LCQ; Thermo Scientific, Waltham, MA, USA). Data is typically collected in automatic mode with an MS scan (2× 500 msec) followed by two MS/MS scans (3× 750 msec) of the two highest-intensity peptides with a dynamic exclusion of 2 and a mass gate of 2.0 daltons. Alternatively, MS/MS data may be collected manually by choosing peaks of interest for fragmentation from the MS scans during the run.

29. We use Turbosequest software (Thermo Scientific) to interpret MS/MS data, allowing for a differential mass increase of 136.2 daltons to Ser and Thr residues, 120.2 daltons to Cys residues that may have been derivatized with DTT, 48.0 daltons to performic acid-oxidized Cys and Trp residues, and 32.0 daltons to performic acid-oxidized Met residues.

4 Notes

1. The antibody HGAC85 was raised against a streptococcal N-linked glycoprotein *(13)* and should be used with caution; samples should be treated with PNGase F to show that any positive antibody signal is not the result of an N-linked sugar. Unlike RL2 and CTD110.6, the blocking solution should be 2% w/v BSA in TBST.

2. Most mammalian cells are more responsive to PUGNAc at physiological levels of Glc, for two possible reasons: PUGNAc may be transported into cells through Glc transporters, and at lower Glc concentrations there is less competition; or O-GlcNAc levels are responsive to extracellular Glc concentrations, and thus at 1 g/L the baseline O-GlcNAc levels are lower.

3. Typically, O-GlcNAc can be detected using 10–15 μg of cell lysate, although this is dependent on the cell type. The best data is obtained when ~30 μg of protein is loaded onto mini- and mid-sized SDS-PAGE gels, and at least ~200 μg on large-format gels. Because O-GlcNAc is concentrated in nuclear proteins, ~15 μg is usually sufficient on mini- and mid-sized gels. When detecting O-GlcNAc on immunoprecipitated proteins, we generally load the immunoprecipitate from ~300 μg of cell lysate.

4. O-GlcNAc is highly labile as a result of chemical instability and high concentrations of hexosaminidases in cell lysates. In addition to protease and phosphatase inhibitors, we generally include the O-GlcNAcase inhibitor PUGNAc at 200 n*M* in all cell extraction buffers.

5. We prefer nitrocellulose membranes for this application and use one produced by Bio-Rad, or PVDF from Millipore. To control for the specificity of GlcNAc-spe-

cific antibodies and lectins, the signal should be competed away by free GlcNAc. For lectin-binding studies, it is advisable to run a third blot and demonstrate that there is no competition of sWGA by a nonspecific sugar such as Gal.

6. GlcNAc solutions should be made fresh. As GlcNAc is hydroscopic, the bottle should be removed from the −20°C freezer and allowed to warm to room temperature before opening.

7. For the antibody CTD110.6, man, Glc, Gal, and GalNAc are appropriate sugars for nonspecific competition. We typically use Glc or Gal, which are less expensive and commonly found in most laboratories.

8. Ascites fluid can contain up to 50 mg/mL of albumin and other proteins. We find that purification of the antigen results in immunoblots of higher quality and is preferable before covalent coupling.

9. Covance Research Products sells CTD110.6 ascites fluid, and Pierce Biotechnology sells CTD110.6 as part of an O-GlcNAc detection kit. Neither company antigen-purifies the antibody before sale.

10. We have used both products to successfully purify CTD110.6.

11. Glc or Gal provides a specificity control. This control is *most* essential for immunofluorescence and immunoblots of immunoprecipitates.

12. Milk proteins are modified by glycans with terminal GlcNAc, and will react with sWGA. Thus, membranes are blocked with BSA that is not glycosylated.

13. WGA recognizes both sialic acid and GlcNac; succinylation blocks the recognition of sialic acid (*see* **Section 2.4, step 5**).

14. sWGA can be stored in PBS:glycerol (1:1) at −20°C for at least 1 year.

15. Other forms of desalting can be used, e.g., TCA precipitation, spin filtration/buffer exchange, or commercial desalting columns such as PD-10 columns (GE Healthcare). However, the 30-cm column described here provides the best resolution between labeled protein and unincorporated [^{3}H]UDP-Gal. The addition of carrier proteins such as BSA (~67 kDa) or cytochrome C (~12.5 kDa) to dilute protein samples will reduce protein loss due to nonspecific adsorption.

16. Bovine milk Gal-T is modified by N-linked glycans that can terminate in GlcNAc. Thus it is necessary to block these with unlabeled Gal.

17. CTD110.6 affinity chromatography can also be performed in 1% (v/v) NP-40 and RIPA cell lysis buffer. Typically, cell lysates are diluted twofold with 50 mM Tris-HCl before loading onto the affinity columns, but for sWGA they should be diluted fivefold.

18. Sepharose (or agarose) is used in the preclearing step to remove any proteins that will nonspecifically bind to the resin. Sepharose (or agarose) coupled to a nonspecific IgM can also be used to preclear. If using CTD110.6 coupled to Sepharose, Sepharose should be used in the preclearing step.

19. For this procedure, we typically couple CTD110.6 to cyanogen bromide-activated Sepharose at 2–3 mg/mL. The covalently coupled antibody is reusable, and less antibody is released into protein extracts which is preferable upstream of MS.

20. A nonspecific IgM provides an ideal specificity control and should be coupled to the same matrix as CTD110.6.

21. Triton X-100 and NP-40 are both compatible with this method.

22. A low-pH wash with free GlcNAc cleanses the column, reducing background in subsequent experiments. It may also be useful for proteins that appear to bind to the CTD110.6 column with high affinity.

23. The pCITE vectors (Novagen, Madison, WI, USA) are designed to work at high efficiency in in vitro transcription–translation reactions.

24. The choice of label depends on the amino acid composition of the protein. [^{35}S]Cys is the most common, but many proteins have only a few Cys residues and another label may be a more appropriate choice.

25. The sample should be spiked with 1–10 pmol of a known phosphorylated and/or glycosylated peptide. An O-GlcNAc-bearing peptide from the basic phospho-protein (BPP) of human cytomegalovirus, PSVPVS(O-GlcNAc)GSAPGR, is commonly used, and is synthesized as previously reported *(54)*. Alternatively, Invitrogen sells a phosphorylated and glycosylated version of the CREB pep-tide TAPTS(O-GlcNAc/phosphate)TIAPG (cat. C33373).

26. The amount of starting material will depend on sample purity, the stoichiometry of O-GlcNAc, and the sensitivity of the mass spectrometer. Using a >90% pure sample with high stoichiometry on a Finnigan LCQ Classic, reasonable data can be collected on picomole amounts of material.

27. To reduce contamination of your sample(s) from the tubes, we recommend these tubes. We also rinse all tubes with 50% acetonitrile before use. Tubes are never autoclaved.

28. Different cells lines have different responsiveness to PUGNAc and DON. At high levels or for extended treatments, both PUGNAc and DON can be toxic to cells.

29. Most cell lines show elevations in O-GlcNAc by 2–4 h with robust increases by 8 h. Cells should not be treated for longer than 18 h, as defects in growth and division have been observed. Decreases in O-GlcNAc levels are seen in 12–18 h with DON, and prolonged treatment causes defects in growth and division.

30. Ovalbumin contains terminal GlcNAc residues. It should react with sWGA and WGA, but *not* with any of the O-GlcNAc-specific antibodies (RL2, CTD110.6, MY95). BSA-GlcNAc can easily be made using aminophenyl-GlcNAc (EY Labs; cat. CG-029-5; Sigma-Aldrich; cat. B6893).

31. Tween-20 can cause an increased background and spottiness in the ECL reaction.

32. Some cross-reactivity with prestained markers may be observed; this is normal.

33. To avoid air bubbles in the column bed, add the resin to water already in the column housing.

34. The volumes described in this protocol are appropriate for 1 mL of ascites fluid. The yield is usually 2–3 mg of purified antibody.

35. This protocol is either performed in the cold room or with cold buffers kept on ice, and fractions are stored in the cold room or on ice to avoid degradation.

36. For an open system, the flow rate can be altered by attaching a narrow-gauge needle to the exit of the column. In a closed system running under gravity, the flow rate can be adjusted by changing the height of the buffer reservoir.

37. For some applications, PBS is not the best dialysis buffer. For instance, for covalent coupling to CNBR-activated Sepharose, the antibody should be dialyzed against 200 m*M* sodium bicarbonate and 500 m*M* NaCl, pH 8.3–8.5.

38. It is possible to determine if all the CTD110.6 has been removed from the flow-through by SDS-PAGE and Coomassie blue staining, ELISA, or Western blot.

39. Typically sWGA reacts best with proteins/peptides that have a cluster of O-GlcNAc sites. GS-1 and GS-2 requires just two GlcNAc residues for maximum binding. Importantly, 0.5 m*M* CaCl$_2$ should be added to buffers when using GS-1 and GS-2.

40. Large amounts of ethanol inhibit Gal-T, and no more than 4 µL should be present in a 500-µL reaction.

41. AMP is included to inhibit any phosphodiesterase activity in protein preparations, which degrades UDP-GlcNAc.

42. There should be a positive control (2 µg ovalbumin) and a blank in addition to the sample(s).

43. UDP, a byproduct of the reaction, is a potent inhibitor of Gal-T. Alkaline phosphatase degrades UDP.

44. For studies where complete labeling of O-GlcNAc is required, such as site-mapping, reactions are chased with cold UDP-Gal.

45. ~2×10^6 dpm should be incorporated into 2 µg of ovalbumin.

46. Most 15-mL polypropylene or polycarbonate tubes can be used for acetone precipitation, although centrifuging faster than 3,000×*g* is *not* recommended.

47. Do not over-dry the pellet, otherwise it will be dehydrated and difficult to resuspend.

48. The size of the β-eliminated product, a sugar alditol, is assessed by size exclusion chromatography on BioGel P4; and finally, the composition of the disaccharide βGal1-4βGlcNAc is confirmed by high performance anion-exchange chromatography (HAPEC) on a CarboPac PA100 column (Dionex, Sunnyvale, CA, USA) *(20, 21)*.

49. For RCA1 affinity chromatography, GlcNAc should be the "nonspecific sugar" wash and Gal should be in the elution buffer.

50. Lysates should be precleared with resin alone or with another nonspecific lectin. ConA affinity chromatography would be ideal, as this would remove many proteins modified by N-linked carbohydrates.

51. Proteins that bind to the nonspecific IgM resin are deemed "not modified by O-GlcNAc." Thus, this is an additional specificity control.

52. IgMs are pentamers and not all the subunits are coupled to the resin, so that a small amount of antibody can leach off the column. The subsequent anti-IgM columns are to trap any CTD110.6 released from the affinity column. This is especially important for samples being prepared for mass spectrometry.

53. For samples to be analyzed by mass spectrometry, Triton X-100 and NP-40 can be omitted from subsequent steps.

54. To concentrate proteins and remove free GlcNAc, proteins are precipitated by adding at least 10 volumes of ice-cold methanol to samples and incubating overnight at −20°C. Proteins are precipitated by centrifuging at >3,000×*g* for

30 min at 4°C and the pellets are air-dried for 5–10 min. Note that acetone precipitates free GlcNAc and should not be used.

55. This is optional, but provides an additional level of confidence.

56. PNGase F is inhibited by SDS. It is essential to add NP-40.

57. Ovalbumin is modified by one N-linked glycan, and treatment with PNGase F should increase its mobility by several kilodaltons on a 10–12% SDS-PAGE gel, a shift which is difficult to resolve on a 7.5% gel. Ideally, samples would be blotted to a membrane and detected with sWGA.

58. The performic acid oxidation step can be performed before or after trypsinization of the protein(s). This step may help denature proteins, increasing the efficiency of the trypsin digestion. Alternatively, proteins can be reduced and alkylated prior to this step. If the alkylation is performed with iodoacetamine, there will be a mass increase of 57.052 daltons per Cys residue.

59. This method can be adapted to map Ser/Thr phosphorylation sites as follows: instead of the phosphatase treatment, samples are acidified to pH 4.5 with TFA and then treated with hexosaminidase (*see* **Section 2.9**). Additionally, the BEMAD solution should be altered to 2% v/v triethylamine, 0.2% v/v NaOH, and 10 m*M* DTT and the reaction should be performed for 5 h at 50°C.

60. Peptides can be bound to the column for longer than 1 h.

61. The LC-MS/MS methodology described here should be used as a guide, and should be optimized for the mass spectrometer available.

Acknowledgments The author acknowledges Prof. Gerald W. Hart (Johns Hopkins University School of Medicine) for comments on this manuscript, and the technical help of Katie Zoey Ho (Johns Hopkins Singapore). NEZ was supported by an A*Star Research grant to Johns Hopkins Singapore.

References

1. Hart, G. W., Haltiwanger, R. S., Holt, G. D., and Kelly, W. G. (1989) Glycosylation in the nucleus and cytoplasm. *Annu. Rev. Biochem.* **58**, 841–874.

2. Lomako, J., Lomako, W. M., and Whelan, W. J. (2004) Glycogenin: the primer for mammalian and yeast glycogen synthesis. *Biochim. Biophys. Acta* **1673**, 45–55.

3. Marchase R, B., Bounelis P, Brumley L. M., Dey N, Browne B, Auger D, Fritz T. A., Kulesza P, and Bedwell D. M. (1993) Phosphoglucomutase in Saccharomyces cerevisiae is a cytoplasmic glycoprotein and the acceptor for a Glc-phosphotransferase. *J. Biol. Chem.* **268**, 8341–8349.

4. Dey, N. B., Bounelis, P., Fritz, T. A., Bedwell, D. M., and Marchase, R. B. (1994) The glycosylation of phosphoglucomutase is modulated by carbon source and heat shock in Saccharomyces cerevisiae. *J. Biol. Chem.* **269**, 27143–27148.

5. Veyna N. A., Jay J. C., Srisomsap C, Bounelis P, and Marchase R. B. (1994) The addition of glucose-1-phosphate to the cytoplasmic glycoprotein phosphoglucomutase is modulated by intracellular calcium in PC12 cells and rat cortical synaptosomes. *J. Neurochem.* **62**, 456–464.

6. Schirmer, J. and Aktories, K. (2004) Large clostridial cytotoxins: cellular biology of Rho/Ras-glucosylating toxins. *Biochim. Biophys. Acta* **1673**, 66–74.

7. West, C. M., Van Der Wel, H., Sassi, S., and Gaucher, E. A. (2004) Cytoplasmic glycosylation of protein-hydroxyproline and its relationship to other glycosylation pathways. *Biochim. Biophys. Acta* **1673**, 29–44.

8. Zachara, N. E. and Hart, G. W. (2004) O-GlcNAc a sensor of cellular state: the role of nucleocytoplasmic glycosylation in modulating cellular function in response to nutrition and stress. *Biochim. Biophys. Acta* **1673**, 13–28.

9. Zachara, N. E. and Hart, G. W. (2006) Cell signaling, the essential role of O-GlcNAc. *Biochim. Biophys. Acta* **1761**, 599–617.

10. O'Donnell, N., Zachara, N. E., Hart, G. W., and Marth, J. D. (2004) Ogt-dependent X-chromosome-linked protein glycosylation is a requisite modification in somatic cell function and embryo viability. *Mol. Cell. Biol.* **24**, 1680–1690.

11. Shafi, R., Iyer, S. P., Ellies, L. G., O'Donnell, N., Marek, K. W., Chui, D., Hart, G. W., and Marth, J. D. (2000) The O-GlcNAc transferase gene resides on the X chromosome and is essential for embryonic stem cell viability and mouse ontogeny. *Proc. Natl. Acad. Sci. USA* **97**, 5735–5739.

12. Hartweck, L. M., Scott, C. L., and Olszewski, N. E. (2002) Free in PMC Two O-linked N-acetylglucosamine transferase genes of Arabidopsis thaliana L. Heynh. have overlapping functions necessary for gamete and seed development. *Genetics* **161**, 1279–1291.

13. Turner, J. R., Tartakoff, A. M., and Greenspan, N. S. (1990) Cytologic assessment of nuclear and cytoplasmic O-linked N-acetylglucosamine distribution by using anti-streptococcal monoclonal antibodies. *Proc. Natl. Acad. Sci. USA* **87**, 5608–5612.

14. Holt, G. D., Snow, C. M., Senior, A., Haltiwanger, R. S., Gerace, L., and Hart, G. W. (1987) Nuclear pore complex glycoproteins contain cytoplasmically disposed O-linked N-acetylglucosamine. *J. Cell Biol.* **104**, 1157–1164.

15. Snow, C. M., Senior, A., and Gerace, L. (1987) Monoclonal antibodies identify a group of nuclear pore complex glycoproteins. *J. Cell Biol.* **104**, 1143–1156.

16. Comer, F. I., Vosseller, K., Wells, L., Accavitti, M. A., and Hart, G. W. (2001) Characterization of a mouse monoclonal antibody specific for O-linked N-acetylglucosamine. *Anal. Biochem.* **293**, 169–177.

17. Matsuoka, Y., Shibata, S., Yasuhara, N., and Yoneda, Y. (2002) MAb MY95, Anti-O-linked N-acetylglucosamine-modified proteins. *Hybrid. Hybridomics* **21**, 233–236.

18. Kamemura, K., Hayes, B. K., Comer, F. I., and Hart, G. W. (2002) Dynamic interplay between O-glycosylation and O-phosphorylation of nucleocytoplasmic proteins: alternative glycosylation/phosphorylation of THR-58, a known mutational hot spot of c-Myc in lymphomas, is regulated by mitogens. *J. Biol. Chem.* **277**, 19229–19235.

19. Ludemann, N., Clement, A., Hans, V. H., Leschik, J., Behl, C., and Brandt, R. (2005) O-glycosylation of the tail domain of neurofilament protein M in human neurons and in spinal cord tissue of a rat model of amyotrophic lateral sclerosis (ALS). *J. Biol. Chem.* **280**, 31648–31658.

20. Roquemore, E. P., Chou, T. Y., and Hart, G. W. (1994) Detection of O-linked N-acetylglucosamine (O-GlcNAc) on cytoplasmic and nuclear proteins. *Methods Enzymol.* **230**, 443–460.

21. Torres, C. R., and Hart, G. W. (1984) Topography and polypeptide distribution of terminal N-acetylglucosamine residues on the surfaces of intact lymphocytes. Evidence for O-linked GlcNAc. *J. Biol. Chem.* **259**, 3308–3317.

22. Khidekel, N., Arndt, S., Lamarre-Vincent, N., Lippert, A., Poulin-Kerstien, K. G., Ramakrishnan, B., Qasba, P. K., and Hsieh-Wilson, L. C. (2003) A chemoenzymatic approach toward the rapid and sensitive detection of O-GlcNAc posttranslational modifications. *J. Am. Chem. Soc.* **125**, 16162–16163.

23. Vocadlo, D. J., Hang, H. C., Kim, E. J., Hanover, J. A., and Bertozzi, C. R. (2003) A chemical approach for identifying O-GlcNAc-modified proteins in cells. *Proc. Natl. Acad. Sci. USA* **100**, 9116–9121.

24. Hayes, B. K., Greis, K. D., and Hart, G. W. (1995) Specific isolation of O-linked N-acetylglucosamine glycopeptides from complex mixtures. *Anal. Biochem.* **228**, 115–122.

25. Greis, K. D., Hayes, B. K., Comer, F. I., Kirk, M., Barnes, S., Lowary, T. L., and Hart, G. W. (1996) Selective detection and site-analysis of O-GlcNAc-modified glycopeptides by beta-elimination and tandem electrospray mass spectrometry. *Anal. Biochem.* **234**, 38–49.
26. Greis, K. D. and Hart, G. W. (1998) Analytical methods for the study of O-GlcNAc glycoproteins and glycopeptides. *Methods Mol. Biol.* **76**, 19–33.
27. Haynes, P. A. and Aebersold, R. (2000) Simultaneous detection and identification of O-GlcNAc-modified glycoproteins using liquid chromatography-tandem mass spectrometry. *Anal. Chem.* **72**, 5402–5410.
28. Zachara, N. E. and Gooley, A. A. (2000) Identification of glycosylation sites in mucin peptides by edman degradation. *Methods Mol. Biol.* **125**, 121–128.
29. Vosseller, K., Hansen, K. C., Chalkley, R. J., Trinidad, J. C., Wells, L., Hart, G. W., and Burlingame, A. L. (2005) Quantitative analysis of both protein expression and serine/threonine post-translational modifications through stable isotope labeling with dithiothreitol. *Proteomics* **5**, 388–398.
30. Wells, L., Vosseller, K., Cole, R. N., Cronshaw, J. M., Matunis, M. J., and Hart, G. W. (2002) Mapping sites of O-GlcNAc modification using affinity tags for serine and threonine post-translational modifications. *Mol. Cell. Proteomics* **1**, 791–804.
31. Chalkley, R. J., and Burlingame, A. L. (2003) Identification of novel sites of O-N-acetylglucosamine modification of serum response factor using quadrupole time-of-flight mass spectrometry. *Mol. Cell. Proteomics* **2**, 182–190.
32. Chalkley, R. J. and Burlingame, A. L. (2001) Identification of GlcNAcylation sites of peptides and alpha-crystallin using Q-TOF mass spectrometry. *J. Am. Soc. Mass. Spectrom.* **12**, 1106–1113.
33. Reason, A. J., Morris, H. R., Panico, M., Marais, R., Treisman, R. H., Haltiwanger, R. S., Hart, G. W., Kelly, W. G., and Dell, A. (1992) Localization of O-GlcNAc modification on the serum response transcription factor. *J. Biol. Chem.* **267**, 16911–16021.
34. Reason, A. J., Blench, I. P., Haltiwanger, R. S., Hart, G. W., Morris, H. R., Panico, M., and Dell, A. (1991) High-sensitivity FAB-MS strategies for O-GlcNAc characterization. *Glycobiology* **1**, 585–594.
35. Reason, A., Dell, A., Morris, H. R., Panico, M., Treisman, R., Marais, R., Hart, G. W., and Haltiwanger, R. S. (1991) Identification of O-GlcNAc attachment sites in transcription factors. *Glycoconjugate J.* **8**, 211.
36. Nandi, A., Sprung, R., Barma, D. K., Zhao, Y., Kim, S. C., Falck, J. R., and Zhao, Y. (2006) Global identification of O-GlcNAc-modified proteins. *Anal. Chem.* **78**, 452–458.
37. Sprung, R., Nandi, A., Chen, Y., Kim, S. C., Barma, D., Falck, J. R., and Zhao, Y. (2005) Tagging-via-substrate strategy for probing O-GlcNAc modified proteins. *J. Proteome Res.* **4**, 950–957.
38. Hart, G. W., Cole, R. N., Kreppel, L. K., Arnold, C. S., Comer, F. I., Iyer, S., Cheng, X., Carroll, J. and Parker, G. J. (2000) Glycosylation of proteins—a major challenge in mass spectrometry and proteomics. In: *Proceedings of the 4th international symposium on mass spectrometry in the health and life sciences* (Burlingame, A., Carr, S., and Baldwin, M., eds), Humana Press, Totowa, New Jersey, pp. 365–382.
39. Bakhtiar, R. and Guan, Z. (2006) Electron capture dissociation mass spectrometry in characterization of peptides and proteins. *Biotechnol. Lett.* **28**, 1047–1059.
40. Dong, D. L. and Hart, G. W. (1994) Purification and characterization of an O-GlcNAc selective N-acetyl-beta-D-glucosaminidase from rat spleen cytosol. *J. Biol. Chem.* **269**, 19321–19330.
41. Haltiwanger, R. S., Grove, K., and Philipsberg, G. A. (1998) Modulation of O-linked N-acetylglucosamine levels on nuclear and cytoplasmic proteins in vivo using the peptide O-GlcNAc-beta-N-acetylglucosaminidase inhibitor O-(2-acetamido-2-deoxy-D-glucopyranosylidene)amino-N-phenylcarbamate. *J. Biol. Chem.* **273**, 3611–3617.
42. Kim, E. J., Perreira, M., Thomas, C. J., and Hanover, J. A. (2006) An O-GlcNAcase-specific inhibitor and substrate engineered by the extension of the N-acetyl moiety. *J. Am. Chem. Soc.* **128**, 4234–4235.

43. Lee, T. N., Alborn, W. E., Knierman, M. D., and Konrad, R. J. (2006) Alloxan is an inhibitor of O-GlcNAc-selective N-acetyl-beta-D-glucosaminidase. *Biochem. Biophys. Res. Commun.* **350**, 1038–1043.
44. Macauley, M. S., Whitworth, G. E., Debowski, A. W., Chin, D., and Vocadlo, D. J. (2005) O-GlcNAcase uses substrate-assisted catalysis: kinetic analysis and development of highly selective mechanism-inspired inhibitors. *J. Biol. Chem.* **280**, 25313–25322.
45. Stubbs, K. A., Zhang, N., and Vocadlo, D. J. (2006) A divergent synthesis of 2-acyl derivatives of PUGNAc yields selective inhibitors of O-GlcNAcase. *Org. Biomol. Chem.* **4**, 839–845.
46. Whitworth, G. E., Macauley, M. S., Stubbs, K. A., Dennis, R. J., Taylor, E. J., Davies, G. J., Greig, I. R., and Vocadlo, D. J. (2007) Analysis of PUGNAc and NAG-thiazoline as transition state analogues for human O-GlcNAcase: mechanistic and structural insights into inhibitor selectivity and transition state poise. *J. Am. Chem. Soc.* **129**, 635–644.
47. Okuyama, R. and Yachi, M. (2001) Cytosolic O-GlcNAc accumulation is not involved in beta-cell death in HIT-T15 or Min6. *Biochem. Biophys. Res. Commun.* **287**, 366–371.
48. Gao, Y., Parker, G. J., and Hart, G. W. (2000) Streptozotocin-induced beta-cell death is independent of its inhibition of O-GlcNAcase in pancreatic Min6 cells. *Arch. Biochem. Biophys.* **383**, 296–302.
49. Gross, B. J., Kraybill, B. C., and Walker, S. (2005) Discovery of O-GlcNAc transferase inhibitors. *J. Am. Chem. Soc.* **127**, 14588–14589.
50. Fiordaliso, F., Leri, A., Cesselli, D., Limana, F., Safai, B., Nadal-Ginard, B., Anversa, P., and Kajstura, J. (2001) Hyperglycemia activates p53 and p53-regulated genes leading to myocyte cell death. *Diabetes* **50**, 2363–2375.
51. James, L. R., Tang, D., Ingram, A., Ly, H., Thai, K., Cai, L., and Scholey, J. W. (2002) Flux through the hexosamine pathway is a determinant of nuclear factor kappaB- dependent promoter activation. *Diabetes* **51**, 1146–1156.
52. Konrad, R. J., Zhang, F., Hale, J. E., Knierman, M. D., Becker, G. W., and Kudlow, J. E. (2002) Alloxan is an inhibitor of the enzyme O-linked N-acetylglucosamine transferase. *Biochem. Biophys. Res. Commun.* **293**, 207–212.
53. Marshall, S., Bacote, V., and Traxinger, R. R. (1991) Complete inhibition of glucose-induced desensitization of the glucose transport system by inhibitors of mRNA synthesis. Evidence for rapid turnover of glutamine:fructose-6-phosphate amidotransferase. *J. Biol. Chem.* **266**, 4706–4712.
54. Greis, K. D., Hayes, B. K., Comer, F. I., Kirk, M., Barnes, S., Lowary, T. L., and Hart, G. W. (1996) Selective detection and site-analysis of O-GlcNAc-modified glycopeptides by beta-elimination and tandem electrospray mass spectrometry. *Anal. Biochem.* **234**, 38–49.
55. Myllyharju, J. and Nokkala, S. (1998) Localization and identification of galactose/N-acetyl-galactosamine and sialic acid-containing proteins in Chinese hamster metaphase chromosomes. *Cell. Biol. Int.* **22**, 85–89.

Chapter 14
Detection of Sumoylated Proteins

Ok-Kyong Park-Sarge and Kevin D. Sarge

Keywords Sumoylation; SUMO-1; SUMO-2; SUMO-3; ubc9; Immuno-precipitation; In vitro modification; HSF1; HSF2

Abstract Small ubiquitin-related modifier (SUMO) is an ubiquitin-like protein that is covalently attached to a variety of target proteins. Unlike ubiquitination, sumoylation does not target proteins for proteolytic breakdown, but is instead involved in regulating a variety of different protein functional properties, including protein–protein interactions and subcellular targeting, to name a few. Protein sumoylation has been particularly well characterized as a regulator of many nuclear processes as well as of nuclear structure, making the characterization of this modification vital for understanding nuclear structure and function. Because sumoylation plays an important role in regulating so many important cellular processes, there has been intense interest in identifying new proteins that are targets of this modification and determining what role sumoylation plays in regulating the protein functions. This chapter presents methodologies for determining whether a particular protein is a substrate of sumoylation, and for identifying the lysine residue(s) where the modification occurs.

1 Introduction

Small ubiquitin-related modifier (SUMO) was discovered as a modifier of mammalian proteins in 1997 *(1, 2)*. SUMO has since been demonstrated to be a modifier of many important proteins, giving this modification a vital role in modulating a large number of important cellular processes *(3–5)*. SUMO proteins are very similar to ubiquitin structurally, but sumoylation does not promote degradation of proteins and instead regulates key functional properties of target proteins. These properties include subcellular localization, protein partnering, and transactivation functions of transcription factors, among others *(3–5)*. Protein sumoylation plays a particularly vital role in regulating many important processes occurring in the nucleus, and although sumoylation can be found on proteins that exist in a number of cellular compartments, most of the sumoylation characterized to date occurs on nuclear proteins *(4, 5)*. Indeed, proteins of the SUMO conjugation machinery have been found to be localized to nuclear pore complexes, in addition to other locations in the nucleus.

SUMO proteins are covalently attached to lysine residues of proteins, which are generally found within the consensus motif ΨKXE, where Ψ is a hydrophobic amino acid and X is any residue. Like ubiquitination, the covalent attachment of SUMO to other proteins involves a series of enzymatic steps (see Fig. 14.1), but the proteins involved are distinct from those in the ubiquitin-conjugation pathway. First, the SUMO proteins have to undergo proteolytic processing near their C-terminal end to form the mature protein, a step that is performed by SUMO proteases (Ulps). These proteases are dual-functional, as they are also responsible for cleaving SUMO groups from substrate proteins by cleaving the isopeptide bonds by which they are joined *(3–5)*. The mature processed SUMO protein is covalently attached via a thioester bond to the Uba2 subunit of the heterodimeric SUMO E1 activating enzyme in an ATP-dependent reaction *(6–9)*. The SUMO moiety is transferred from the E1 to ubc9, the SUMO E2 enzyme, which then binds to the ΨKXE consensus sequence in target proteins and forms an isopeptide bond between the ε-amino group of the lysine within this sequence and the carboxyl group of the C-terminal glycine of the SUMO polypeptide *(10–13)*. SUMO E3 proteins have been identified that enhance the efficiency of SUMO attachment by interacting with both ubc9 (the E2 enzyme) and the substrate, thereby acting as bridging factors *(3–5)*. Vertebrate cells contain three SUMO paralogs. SUMO-2 and SUMO-3 are very similar to each other in sequence and have approximately

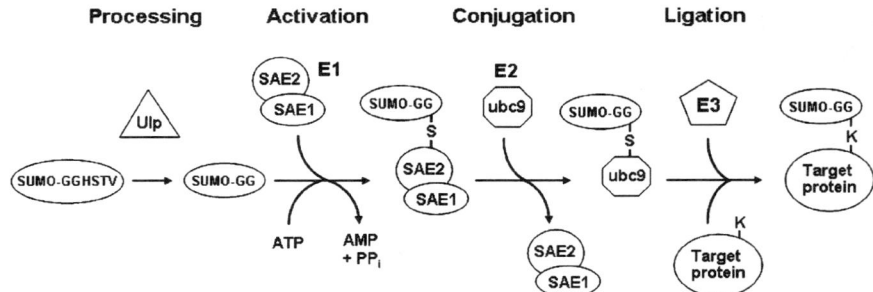

Fig. 14.1 The SUMO conjugation pathway. After they are translated, SUMO proteins must first be processed by a SUMO protease, such as Ulp1, which removes four C-terminal residues so that the mature form ends with a glycine. These proteases are also responsible for removing SUMO groups from proteins. This mature form is then activated in an ATP-dependent manner by forming a thioester bond with a cysteine residue in the SAE2 subunit of the heterodimeric E1-activating enzyme. Following this activation step, the SUMO moiety is transferred to the E2-conjugating enzyme, ubc9. In the final step, SUMO is transferred in a ligation reaction from ubc9 to a substrate protein, forming an isopeptide bond between the terminal glycine of SUMO and the ε-amino group of a lysine in the target protein. The efficiency of sumoylation of some proteins is enhanced by SUMO ligase E3 proteins, via their ability to bind both ubc9 and the target protein, thereby increasing the efficiency of SUMO transfer

50% sequence identity with SUMO-1, which is the best characterized of the three vertebrate SUMO proteins.

In this chapter, we describe two different experimental approaches for determining whether a specific protein is sumoylated. One method employs immunoprecipitation of the protein of interest, either an endogenous or a transfected epitope-tagged protein, followed by Western blotting with SUMO antibodies. The second method involves incubating the protein, either as a ^{35}S-labeled in vitro translation product or a purified recombinant protein, with a reconstituted in vitro sumoylation enzymatic reaction followed by sodium dodecyl sulfate (SDS) polyacrylamide gel electrophoresis (PAGE) and autoradiography or Western blotting, respectively, to look for the appearance of higher molecular weight bands indicative of sumoylation. By comparing wild-type protein constructs with those containing nonsumoylatable arginine substitutions of candidate target lysine residues, these protocols can also allow identification of those lysine residues where SUMO attachment actually occurs in a given protein. This information then provides the critical reagents for testing the functional consequences of blocking sumoylation of that particular protein. To illustrate the types of data that can be obtained using these methodologies, we present figures showing the results of immunoprecipitation and in vitro sumoylation analyses of a transcription factor we study, called HSF2.

2 Materials

2.1 Detection of Sumoylated Proteins by Immunoprecipitation Analysis

1. Cells expressing the protein of interest.
2. Phosphate-buffered saline (PBS): 137 mM NaCl, 2.7 mM KCl, 4.3 mM Na$_2$HPO$_4$, and 1.5 mM KH$_2$PO$_4$.
3. PBS containing 0.5% (v/v) Nonidet P40.
4. Lysis solution: 0.15 M Tris-HCl, pH 6.7, 5% SDS, and 30% glycerol.
5. N-ethylmaleimide.
6. Complete protease inhibitor (Roche, Indianapolis, IN, USA).
7. Protein G-Sepharose or protein A-Sepharose.
8. Primary antibody capable of immunoprecipitating the protein of interest (or against the epitope tag if you are analyzing a transfected tagged protein), species-matched nonspecific IgG, and antibodies against SUMO-1, SUMO-2, or SUMO-3 (Invitrogen, Carlsbad, CA, USA).
9. 4× SDS sample buffer: 250 mM Tris, pH 6.8, 20% (v/v) glycerol, 4% SDS, 0.01% bromophenol blue, and 50 mM β-mercaptoethanol.
10. Polyacrylamide gel electrophoresis equipment and solutions for SDS-PAGE.
11. Reagents for immunoblotting and detection (nonfat dried milk for blocking, and ECL or another detection system).

2.2   Detection of Sumoylated Proteins by In Vitro Sumoylation Analysis

1. A plasmid containing an open-reading frame of the protein of interest oriented to be expressed from a T7 promoter in the vector.
2. pGEX-SUMO-1, pQE30-SUMO-1, pGEX-Ubc9, and pGEX-SAE2/SAE1 bicistronic expression constructs (*see* **Note 1**) or purified SUMO, ubc9, SAE1/SAE2 (LAE Biotech International, Rockville, MD, USA).
3. Ampicillin and LB media.
4. Isopropyl-β-D-thio-galactopyranoside (IPTG).
5. French press (e.g., GlenMills, Clifton, NJ, USA) to break cells.
6. phenylmethylsulfonyl fluoride (PMSF).
7. Glutathione-agarose and Ni-NTA-agarose.
8. In vitro translation kit (e.g., TNT T7 Quick for PCR DNA; Promega, Madison, WI, USA).
9. PAGE equipment and SDS-PAGE solutions.
10. 10× sumoylation buffer: 500 mM Tris-HCl (pH 7.6), 500 mM KCl, 50 mM MgCl$_2$, 10 mM DTT, and 10 mM ATP.
11. Whatman paper, X-ray film, and cassettes for detection of ^{35}S-labeled proteins in a sumoylation reaction.

3   Methods

3.1   Detection of Sumoylated Proteins by Immunoprecipitation Analysis

In this protocol, proteins to be tested for sumoylation are immunoprecipitated using lysis buffers designed to block the action of desumoylating enzymes (*see* **Note 2**) and then subjected to Western blot analysis using anti-SUMO antibodies to look for the appearance of a band with a size consistent with a sumoylated form of the protein. Although the theoretical molecular weight of the SUMO proteins is ~11 kDa, the size increase on SDS-PAGE gels for each SUMO added is typically in the range of 15–17 kDa. In the case of a protein with multiple sumoylation sites, or where SUMO-chains form on a lysine target site (*see* **Note 3**), multiples of this size increase are expected, sometimes yielding very large shifts in mobility. Multiple bands representing different sumoylation states of the protein are also possible. This approach can be used to analyze endogenous proteins, or transfected proteins containing an epitope tag, for example FLAG or myc (*see* **Note 4**). The transfection approach can also be used to determine the lysine residue(s) to which the SUMO group is attached, by comparing the sumoylation of a wild-type construct to those in which candidate lysines have been changed to nonsumoylatable arginines. Two different lysis conditions are described, one using SDS to inhibit desumoylase enzymes and the other containing *N*-ethylmaleimide, a chemical inhibitor of these enzymes.

Cultured cells are grown in media appropriate for the cell line or primary cell type, and typically at least 1×10^6 cells are needed for each immunoprecipitation.

1. For harvesting place the plate of cells on ice, remove the medium by aspiration and add 1 mL of ice-cold PBS to the plate. Scrape the cells off the plate with a cell scraper and transfer to a 1.5-mL microcentrifuge tube.

2. Collect the cells by centrifugation at $13,000\times g$ for 30 sec at room temperature, and remove the supernatant by aspiration.

3. Lyse the cells in 150 µL of lysis solution, which is then diluted 1:10 in PBS/0.5% NP-40 plus complete protease inhibitor (Roche) and centrifuged ($16,000\times g$, 10 min, 4°C) to remove cellular debris. Sonication can be done prior to the centrifugation step if the lysate is highly viscous. Alternatively, the cells can be lysed in any standard immunoprecipitation lysis buffers (e.g., RIPA) which is known to extract the protein of interest, providing that the solution is supplemented with 20 m*M* *N*-ethylmaleimide (desumoylase inhibitor, freshly dissolved).

4. While the cell lysate is being centrifuged, prepare 30 µL of 50% protein G-Sepharose (or protein A-Sepharose if this is preferable for the particular antibody) following the manufacturer's instructions. After the protein G-Sepharose has been prepared, add the cell lysate to it and rotate at 4°C for 30 min to pre-clear the lysate.

5. Pellet the protein G-Sepharose by centrifugation ($16,000\times g$, 20 sec, 4°C) and transfer the supernatant to a new tube. At this point take 30 µL of cell lysate, place it in a separate tube with 10 µL of 4× SDS-PAGE gel loading buffer, and label "input."

6. Divide the remainder of each lysate into two equal amounts in separate 1.5-mL microcentrifuge tubes. To one of the tubes add a sufficient amount of primary antibody, and to the other add a species-matched nonspecific IgG. Place the samples on a rotator at 4°C for 1 h, and then add 20 µL of PBS-washed protein G-Sepharose and rotate at 4°C for 3 h.

7. Collect the beads by centrifugation ($16,000\times g$, 10 sec, 4°C) and discard the supernatant.

8. Wash the beads 4× with PBS, 0.5% NP-40 plus complete protease inhibitors, or another lysis buffer if chosen, collecting the beads by centrifugation after each wash ($16,000\times g$, 10 sec, 4°C). Add 30 µL of 2× SDS-PAGE gel loading buffer to the beads after removing the supernatant from the final wash.

9. Separate the immunoprecipitated proteins by SDS-PAGE.

10. Transfer proteins to a nitrocellulose or nylon membrane and subject to Western blotting using antibodies against SUMO-1, SUMO-2, or SUMO-3 (*see* **Section 2.1.8**). This methodology can also be used to examine the sumoylation state of an epitope-tagged version of the protein of interest being expressed in transfected cells (*see* **Note 4**). The results of an immunoprecipitation to examine sumoylation of the protein HSF2 are shown in Fig. 14.2.

Fig. 14.2 Detection of sumoylation by
immunoprecipitation followed by SDS-PAGE and
SUMO-1 Western blot. The protein HSF2 was
immunoprecipitated from extracts of HeLa cells,
followed by Western blotting using anti-SUMO-1
antibodies. The positions of molecular weight
standards are indicated on the left side of the panel.
This figure is reproduced from our published work
(15) with permission

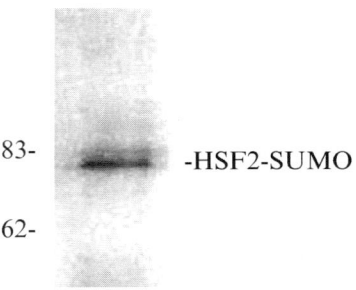

83-   -HSF2-SUMO

62-

3.2   In Vitro Sumoylation Assay

In this protocol, the protein of interest is translated in vitro (typically with
[^{35}S]methionine incorporation) and then incubated in a reaction containing the
SUMO E1 and E2 enzymes and SUMO-1, followed by SDS-PAGE and autoradiography to determine whether a lower-mobility band appears that would be consistent with a sumoylated form of the target protein. Because of the high concentrations
of SUMO E1 and E2 enzymes used in this assay the need for SUMO E3 proteins
is diminished, and thus sumoylation can be detected without their addition.
Performing the in vitro reaction using mutants of the protein in which candidate
lysine residues for sumoylation are replaced by nonsumoylatable arginines can be
used to determine the site(s) at which SUMO attachment is occurring (*see* Fig.
14.4). The expression and purification of the recombinant proteins required for the
in vitro sumoylation assay is described in **Section 3.2.1** and the protocol for performing the assay itself in **Section 3.2.3**.

3.2.1   Expression of Recombinant SUMO, ubc9 (SUMO E2), and SAE1/
SAE2 (SUMO E1) Heterodimer

The SUMO proteins are expressed and purified as fusion proteins with a GST or
6×His affinity tag at the N-terminal end, which does not need to be removed prior
to using the protein for the sumoylation reaction. The size difference of GST-
SUMO compared with 6×His-SUMO can also provide a useful control for the
sumoylation reaction, as it yields a predictable difference between the sizes of
sumoylation products (Fig. 14.3). SUMO E1 is a heterodimer of SAE1/SAE2, and
is active when expressed in *Escherichia coli* from a bicistronic construct of GST-
SAE2 and untagged SAE1; the two proteins form a complex that can be purified
using glutathione-agarose affinity chromatography *(14)*. SUMO E2 can also be
expressed and appears to be more active when the GST tag is removed by thrombin
cleavage, at least in our hands. The following is the general protocol for expressing
and purifying these recombinant proteins for the in vitro sumoylation assay. Once
purified, these proteins can be stored at −80°C for extended periods of time.

Fig. 14.3 Analysis of SUMO-1
modification by reconstituted in vitro
sumoylation reactions. **a** In vitro-translated
^{35}S-labeled HSF2 protein was incubated
with HeLa cell cytosol (as a source of E1),
Ubc9, SUMO-1, SUMO-2, or various
combinations of each of these, and then
subjected to SDS-PAGE followed by
autoradiography. The positions of
unmodified and SUMO-modified HSF2
are indicated to the right of the panel.
b In vitro-translated ^{35}S-labeled HSF2
protein was subjected to the in vitro
SUMO-1 modification assay using either
6×His-SUMO-1 or GST-SUMO-1 as
the SUMO-1 substrate for the reaction.
This figure is reproduced from our
published work *(15)* with permission

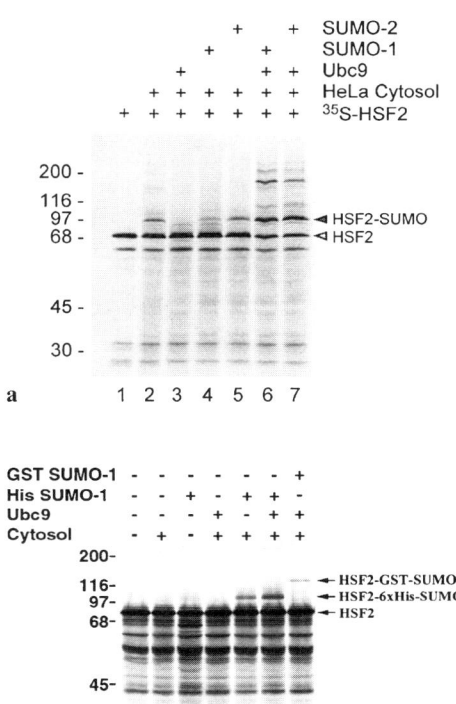

1. Transform the expression construct into *E. coli* DH10B cells using standard molecular biology methods.
2. Plate the cells on LB plates containing ampicillin and incubate overnight at 37°C.
3. Select single colonies and grow overnight at 37°C in LB media containing ampicillin.
4. Inoculate individual 1-L cultures of LB containing ampicillin with 5-mL aliquots of overnight culture and grow the cells to an $OD_{600\,nm}$ of 0.6–0.8.
5. Induce expression by adding 1 m*M* IPTG for 3 h.
6. Harvest the cells by centrifugation (4,000×*g*, 5 min, 4°C), resuspend the pellet in PBS at 4°C, then repellet the cells. At this point the cell pellet can be extracted or stored at −80°C.
7. To extract protein from the cells, resuspend the cells in PBS at 4°C with PMSF at a final concentration of 1 m*M*.
8. Pass the cells through a French press at 10,000 to 14,000 psi, and repeat to ensure complete lysis.
9. Centrifuge the cell lysate (30,000×*g*, 1 h, 4°C) and retain the supernatant.

10. Purify proteins using standard nondenaturing affinity chromatography techniques suitable for their fusion tags (glutathione agarose for GST-fusion proteins, Ni-NTA-agarose for 6×His-fusion proteins).

11. Check the purified proteins by Coomassie Blue staining of an SDS-PAGE gel (GST-SUMO1 = 38 kDa, GST-Ubc9 = 44 kDa, and 6×His-SUMO1 = 14 kDa). For the in vitro sumoylation assays, the GST-Ubc9 needs to be thrombin-cleaved to remove the GST-tag; this can be done following standard protocols (e.g., RECOMT Thrombin CleanCleave Kit; Sigma-Aldrich, St. Louis, MO, USA).

12. Check the thrombin cleavage by Coomassie Blue staining of an SDS-PAGE gel (Ubc9 = 18 kDa).

3.2.2 In Vitro Sumoylation of Rabbit Reticulocyte System-Translated Proteins

The in vitro sumoylation assay uses a [^{35}S]methionine-radiolabeled protein as the substrate. We generate radiolabeled proteins using the Promega TNT T7 Quick for PCR DNA kit following the manufacturer's instructions. Described below is the SUMO modification procedure to be utilized with radiolabeled translated proteins.

1. Translate fresh target protein before each sumoylation assay, and place on ice until step 3.

2. For each sumoylation reaction, prepare a mixture containing the following (on ice): 1 μL of 10× sumoylation buffer, 0.4 U of creatine phosphokinase, 10 mM creatine phosphate, 0.6 U of inorganic pyrophosphatase, 100 ng of purified SAE1/SAE2 heterodimer (E1), 400 ng of purified ubc9 (E2), 1 μg of purified 6×His-SUMO-1, and H$_2$O to 10 μL. As a negative control, make another mix identical to that above but lacking SUMO protein. If desired, a third mix can be made in which GST-SUMO-1 is added instead of 6×His-SUMO.

3. Set up three reactions (no SUMO control, +6×His-SUMO, and +GST-SUMO) by adding 2 μL of in vitro-translated protein to the 10-μL reaction mix, and then incubate at 37°C for 1 h.

4. Terminate the reaction by adding 12 μL of 2× SDS-PAGE loading buffer. Store at −20°C until gel electrophoresis.

5. Boil the samples for 5 min and separate on an SDS-PAGE gel. After electrophoresis, the gel is fixed for 10 min in SDS-PAGE fixing solution (50% [v/v] methanol and 10% [v/v] acetic acid), dried on Whatman paper, and finally placed on X-ray film. ^{35}S emissions have a low energy, and therefore if possible it is not advisable to leave plastic wrap between the dried gel and the X-ray film as this can increase exposure times.

6. One important experiment to give confidence in this in vitro assay is a reconstitution test where each of the required components (E1, ubc9, or

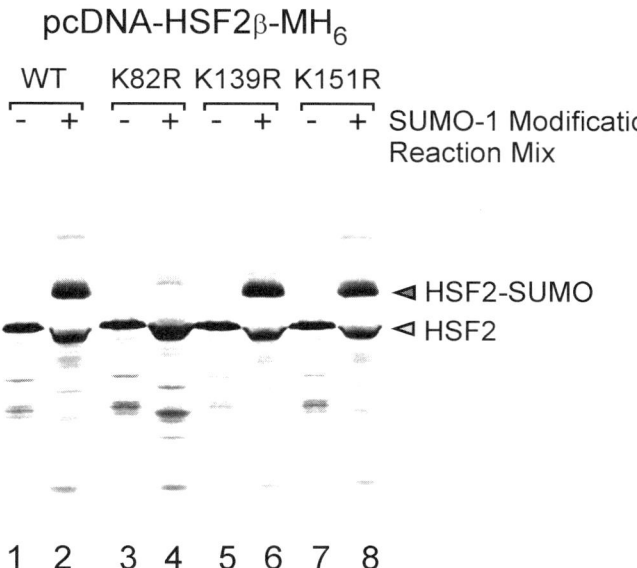

Fig. 14.4 In vitro-translated ^{35}S-labeled wild-type HSF2 protein and the HSF2 SUMO-1 consensus site mutants K82R, K139R, and K151R were used as substrates in in vitro SUMO-1 modification reactions. The positions of unmodified and SUMO-modified HSF2 proteins are indicated to the right of the panel. This figure is reproduced from our published work *(15)* with permission

SUMO-1) is individually left out of the reaction and the results are compared with those of a complete reaction (Fig. 14.3a). As shown in Fig. 14.3a, such an experiment can also reveal the relative efficiency of sumoylation of your target protein by different SUMO proteins (e.g., SUMO-1 vs. SUMO-2). Another experiment that increases confidence that your protein is indeed being sumoylated, rather than targeted by some other modification, is to compare the results of using 6×His-SUMO or GST-SUMO as donor, whose sumoylated products show a predictable size difference (Fig. 14.3b).

7. To identify the sumoylation site in the protein of interest, site-directed mutagenesis is done to replace candidate lysine residues in the sumoylation consensus sequence (ΨKXE) by nonsumoylatable arginine residues, using protocols such as the Quickchange Site-Directed Mutagenesis Kit (Stratagene, La Jolla, CA, USA). The extent of sumoylation of the wild-type compared with the mutant protein can be compared using either the transfection/immunoprecipitation approach or the in vitro sumoylation assay. Results of such an experiment using the in vitro sumoylation assay are shown in Fig. 14.4.

4 Notes

1. These have been constructed by several investigators; contact information can be obtained from the authors.
2. SUMO-modified proteins are highly susceptible to SUMO proteases. The SDS in the lysis buffer described in this protocol inactivates these proteases, allowing easier detection of sumoylated proteins. However, a common complication with this method is that the lysate tends to be very viscous and sticky due to genomic DNA. This problem is remedied by brief sonication, which shears the DNA and makes the samples easier to manipulate. SUMO proteases can also be inhibited by the addition of the isopeptidase inhibitor N-ethylmaleimide (20 mM) to standard lysis buffers such as NP-40 lysis buffer (50 mM Tris-HCl, pH 8.0, 150 mM NaCl, and 1% NP-40) if lysis in a buffer other than SDS buffer is desirable.
3. The size of putative sumoylated forms of a protein on SDS-PAGE will depend on how many different SUMO attachment sites the protein has. In addition, SUMO-2 and SUMO-3 have been reported to form polymeric chains reminiscent of ubiquitin, which could result in large increases in size for a sumoylated protein on SDS-PAGE compared with the nonsumoylated form *(14)*. Thus, it is possible to observe bands that are multiples of the ~15–17 kDa size of each SUMO unit, as well as multiple bands representing different sumoylation states.
4. Investigating sumoylation may also be done using cells transfected with fusion-tagged plasmid constructs for the protein thought to be sumoylated, with immunoprecipitation utilizing antibodies to the fusion tag available from commercial sources (for example GFP, FLAG, Myc, or 6×His tags). In these types of experiments, it is advisable to co-transfect the cells with a SUMO expression construct (often this is epitope-tagged) to ensure that sufficient SUMO protein is present in the transfected cells to allow efficient sumoylation of the target protein.

Acknowledgments We are very grateful to Mike Matunis (Johns Hopkins), Ron Hay (University of St. Andrews), and Chris Lima (Sloan Kettering Institute) for providing constructs and reagents.

References

1. Mahajan, R., Delphin, C., Guan, T., Gerace, L., and Melchior, F. (1997). A small ubiquitin-related polypeptide involved in targeting RanGAP1 to nuclear pore complex protein RanBP2. *Cell* **88**, 97–107.
2. Matunis, M. J., Wu, J., and Blobel, G. (1998). SUMO-1 modification and its role in targeting the Ran GTPase-activating protein, RanGAP1, to the nuclear pore complex. *J. Cell Biol.* **140**, 499–509.
3. Bossis, G. and Melchior, F. (2006) SUMO: regulating the regulator. *Cell Div.* Jun 29, 1–13.

4. Kerscher, O., Felberbaum, R., and Hochstrasser, M. (2006) Modification of proteins by ubiquitin and ubiquitin-like proteins. *Annu. Rev. Cell Dev. Biol.* **22**, 159–180.
5. Hay, R.T. (2005) SUMO: a history of modification. *Mol. Cell* **18**, 1–12.
6. Johnson, E.S., Schwienhorst, I., Dohmen, R.J., and Blobel, G. (1997) The ubiquitin-like protein Smt3p is activated for conjugation to other proteins by an Aos1p/Uba2p heterodimer. *EMBO J.* **16**, 5509–5519.
7. Desterro, J.M., Rodriguez, M.S., Kemp, G.D., and Hay, R.T. 1999. Identification of the enzyme required for activation of the small ubiquitin-like protein SUMO-1. *J. Biol. Chem.* **274**, 10618–10624.
8. Gong, L., Li, B., Millas, S., and Yeh, E.T. (1999) Molecular cloning and characterization of human AOS1 and UBA2, components of the sentrin-activating enzyme complex. *FEBS Lett.* **448**, 185–189.
9. Okuma, T., Honda, R., Ichikawa, G., Tsumagari, N., and Yasuda, H. (1999) In vitro SUMO-1 modification requires two enzymatic steps, E1 and E2. *Biochem. Biophys. Res. Commun.* **254**, 693–698.
10. Desterro, J. M., Thomson, J., and Hay, R. T. (1997) Ubch9 conjugates SUMO but not ubiquitin. *FEBS Lett.* **417**, 297–300.
11. Johnson, E.S. and Blobel, G. (1997) Ubc9p is the conjugating enzyme for the ubiquitin-like protein Smt3p. *J. Biol. Chem.* **272**, 26799–26802.
12. Rodriguez, M.S., Dargemont, C., and Hay, R.T. (2001) SUMO-1 conjugation in vivo requires both a consensus modification motif and nuclear targeting. *J. Biol. Chem.* **276**, 12654–12659.
13. Sampson, D.A., Wang, M., and Matunis, M.J. (2001) The small ubiquitin-like modifier-1 (SUMO-1) consensus sequence mediates Ubc9 binding and is essential for SUMO-1 modification. *J. Biol. Chem.* **276**, 21664–21669.
14. Tatham, M.H., Jaffray, E., Vaughan, O.A., Desterro, J.M., Botting, C.H., Naismith, J.H., and Hay, R.T. (2001) Polymeric chains of SUMO-2 and SUMO-3 are conjugated to protein substrates by SAE1/SAE2 and Ubc9. *J. Biol. Chem.* **276**, 35368–35374.
15. Goodson, M.L., Hong, Y., Rogers, R., Matunis, M.J., Park-Sarge, O.K., and Sarge, K.D. (2001) Sumo-1 modification regulates the DNA binding activity of heat shock transcription factor 2, a promyelocytic leukemia nuclear body associated transcription factor. *J. Biol. Chem.* **276**, 18513–18518.

Chapter 15
Detection of the Nuclear Poly(ADP-ribose)-Metabolizing Enzymes and Activities in Response to DNA Damage

Jean-Christophe Amé, Antoinette Hakmé, Delphine Quenet, Elise Fouquerel, Françoise Dantzer, and Valérie Schreiber

Keywords Poly(ADP-ribose); Poly(ADP-ribose) polymerases; Poly(ADP-ribose) glycohydrolase; DNA damage; Genome integrity; Western blotting; Immunofluorescence

Abstract Poly(ADP-ribosyl)ation is a posttranslational modification of proteins in higher eukaryotes mediated by poly(ADP-ribose) polymerases (PARPs) that is involved in many physiological processes such as DNA repair, transcription, cell division, and cell death. Biochemical studies together with PARP-1- or PARP-2-deficient cellular and animal models have revealed the redundant but also complementary functions of the two enzymes in the surveillance and maintenance of genome integrity. Poly(ADP-ribose) is degraded by the endo- and exo-glycosidase activities of poly(ADP-ribose) glycohydrolase (PARG). In this chapter, biochemical and immunofluorescence methods are described for detecting and assaying PARPs and PARG.

1 Introduction

Poly(ADP-ribosyl)ation is a posttranslational modification of proteins mediated by poly(ADP-ribose) polymerases (PARPs). It occurs in higher eukaryotes and is involved in many physiological processes such as DNA repair, transcription, cell division, and cell death. The widespread subcellular localisations of the 17 members of the PARP family underline the probable occurrence of this modification all over the cell (*1–3*). Poly(ADP-ribose) is generated by polymerisation of ADP-ribose moieties from NAD^+ on acceptor proteins, either PARPs themselves (automodification) or heterologous substrates (heteromodification). PARP-1, the founding member of the PARP family, is a DNA break-sensing molecule that translates the presence of single-strand DNA interruptions to several outcomes: poly(ADP-ribosyl)ation of histones, mainly H1, leading to chromatin decondensation and accessibility of the damaged site; recruitment of repair factors such as the base excision repair/single strand break repair factor XRCC1, which has a strong affinity for poly(ADP-ribose); and informing the cell about the severity of the damage, reflected by the amount of poly(ADP-ribose) produced. From this will depend the fate of the damaged cell, either survival after repair of the damaged DNA or death. Another PARP, PARP-2, possesses a DNA-dependent activity. Biochemical studies together with PARP-1- or PARP-2-deficient cellular and animal models have revealed the redundant but also complementary functions of the two enzymes in the surveillance and maintenance of genome integrity *(4–8)*.

Poly(ADP-ribosyl)ation is a transient modification of proteins, since poly(ADP-ribose) is efficiently and rapidly degraded by the endo- and exo-glycosidase activities of poly(ADP-ribose) glycohydrolase (PARG). The PARG gene encodes a nuclear enzyme (PARG[110]) as well as smaller isoforms showing cytoplasmic and mitochondrial localisations *(9, 10)*. PARG[110] is the most active isoform detected in a cell extract, probably because it has to face the huge amount of PAR produced by PARP-1 during the DNA damage response. Many studies have revealed the importance of PARP and PARG in the control of poly(ADP-ribose) levels regulating the balance between life-and-death in response to DNA injury *(11, 12)*.

2   Materials

2.1   Detection of PARPs and PARG by SDS-PAGE and Western Blotting

1. Cell extract prepared by harvesting 10^6–10^7 cells and washing 2× in phosphate-buffered saline (*see* **Section 2.2.5**) prior to suspension in 100 μL of sodium dodecyl sulfate (SDS) polyacrylamide gel electrophoresis (PAGE) sample buffer (*see* **Section 2.1.9**); alternatively, any other cell extract can be used.
2. Separating gel buffer (4×): 1.5 M Tris-HCl, pH 8.8, 0.4% (w/v) SDS. Store at 4°C.
3. Stacking gel buffer (4×): 0.5 M Tris-HCl, pH 6.8, 0.4% (w/v) SDS. Store at 4°C.
4. Acrylamide/bis solution (40% acrylamide, 37.5:1 ratio) and N,N,N,N'-tetramethyl-ethylenediamine (TEMED) (Bio-Rad, Marnes-la-Coquette, France). Store at 4°C.
5. Ammonium persulfate solution: 10% (w/v) in dH_2O, freeze in aliquots at −20°C.
6. H_2O-saturated isobutanol: shake equal volumes of dH_2O and isobutanol in a glass bottle and allow to separate. Store at room temperature. Use the top isobutanol layer.
7. Gel running buffer (5×): 30 g/L Tris-base, 144 g/L glycine, 5 g/L SDS. The pH is not adjusted. Store at room temperature.
8. Sample buffer: 50 mM Tris-HCl, pH 6.8, 6 M urea, 3% (w/v) SDS, 6% (v/v) β-mercaptoethanol. Add 3 mg of bromophenol blue to 100 mL. Aliquot and store at −20°C.

9. Prestained protein molecular weight markers (e.g. Bio-Rad, all blue Precision Plus).

10. Blotting transfer buffer: prepare a 10× stock solution containing 30 g/L Tris-base, 144 g/L glycine, and 10 g/L SDS. The pH is not adjusted but should be 8.3. Store at room temperature. For 1 L of 1× transfer buffer, mix 100 mL of 10× stock buffer, 700 mL of dH_2O and 200 mL of 95% ethanol.

11. Blotting system: we use a Mini Trans-Blot Electrophoretic Transfer Cell (Bio-Rad).

12. Nitrocellulose transfer membranes (e.g. Schleicher and Schuell Protan BA 83, 0.2 µm porosity; Whatman, Dassel, Germany) and Whatman 3 MM filter paper sheets.

13. TBS buffer (10×): Tris-base 60.5 g/L, adjust to pH 8.0 with 37% HCl, add NaCl to 87.66 g/L.

14. TBS-Tween buffer (1×): 100 mL of 10× TBS buffer, 3 mL of Tween-20, complete with dH_2O to 1 L.

15. Primary antibodies: anti-PARP-1 rabbit polyclonal (#ALX-210-897-R100; Alexis Biochemicals, Lausen, Switzerland), anti-PARP-2 rabbit polyclonal (#ALX-210-896-R100; Alexis Biochemicals), and anti-PARG polyclonal raised against the catalytic domain of bovine PARG *(9)*.

16. Secondary antibody: horseradish peroxidase (HRP)-conjugated goat anti-rabbit (Sigma-Aldrich).

17. Blocking buffer: 5% (w/v) nonfat dry milk in TBS-Tween.

18. Detection reagents: enhanced chemiluminescence (ECL Plus Western Blotting Detection System; Amersham GE Healthcare, Orsay, France) and detection film (BioMax MR; Kodak, Rochester, NY, USA; or Hyperfilm ECL; Amersham GE Healthcare).

2.2 Detection of PARPs and PARG by Immunofluorescence Microscopy

1. Six-well culture plates or 35-mm diameter plastic dishes, coverslips (0.17-mm thickness), and microscope slides.

2 Dulbecco's Modified Eagle's Medium (DMEM) (Gibco/BRL, Bethesda, MD, USA) supplemented with 10% (v/v) fetal bovine serum (FBS; AdGenix, Voisins le Bretonneux, France) and 1% (w/v) gentamicin (Gibco/BRL).

3 Trypsin (0.05% w/v) and EDTA (0.53 m*M*) solutions (Gibco/BRL).

4. Ethanol 70% (v/v): mix 700 mL of 100% ethanol with 300 mL of dH_2O.

5. Dulbecco's phosphate-buffered saline (PBS): for a 10× stock solution, dissolve 96 g of premixed PBS powder (#D5652; Sigma-Aldrich) in 1 L of dH_2O. For 1× PBS, dilute 100 mL of 10× PBS in 900 mL of dH_2O and sterilize by autoclaving.

6. PBS-Triton: dilute 5 mL of 10× PBS and 50 µL of Triton X-100 in dH_2O to 50 mL. Store these solutions at 4°C.

7. Antibody dilution solution: to 5 mL of PBS-Triton, add 500 μL of bovine serum albumin (BSA) solution (10 mg/mL).
8. Primary antibodies: anti-PARP-1 rabbit polyclonal (#ALX-210-897-R100; Alexis Biochemicals), C_{2-10} anti-PARP-1 mouse monoclonal (#ALX-804-210-R050, Alexis Biochemicals), anti-PARP-2 rabbit polyclonal (#ALX-210-303-R100, Alexis Biochemicals), and anti-PARG polyclonal (9).
9. Secondary antibodies: goat anti-mouse (GAM) or goat anti-rabbit (GAR), both conjugated to Alexa Fluor 488 or Alexa Fluor 568 (Molecular Probes; Invitrogen, Cergy Pontoise, France).
10. Formaldehyde solution (2% w/w): just before use, add 2.7 mL of 37% (w/w) formaldehyde solution (stored in the dark at room temperature) to 47.3 mL of PBS. To make 1% formaldehyde–0.1% Triton solution, mix 1.35 mL of 37% formaldehyde with 48.65 mL of PBS containing 0.1% Triton X-100.
11. 4′,6-Diamidino-2-phenylindole (DAPI) stock solution: 1 μg/μL DAPI (#D9542 Sigma-Aldrich) in PBS containing 50% (v/v) glycerol. Store at −20°C. Dilute 1:20,000 in PBS to give a working solution.
12. Mounting solution containing antifading reagent (Mowiol/Dabco): add 2.4 g of MOWIOL 4-88 (Calbiochem; VWR, Fontenay sous Bois, France) to 6 g of glycerol, stir to mix. Add 6 mL of dH_2O, and stir at room temperature for several hours. Add 12 mL of 0.2 M Tris-Cl, pH 8.5, and heat to 50°C for 10 min with occasional mixing. Clarify by centrifugation at 5,000×g for 15 min. Add 1,4,-diazobicyclo-[2.2.2]-octane (DABCO; Sigma-Aldrich) to 2.5% (w/v). Aliquot into airtight containers and store at −20°C.
13. Fluorescence microscope with oil-immersion objectives and a digital CCD camera (e.g. ORCA ER; Hamamatsu Photonics, Massy, France).

2.3   Detection of PAR Synthesis by Western Blotting

1. Cell extracts and materials for SDS-PAGE and Western blots, as described in **Section 2.1**.
2. Mouse monoclonal anti-PAR antibody 10H ($IgG_3κ$) (hybridoma supernatant provided by Dr. M. Miwa and Dr. T. Sugimura, Tokyo, Japan), rabbit polyclonal anti-PAR antibody LP96-10 (Pharmingen; BD Biosciences, Le Pont de Claix, France).
3. Horseradish peroxidase (HRP)-conjugated goat anti-rabbit and HRP-conjugated goat anti-mouse secondary antibodies (Sigma-Aldrich).

2.4   Detection of PAR by Immunofluorescence Microscopy

1. For preparation of cells, *see* **Section 2.2**.
2. DNA-damaging agents:

(a) H_2O_2: prepare a 100× working solution by diluting 10 µL of 8.8 M stock H_2O_2 (Sigma-Aldrich; stored at 4°C) with 870 µL of dH_2O.

(b) 1-methyl-3-nitro-1-Nitrosoguanidine (MNNG): prepare a stock solution dissolved at 2 M in DMSO. Light sensitive, store in the dark at −80°C. Add to the culture medium to a final concentration of 50 µM. *MNNG is carcinogenic, manipulate with precaution following the NIH Guidelines for the Laboratory Use of Chemical Carcinogens.*

(c) 1-methyl-1-Nitrosourea (MNU): dissolve at 2 M in DMSO and store at −20°C. Add to the culture medium to a final concentration of 2 mM. *MNU is carcinogenic, manipulate with precaution following the NIH Guidelines for the Laboratory Use of Chemical Carcinogens.*

(d) Methyl methane sulfonate (MMS): dissolve in dH_2O and store at 4°C. Add to the culture medium to a final concentration of 200 µM. *MMS is carcinogenic, manipulate with precaution following the NIH Guidelines for the Laboratory Use of Chemical Carcinogens.*

(e) Ionizing radiation: we use a Pantak Seifert X-ray system (Ahrensburg, Germany) and a UNIDOS E Universal Dosemeter (PTW, La Ville du Bois, France).

3. PBS, 10× and 1× solutions (*see* **Section 2.2.5**).
4. PBS-Tween solution: dilute 100 mL of 10× PBS and 500 µL of Tween-20 in 899.5 mL of dH_2O. Store at 4°C.
5. Antibody dilution solution (PBS-Tween-BSA): to 5 mL of PBS-Tween, add 500 µL of BSA solution (10 mg/mL).
6. Primary antibodies: as described in **Section 2.3.2**.
7. Secondary antibodies: as described in **Section 2.2.9**.
8. 100% methanol and acetone at −20°C.
9. DAPI solution: as described in **Section 2.2.11**.
10. Mounting solution: as described in **Section 2.2.12**.

2.5 Assay of PARP Activity in Solution

1. Glass assay tubes: 5 mL.
2. Glass microfibre filters: 25-mm diameter (Whatman, Maidstone, England) or equivalent, with a 25-mm-diameter vacuum filtration unit.
3. Precipitation buffer: 5% (w/v) trichloroacetic acid, 1% (w/v) sodium pyrophosphate prepared from 100% and 5% stock solutions, respectively. Store at 4°C.
4. Ethanol 95% (v/v), kept on ice.
5. Purified PARP-1 (#ALX-201-063; Alexis Biochemicals).
6. DNase I-treated DNA (#ALX-840-040-C010; Alexis Biochemicals).
7. [^{32}P]NAD$^+$ (1,000 Ci/mmol, 10 µCi/µL; GE HealthCare).

8. PARP activity incubation mixture (10×): 500 mM Tris-Cl, pH 8, 40 mM MgCl$_2$, 1 M NaCl, 10 mM DTT, 2 µg (60 pmol) DNase I-treated DNA, 1 µg/µL BSA, 4 mM NAD, and 1 µCi [^{32}P]NAD$^+$.

9. Liquid scintillation cocktail for aqueous and nonaqueous samples, high flash-point (e.g. Ultima Gold; Packard BioSciences, Groningen, The Netherlands) and liquid scintillation counter.

2.6   Detection of PARG Activity

2.6.1   Large-scale Synthesis of ^{32}P-Labelled PAR

1. Siliconized tubes: to prevent unspecific binding of PAR, 10-mL Sarstedt screw-cap tubes (round base) are siliconized by adding 1 mL of Sigmacote (Sigma-Aldrich) and vortexing. Remove the solution and let the tubes air-dry under a chemical hood.

2. PARP activity buffer (10×): 500 mM Tris-Cl, pH 8.0, 40 mM MgCl$_2$, 2 mM DTT, 500 µg/mL BSA, 20 µg/mL DNase I-activated DNA, and 4 mM NAD$^+$. Can be kept in 200-µL aliquots for more than a year at −20°C.

3. [^{32}P]NAD$^+$ (1,000 Ci/mmol, 10 µCi/µL; GE HealthCare).

4. Purified PARP-1 (#ALX-201-063; Alexis Biochemicals).

5. DNase I solution: 10 mg/mL DNase I (Roche Diagnostics, Mannheim, Germany) in dH$_2$O.

6. Proteinase K solution: 10 mg/mL proteinase K (Merck) in 20% (w/v) SDS.

7. 0.2 M NaOH and 40 mM EDTA solution, pH 8.0.

8. 0.2 M HCl.

9. Phenol/CHCl$_3$/isoamyl alcohol (25:24:1) mixture (Roth, Karlsruhe, Germany).

10. 3 M potassium acetate, pH 4.8.

11. Isopropyl alcohol 100% (Fluka-Sigma-Aldrich).

12. 80% Ethanol.

13. TE buffer: 10 mM Tris-HCl and 1 mM EDTA, pH 8.0.

2.6.2   Assay of PARG Activity in Solution

1. The reaction is performed in 5-mL glass hæmolysis tubes.

2. ^{32}P-labelled PAR, see **Section 2.6.1**.

3. Either purified recombinant PARG *(13)* or a cell lysate in which PARG activity is to be evaluated.

4. PARG activity buffer (10×): 500 mM Tris-HCl, pH 8.0, 20 mM MgCl$_2$, 10 mM DTT, and 1 M NaCl. Store in small aliquots at −20°C.

5. Precipitation buffer: see **Section 2.5.3**.

6. Glass fibre filters and LSC-cocktail: see **Section 2.5.2**.

2.6.3 Detection of PARG Activity by Zymography

1. Materials for SDS-PAGE gels: *see* **Section 2.1**.
2. ^{32}P-labelled-automodified-PARP-1 (10^6 dpm): *see* **Section 2.6.1**.
3. Zymogram renaturation buffer: $50\,mM$ sodium phosphate buffer, pH 7.5 (prepare a $1\,M$ stock solution by mixing $16\,mL$ of $2\,M$ NaH_2PO_4 with $84\,mL$ of $2\,M$ Na_2HPO_4 and $100\,mL$ of dH_2O), $50\,mM$ NaCl, 10% (v/v) glycerol, 1% (v/v) Triton X-100, and $10\,mM$ β-mercaptoethanol. Prepare $1\,L$ just before use, keep at room temperature.
4. Gel destaining buffer: 40% (v/v) methanol and 30% (v/v) acetic acid in dH_2O, store at room temperature.
5. Heated vacuum gel dryer (e.g. Bio-Rad model 583 or equivalent).

3 Methods

3.1 Detection of PARPs and PARG by SDS-PAGE and Western Blotting

1. The following method assumes the use of a 11×10-cm gel system, but is easily adaptable to other formats. It is important that the glass plates are cleaned before use with a soft detergent (Teepol) and rinsed extensively with dH_2O, then with 95% ethanol, and air-dried.
2. Prepare a 1-mm thick, 10% gel by mixing $5\,mL$ dH_2O with $2.5\,mL$ of 4× separating gel buffer, $2.5\,mL$ of acrylamide/bis solution, $68\,\mu L$ of ammonium persulfate solution, and $14\,\mu L$ of TEMED. Pour the gel, leaving space for a stacking gel, and carefully overlay with dH_2O-saturated isobutanol. The gel should polymerize in 10–15 min.
3. Pour off the isobutanol and rinse the top of the gel 2× with dH_2O.
4. Prepare the stacking gel by mixing $3.2\,mL$ of dH_2O with $1.25\,mL$ of 4× stacking gel buffer, $500\,\mu L$ of acrylamide/bis solution, $40\,\mu L$ of ammonium persulfate solution, and $8\,\mu L$ of TEMED. Pour the stacking gel and insert the comb. The stacking gel should polymerize in about 10 min.
5. Prepare the running buffer by mixing $200\,mL$ of 5× running buffer with $800\,mL$ of dH_2O.
6. After polymerization of the stacking gel, the comb is carefully removed and the wells are washed with 1× running buffer using a 3-mL syringe with a 22-gauge needle.
7. Add running buffer to the lower chamber first (avoid any bubbles that could be trapped between the glass plates on the bottom of the gel), then to the upper gel unit.
8. Mix samples containing 10–100 μg of total protein 1:4 (v:v) with sample buffer, heat 4 min at 95°C, and load 10 to 25 μL into sample wells.

9. Load prestained molecular weight markers into one well.
10. Complete assembly of the gel unit and connect the power supply. The gel is run at 12 V/cm for 90 min at room temperature or until the dye front reaches the bottom of the gel.
11. Prepare a tray of transfer buffer large enough to lay out a transfer cassette with its pieces of Scotch-brite pads and with two sheets of Whatman 3 MM paper submerged on one side. Cut a sheet of nitrocellulose membrane just larger than the size of the gel (separating and stacking), wet and soak in the transfer buffer.
12. Disconnect the gel unit from the power supply and disassemble. Set up the transfer sandwich from bottom to top: one side of the transfer cassette, a piece of Scotch-brite pad, a sheet of Whatman 3 MM paper, the nitrocellulose membrane, the gel (ensuring that no bubbles are trapped between the membrane and the gel), a sheet of Whatman 3 MM paper, and a piece of Scotch-brite pad. The transfer cassette is then carefully closed.
13. Place the cassette into the transfer tank such that the nitrocellulose membrane is between the gel and the anode (+). Fill up the tank with cold transfer buffer and add a magnetic stir-bar. The transfer is performed in a cold chamber (4°C) with the tank placed on top of a magnetic stirrer; make sure that the stir-bar rotates freely.
14. Put the lid on the tank and activate the power supply. The transfer is accomplished at 90 V (250 mA) for 1.5 h.
15. Once the transfer is complete, take out the cassette from the tank and carefully disassemble it, with the top Scotch-brite pad and sheets of 3 MM paper removed. Do not forget to note the orientation of the membrane. The gel can be discarded and the excess nitrocellulose cut off. The coloured molecular weight markers should be clearly visible on the membrane.
16. Incubate the nitrocellulose membrane in 20 mL of TBS-Tween for 5 min, then in 20 mL of blocking buffer for 20 min at room temperature on a rocking platform.
17. Discard the blocking buffer and replace by a minimum volume (10 mL) of blocking buffer containing anti-PARP-1 antibody (working dilution 1:4,000), anti-PARP-2 antibody (working dilution 1:4,000), or anti-PARG antibody (working dilution 1:5,000). Incubate at room temperature for 2 h or at 4°C overnight on a rocking platform.
18. Remove the primary antibody; the diluted antibody can be kept at −20°C for further use. Wash the membrane 3× 10 min each with 20 mL of TBS-Tween.
19. The HRP-conjugated goat anti-rabbit secondary antibody is freshly prepared for each experiment as a 1:20,000-fold dilution in blocking buffer, and added to the membrane for 2 h at room temperature on a rocking platform.
20. Discard the secondary antibody and wash the membrane 3× 10 min each with TBS-Tween at room temperature.
21. After removing the excess of buffer, lay the membrane on a clean glass plate. Mix 1 mL of ECL reagent A and 25 μL of reagent B and spread evenly on the membrane in a dark chamber. After 2–3 min incubation remove excess ECL

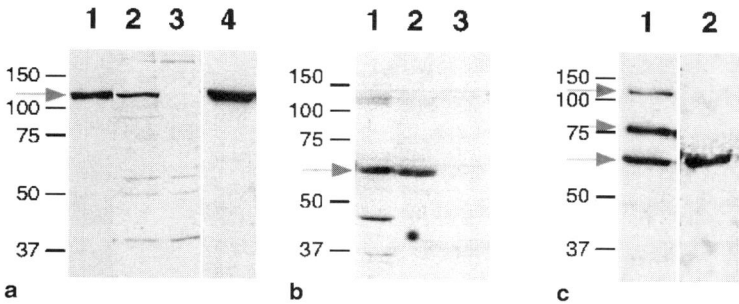

Fig. 15.1 Detection of PARP-1, PARP-2, and PARG by Western blotting. Total protein extracts were separated in a 10% SDS-PAGE gel, transferred to a nitrocellulose membrane, and probed with antibodies against (**a**) PARP-1, (**b**) PARP-2, or (**c**) PARG. The extracts (corresponding to 2×10^5 cells) are from: (**a**) *lane 1*, HeLa cells; *lane 2*, 3T3 MEFs cells; *lane 3*, PARP-1$^{-/-}$ MEFs cells; *lane 4*, MRC5 cells; (**b**) *lane 1*, HeLa cells; *lane 2*, 3T3 MEFs cells; *lane 3*, PARP-2$^{-/-}$ MEFs cells; and (**c**) *lane 1*, HeLa cells; *lane 2*, mitochondria from rat liver. Notice that in (**c**), the antibody directed against PARG detects various isoforms of the enzyme in HeLa cells (*9*)

solution from the blot with Kim-Wipes, then place on Whatman 3 MM paper and wrap in Saran Wrap.

22. Place the blot in an X-ray film cassette with film for a suitable exposure time, typically a few minutes. The film is developed using an automatic film developer (e.g. X-OMAT; Eastman Kodak). An example of results is shown in Fig. 15.1.

3.2 Detection of PARPs and PARG by Immunofluorescence Microscopy

1. Sterilize coverslips by dipping in 70% ethanol, then in 100% ethanol and let stand in a petri dish under the hood until dry.
2. Cells: set up 6-well culture plates or 35-mm diameter Petri dishes with sterile coverslips and seed 40,000–50,000 cells in 2 mL of medium per well.
3. The next day, wash the cells twice with ice-cold PBS, then fix for 15 min at 4°C with 2 mL of fixation solution. The choice of fixation depends on both the antigen and the antibody involved; for PARP-1, PARP-2, and PARG, the fixation can be realized with either 2% formaldehyde solution or 1% formaldehyde, 0.1% Triton X-100 (*see* Fig. 15.2).
4. Discard the fixation solution (into a specific waste container) and wash the cells 3× 10 min each with ice-cold PBS-Triton-BSA.
5. Dilute the primary antibodies in PBS-Triton-BSA: 1:2,000 for anti-PARP-1, 1:400 for anti-PARP-2, and 1:4,000 for anti-PARG Cter- or Nter-. Remove the PBS-Triton from the cells and add 50 μL of diluted antibody, then gently place

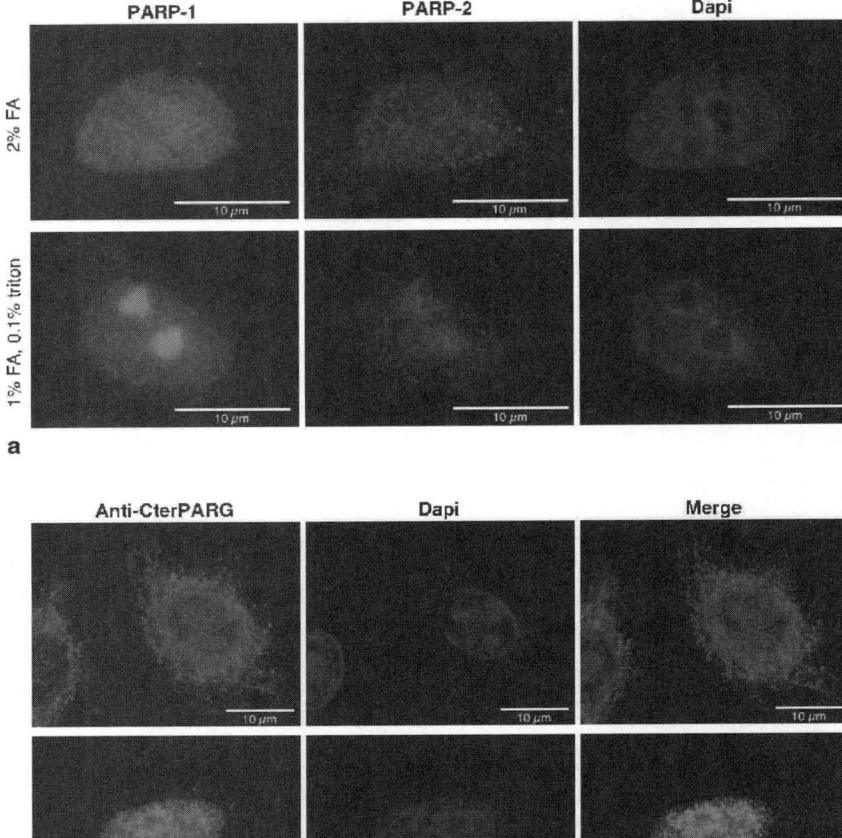

Fig. 15.2 Detection of PARP-1, PARP-2, and PARG by immunofluorescence microscopy. **a** HeLa cells were probed with an anti-PARP-1 monoclonal antibody (C_{2-10}) revealed by green fluorescence, and an anti-PARP-2 rabbit polyclonal antibody revealed by red fluorescence, following fixation with 2% formaldehyde (*FA*; *upper panels*) or with 1% formaldehyde, 0.1% Triton X-100 (*lower panels*). DNA is stained with DAPI (*right panels*). Note the influence of the type of fixation on the detection of PARP-1 and PARP-2: nucleolar PARP-1 and PARP-2 are revealed using 1% formaldehyde, 0.1% Triton X-100 *(14)*. **b** HeLa cells were probed with an anti-CterPARG rabbit polyclonal antibody revealed by red fluorescence (*upper panels*), or with an anti-NterPARG rabbit polyclonal antibody revealed by green fluorescence (*lower panels*). DNA is stained with DAPI (*middle panels*) and merged with the antibody fluorescence in the *right panels*. Note the mixed nuclear and mitochondrial localization observed with the anti-CterPARG antibody, which recognizes various isoforms of the enzyme, in contrast to the uniquely nuclear signal observed with the anti-NterPARG antibody, which recognizes the nuclear full-length PARG. To view this figure in color, see COLOR PLATE 7

a plastic cover over the coverslip to ensure a uniform distribution of liquid over the cells and prevent dehydration.

6. Place the culture plates or Petri dishes in a plastic box containing a wet paper towel and incubate overnight at 4°C (*see* **Note 1**).

7. Add 2 mL of PBS-Triton, gently remove the plastic cover, and wash the coverslips 3× with PBS-Triton X at room temperature.

8. Dilute the secondary antibody in PBS-Tween-BSA. For a mouse monoclonal primary antibody, use a goat anti-mouse secondary antibody conjugated to Alexa Fluor 488 for green fluorescence or to Alexa Fluor 568 for red fluorescence. For a rabbit primary antibody, use a goat anti-rabbit secondary antibody conjugated to Alexa Fluor 488 for green fluorescence or to Alexa Fluor 568 for red fluorescence. Dilutions of 1:1,500 are recommended for the secondary antibodies. Incubate for 2 h at room temperature in a humid chamber in the dark with a plastic cover over the coverslip.

9. Wash 2× 10 min in PBS-Triton, then 1× in PBS, then stain DNA with PBS containing DAPI for 15 min. Wash again in PBS. During all these steps, avoid direct bright light and keep the samples in the dark during the incubations to avoid premature fading of the signal.

10. Label microscope slides, rapidly rinse the coverslip in dH$_2$O, and mount it (inverted) on a slide using a drop (35 µL) of Mowiol mounting solution. Store in the dark at room temperature for several hours to dry, then at −20°C to minimize background fluorescence.

11. Observe the slides under a fluorescence microscope.

3.3   Detection of PAR by Western Blotting

In order to trigger activation of PARP, cells may be treated by a DNA-damaging agent before lysis. Cells are harvested directly in sample buffer in order to shorten as much as possible the time between PARP activation and cell lysis and to minimize PAR degradation by the very active endogenous PARG enzyme.

1. Samples that have been separated by SDS-PAGE are transferred to nitrocellulose membranes and incubated in blocking buffer as described in **Section 3.1**, **steps 1–16**.

2. The blocking buffer is discarded and replaced by a minimum volume (10 mL) of blocking buffer containing anti-PAR antibody. Use a 1:1,500 dilution for the rabbit polyclonal LP96-10 and a 1:500 dilution for the monoclonal 10H (IgG$_3$κ). Incubate at room temperature for 2 h or at 4°C overnight on a rocking platform.

3. The primary antibody is then removed; it can be kept for further use. The membrane is washed 3× 10 min each with 20 mL of TBS-Tween at room temperature.

4. The secondary antibody is freshly diluted in blocking buffer at 1:20,000 for HRP-conjugated goat anti-rabbit antibody or 1:500 for HRP-conjugated goat

Fig. 15.3 Detection of poly(ADP-ribose) synthesis by
Western blotting. Total proteins from 2×10^5 primary
human fibroblasts not treated (*lane 1*) or treated with
$500\,\mu$M H_2O_2 (*lane 2*) were separated on a 10%
SDS-PAGE gel, transferred to a nitrocellulose
membrane, and probed with the anti-PAR antibody,
LP96-10

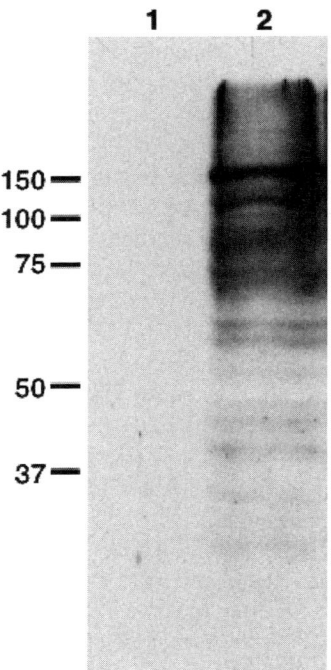

anti-mouse $IgG_3\kappa$ antibody, and incubated with the membrane for 2 h at room
temperature on a rocking platform.
5. Washing, signal development and detection are as described in **Section 3.1,
steps 20–22**. An example of results is shown in Fig. 15.3.

3.4   Detection of PAR by Immunofluorescence

1. Prepare cells grown on coverslips as described in **Section 3.2, steps 1** and **2**.
2. Damaging DNA:

   (a) With H_2O_2: wash the cells 2× with 2 mL of PBS at room temperature. Add
   to the cells 2 mL of PBS containing 10 μL of freshly diluted 100× H_2O_2
   ($500\,\mu M$) and incubate for 10 min at 37°C. If required, replace with fresh
   medium and incubate the cells further at 37°C for 0 to 40 min.
   (b) With a monofunctional alkylating agent: replace the medium with medium
   containing a suitable dilution (*see* **Section 2.4.2**) of MNNG, MNU, or
   MMS and incubate for 10–30 min at 37°C, then replace with fresh medium
   and incubate further at 37°C for 0 to 60 min. *Discard media containing
   alkylating agents in an appropriate flask for elimination according to the
   NIH Guidelines for the Laboratory Use of Chemical Carcinogens.*

    (c) Using ionizing radiation (X-rays): place Petri dishes (with the lid removed) on a horizontally rotating tray to produce a homogeneous irradiation under the beam area of the X-ray source. Detection of PAR is easier when the dishes are placed on an ice tray for irradiation, because this minimizes PAR degradation by PARG. Adjust the settings of the source and the distance from the cells to deliver ~2 Gy in 3 min (*see* **Note 2**). Irradiation is stopped at the desired dose measured by a dosimeter, and the cells are left at room temperature for 5–10 min to allow PAR synthesis.

3. Place the culture plates or Petri dishes on ice in a tray, remove the medium by aspiration, and rinse the cells 2× with ice-cold PBS.
4. Fix the cells with methanol:acetone (1:1, prepared just before use and cooled at −20°C) for a minimum of 20 min on ice (*see* **Note 3**).
5. Aspirate the methanol/acetone and wash the cells 3× 10 min with 2 mL of ice-cold PBS-Tween.
6. Dilute the primary antibody in PBS-Tween-BSA. The monoclonal anti-PAR antibody 10 H is used at a 1:400 dilution and LP96-10 at a 1:1,500 dilution. Remove the PBS-Tween from the cells and add 50 μL of diluted antibody, then gently place a plastic cover over the coverslip to ensure a uniform distribution of liquid over the cells and prevent dehydration.
7. Place the culture plates or dishes in a plastic box containing a wet paper towel and incubate overnight at 4°C (*see* **Note 1**).
8. Add 2 mL of PBS-Tween, gently remove the plastic cover and wash 3× with PBS-Tween at room temperature.
9. Dilute the secondary antibodies in PBS-Tween-BSA. For the primary antibody 10 H, use goat anti-mouse secondary antibody conjugated to Alexa Fluor 488 for green fluorescence or to Alexa Fluor 568 for red fluorescence. For LP96-10, use goat anti-rabbit secondary antibody conjugated to Alexa Fluor 488 for green fluorescence or to Alexa Fluor 568 for red fluorescence. Dilutions of 1:1,500 are recommended. Incubate for 2 h at room temperature in a humid chamber in the dark with a plastic cover over the coverslip.
10. Wash 2× 10 min in PBS-Tween, then once in PBS, then stain DNA with PBS containing DAPI for 15 min. Wash again in PBS. During all these steps avoid direct bright light and keep the samples in the dark during the incubations to avoid premature fading of the signal.
11. Mount the coverslips on slides as described in **Section 3.2**, **step 10**. An illustration of PAR detection by immunofluorescence in cells treated with H_2O_2 is shown in Fig. 15.4.

3.5   *Assay of PARP Activity in Solution*

This protocol has been developed to measure in vitro activity of either purified PARP-1 or PARP-2, or in cell lysates containing 10 to 100 μg of total protein.

| 10H | LP96-10 | Dapi | Merge |

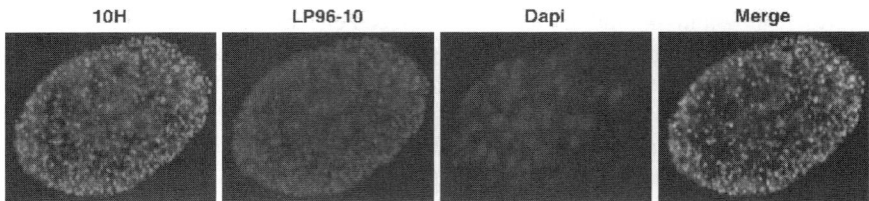

Fig. 15.4 Detection of PAR by immunofluorescence microscopy. 3T3 cells were exposed to H_2O_2 (500 μM in PBS) for 10 min followed by incubation for 10 min in culture medium at 37°C. After fixation in methanol:acetone, PAR was revealed by the monoclonal anti-PAR antibody 10 H or by the rabbit anti-PAR antibody LP96-10. DNA was stained with DAPI. Colocalization of signals is shown in the merged image. Images were captured using a Zeiss Axioplan microscope (×100 objective) with an Olympus DP50 camera. Image width is 15 μm. To view this figure in color, see COLOR PLATE 8

1. The reaction should be done in triplicate. In a glass tube, mix 10 μL of PARP activity incubation mixture with 90 μL of dH_2O.
2. Add 200 ng of PARP-1 or up to 10 μL of cell extract, and immediately incubate the reaction for 10 min at 25°C.
3. Stop the reaction by adding 4 mL of ice-cold precipitation buffer.
4. Filter the reaction solution on a glass microfibre filter using a vacuum filtering system. Poly(ADP-ribose) will be retained on the filters.
5. Wash the filters twice with 4 mL of ice-cold precipitation buffer and once with ice-cold 95% (v/v) ethanol.
6. Air-dry the filters and transfer them into scintillation counting vials, add 10 mL of scintillation cocktail, and measure radioactivity in a liquid scintillation counter. The radioactivity retained on the filter represents the PAR synthesized during the enzymatic reaction.

3.6 Detection of PARG Activity

3.6.1 Large-Scale Synthesis of ^{32}P-Labelled PAR

The detection of PARG activity in solution or by zymography requires a large amount of ^{32}P-labelled poly(ADP-ribose), whose synthesis is described here.

1. For a 2 mL reaction, add into a siliconized 10-mL Sarstedt tube 1,745 μL of dH_2O, 200 μL of 10× PARP activity buffer, 50 μCi (5 μL) of [^{32}P]NAD$^+$, and 20 μg of PARP-1. Gently vortex and incubate for 1 h at 25°C.
2. Add 4 μL of 1 M $CaCl_2$ and 4 μL of a 10 mg/mL solution of DNase I, mix, and incubate for 1 h at 37°C.

3. Add 935 µL of dH$_2$O, 15 µL of 20% (w/v) SDS, and 50 µL of a 10 mg/mL solution of proteinase K. Mix and incubate for 4 h or overnight at 37°C.

4. Remove the amino acid residues covalently bound to PAR by adding 2 mL of 0.2 M NaOH and 40 mM EDTA, pH 8.0. Mix and incubate 1 h at 37°C, then neutralize by adding 2 mL of 0.2 M HCl.

5. Add 500 µL of phenol/CHCl$_3$/isoamyl alcohol (25/24/1), vortex, and centrifuge at 3,000×g. Carefully transfer the upper phase to a new siliconized tube and repeat the extraction until the organic/aqueous interphase is clean.

6. Precipitate PAR by adding to the aqueous phase 300 µL of 3 M potassium acetate and 3.5 mL of isopropyl alcohol, mix, and let the PAR precipitate for 1 h at −20°C. Centrifuge 30 min at 6,000×g at 4°C and wash the pellet twice with 3 mL of ice cold 80% ethanol. After centrifugation for 30 min at 6,000×g at 4°C, air-dry the pellet by leaving the tube open on the bench. The pellet of ^{32}P-labelled PAR is resuspended in TE buffer.

3.6.2 Detection of PARG Activity in solution

1. The reaction should be done in triplicate. In a glass tube, incubate the following mixture for 10 min at room temperature: 10 µL of 10×PARG activity buffer, 10 µL of 32(superscript)P-labelled-PAR (20,000 dpm), an appropriate dilution of purified PARG (0.1 to 50 ng) or of a cell protein extract (up to 10 µL), and dH2(subscript)O to 100 µL.

2. Stop the reaction, and filter an aliquot as described in Section 3.5, steps 3–6. The radioactivity retained on the filter represents undigested PAR.

3.6.3 Detection of PARG Activity by Zymography

Zymography is an electrophoretic method for detecting a specific enzymatic activity. This method has been adapted to qualitatively detect PARG activity within an SDS-PAGE gel co-polymerized with ^{32}P-labelled PAR. After electrophoresis of purified PARG- or of PARG-containing protein extracts, the gel is incubated in a renaturation solution with several changes and the degradation of the radioactive PAR is revealed by autoradiography as a white band on a dark background.

1. Prepare an SDS-PAGE gel (*see* **Section 3.1**), including in the separating gel mixture 250 µL (10^6 dpm) of ^{32}P-labelled PAR (*see* **Section 3.6.1**) (no radioactive PAR is required in the stacking gel). *Because the gel is now radioactive, all operations should be performed in a controlled area or a specific room dedicated for use of radioactivity.*

2. Transfer the entire gel (stacking plus separating gel) to a clear box containing 200 mL of zymogram renaturation buffer.

3. Renature under agitation at room temperature. After 3 h, carefully discard the solution (*radioactive, collect in a dedicated ^{32}P waste bottle*) and replace with

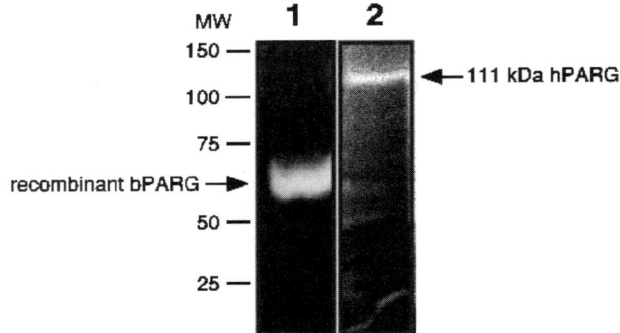

Fig. 15.5 Analysis of PARG activity by zymography. Activity gel assays for PARG were done by casting polyacrylamide gels containing PARP-1 automodified with ^{32}P-labelled PAR. PARG activity was detected following autoradiography as a clear band on a black background. *Lane 1,* 50 ng of recombinant 65-kDa PARG; *lane 2,* 30 μg of total protein from a cleared lysate of HeLa cells

200 mL of fresh renaturation solution. Repeat this step three times. The last incubation is performed at 37°C for 3 h.

4. Remove the renaturation solution and incubate the gel for 1 h in 200 mL of destaining buffer at room temperature under agitation.

5. Gently transfer the gel onto a plastic nonporous film such as Saran Wrap and cover with a sheet of Whatman 3 MM filter paper. The gel is dried for 1 h in a heated vacuum gel dryer with the temperature set at 80°C.

6. Expose the zymogram to Kodak BioMax MS film in a film cassette at −70°C for a few hours and develop the autoradiogram using an automatic film developer. An example is shown in Fig. 15.5.

4 Notes

1. Shorter incubation at higher temperature also work, but may give a weaker signal.

2. For a Pantak Seifert X-ray system, settings are 100 kV and 4.5 mA with the metallic shutter removed.

3. Paraformaldehyde fixation is not recommended for PAR detection, as it can damage DNA independently of the damaging agent previously used and this could inadvertently induce PAR synthesis through PARP activation.

References

1. Kraus, W.L. and Lis, J.T. (2003) PARP goes transcription. *Cell* **113**, 677–683.
2. Amé, J.C., Spenlehauer, C., and de Murcia, G. (2004) The PARP superfamily. *Bioessays* **26**, 882–893.

3. Schreiber, V., Dantzer, F., Amé, J.C., and de Murcia, G. (2006) Poly(ADP-ribose): novel functions for an old molecule. Nat. Rev. Mol. Cell. Biol. 7, 517–528.
4. Ménissier de Murcia, J., Niedergang, C., Trucco, C., Ricoul, M., Dutrillaux, B., Mark, M., Olivier, F.J., Masson, M., Dierich, A., LeMeur, M., Walztinger, C., Chambon, P., and de Murcia, G. (1997) Requirement of poly(ADP-ribose)polymerase in recovery from DNA damage in mice and in cells. Proc. Natl. Acad. Sci. USA 94, 7303–7307.
5. Wang, Z.Q., Stingl, L., Morrison, C., Jantsch, M., Los, M., Schulze-Osthoff, K., and Wagner, E.F. (1997) PARP is important for genomic stability but dispensable in apoptosis. *Genes Dev.* **11**, 2347–2358.
6. Masutani, M., Nozaki, T., Nishiyama, E., Shimokawa, T., Tachi, Y., Suzuki, H., Nakagama, H., Wakabayashi, K., and Sugimura, M. (1999) Function of poly(ADP-ribose) polymerase in response to DNA damage: gene-disruption study in mice. *Mol. Cell. Biochem.* **193**, 149–152.
7. Schreiber, V., Amé, J.C., Dolle, P., Schultz, I., Rinaldi, B., Fraulob, V., Menissier-de Murcia, J., and de Murcia, G. (2002) Poly(ADP-ribose) polymerase-2 (PARP-2) is required for efficient base excision DNA repair in association with PARP-1 and XRCC1. *J. Biol. Chem.* **277**, 23028–23036.
8. Ménissier de Murcia, J., Ricoul, M., Tartier, L., Niedergang, C., Huber, A., Dantzer, F., Schreiber, V., Amé, J.C., Dierich, A., LeMeur, M., Sabatier, L., Chambon, P., and de Murcia, G. (2003) Functional interaction between PARP-1 and PARP-2 in chromosome stability and embryonic development in mouse. *EMBO J.* **22**, 2255–2263.
9. Amé, J.C., Jacobson, E.L., and Jacobson, M.K. (1999) Molecular heterogeneity and regulation of poly(ADP-ribose) glycohydrolase. *Mol. Cell. Biochem.* **193**, 75–81.
10. Meyer-Ficca, M.L., Meyer, R.G., Coyle, D.L., Jacobson, E.L., and Jacobson, M.K. (2004) Human poly(ADP-ribose) glycohydrolase is expressed in alternative splice variants yielding isoforms that localize to different cell compartments. *Exp. Cell. Res.* **297**, 521–532.
11. Andrabi, S.A., Kim, N.S., Yu, S.W., Wang, H., Koh, D.W., Sasaki, M., Klaus, J.A., Otsuka, T., Zhang, Z., Koehler, R.C., Hurn, P.D., Poirier, G.G., Dawson, V.L., and Dawson, T.M. (2006) Poly(ADP-ribose) (PAR) polymer is a death signal. *Proc. Natl. Acad. Sci. USA* **103**, 18308–18313.
12. Gagné, J.P., Hendzel, M.J., Droit, A., and Poirier, G.G. (2006) The expanding role of poly(ADP-ribose) metabolism: current challenges and new perspectives. *Curr. Opin. Cell. Biol.* **18**, 145–151.
13. Lin, W., Amé, J.C., Aboul-Ela, N., Jacobson, E.L., and Jacobson, M.K. (1997) Isolation and characterization of the cDNA encoding bovine poly(ADP-ribose) glycohydrolase. *J. Biol. Chem.* **272**, 11895–11901.
14. Méder, V.S., Boeglin, M., de Murcia, G., and Schreiber, V. (2005) PARP-1 and PARP-2 interact with nucleophosmin/B23 and accumulate in transcriptionally active nucleoli. *J. Cell Sci.* **118**, 211–222.

Chapter 16
Purification and Analysis of Variant and Modified Histones Using 2D PAGE

George R. Green and Duc P. Do

Keywords Histone variants; Histone modification; Phosphorylation; Acetylation; Ubiquitylation; Two-dimensional polyacrylamide gel electrophoresis; Electroelution; Western analysis; Triton X-100; SDS-PAGE

Abstract Two-dimensional (2D) polyacrylamide gel electrophoresis (PAGE) systems employing combinations of acetic acid/urea (AU), acetic acid/urea/Triton X-100 (AUT) and sodium dodecyl sulfate (SDS) gel formulations are uniquely effective for resolution of histone variants and their modified derivatives. Coupled with Western transfer methods using modification-specific antibodies and recent advances in mass spectrometry, 2D PAGE emerges as a versatile tool for histone purification and analysis. This chapter describes 2D PAGE gel systems appropriate for histone proteins, including detailed procedures for designing, running, and staining gels. Methods for electrophoretic transfer of histones from AUT×SDS and AUT×AU 2D gels are described and evaluated. Alternatively, methods are provided for obtaining highly purified protein samples from fixed and stained gels via electroelution of proteins from specific gel spots.

1 Introduction

Histones, the protein building blocks of eukaryotic chromatin, have been substantially conserved during evolution, a consequence of their central role in chromatin structure. Variation in histone structure and function does occur, however, as the product of two fundamentally different processes. First, diverse gene families within and between species produce both homomorphic and heteromorphic histone variants, with long-term adaptive consequences for chromatin structure and function (1–5). Second, histones may be altered postsynthetically by processes such as phosphorylation, acetylation, methylation, and ubiquitinylation to cause transient changes in chromatin structure and function, as might occur during activation of a gene or condensation of a chromosome (6–9). Variant and modified histones are therefore of central interest in modern chromatin research.

This chapter describes methods for purification and analysis of histone variants and their derivative forms using high-resolution two-dimensional (2D) polyacrylamide gel electrophoresis (PAGE). Commonly used 2D PAGE systems combine methods that differ in separation principle, providing unparalleled resolution for analytical purposes with sufficient capacity for preparation of highly pure histone fractions. First-dimension AUT gels contain acetic acid, urea, and the nonionic detergent Triton X-100 (TX100), which binds differentially to homomorphic histone variants, slowing their electrophoretic migration and facilitating resolution *(10)*. 2D gels are of two types: sodium dodecyl sulfate (SDS) gels, which resolve histones according to their molecular mass *(11,12)*, and AU gels, which resolve histones as a function of their charge/mass ratio *(13,14)*. Careful selection from available 2D PAGE formats and formulations yields precise 2D displays of differentially migrating histone variants and their modified forms (Fig. 16.1; *see* **refs.** *(15–28)* for examples).

Once resolved by 2D PAGE, the amount of histone protein contained in a single stained gel spot is often sufficient for direct in-gel peptide mapping *(19, 20, 22, 23, 25–28, 29)* or, using material eluted from the gel *(30)*, for amino acid analysis *(31)*, phosphoamino acid analysis *(19, 20, 22, 26)*, and tryptic peptide mapping *(19)*. Alternatively, conventional Western analysis using antibodies developed to recognize specific histone modifications may be used to probe the 2D PAGE display *(32–35)*. Coupled with recent advances in mass spectrometry *(35–46)*, high resolution 2D PAGE and Western analysis emerge as potent tools for analysis of histone variants and their modified forms. This chapter provides detailed procedures for designing, constructing, and running 2D PAGE gels, together with methods for staining, Western analysis, electroelution, and quantification of single protein spots from fixed and stained gels.

2 Materials

2.1 Sample Preparation

1. All-purpose buffer (APB): $0.15 M$ NaCl and $10 mM$ Tris-HCl, pH 8.0. Store at 4°C. APB may be prepared as a 10× stock solution.
2. $0.4 M$ H_2SO_4. Store at 4°C.
3. 100% w/v trichloroacetic acid (TCA) (Sigma-Aldrich, St. Loius, MO, USA; cat. T6399). Store at 4°C.
4. Acetone (Fisher, Pittsburgh, PA; cat. A929). Store at room temperature.
5. Acetic acid/urea loading buffer (AULB): $8 M$ urea (Sigma-Aldrich; cat. U15), 5% v/v acetic acid, 5% v/v β-mercaptoethanol, 0.2 mg/mL crystal violet. Store at room temperature.

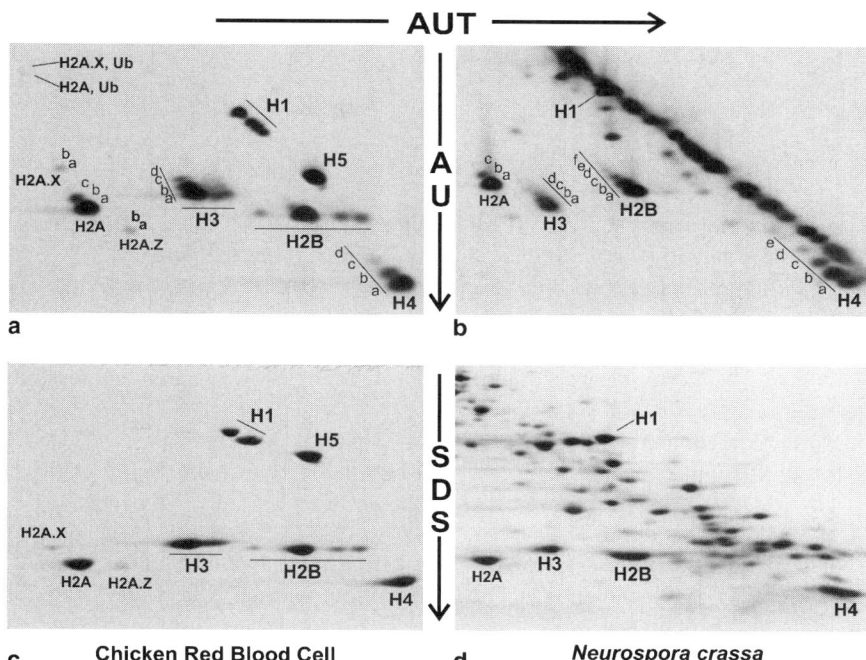

Fig. 16.1 Examples of variant and modified histones resolved by AUT×AU and AUT×SDS 2D PAGE. Nuclei were prepared from chicken red blood cells nuclei *(27)* and *Neurospora crassa* *(49)* and histones were extracted and prepared for electrophoresis as described in **Section 3.1**. The histones were resolved in the first (horizontal) dimension using a 6 *M* urea AUT gel (Sections 3.2 and 3.3), followed in the second (vertical) dimension by either an AU gel (**a** and **b**; **Section 3.4**) or an SDS gel (**c** and **d**; **Section 3.5**). Midi gel formats were used for both dimensions. The gels were stained with Coomassie blue R-250, dried between transparent cellophane sheets, and scanned with an HP Scanjet 5550C scanner (**Section 3.6**). Heteromorphic histone variants in the H1 and H2A histone classes and homomorphic variants in the H3 and H2B histone classes characterize the histone pattern obtained from chicken red blood cell nuclei (**a** and **c**). Modified core histone species, indicated by the *lower case type*, are observed in both species but are especially abundant in *Neurospora* (**b** and **d**)

2.2   Gel Equipment and Formats

1. Mini gel electrophoresis: Biorad Mini-Protean 3 vertical gel apparatus.
2. Midi gel electrophoresis: Hoeffer SE 600 equipped with divider plates.
3. Maxi gel electrophoresis: Hoeffer SE 660 equipped with divider plates.

4. BioRad Mini-Blot Transfer Cell.
5. Electrophoretic gel elution apparatus *(30)*.

2.3 AUT 1D Gels and AU 2D Gels

1. Ammonium persulfate (APS; Fisher). Store desiccated at room temperature.
2. Dry urea and $10M$ urea (Sigma-Aldrich). Store at room temperature.
3. Stacking gel buffer: $6M$ urea (Sigma-Aldrich), $10mM$ potassium acetate and 0.5% v/v acetic acid. Store at room temperature.
4. Acrylamide for AU and AUT gels: 60% w/v acrylamide (Sigma-Aldrich; cat. A8887), 0.4% w/v bisacrylamide (Sigma-Aldrich). Store at room temperature. *Caution: Acrylamide is a neurotoxin.*
5. 10% v/v Triton X-100 (TX100) (Acros Organics, Morris Plains, NJ, USA). Store at room temperature over mixed-bed deionizing resin (Sigma-Aldrich; cat. M8157).
6. Glacial acetic acid.
7. Riboflavin phosphate (Sigma-Aldrich), 1.0 mg/mL in water. Store at 4°C.
8. Tetramethylethylenediamine (TEMED) (Sigma-Aldrich). Store desiccated at room temperature.
9. Separating gel overlay buffers contain 0.02% w/v crystal violet dissolved in the same reagents used to compose the corresponding separating gel except acrylamide, bisacrylamide, APS, and TEMED, for which H_2O is substituted. Store at room temperature.
10. Stacking gel overlay buffer: 5% v/v acetic acid, 5 mg/mL protamine sulfate (Sigma-Aldrich), 20 mg/mL β-mercaptoethylamine (Sigma-Aldrich), and 0.02% w/v crystal violet. Store at room temperature.
11. AU/AUT gel running buffer (AURB): 5% v/v acetic acid. Store at room temperature.

2.4 AUT×AU 2D Gels

1. Lane-locator stain: 0.01% w/v Coomassie blue R-250 (CB) (Sigma-Aldrich), freshly prepared from a 0.1% w/v CB stock dissolved in 95% v/v ethanol.
2. Stacking gel overlay buffer: 5% v/v acetic acid, 5 mg/mL protamine sulfate (Sigma-Aldrich), and 20 mg/mL β-mercaptoethylamine (Sigma-Aldrich). Store at room temperature.
3. Agarose sealant: for 100 mL, make freshly by dissolving 0.5 g of protamine sulfate (Sigma-Aldrich P4020) in 94 mL of 0.02% w/v crystal violet. Add 1 g of agarose (Sigma-Aldrich A9539) and heat in a microwave oven to dissolve the agarose. Remove the mixture from the microwave oven; add 2.0 g of β-mercaptoethylamine (Sigma-Aldrich) and mix to dissolve. Then add 5 mL of glacial acetic acid, mix well, and use immediately to seal the 1D AUT gel to the surface of the 2D AU gel.

2.5   AUT×SDS 2D Gels

1. Ammonium persulfate: *see* **Section 2.3.1**
2. Separating gel buffer: 1.5 M Tris-HCl, pH 8.8. Store at room temperature.
3. Stacking gel buffer: 0.5 M Tris-HCl, pH 6.8. Store at room temperature.
4. Acrylamide for SDS separating gels: 30% w/v acrylamide (Sigma-Aldrich), 0.4% w/v bisacrylamide (Sigma-Aldrich). Store at room temperature. *Caution: acrylamide is a neurotoxin.*
5. Acrylamide for SDS stacking gels: 30% w/v acrylamide, 0.8% w/v bisacrylamide. Store at room temperature. *Caution: acrylamide is a neurotoxin.*
6. 10 % w/v SDS (Fisher; cat. BP166).
7. TEMED (Sigma-Aldrich). Store desiccated at room temperature.
8. SDS running buffer (SDSRB): 25 mM Tris base, 192 mM glycine, 0.1% w/v SDS. Store at room temperature.
9. SDS preequilibration buffer: 50 mM Tris-HCl, pH 6.8, and 20 µg/mL bromophenol blue.
10. SDS equilibration buffer: 50 mM Tris-HCl, pH 6.8, 0.2% w/v SDS, 0.2% w/v β-mercaptoethanol, 20% v/v glycerol, and 20 µg/mL bromophenol blue.

2.6   Gel Staining, Drying, and Scanning

1. Coomassie blue gel stain: 0.05% w/v Coomassie blue R-250 (Sigma-Aldrich), 40% v/v ethanol, 5% v/v acetic acid.
2. Gel destain: 20% v/v ethanol, 0.5% v/v acetic acid.
3. Gel drying solution: 50% v/v methanol, 2% v/v glycerol.
4. Plexiglas gel drying frames.
5. Cellophane gel drying sheets (Idea Scientific, Minneapolis, MN, USA; cat. 1080).

2.7   Electroelution of Proteins from Fixed and Stained Gel Spots

1. Elution buffer: 30 mM Tris-HCl and 0.1% w/v SDS, pH 8.8.
2. Eluter preequilibration buffer: 10 mM Tris-HCl, pH 6.8.
3. Eluter equilibration buffer: 10 mM Tris-HCl, pH 6.8, 0.2% w/v SDS, 0.2% v/v β-mercaptoethanol, and 20% v/v glycerol.
4. Recovery zone buffer: 10 mM Tris-HCl, pH 6.8, and 0.1% w/v SDS.
5. Gel-slice sealing agarose: 0.5% w/v agarose (Sigma-Aldrich A9539), 10 mM Tris-HCl, pH 6.8, and 0.1% w/v SDS. Melt the agarose by heating in a microwave oven just prior to use.

2.8 Recovery and Quantification of Electroeluted Proteins

1. Ion-exchange extraction buffer: 90% v/v acetone, 5% v/v acetic acid, 5% v/v triethylamine (Fluka-Sigma-Aldrich). Store at 4°C.
2. Acetone (Fisher). Store at room temperature.

2.9 Electrophoretic Transfer for Western Analysis

1. SDS gel transfer buffer: 25 mM Tris, 192 mM glycine, 20% v/v methanol, and 0.05% w/v SDS. Store at 4°C.
2. AU gel transfer buffer: 0.7% v/v acetic acid. Store at 4°C.
3. Ponceau S stain: 0.1% w/v Ponceau S in 5% v/v acetic acid.

3 Methods

3.1 Sample Preparation

Histones should be prepared from biological materials using methods that preserve their structure and postsynthetic modifications. They may be extracted from subcellular fractions such as nuclei or chromatin, and in some cases directly from whole cells. Methods for preparing histone-rich subcellular fractions are specific to the material being studied, and will not be covered here. Beginning with an appropriate preparation, histones may be extracted for electrophoretic analysis using the following procedure:

1. Extract histones at 4°C by adding an equal volume of ice-cold $0.4\,M$ H_2SO_4 to chromatin, nuclei, or other biological preparation dissolved or suspended in APB or an equivalent solution. Mix thoroughly and incubate on ice for 24 h with occasional mixing.
2. Remove acid-insoluble material by centrifuging at 10,000×g for 10 min at 4°C.
3. Transfer the clear supernatant to a fresh centrifuge tube, add ice cold 100% TCA to a final concentration of 20% TCA (1/4 volume), mix thoroughly, and incubate on ice for 1 h to precipitate the histones.
4. Collect the precipitated histones by centrifugation at 10,000×g for 10 min at 4°C, discard the supernatant, and resuspend the histone pellet in ice-cold acetone, using a plastic transfer pipette and a bath sonicator to fully disperse the precipitated protein.
5. Collect the histone precipitate by centrifugation at 10,000×g for 10 min at 4°C, discard the supernatant, and wash the pellet by suspension in cold acetone followed by centrifugation at 10,000×g for 10 min at 4°C.
6. Thoroughly drain the supernatant, dry the pellet briefly at room temperature, and dissolve in AULB to a concentration of 1–10 mg protein/mL.

3.2   Gel Equipment and Formats

Commercially available gel electrophoresis apparatuses are available to accommodate three vertical gel formats, Mini gels, Midi gels, and Maxi gels (Table 16.1). Use thinner gel formats for analytical applications and to minimize sample consumption, and thicker formats to maximize protein capacity for subsequent protein recovery by electroelution. Use longer gel formats to increase resolution and protein capacity. The 1D gel must be thinner by one spacer size than the corresponding 2D gel so that it will fit between the 2D gel plates. Thus, a 0.75-mm 1D gel would be combined with a 1.0-mm 2D gel, a 1.0-mm 1D gel would be combined with a 1.5-mm 2D gel, and a 1.5-mm 1D gel would be combined with a 2.0-mm 2D gel. The use of gel divider plates to create a "club sandwich" containing two vertical slab gels conveniently increases the carrying capacity of the PAGE apparatus and creates a wider space between the outer plates to facilitate delivery of liquid acrylamide into the gel mold. Typical 2D gel formats include:

1. For sample screening: Use a Mini 1D gel and apply two 1D lanes to a single Midi 2D gel to increase the capacity of the apparatus and speed sample analysis. In addition to increased analytical capacity, this system consumes less reagents, sample, and effort than other methods, especially when 0.75-mm-thick 1D gels are employed in the first dimension followed by 1.0-mm-thick 2D gels.
2. Analytical gels: Use a Midi 1D gel followed by a Midi 2D gel to obtain routine high-resolution 2D gels, using gel formulations that have been adjusted to assure optimal resolution of the specific sample being studied. Each sample will require one full slab gel for the 2D analysis, so it is usually convenient to run multiple 2D gels in parallel using multiple gel apparatuses fitted with gel divider plates. Typically, 12 to 24 2D gels may be processed simultaneously by an adequately stocked facility.
3. High capacity and high-resolution analytical gels: The capacity of any 2D PAGE system may be increased by using thicker gels or by running longer gels to

Table 16.1 Dimensions, volumes, and properties of gel formats

Gel type	Gel thickness (mm)	Mini[a] (mL)	Midi[b] (mL)	Maxi[c] (mL)
Separating gel	0.75	3.7	13.5	24
	1.0	5.0	18	32
	1.5	7.5	27	48
Stacking gel	0.75	0.8	2.9	2.9
	1.0	1.1	3.8	3.8
	1.5	1.6	5.7	5.7
Resolution	Similar	Good	Better	Best
Capacity	Increases	Low	Moderate	High
Reagent costs	Increases	Low	Moderate	High
Effort	Similar	Less	Moderate	More

Volumes exceed the actual volumes by ~10%
[a]Height × width = 8.5×8 cm; separating gel = 6 cm, stacking gel = 1.5 cm
[b]Height × width = 16×14 cm; separating gel = 12 cm, stacking gel = 3 cm
[c]Height × width = 24×14 cm; separating gel = 20 cm, stacking gel = 3 cm

increase separation between protein spots. In addition, use of longer gels will increase resolution of protein spots that are hard to separate, such as the acetylated forms of H3. High capacity, high-resolution 2D systems consist of a Midi 1D gel followed by a Maxi 2D gel. To increase resolution in the first dimension, the 1D gel may be run as a Maxi gel and cut into two pieces for subsequent 2D analysis on Midi or Maxi gels. High capacity, high-resolution 2D systems are especially useful for analysis and recovery of modified or variant histones that resolve poorly or are present in small amounts. However, these gels also require special gel equipment, are more laborious, and consume more sample and reagents than mini and midi formats.

Optimal resolution of histone subtypes and their modified derivatives requires careful selection of the appropriate gel systems. Commonly used 2D PAGE gel systems for histone analysis employ AUT in the first dimension, but differ in the urea/TX100 ratio used, which can be altered to optimize histone resolution. Increasing the urea concentration in the AUT gel reduces the binding of TX100 to histones, effectively increasing their electrophoretic mobility. Since histone variants bind TX100 with differential affinities, careful adjustment of the TX100/urea ratio in the AUT dimension can permit resolution of proteins that are difficult to separate. Formulations for AUT gels with differing TX100/urea ratios (including AU gels) are provided in Table 16.2. Procedures for casting and running AU and AUT gels are provided in **Section 3.3**.

The 2D gel may consist of either an AU gel (Fig. 16.1a, b) or an SDS gel (Fig. 16.1c, d). AUT×AU 2D systems, described in Sections 3.3 and 3.4, are best for separating core histone variants and their charge-altered derivatives. AUT×SDS 2D systems, described in **Section 3.5**, are useful for resolving heteromorphic histone variants and histones that do not bind TX100 (primarily H1 histone variants).

Recovery of specific protein spots from 2D gels may be accomplished by electroelution of gel slices from fixed and stained gels using the apparatus described in **ref.** *(30)* (*see* **Section 3.7**). Alternatively, proteins may be transferred from the 2D gel before staining using commercially available electroblot equipment (e.g., BioRad Mini-Blot Transfer Cell) and Western blot procedures (*see* **Section 3.9**).

Table 16.2 Formulation of AUT and AU gels

For 100 mL	Final [urea] (M)	Urea 10M (mL)	APS (mg)	H$_2$O (mL)	Acryl[a] (mL)	Triton 10% (mL)	Acetic acid (mL)	Urea (g)	TEMED (mL)
AUT	4	40	100	30	20	4.0	5.0	—	1.0
AUT	5	50	100	20	20	4.0	5.0	—	1.0
AUT	6	60	100	10	20	4.0	5.0	—	1.0
AUT	7	70	100	—	20	4.0	5.0	—	1.0
AUT	8	60	100	—	20	4.0	5.0	12	1.0
AU	8.4	64	100	—	20	—	5.0	12	1.0

[a]60% acrylamide, 0.4% bisacrylamide mixture

3.3   AUT 1D Gels and AU 2D Gels

This section is adapted from **refs.** *(10, 14, 47)*.

1. Design an appropriate AUT×AU 2D system (*see* **Section 3.2**) by selecting appropriate gel formats (Table 16.1) and gel recipes (Table 16.2).
2. Assemble the gel molds and estimate the required volume of gel solution (*see* **Note 1**).
3. Using a gel recipe from Table 16.2, mix all the reagents except TEMED in the order listed from left to right. If dry urea is included in the recipe, stir gently until the urea is fully dissolved (*see* **Note 2**).
4. Add TEMED and mix the gel solution quickly and thoroughly.
5. Pour the gel solution into the gel molds allowing approximately 2.5 cm for the stacking gel, overlay with water, and polymerize for 1 h.
6. Pour off the water overlay solution and replace with separating gel overlay buffer. Allow the gel to polymerize for an additional hour.
7. Assemble the gel apparatus and then fill the upper and lower reservoir with AURB.
8. Pre-electrophorese the gel using a constant voltage of 100 V until the crystal violet tracking dye reaches the bottom of the gel.
9. Remove the gel mold from the apparatus and remove the gel overlay buffer in preparation for casting the stacking gel.
10. For 100 mL of AUT/AU stacking gel solution, mix 85 mL of 6.0 M urea/10 mM K$^+$ acetate and 15 mL of 60%/0.4% acrylamide/bisacrylamide.
11. Add 500 µL of riboflavin phosphate and 500 µL of TEMED, mix well, and pour into the gel mold.
12. For AUT 1D gels, insert an appropriate well-forming comb into the stacking gel solution.
13. For AU 2D gels, fill with stacking gel solution to 1 cm below the top of the glass plates and overlay with water to create a slot to receive the 1D gel.
14. Place the stacking gel in close contact with a bright fluorescent light source (*see* **Note 3**) and polymerize for 1 h.
15. Remove the gel-forming comb (1D gels) or H$_2$O overlay (2D gels), rinse with AURB, and fill the sample wells (1D gels) or 1D gel slot (2D gels) with stacking gel overlay buffer.
16. For 1D AUT gels, layer the samples into the wells beneath the stacking gel overlay, assemble the gel apparatus, fill with AURB, and run at 100 V until the crystal violet tracking dye reaches the bottom of the gel.
17. For 2D AU gels, *see* **Section 3.4**.

3.4   AUT×AU 2D Gels

This section is adapted from **refs.** *(25,47)*.

1. Remove the 1D AUT gel from the gel mold and soak in lane-locator stain for the minimal amount of time needed to make the sample lanes visible.
2. Wash the gel briefly with H_2O, then use a Plexiglas gel cutter (UVP, Upland, CA, USA; cat. 85-0002-01) to accurately excise the individual gel lanes.
3. Insert the excised 1D gel lanes between the glass plates of the pre-electrophoresed 2D AU gel, and use a gel spacer to press the 1D gel into contact with the surface of the 2D stacking gel.
4. Remove residual stacking gel overlay buffer and replace with freshly prepared agarose sealant, taking care to seat the 1D gel tightly against the 2D stacking gel.
5. Assemble the gel apparatus, fill with AURB, and run at 100 V constant voltage until the crystal violet tracking dye reaches the bottom of the gel.

3.5   AUT×SDS 2D Gels

This section is adapted from **refs. (11, 15, 47)**.

1. Remove the 1D AUT gel from the gel mold, stain, and destain (**Section 3.6**).
2. Wash the destained 1D gel with tap water and then incubate with gentle shaking in preequilibration buffer for 1–24 h.
3. Assemble the gel molds and estimate the required volume of gel solution (*see* **Note 1**).
4. For 100 mL of SDS separating gel solution, dissolve 50 mg of APS in 24 mL of H_2O and then add 50 mL of acrylamide solution (30% acrylamide, 0.4% bisacrylamide), 25 mL of 1.5 M Tris-HCl, pH 8.8, and 1.0 mL of 10% SDS, and mix thoroughly (*see* **Note 2**).
5. Add 50 μL of TEMED and mix the gel solution thoroughly.
6. Pour the gel solution into the gel molds allowing approximately 2.5 cm for the stacking gel, overlay with water, and polymerize for 1 h.
7. Pour off the water overlay solution in preparation for casting the stacking gel.
8. For 100 mL of SDS stacking gel solution, dissolve 50 mg of APS in 54 mL of H_2O and then add 25 ml of 0.5 M Tris-HCl, pH 6.8, 20 mL of acrylamide (30% acrylamide, 0.8% bisacrylamide), and 1.0 mL of 10% SDS with good mixing.
9. Add 50 μL of TEMED, mix well, and pour into the gel mold to 1.0 cm below the top of the glass plates.
10. Overlay with H_2O and polymerize for 1 h.
11. Use a Plexiglas gel cutter to accurately excise the individual gel lanes from the preequilibrated fixed and stained 1D gel.
12. Incubate the excised gel lanes in equilibration buffer for 2 h, changing buffer every 30 minutes.
13. Insert an excised 1D gel lane between the glass plates of an SDS 2D gel, using a gel spacer to press the 1D gel into close contact with the surface of the stacking gel.
14. Assemble the gel apparatus, fill with SDSRB, and run at a constant current of 20 mA for each 1 mm of gel thickness until the bromophenol blue tracking dye reaches the bottom of the gel.

3.6 Gel Staining, Drying, and Scanning

1. Remove the gel from the gel mold, place in a plastic container, and add CB stain to a level sufficient to completely wet the gel.
2. Incubate at room temperature with gentle shaking for 16–24 h.
3. Remove the stain, rinse the gel briefly with tap water, add destain and incubate at room temperature with gentle shaking for several hours.
4. Repeat the destaining procedure as required to fully destain CB from the gel background.
5. To dry the gel, soak in gel drying solution for several hours with two changes of the drying solution to fully equilibrate the gel.
6. Soak a piece of transparent gel-drying cellophane in gel-drying solution and laminate the cellophane to a Plexiglas plate with the aid of a gel squeegee (*see* Fig. 16.2a).
7. Cover the gel with a second piece of wet cellophane and remove as much gel-drying solution as possible with the gel squeegee.
8. Attach a gel-drying frame to the Plexiglas plate using binder clamps.
9. Incubate at room temperature until fully dry, usually 16–24 h depending on gel thickness, ambient temperature, and humidity.
10. Remove the dried gel from the drying apparatus and cut away excess cellophane with scissors or a paper cutter.
11. Obtain a scanned image of the dried gel using a conventional flat-bed scanner.

3.7 Electroelution of Proteins from Fixed and Stained Gel Spots

This section is adapted from **refs. (30, 47)**.

The eluter is used for multiple elution, concentration, and quantification of proteins from individual Coomassie blue-stained gel slices. Eluted proteins are

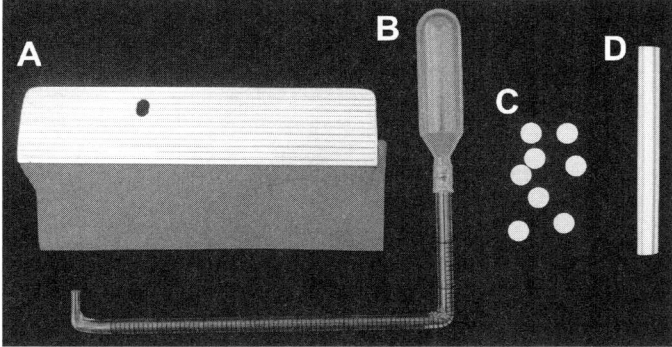

Fig. 16.2 Special tools used for gel drying and electroelution. Gel squeegee (**A**), bubble extractor (**B**), paper discs (**C**), and gel spot cutter (**D**). Details of construction and use are provided in the text

collected in volumes of less than 30 µL with efficiencies of approximately 90%. The eluter model described here is capable of eluting protein gel slices from 12 separate sample wells, each containing two to three gel slices, in about 1 h. Following elution, the eluter may be used for passive or electrophoretic dialysis of the eluted proteins to remove SDS and change solvent constituents. Alternatively, eluted proteins may be recovered by ion-pair extraction to yield highly pure acetone precipitates ((48); *see* **Section 3.8**). Protein-bound Coomassie blue dye is quantitatively recovered during electroelution, permitting direct spectrophotometric quantification of the recovered proteins.

The procedure for electroelution of proteins from gel spots stained with Coomassie blue R250 is as follows:

1. Cut 1 mm off each edge of a 5.4-cm-long piece of flat, dry dialysis tubing (MW cutoff 3,500; dry cylinder diameter 34.0 mm; Spectrophor 132725; Spectrum Industries, Los Angeles, CA, USA) and wet the tubing under hot tap water to produce two sheets of dialysis membrane. Extra dialysis membrane sheets may be stored at 4°C in eluter running buffer.
2. Assemble the eluter cassette (for details *see* **ref. (30)**), positioning a single dialysis sheet as shown in Fig. 16.3. Rinse the assembled cassette with hot tap water and deionized water and insert into the lower reservoir.
3. Fill the lower reservoir with elution buffer and remove air bubbles from below each well using the bubble extractor shown in Fig. 16.2b, constructed from a 1-mL plastic serological pipette attached to a bulb cut from a plastic transfer pipette (Fisher; cat. 13-711-9A).
4. Flatten one end of a short plastic soda straw (Fig. 16.2d) to form an ovoid cutting edge, and cut specific gel slices from a fixed and stained gel using gentle pressure and twisting of the straw to excise the gel slice.
5. Incubate the gel slices in eluter preequilibration buffer at room temperature for 60 min.

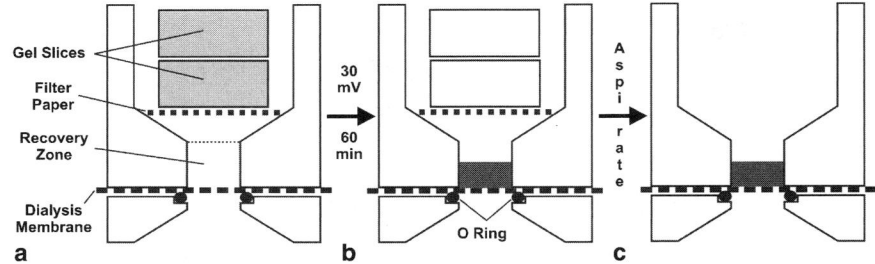

Fig. 16.3 Electroelution of proteins from fixed and stained gel slices. A detail of one well from the 12-well electrophoretic eluter *(30)* diagrams the progressive stages of electroelution (**a–c**). The eluter is assembled and equilibrated gel slices are placed in the wells (**a**) (*see* **Section 3.7**). A constant voltage of 30 V is applied for 1 h, eluting the proteins from the gel slices into the sample well recovery zone (**b**). Elution buffer, gel slices, and filter discs are removed, yielding the eluted protein in a volume of approximately 30 µL (**c**)

6. Incubate the gel slices in eluter equilibration buffer at room temperature for 60 min.

7. Fill the lower section of the sample wells with 100 μL of recovery-zone buffer.

8. Use a standard paper punch to cut 7-mm-diameter discs from Whatmann Type 1 filter paper (Fig. 16.2c), soak the discs in recovery-zone buffer, and insert into the wells as shown in Fig. 16.3. The surfaces of the filter discs should sit at the surface of the recovery zone buffer, just wetted by it and not trapping air bubbles; if bubbles are present repeat **step 7**, increasing the volume of recovery-zone buffer. Carefully remove excess recovery-zone buffer using a Hamilton syringe attached to an aspirator.

9. Blot residual equilibration buffer from the surfaces of the equilibrated gel slices and position them in the sample wells, as shown in Fig. 16.3a. Use 75 μL of freshly melted gel slice-sealing agarose to bond the gel slices to the filter discs and allow 10 min for the agarose to cool and gel.

10. Carefully fill the upper reservoir of the eluter cassette with elution buffer, taking care not to disturb the gel slices.

11. Insert the upper electrode holder and connect the eluter to a power supply with the negative electrode leading to the upper reservoir and the positive electrode leading to the lower reservoir. Elute at 30 V for 1 h until all Coomassie Blue has eluted from the gel slices and entered the recovery zone.

12. The eluted protein sample will be accurately marked by the coeluting CB dye, which should form a boundary with the elution buffer as shown in Fig. 16.3b. Take care not to disturb or remove this material during removal of the elution buffer, gel slices, and filter paper discs.

13. Remove elution buffer from the upper reservoir using a plastic transfer pipette attached to an aspirator, then carefully remove residual elution buffer from the sample wells using a Hamilton syringe attached to an aspirator, exposing the remaining gel slices and filter paper discs.

14. Use forceps to remove the vacated gel slices and filter paper discs, then use the Hamilton syringe to carefully remove residual buffer from the sample wells and recovery zone leaving the eluted protein samples, marked by CB dye, at the bottom of the recovery zone.

15. Remove the eluted proteins from the recovery zones using a plastic-tipped micropipettor, taking care not to puncture the dialysis membrane.

3.8 Recovery and Quantification of Electroeluted Proteins

This section is adapted from **refs.** *(30, 47, 48)*.

1. Following electrophoretic elution, SDS and some nonprotein constituents may be removed from the eluted protein by dialysis of the sample while still in the eluter cassette. Note that CB dye will not be removed by this procedure. Replace the lower reservoir buffer with an appropriate dialysate, add a stir bar, and mix the

lower reservoir buffer with a magnetic stirrer. Dialysis time will depend on the sample, and dialysate constituents and should be tested for specific applications.

2. For removal of CB dye, SDS, and other nonprotein constituents and recovery of the eluted protein sample as an acetone precipitate, use the ion-pair extraction procedure described by Henderson et al. *(48)*. The ion-pair extraction solvent is especially efficient in the removal of SDS, which accumulates at high concentrations in the recovery zone *(30)*. Transfer the eluted sample to a 1.5-mL microcentrifuge tube, add 20 volumes of ice cold ion-pair extraction buffer and mix thoroughly. Incubate the samples at −80°C for 24 h, warm to 4°C, and centrifuge for 5 min at 4°C and 16,000×*g*. Carefully decant the clear blue supernatant and save for spectrophotometric quantification (*see* **Section 3.8.3**). Wash the pellet once with cold ion-pair extraction buffer and once with cold acetone, then air dry. Tightly cap the microcentrifuge tube for storage. The sample typically appears a nearly invisible speck stuck to the bottom of the tube; note its appearance to assure recovery during subsequent analytical procedures.

3. For quantification of an eluted protein sample immediately after elution, dilute to an appropriate volume with H_2O and measure the absorbance at 600 nm using a conventional spectrophotometer. For protein quantification after ion-pair extraction, read the absorbance of the clear blue ion-pair extraction supernatant obtained in **step 2** in the same way. Standard curves for either procedure may be created by running different amounts of a reference protein in the same gel system, with parallel staining and elution procedures.

3.9   Electrophoretic Transfer for Western Analysis

Conventional electroblotting methods using nitrocellulose or nylon membranes may be used to transfer proteins from unstained 2D gels run in either the AUT×AU or AUT×SDS format, using acetic acid-based systems for AUT×AU gels and SDS-based transfer systems for AUT×SDS gels. The efficiency of histone transfer to nitrocellulose versus nylon membranes appears to be similar, but the choice of membrane material may affect other aspects of the Western procedure and the membrane should be tested for specific applications to assure optimal results. Transfer from AUT×AU gels using 0.7% acetic acid for the transfer buffer is very efficient and especially appropriate for highly basic histone proteins. Conversely, we find that transfer of histones from AUT×SDS gels using SDS-based systems can be both selective and inefficient. For example, when 20% methanol is used in the SDS transfer buffer, chicken red blood cell histone H5 is quantitatively retained in the gel, with essentially no electrophoretic transfer. In addition, loss of core histones, especially histones H2A and H4, occurs during electrophoretic transfer in SDS-based systems, leading in many cases to poor recovery of certain histone subtypes. These circumstances require optimization of SDS-based transfer system conditions and elution buffers to obtain effective electrophoretic transfer of histones of interest.

Methods for electrophoretic transfer from AUT×AU and AUT×SDS gels are similar, except that different buffers are used (*see* **Section 2.9**) and the directions of protein migration are reversed: in SDS-based transfers, histones migrate toward the positive electrode, whereas in acetic acid-based transfers they migrate toward the negative electrode. The general procedure is as follows:

1. Soak the unstained gel in gel transfer buffer for 30 min. If a nitrocellulose membrane is used, wash the membrane with gel transfer buffer before installing. If a nylon membrane is used, wash in 100 % methanol for 10 sec followed by a brief wash in gel transfer buffer prior to use.
2. Assemble the transfer apparatus according to the manufacturer's instructions, insert the prechilled cooling module provided with the apparatus, and add cold gel transfer buffer.
3. Connect the apparatus to a power supply. For acetic acid-based transfers, orient the AU gel toward the positive electrode and the blotting membrane toward the negative electrode. For SDS-based transfers, orient the SDS gel toward the negative electrode and the blotting membrane toward the positive electrode.
4. Electroblot at 100 V for 60 min.
5. Remove the membrane from the assembly and wash briefly in H_2O. The blot may be stained by soaking in Ponceau S stain followed by brief destaining in 0.5% acetic acid, or used immediately for Western analysis.

4 Notes

1. Use clean, dry components for the gel apparatus. Glass plates should be washed in hot detergent (Alconox), thoroughly rinsed with hot water, and dried in a rack that holds them vertically. If properly washed and rinsed, additional treatments with nitric acid, ethanol, or distilled water are not necessary.
2. To obtain consistent gel polymerization kinetics, gel reagents should be at room temperature before adding TEMED and pouring the gel. Mixing and pouring of gel solutions are facilitated by tricorner plastic beakers of appropriate size, using two beakers to efficiently mix reagents by pouring back and forth between two beakers. The pointed lips of the tricorner beaker allow the gel solution to be poured directly from the beaker into the gel mold without the aid of pipettes or syringes.
3. The fluorescent light source for polymerizing stacking gels must provide an intense white light source within a few centimeters of the gel. We use a 20-W portable "light stick" (Lights of America Model 7020), which is inexpensive, easily stored when not in use, and available through most local home supply stores. The light stick fits snuggly between the two gel sandwiches held by a single Midi gel casting stand, simultaneously providing intense light to both sandwiches at the height of the stacking gel without additional support. Up to three casting stands holding 6 gel sandwiches and, with divider plates, up to 12 slab gels may be simultaneously polymerized with one li ght stick.

Acknowledgments   We thank Eric Selker, Kristina Smith, and Keyur Adhvaryu for providing *Neurospora* strains and assistance with isolating nuclei and histones from this species.

References

1. Ausio, J. (2006). Histone variants—the structure behind the function. *Brief. Funct. Genomics Proteomics* **5**, 228–243.
2. Hake, S.B. and C.D. Allis (2006). Histone H3 variants and their potential role in indexing mammalian genomes: The "H3 barcode hypothesis." *Proc. Natl. Acad. Sci. USA* **103**, 6428–6435.
3. Kamakaka, R.T. and S. Biggins (2005). Histone variants: Deviants? *Genes Dev.* **19**, 295–316.
4. Pusarla, R.H. and P. Bhargava (2005). Histones in functional diversification: Core histone variants. *FEBS J.* **272**, 5149–5168.
5. Mizzen, C.A. (2004). Purification and analyses of histone H1 variants and H1 posttranslational modifications. *Methods Enzymol.* **375**, 278–297.
6. Kouzarides, T. (2007). Chromatin modifications and their function. *Cell* **128**, 693–705.
7. Mersfelder, E.L. and M.R. Parthun (2006). The tale beyond the tail: Histone core domain modifications and the regulation of chromatin structure. *Nucleic Acids Res.* **34**, 2653–2662.
8. Henikoff, S. (2005). Histone modifications: Combinatorial complexity or cumulative simplicity? *Proc. Natl. Acad. Sci. USA* **102**, 5308–5309.
9. Green, G.R. (2001). Phosphorylation of histone variant regions in chromatin: Unlocking the linker? *Biochem. Cell Biol.* **79**, 275–287.
10. Zweidler, A. (1978). Resolution of histones by polyacrylamide gel electrophoresis in presence of nonionic detergents. *Methods Cell Biol.* **17**, 223–333.
11. Laemmli, U.K. (1970). Cleavage of structural proteins during the assembly of the head of bacteriophage T4. *Nature* **227**, 680–685.
12. Panyim, S. and R. Chalkley (1971). The molecular weights of vertebrate histones exploiting a modified sodium dodecyl sulfate electrophoretic method. *J. Biol. Chem.* **246**, 7557–7560.
13. Reisfeld, R.A., U.J. Lewis, and D.E. Williams (1962). Disk electrophoresis of basic proteins and peptides on polyacrylamide gels. *Nature* **195**, 281–283.
14. Panyim, S. and R. Chalkley (1969). High resolution acrylamide gel electrophoresis of histones. *Arch. Biochem. Biophys.* **130**, 337–346.
15. Savic, A. and D. Poccia (1978). Separation of histones from contaminating ribosomal proteins by two-dimensional gel electrophoresis. *Anal. Biochem.* **88**, 573–579.
16. Poccia, D., J. Salik, and G. Krystal (1981). Transitions in histone variants of the male pronucleus following fertilization and evidence for a maternal store of cleavage-stage histones in the sea urchin egg. *Dev. Biol.* **82**, 287–296.
17. Salik, J., L. Herlands, H.P. Hoffmann, and D. Poccia (1981). Electrophoretic analysis of the stored histone pool in unfertilized sea urchin eggs: Quantification and identification by antibody binding. *J. Cell Biol.* **90**, 385–395.
18. Poccia, D., T. Greenough, G.R. Green, E. Nash, J. Erickson, and M. Gibbs (1984). Remodeling of sperm chromatin following fertilization: Nucleosome repeat length and histone variant transitions in the absence of DNA synthesis. *Dev. Biol.* **104**, 274–286.
19. Green, G.R. and D.L. Poccia (1985). Phosphorylation of sea urchin sperm H1 and H2B histones precedes chromatin decondensation and H1 exchange during pronuclear formation. *Dev. Biol.* **108**, 235–245.
20. Poccia, D.L, M.V. Simpson, and G.R. Green (1987). Transitions in histone variants during sea urchin spermatogenesis. *Dev. Biol.* **121**, 445–453.
21. Green, G.R. and D.L. Poccia (1988). Interaction of sperm histone variants and linker DNA during spermiogenesis in the sea urchin. *Biochemistry* **27**, 619–625.
22. Green, G.R. and D.L. Poccia (1989). Phosphorylation of sea urchin histone CS H2A. *Dev. Biol.* **134**, 413–419.

23. Vodicka, M., G.R. Green, and D.L. Poccia (1990). Sperm histones and chromatin structure of the "primitive" sea urchin *Eucidaris tribuloides*. *J. Exp. Zool.* **256**, 179–188.

24. Poccia, D., W. Pavan, and G.R. Green (1990). 6DMAP inhibits chromatin decondensation but not sperm histone kinase in sea urchin male pronuclei. *Exp. Cell Res.* **188**, 226–234.

25. Green, G.R., L.C. Gustavsen, and D.L. Poccia (1990). Phosphorylation of plant H2A histones. *Plant Phys.* **93**, 1241–1245.

26. Green, G.R., J.C. Patel, N.B. Hecht, and D.L. Poccia (1991). A complex pattern of H2A phosphorylation in the mouse testis. *Exp. Cell Res.* **195**, 8–12.

27. Green, G.R., R.R. Ferlita, W.F. Walkenhorst, and D.L. Poccia (2001). Linker DNA destabilizes condensed chromatin. *Biochem. Cell Biol.* **79**, 349–363.

28. Wyatt, H.R., H. Liaw, G.R. Green, and A.J. Lustig (2003). Multiple roles for *Saccharomyces cerevisiae* histone H2A in telomere position effect, Spt phenotypes and double-strand-break repair. *Genetics* **164**, 47–64.

29. Cleveland, D.W., S. G. Fischer, M. W. Kirschner, and U. K. Laemmli (1977). Peptide mapping by limited proteolysis in sodium dodecyl sulfate and analysis by gel electrophoresis. *J. Biol. Chem.* **252**, 1102–1106.

30. Green, G.R., D. Poccia, and L. Herlands (1982). A multisample device for electroelution, concentration, and dialysis of proteins from fixed and stained gel slices. *Anal. Biochem.* **123**, 66–73.

31. Green, G.R., D.G. Searcy, and R.J. DeLange (1983). Histone-like protein in the archaebacterium *Sulfolobus acidocaldarius*. *Biochim. Biophys. Acta* **741**, 251–257.

32. Waterborg, J.H. and R.E. Harrington (1987). Western blotting of histones from acid-urea-Triton and sodium dodecyl sulfate-polyacrylamide gels. *Anal. Biochem.* **162**, 430–434.

33. Delcuve, G.P. and J.R. Davie (1992). Western blotting and immunochemical detection of histones electrophoretically resolved on acid-urea-triton and sodium dodecyl sulfate-polyacrylamide gels. *Anal. Biochem.* **200**, 339–341.

34. Thiriet, C. and P. Albert (1995). Rapid and effective Western blotting of histones from acid-urea-Triton and sodium dodecyl sulfate polyacrylamide gels: Two different approaches depending on the subsequent qualitative or quantitative analysis. *Electrophoresis* **16**, 357–361.

35. Albert, P. and C. Redon (1998). Efficient antibody generation using histone H1 subfractions purified from Western blots. *Anal. Biochem.* **261**, 87–92.

36. Gorg, A., W. Weiss, and M.J. Dunn (2004). Current two-dimensional electrophoresis technology for proteomics. *Proteomics* **4**, 3665–3685.

37. Goshe, M. B. (2006). Characterizing phosphoproteins and phosphoproteomes using mass spectrometry. *Brief. Funct. Genomics Proteomics* **4**, 363–376.

38. Domon, B. and R. Aebersold (2006). Mass spectrometry and protein analysis. *Science* **312**, 212–217.

39. Wittmann-Liebold, B., H.R. Graack, and T. Pohl (2006). Two-dimensional gel electrophoresis as tool for proteomics studies in combination with protein identification by mass spectrometry. *Proteomics* **6**, 4688–4703.

40. Garcia, B.A., J. Shabanowitz, and D.F. Hunt (2007). Characterization of histones and their post-translational modifications by mass spectrometry. *Curr. Opin. Chem. Biol.* **11**, 66–73.

41. Garcia, B.A., S.B. Hake, R.L. Diaz, M. Kauer, S. A. Morris, J. Recht, J. Shabanowitz, N. Mishra, B. D. Strahl, C. D. Allis, and D. F. Hunt (2007). Organismal differences in post-translational modifications in histones H3 and H4. *J. Biol. Chem.* **282**, 7641–7655.

42. Wisniewski, J.R., A. Zougman, S. Krüger, and M. Mann (2007). Mass spectrometric mapping of linker histone H1 variants reveals multiple acetylations, methylations, and phosphorylation as well as differences between cell culture and tissue. *Mol. Cell. Proteomics* **6**, 72–87.

43. Bonenfant, D., M. Coulot, H. Towbin, P. Schindler, and J. Oostrum (2006). Characterization of histone H2A and H2B variants and their post-translational modifications by mass spectrometry. *Mol. Cell. Proteomics* **5**, 541–552.

44. Beck, H.C., E.C. Nielsen, R. Matthiesen, L.H. Jensen, M. Sehested, P. Finn, M. Grauslund, A.M. Hansen, and O.N. Jensen (2006). Quantitative proteomic analysis of post-translational modifications of human histones. *Mol. Cell. Proteomics* **5**, 1314–1325.

45. Boyne, M.T., J.J. Pesavento, C.A. Mizzen, and N.L. Kelleher (2006). Precise characterization of human histones in the H2A gene family by top down mass spectrometry. *Proteome Res.* **5**, 248–253.

46. Burlingame, A.L., X. Zhang, and R.J. Chalkley (2005). Mass spectrometric analysis of histone posttranslational modifications. *Methods* **36**, 383–394.

47. Poccia, D.L. and G.R. Green (1986). Nuclei and chromosomal proteins. *Methods Cell Biol.* **27**, 153–174.

48. Henderson, L.E., S. Oroszlan, and W. Konigsberg (1979). A micromethod for complete removal of dodecyl sulfate from proteins by ion-pair extraction. *Anal. Biochem.* **93**, 153–157.

49. Baum, J.A. and N.H. Giles (1985). Genetic control of chromatin structure 5′ to the qa-x and qa-2 genes of *Neurospora*. *J. Mol. Biol.* **182**, 79–89.

Chapter 17
Quantification of Redox Conditions in the Nucleus

Young-Mi Go, Jan Pohl, and Dean P. Jones

Keywords Nuclear redox state; Thioredoxin-1; Thioredoxin reductase; Glutathione; Protein S-glutathionylation; Nuclear proteins; Redox Western blot; Biotinylated iodoacetamide (BIAM); Isotope-coded affinity tags (ICAT) analysis

Abstract Many nuclear proteins contain thiols, which undergo reversible oxidation and are critical for normal function. These proteins include enzymes, transport machinery, structural proteins, and transcription factors with conserved cysteine in zinc fingers and DNA-binding domains. Uncontrolled oxidation of these thiols causes dysfunction, and two major thiol-dependent antioxidant systems provided protection. The redox states of these systems, including the small redox active protein thioredoxin-1 (Trx1) and the abundant, low molecular weight thiol antioxidant glutathione (GSH), in nuclei provide means to quantify nuclear redox conditions. Redox measurements are obtained under conditions with excess thiol-reactive reagents. Here we describe a suite of methods to measure nuclear redox state, which include a redox Western blot technique to quantify the redox state of Trx1, a biotinylated iodoacetamide (BIAM) method for thioredoxin reductase-1 (TrxR1), GSH redox measurement using total protein S-glutathionylation, and a redox isotope-coded affinity tag (ICAT) method for measuring oxidation of specific cysteines in high-abundance nuclear proteins.

1 Introduction

The major intracellular thiol/disulfide systems, including GSH/diglutathione (GSSG) and Trx1, control diverse cellular events through discrete redox pathways that are responsive to oxidative stress and function in redox signaling (*1*). GSH and Trx1 are components of redox control systems present in both cytoplasm and nuclei (*2*).

GSH is important in the regulation of nuclear matrix organization (*3*), maintenance of cysteine residues on zinc-finger DNA-binding motifs in a reduced and functional state (*4*), chromosome consolidation (*5*), DNA synthesis (*6*), DNA protection from oxidative stress (*7*), and protection of DNA-binding proteins (*8*). Trx1 is a redox protein found in the cytoplasm and nucleus with many central redox signaling and control functions. Trx1 can be imported to the nucleus from the cyto-plasm during various forms of oxidative stress *(9, 10)*. Trx1 in the nucleus controls transcription factor binding to DNA. The activation of transcription factors includ-ing NF-κB, AP-1, Nrf-2, glucocorticoid receptor, and p53 under oxidative stress is associated with the redox state of thiol/disulfide in the nuclei and is functionally important in critical cell processes, including those resulting in mutation of DNA and causing cell death. Thus, maintenance of a relatively reduced nuclear redox state is critical for transcription factor binding in transcriptional activation in response to oxidative stress.

Earlier studies indicating a sequestered nuclear GSH pool were controversial because of possible redistribution artifacts *(11–13)*. In addition, artifacts can occur due to oxidation or reduction reactions during sample processing and analysis. Recently, approaches have been introduced to minimize these artifacts, and the results show that the redox states of nuclear proteins can vary independently of cytoplasmic proteins. Four methods are presented below for analysis of nuclear thiol/disulfide redox states. The first is a specific redox Western blot that measures the redox state of the central redox protein, Trx1. This protein has been extensively characterized and the method provides a means to quantify the steady-state redox potential in the nucleus. The second is a more general but semiquantitative method (BIAM-blot) which is useful for many proteins with oxidizable thiols, such as thioredoxin reductase-1 (TrxR1) and redox factor-1 (REF-1; apurinic/apyrimidinic (AP) endonuclease-1). The third gives a method to measure GSH redox state in nuclei based upon total protein S-glutathionylation. The fourth provides a method to study oxidation of specific cysteines in high-abundance nuclear proteins. Each of these procedures relies upon the reaction of thiols with iodoacetic acid or maleim-ides under conditions where disulfides do not react.

2 Materials

2.1 *Redox Western Blot for Quantification of Nuclear Thioredoxin-1 Redox State*

2.1.1 Cell Culture, Lysis, Fractionation, and Derivatization for Trx1 Redox State

1. Dulbecco's Modified Eagle's Medium (DMEM; Mediatech Inc, Herndon, VA, USA) supplemented with 10% fetal bovine serum (FBS; Mediatech).
2. Nonidet P-40 (Sigma-Aldrich, St. Louis, MO, USA).
3. Hypotonic buffer: $10\,mM$ Hepes, pH 7.8, $10\,mM$ KCl, $2\,mM$ $MgCl_2$, $0.1\,mM$ EDTA, $0.2\,mM$ NaF, and $0.2\,mM$ Na_3VO_4, with freshly added protease inhibitors (leupeptin, aprotinin, pepstatin, and phenylmethylsulfonyl fluoride [PMSF], all at $1\,mM$).

4. Guanidine-HCl lysis buffer (G-lysis buffer): $6\,M$ guanidine, $50\,mM$ Tris-HCl, $3\,mM$ EDTA, pH adjusted to 8.3 with 1 N NaOH.

5. G-lysis buffer with IAA: iodoacetic acid (IAA; Sigma-Aldrich) is dissolved in G-lysis buffer at $50\,mM$ and pH is adjusted to 8.3 with 1 N NaOH.

6. G-25 Microspin columns (GE Healthcare, Buckingamshire, UK) are prepared prior to use by centrifugation at $1,100\times g$ for 2 min.

2.1.2 Redox Western Blotting

1. 30% acrylamide/bis solution (Bio-Rad, 37.5:1) and N,N,N,N'-tetramethyl-ethylenediamine (TEMED; Bio-Rad, XX).

2. Ammonium persulfate: 1.5% (w/v) solution in water.

3. Separating gel buffer (5×): $1.5\,M$ Tris-HCl, pH 8.8.

4. Stacking gel buffer (5×): $0.5\,M$ Tris-HCl, pH 6.8.

5. Running buffer (5×): $125\,mM$ Tris base and $960\,mM$ glycine.

6. Transfer buffer: $25\,mM$ Tris base, $192\,mM$ glycine, and 20% v/v methyl alcohol.

7. Nitrocellulose or PVDF transfer membrane.

8. Blocking buffer: dilute Odyssey blocking buffer (LI-COR, Lincoln, Nebraska, USA) with phosphate-buffered saline (PBS) with 0.05% v/v Tween-20 (PBS-T) (1:1).

9. Primary antibody: anti-Trx (BD Biosciences, San Jose, CA, USA).

10. Secondary antibody: Alexa Fluor 680-goat anti-mouse IgG (H+L) (Molecular Probes, Eugene, OR, USA).

11. Wash buffer (PBST): 0.05% v/v Tween-20 in 1× PBS.

2.2 Redox State Measurement for Thioredoxin Reductase-1 by BIAM-Labeling

2.2.1 Buffers

1. Lysis buffer A: $50\,mM$ Bis-Tris-HCl, pH 6.5, 0.5% (v/v) Triton X-100, 0.5% deoxycholate, 0.1% SDS, $150\,mM$ NaCl, $1\,mM$ EDTA, leupeptin, aprotinin, and $0.1\,mM$ PMSF.

2.2.2 Redox States for BIAM-Labeled Proteins

1. Biotin-iodoacetamide (N-[biotinoyl]-N'-[iodoacetyl]ethylenediamine, BIAM) and biotin-maleimide (N^{α}-[3-maleimidylpropionyl]biocytin) (both from Molecular Probes): dissolve in $50\,mM$ Tris-HCl, pH 6.8, to make a $0.1\,M$ stock solution of each.

2. Iodoacetamide (IAM) (Sigma-Aldrich): dissolve in 50 mM Tris-HCl, pH 6.8, to give a 5mM solution.
3. Rabbit anti-TrxR1 antibody (LabFrontier, Seoul, Korea) for Western blot.
4. Mouse anti-TrxR1 antibody (LabFrontier) for immunoprecipitation.
5. Rabbit anti-Ref1 antibody (Santa Cruz Biotechnology, Santa Cruz, CA, USA).
6. Protein G sepharose (Sigma-Aldrich).
7. Streptavidin-agarose (Sigma-Aldrich).

2.3   Measurement of GSH Redox State Based upon Protein S-Glutathionylation

1. Trichloroacetic acid (TCA) solution: 100% (w/v) TCA in water.
2. IAA dissolved in water at 7.4 mg/mL (w/v).
3. Dansyl chloride dissolved in acetone at 20 mg/mL (w/v).
4. Perchloric acid (PCA) 70% (Fisher Scientific), boric acid (Sigma-Aldrich), and gamma-glutamylglutamate (Sigma-Aldrich) dissolved in water at 10% (v/v), 0.2 M, and 10 µM concentrations, respectively.

2.4   Measurement of Redox State of Most Abundant Nuclear Proteins Using the Redox ICAT Reagent

1. ICAT assay kit (Applied Biosystems, Foster City, CA, USA).

3   Methods

3.1   Redox Western Blot for Quantification of Nuclear Thioredoxin-1 Redox State

The Trx1 Redox Western blot procedure is based upon that of Holmgren and Fagerstedt (14) in which the redox form of *Escherichia coli* Trx can be separated by gel electrophoresis under native, nonreducing condition by treating samples with iodoacetic acid, which introduces one negative charge for each thiol present. Although the protein contains five cysteines, the active site is most readily oxidized. The procedure described including fraction and nuclear isolation is optimized to measure redox in both nuclear and cytosolic fractions. Detailed characterization and identification of bands in gels is available (15).

3.1.1 Isolation of Nuclei

Nuclei are isolated based on the procedure of Janssen and Sen *(16)*, with modifications to allow redox measurements of Trx1 via the Redox Western blot.

1. Cells are washed 2× with HBSS and lysed in hypotonic buffer (*see* **Section 2.1.1.3**).
2. 50 mM IAA is added to the hypotonic buffer and the pH is adjusted to 7.8.
3. The suspension is incubated on ice for 5 min, and NP-40 is added to a final concentration of 0.6% (v/v).
4. Following centrifugation at 16,100×g for 2 min, the pellet (nuclei) and the supernatant (cytosol) are separated and analyzed as described below (*see* **Note 1**).

3.1.2 Derivatization of Protein Thiols with IAA

1. Cells (10^6 cells/well in 6-well plates) are washed with cold PBS, and the nuclear fraction is isolated and resuspended in 200 µL of G-lysis buffer supplemented with 50 mM IAA and protease inhibitors (PMSF, aprotinin, leupeptin, and pepstatin, 1 mM each).
2. Prepare cells from two extra wells for positive controls, including treatment with DTT (5 mM for 30 min) and with H_2O_2 (2 mM for 30 min) to fully reduce and oxidize proteins, respectively.
3. Prepare G-25 columns in advance by centrifuging them (1,100×g for 2 min) to remove preservation buffer while the samples are incubated at 37°C for IAA labeling.
4. After 30 min incubation at 37°C, the cell lysate is spun briefly to send down all the solution in the tube, and transferred to a G-25 column, which is centrifuged at 1,100×g for 2 min at room temperature to remove excess IAA.
5. Samples can be stored at −20°C or further analyzed by redox Western blotting.

3.1.3 Redox Western Blotting

1. Make a 1-mm-thick, nonreducing 15% acrylamide separating gel. Do not layer butyl alcohol to make the surface flat, but instead use dH$_2$O. Make a 6% polyacrylamide stacking gel.
2. Prepare from each lysate a sample containing 30 to 50 µg of protein and mix with 5× gel loading buffer. Samples do not need to be boiled before loading onto the gel.
3. Run the gel at 150 V for 1–1.5 h until the dye line runs out of the gel. The gel is then electroblotted to nitrocellulose or PVDF membrane (0.2-µm pore size) at 100 V for 1 h at 4°C. Block in blocking buffer for 1 h.
4. Prepare primary antibody for Trx1 (10 µL in 10 mL of blocking buffer (1:1,000), add to the membrane, and incubate on a rocker overnight at 4°C (*see* **Note 2**).

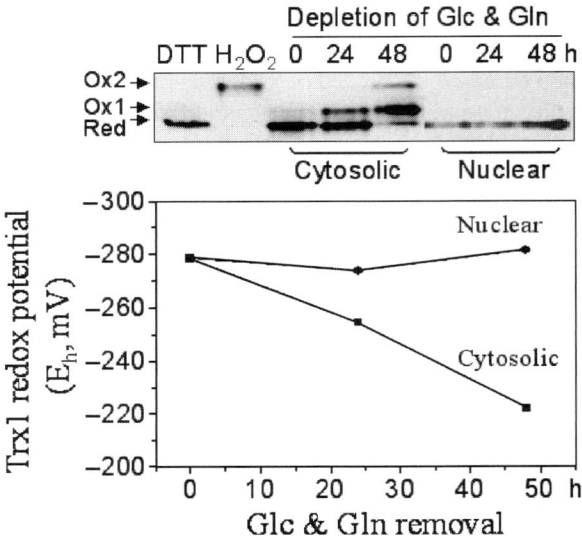

Fig. 17.1 Glucose (*Glc*) and glutamine (*Gln*) depletion cause selective oxidation of cytosolic Trx1 without oxidation of nuclear Trx1. Confluent HT29 cells were cultured in complete medium, washed twice with PBS, and then exposed to Glc- and Gln-free medium for the indicated time. As controls, cells were treated with DTT (for reduction) or H_2O_2 (for oxidation) for 30 min. Redox changes in nuclear and cytosolic Trx1 were analyzed by the methods described in a previous report *(28)*. The *bottom graph* shows changes in redox potential (E_h) of nuclear (*filled circle*) and cytosolic (*filled square*) Trx1 calculated using the Nernst equation ($E_h = E_0 + RT/2F \ln [\text{Trx1}^{\text{ox}}]/[\text{Trx}^{\text{Red}}]$, $E_0 = -240$ mV at pH 7.0)

5. Wash the membrane 3× 5 min with PBST.
6. Prepare the secondary antibody (1:5,000 in 10 mL of PBST), add to the membrane, and incubate on a rocker for 1 h in the dark at room temperature.
7. Wash the membrane 3× 5 min with PBST in the dark, and air-dry.
8. The redox state of Trx1 will be seen upon scanning the membrane. Trx1 is separated into three bands, which represent the most oxidized form (upper band), a mixture of oxidized and reduced forms (middle band), and the most reduced form (lower band), respectively (*see* **Notes 2** and **3**).

An example of a redox Western blot for Trx1 is shown in Fig. 17.1 (*see* **Note 4**).

3.2 Redox State Measurement for Thioredoxin Reductase-1 by BIAM-Labeling

The Trx system is composed of Trx, thioredoxin reductase (TrxR), NADPH, and Trx peroxidases/peroxiredoxins. Trx reduces protein disulfides directly and serves as a reductant for the peroxiredoxins *(17)*. The oxidized form of Trx is reduced by catalytic activity of TrxR using an electron from reduced NADPH *(18)*. TrxR1 is an isoform of TrxRs including *E. coli* TrxR, human TrxR2. The sequence of the catalytic site of TrxR is -Cys-Val-Asn-Val-Gly-Cys- and is located in the FAD domain of these enzymes *(19)*. Furthermore, TrxR1 has a C-terminal selenocysteine residue that is required for catalytic activity but is not part of the conserved active site *(20)*. Since TrxRs are known to reduce oxidized Trx, alterations in TrxR activity may regulate Trx activity. TrxR1 is localized in cytoplasm and nuclei while TrxR2 is predominantly found in mitochondria *(21)*. In this section, we describe an assay for measuring nuclear TrxR1 redox state using the BIAM-labeling technique; the thiol-reactive biotin iodoacetamide and biotin maleimide derivatives can be used for this technique. This method is based on the procedure of Kim et al. *(22)*, with modifications to measurement of redox states of proteins containing thiols in their catalytic site, such as TrxR and redox factor-1 (Ref-1), which is also regulated by Trx1 (*see* **Note 5**).

Fig. 17.2a shows an example of measuring changes in Ref-1 redox state in the cytoplasm and nucleus after removal of glucose and glutamine from the growth medium, and Fig. 17.2b shows measurement of the redox state of mitochondrial TrxR2 after Glc and Gln removal using the BIAM-labeling method.

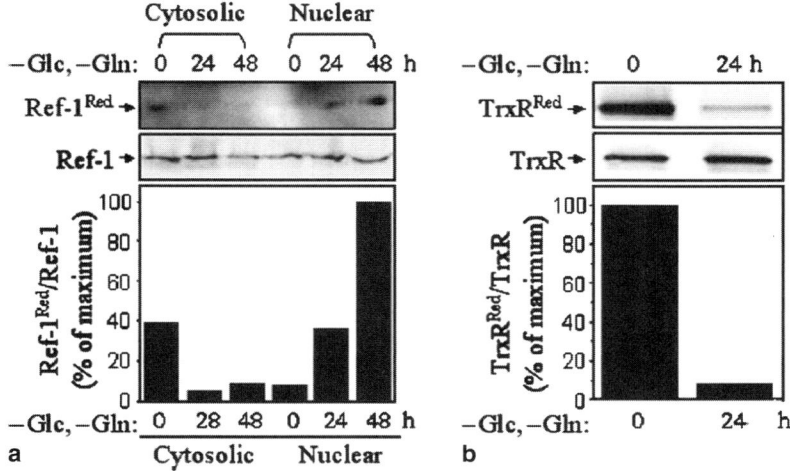

Fig. 17.2 Effect of Glc and Gln depletion on cytosolic and nuclear Ref-1 and TrxR. **a** HT29 cells treated with Glc- and Gln-free medium for the indicated time were analyzed for the reduced form (**a**, *top*) of Ref-1 and total Ref-1 protein (**a**, *middle*) after cytosolic and nuclear fractionation. **b** Cells treated with Glc- and Gln-free medium for 24 h were analyzed for the reduced form (**b**, *top*) of TrxR and total TrxR (**b**, *middle*). After treatment with Glc- and Gln-free medium, cells were labeled with BIAM, washed with PBS, lysed, and immunoprecipitated with an antibody specific for Ref-1 (**a**) or TrxR (**b**). Immunocomplexes were separated by SDS-PAGE, blotted, and probed with streptavidin (*top panels* in **a** and **b**). Western blotting results using cell lysates from the same samples probed with Ref-1 (**a**) or TrxR (**b**) antibody show the total amount of Ref-1 (**a**, *middle panel*) and TrxR (**b**, *middle panel*). The bottom graphs show the percent of reduced Ref-1 and TrxR normalized to total Ref-1 (**a**) and TrxR (**b**), respectively

1. Cells (1–2×10^7 cells in a 10-cm plate) after treatment (e.g., in glucose [glc]- and glutamine [gln]-deficient media) are washed with cold PBS, fractionated, and the nuclear fraction is lysed with 1 mL of lysis buffer A containing 10 μM BIAM. Control cells that were not treated with glc- and gln-deficient media are lysed and labeled with BIAM in parallel.

2. After incubation for 10 min at 37°C in the dark, the labeling reaction is stopped by adding IAM to 5 mM.

3. TrxR1 is precipitated from the reaction mixture by adding rabbit antibodies to TrxR1 and protein G-Sepharose (40 μL/sample). Alternatively, to immunoprecipitate all BIAM-labeled proteins, streptavidin-agarose can be used instead of TrxR1 antibody.

4. The immunocomplex of BIAM-labeled TrxR1 (when using TrxR1 antibody) or all BIAM-labeled proteins (when using streptavidin-agarose) is washed 3× with 1 mL of ice-cold lysis buffer A.

5. Add 40 μL of 2× gel loading buffer to each sample, heat at 95°C for 10 min, separate by SDS-PAGE, and prepare a Western blot on a nitrocellulose or PVDF membrane following a standard protocol.

6. Probe the membrane with streptavidin-Alexa Fluor 680 conjugate at 4°C for 1 h (if using TrxR1 antibody) or overnight (if using another primary antibody, e.g., NF-κB p50 or p65), depending on the protein to be examined.

7. Wash the membrane 3× with PBS-T and scan on the Odyssey scanner to detect reduced and BIAM-labeled TrxR1. If the membrane was probed with an antibody other than TrxR1, wash it 3× with PBST and add secondary antibody (1:5,000 dilution) appropriate for the primary antibody used, wash 3× with PBST, and scan to examine reduced and BIAM-labeled proteins.

3.3 Measurement of GSH Redox State Based upon Protein S-Glutathionylation

Protein thiol groups can be modified by oxidants. The formation of the disulfide bond between protein cysteine thiols and low molecular weight thiols such as cysteine and GSH, generally referred to as S-thiolation, generates mixed disulfides. Because GSH is predominant low molecular weigh thiol in cellular system, glutathionylation of protein that indicates the formation of mixed disulfide between proteins and glutathione (Pr-SSG) is the main form of protein S-thiolation. Glutathionylation of proteins can be found in normal tissue (23, 24), plasma (25), and cells (26), however, its amount can go up under oxidative stress. Therefore, increased amount of protein glutathionylation is often considered as a marker for cells exposed to oxidative stress. The procedures to measure total Pr-SSG are described below.

   Figure 17.3 shows an example of measuring protein glutathionylation (glutathione–protein mixed disulfides [PrSSG]) to examine the effect of glucose and glutamine depletion on changes in redox state of the glutathione system in

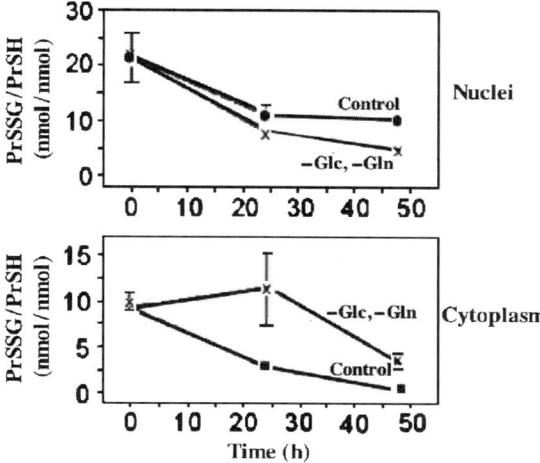

Fig. 17.3 Effect of Glc and Gln depletion on nuclear protein S-glutathionylation/protein thiol (*top*) and cytosolic S-glutathionylation/protein thiol (*bottom*). As controls, cells treated with complete medium are shown by *filled circles* (*top*, nucleus) and *squares* (*bottom*, cytosolic)

nuclei and cytoplasm. The PrSSG was normalized with respect to the total amount of thiols (*see* **Note 6**).

1. Cells (2×10^7 cells in a 10-cm plate) are washed 2× with cold PBS, nuclei are isolated (*see* **Section 3.1.1**) and transferred to an Eppendorf tube. Ice-cold 25% (w/v) TCA is added and the samples are incubated on ice for 10 min.
2. Centrifuge at $16,100 \times g$ for 5 min and save the pellet.
3. Wash pellet with 200 μL of cold 25% TCA.
4. Centrifuge at $16,100 \times g$ for 5 min.
5. Resuspend the pellet in 0.2 mL of 0.1 M NaOH.
6. Sonicate briefly on ice to resuspend proteins (we use six pulses of 2 sec each at 30% power in a probe-type ultrasonic cell disrupter; VirTis, Gardiner, NY, USA). Save a sample to measure the amount of protein (*see* **Note 7**).
7. Transfer 0.125 mL of suspension to a fresh tube containing 0.125 mL of 0.1 M sodium phosphate buffer, pH 6.0, and 5 mM DTT.
8. Incubate at room temperature for 30 min.
9. Add 0.25 mL of PCA.
10. Centrifuge at $16,100 \times g$ for 5 min.
11. Transfer 300 μL of supernatant to a fresh tube.
12. Add 60 μL of IAA and vortex.
13. Adjust pH to 9.0 with 1 M KOH.
14. Incubate at room temperature for 20 min.
15. Prepare the dansyl chloride solution and add 300 μL to each sample.
16. Incubate for 24 h at room temperature in the dark.

17. Add $500\,\mu L$ of chloroform to each sample to extract acetone and free dansyl chloride from the aqueous phase.
18. Centrifuge at 16,100 rpm for 2 min to separate the phases.
19. Take samples for high-performance liquid chromatography (HPLC) analysis (*see* **Note 8**) or store at 4°C.
20. Protein glutathionylation (Pr-SSG) can be evaluated by normalization with respect either to the total protein or the total amount of thiols present in the nuclear fraction (*see* **Note 9**).

3.4   Measurement of Redox State of Most Abundant Nuclear Proteins Using the Redox ICAT Reagent

The ICAT reagent (Applied Biosystems) allows labeling of thiols with heavy (^{13}C) and light (^{12}C) forms of biotin-labeled iodoacetamide to quantify the redox state of specific proteins in response to oxidative stress *(27)*. The technique is based on the original ICAT procedure with modifications; in comparison with the original method used to define changes in expression levels of proteins in response to treatment, the redox ICAT method examines the redox state of specific thiols in proteins. Proteins can be present as either thiols (reduced), internal disulfides (oxidized), or as disulfides with GSH, cysteine (thylation), or other thiol-containing chemicals or proteins (modified thiols). The original ICAT approach does not distinguish whether thiols or disulfides are present in proteins because the heavy and light reagents are added after complete reduction of disulfides by tris(2-carboxyethyl) phosphine (TCEP). This original procedure is modified here by addition of heavy ICAT reagent (H) to react with thiols present in the proteins before reduction, and then addition of light ICAT reagent (L) after reduction of protein disulfides to thiols. Thus both H- and L-labeled proteins are present in the same sample. A higher H:L ratio in the control compared with the experimental sample indicates thiol oxidation by the experimental treatment. Table 17.1 shows an example of measuring the ratio of reduced (H) and oxidized (L) thiols in different proteins present in nuclei.

3.4.1   Cell Lysis and Protein Collection

1. Cells ($1-2\times10^7$ cells in a 10-cm plate) after treatment (e.g., in Glc- and Gln-deficient medium, or with Glc and Glc oxidase to generate H_2O_2 at a physiological level) are washed with cold PBS, and nuclei are isolated (*see* **Section 3.1.1**) and pelleted by centrifugation.
2. Add $80\,\mu L$ of 50 mM Tris, pH 8.5, and 0.1% (w/v) SDS to the nuclear pellet and measure the amount of protein in the nuclear fraction. Prepare $100\,\mu g$ of nuclear proteins in $80\,\mu L$ of denaturation buffer in a microcentrifuge tube.

Table 17.1 Ratio of reduced (H) to oxidized (L) fraction of cysteines in nuclear proteins.

Protein	Peptide sequence	H:L Ratio	
		Control	Glc/Glc oxidase
14-3-3 protein zeta	DICNDVLSLLEK	3.25	1.50
Nuclear ribonucleo-protein D	FGEVVDCTLK	2.51	1.88
	GFCFITFK	2.42	2.23
U2 small nuclear ribonucleoprotein	YCDPDSYHR	2.07	0.93
RNA-dependent helicase p72	CTYLVLDEADR	1.11	1.19
Annexin A2	GLGTDEDSLIEIICSR	6.54	2.24
mRNA splicing factor SRp30C	EAGDVCYADVQK	8.48	7.13
Nuclear ribonucleo-protein K	GSDFDCELR	3.19	2.49
Far upstream element-binding protein 1	CQHAAEIITDLLR	3.43	2.71
Nuclear ribonucleo-protein A1	YHTVNGHNCEVR	11.14	6.52

The ratio of reduced (H) to oxidized (L) fraction of cysteines in nuclear proteins determined by ICAT and nano-liquid chromatography-tandem mass spectrometry (LC/MS/MS) analysis. An equal amount (100 µg) of the nuclear fraction of cultured human HT-29 (colon cancer) cells was treated without (control) or with glucose/glucose oxidase (Glc/Glc oxidase) to generate H_2O_2 for 1 h at 37°C. The samples were then incubated with reagents for redox ICAT as described here, and subjected to nano-LC/MS/MS analysis. Shown are proteins selected by the detection of identical peptides in both samples and the H:L ratios. A higher H:L ratio in the control than after Glc/Glc oxidase treatment indicates that this treatment caused thiol oxidation

3.4.2 ICAT Labeling

1. To the nuclei in denaturation buffer, add 20 µL of heavy ICAT reagent (H).
2. Vortex-mix and pulse-centrifuge, incubate for 60 min at 37°C.
3. Add 2 µL of TCEP, incubate for 20 min at 37°C.
4. Add 25 µL of 10% TCA, place on ice for 30 min.
5. Centrifuge at 16,100×g for 10 min, remove the supernatant.
6. Wash the pellet with 100% acetone.
7. Centrifuge at 16,100×g for 10 min, remove the supernatant.
8. Resuspend the pellet in 80 µL of denaturation buffer.
9. Add 20 µL of light ICAT reagent (L).
10. Vortex and pulse-centrifuge, incubate for 60 min at 37°C.

3.4.3 Trypsinization

1. Dissolve the trypsin included in the ICAT assay kit in 200 µL of MilliQ H_2O.
2. Add 200 µL of trypsin to samples.
3. Vortex and pulse-centrifuge, incubate for 12–16 h at 37°C, vortex and pulse-centrifuge.

3.4.4 Cation-Exchange

1. Place the sample in a 3-mL tube.

2. Add 2 mL of Cation-Exchange Buffer (CEB)-Load to the sample at 1 drop/sec.
3. Vortex and pulse-centrifuge.
4. Check the pH; it should be 2.5–3.3, if not, add more CEB-Load until the pH is correct.
5. Assemble a cartridge holder with a cation-exchange cartridge.
6. Equilibrate the cartridge by injecting 2 mL of CEB-Load at 1 drop/sec (*see* **Note 10**).
7. Inject the sample onto the cartridge slowly, wash with 1 mL of CEB-Load.
8. Elute peptides with 500 µL of CEB-Elute, collect the flow-through (*see* **Note 11**).

3.4.5   Cleaning and Storing the Cation-Exchange Cartridge

Inject 1 mL of CEB to clean the cartridge, inject 2 mL of CEB before storing the cartridge at 4°C.

3.4.6   Purifying Peptides and Cleaving Biotin

1. Insert an avidin cartridge into the cartridge holder.
2. Equilibrate the cartridge by injecting 2 mL of Affinity Buffer (AB)-Elute, discard waste.
3. Inject 2 mL of AB-Load, discard waste.
4. Neutralize samples by adding 500 µL of AB, vortex, and pulse-centrifuge.
5. Check the pH; it should be 7, if not add more AB until it is correct, vortex, and pulse-centrifuge.
6. Slowly inject the sample onto the cartridge (1 drop/sec).
7. Collect the flow-through; this fraction contains unlabeled protein fragments and can be analyzed at a later time if necessary.
8. Inject an additional 500 µL of AB-Load and collect in the same tube as **step 7**, or in a fresh tube if storing for later analysis.
9. Inject 1 mL of AB (first wash) and discard the flow-through.
10. Inject 1 mL of AB (second wash); collect the first 500 µL but discard the latter 500 µL.
11. Inject 1 mL of MilliQ H_2O and discard the flow-through.

3.4.7   Eluting ICAT-labeled Samples

1. Inject 800 µL of AB-Elute.
2. Discard the initial 50 µL.
3. Collect the remaining 750 µL in a tube.
4. Vortex and pulse-centrifuge.

3.4.8   Cleaning and Storing the Avidin Cartridge

1. Inject 2 ml of AB to clean the cartridge, inject 2 ml of AB and store at 4°C (*see* **Note 12**).

3.4.9   Cleaving Biotin

1. Evaporate the samples in a SpeedVac.
2. In another tube, mix cleaving agents A (95 µL) and B (5 µL).
3. Add 95 µL of cleaving reagent solution to the sample, vortex and pulse-centrifuge, incubate for 2 h at 37°C, vortex and snap-centrifuge.
4. Evaporate in a SpeedVac.
5. Send for analysis by mass spectrometry.

4   Notes

1. If cell lysis is unsatisfactory (e.g., due to too high or too low NP-40 in the lysis buffer), test different concentrations of NP-40 and measure the distribution of a marker protein for nuclei (e.g., histone H2B).
2. Not all antibodies recognize both the S-carboxylmethyl- and SS- forms of Trx1 equivalently. Always have positive controls including 5 mM DTT and 5 mM H_2O_2 treatments for identification. Use commercially available Trx1 protein at a sufficient concentration to visualize by Coomassie Blue staining and compare with a Western blot (*15*).
3. Other bands may be present due to other modifications, e.g., Cys73 can be modified by S-glutathionylation and this will alter the mobility. If this poses unacceptable complications for interpretation of the results after treatment with IAA, remove excess IAA with a G25 spin column, reduce with 2 mM DTT, remove excess DTT with a G25 spin column, and treat with 50 mM iodoacetamide. Iodoacetamide does not introduce a negative charge and the modified protein migrates with the disulfide form. To check for autoxidation and incomplete derivatization, vary the concentration of IAA in the lysis buffer (10, 20, 50, and 100 mM) and adjust the pH to 8.3.
4. The integrated intensities of oxidized and reduced Trx1 can be used in the Nernst equation (*15*) to calculate E_h (redox potential) values with E_0 (midpoint potential) = −240 mV at pH 7.0 and 25°C. Linearity of detection is important for redox state calculations; the Odyssey system shows better sensitivity and linearity than others that we have used.
5. In addition to TrxR1, the redox states of other nuclear proteins including Ref-1, NF-kB p65, and p50 subunits can be measured using this method.
6. Ellman's reagent (5,5′-dithio-bis[2-nitrobenzoic acid], DTNB) was used to measure the thiol content of proteins from the cytosolic and nuclear fractions

by protein precipitation, complete denaturation in TCA, and resolubilization *(25)*. Samples (50 µL) were added to a cuvette containing 1 mL of 0.05 M Tris-Cl and 0.05 M EDTA, pH 8.3, and the A_{412} was measured. Ellman's reagent (25 µL, 5 mM in methanol) was then added to each cuvette, and was thoroughly mixed and incubated at room temperature for 20 min before remeasuring the A_{412}. The increase in A_{412} after addition of DTNB was used to determine the concentration of thiols in the sample.

7. Protein pellets do not resuspend well without sonication. Samples to measure protein can be stored at −20°C.

8. HPLC analysis for GSH/GSSG: the derivatized samples are separated on a Supercosil LC-NH2 column (5 µm: 4.6×25 cm; Supelco, Bellefunk, PA, USA) with a Waters 2690 HPLC and autosampler system (Waters, Milford, MA, USA). Detection is by fluorescence using bandpass filters (305–395 nm excitation, 510–650 nm emission; Gilson Medical Electronics, Middletown, WI, USA). The GSH and GSSG are quantified by integration relative to an internal standard and expressed as nanomoles per milligram protein.

9. The fractionation procedure may cause an artifactual increase or decrease of the Pr-SSG level. The nuclear fraction should be used for analysis without freezing or saving at 4°C.

10. All equilibration, load, wash, and elution steps are carried out at 1 drop/sec.

11. Wash the needle and syringe 3× with MilliQ H_2O between samples.

12. Cartridges can be used 50 times.

References

1. Jones, D.P. (2006) Redefining oxidative stress. *Antioxid. Redox Signal.* **8**, 1865–1879.
2. Hansen, J.M., Go, Y.M., and Jones, D.P. (2006) Nuclear and mitochondrial compartmentation of oxidative stress and redox signaling. *Annu. Rev. Pharmacol. Toxicol.* **46**, 215–234.
3. Dijkwel, P.A. and Wenink, P.W. (1986) Structural integrity of the nuclear matrix: differential effects of thiol agents and metal chelators. *J. Cell Sci.* **84**, 53–67.
4. Klug, A. and Rhodes, D. (1987) Zinc fingers: a novel protein fold for nucleic acid recognition. *Cold Spring Harb. Symp. Quant. Biol.* **52**, 473–482.
5. De Capoa, A., Ferraro, M., Lavia, P., Pelliccia, F., and Finazzi-Agro, A. (1982) Silver staining of the nucleolus organizer regions (NOR) requires clusters of sulfhydryl groups. *J Histochem. Cytochem.* **30**, 908–911.
6. Suthanthiran, M., Anderson, M.E., Sharma, V.K., and Meister, A. (1990) Glutathione regulates activation-dependent DNA synthesis in highly purified normal human T lymphocytes stimulated via the CD2 and CD3 antigens. *Proc. Natl. Acad. Sci. USA* **87**, 3343–3347.
7. Sandstrom, B.E. and Marklund, S.L. (1990) Effects of variation in glutathione peroxidase activity on DNA damage and cell survival in human cells exposed to hydrogen peroxide and t-butyl hydroperoxide. *Biochem. J.* **271**, 17–23.
8. Sen, C.K. and Packer, L. (1996) Antioxidant and redox regulation of gene transcription. *FASEB J.* **10**, 709–720.
9. Hirota, K., Murata, M., Sachi, Y., Nakamura, H., Takeuchi, J., Mori, K., and Yodoi, J. (1999) Distinct roles of thioredoxin in the cytoplasm and in the nucleus. A two-step mechanism of redox regulation of transcription factor NF-kappaB. *J. Biol. Chem.* **274**, 27891–27897.

10. Wei, S.J., Botero, A., Hirota, K., Bradbury, C.M., Markovina, S., Laszlo, A., Spitz, D.R., Goswami, P.C., Yodoi, J., and Gius, D. (2000) Thioredoxin nuclear translocation and interaction with redox factor-1 activates the activator protein-1 transcription factor in response to ionizing radiation. *Cancer Res.* **60**, 6688–6695.

11. Bellomo, G., Vairetti, M., Stivala, L., Mirabelli, F., Richelmi, P., and Orrenius, S. (1992) Demonstration of nuclear compartmentalization of glutathione in hepatocytes. *Proc. Natl. Acad. Sci. USA* **89**, 4412–4416.

12. Briviba, K., Fraser, G., Sies, H., and Ketterer, B. (1993) Distribution of the monochlorobimane-glutathione conjugate between nucleus and cytosol in isolated hepatocytes. *Biochem. J.* **294**, 631–633.

13. Cotgreave, I.A. (2003) Analytical developments in the assay of intra- and extracellular GSH homeostasis: specific protein S-glutathionylation, cellular GSH and mixed disulphide compartmentalisation and interstitial GSH redox balance. *Biofactors* **17**, 269–277.

14. Holmgren, A. and Fagerstedt, M. (1982) The in vivo distribution of oxidized and reduced thioredoxin in Escherichia coli. *J. Biol. Chem.* **257**, 6926–6930.

15. Watson, W.H., Pohl, J., Montfort, W.R., Stuchlik, O., Reed, M.S., Powis, G., and Jones, D.P. (2003) Redox potential of human thioredoxin 1 and identification of a second dithiol/disulfide motif. *J. Biol. Chem.* **278**, 33408–33415.

16. Janssen, Y.M. and Sen, C.K. (1999) Nuclear factor kappa B activity in response to oxidants and antioxidants. *Methods Enzymol.* **300**, 363–374.

17. Padgett, C.M. and Whorton, A.R. (1995) S-nitrosoglutathione reversibly inhibits GAPDH by S-nitrosylation. *Am. J. Physiol.* **269**, C739–749.

18. Stadtman, T.C. (2002) Discoveries of vitamin B12 and selenium enzymes. *Annu. Rev. Biochem.* **71**, 1–16.

19. Gasdaska, P.Y., Gasdaska, J.R., Cochran, S., and Powis, G. (1995) Cloning and sequencing of a human thioredoxin reductase. *FEBS Lett.* **373**, 5–9.

20. Mustacich, D. and Powis, G. (2000) Thioredoxin reductase. *Biochem. J.* **346**, 1–8.

21. Soini, Y., Kahlos, K., Napankangas, U., Kaarteenaho-Wiik, R., Saily, M., Koistinen, P., Paaakko, P., Holmgren, A., and Kinnula, V.L. (2001) Widespread expression of thioredoxin and thioredoxin reductase in non-small cell lung carcinoma. *Clin. Cancer Res.* **7**, 1750–1757.

22. Kim, J.R., Lee, S.M., Cho, S.H., Kim, J.H., Kim, B.H., Kwon, J., Choi, C.Y., Kim, Y.D., and Lee, S.R. (2004) Oxidation of thioredoxin reductase in HeLa cells stimulated with tumor necrosis factor-alpha. *FEBS Lett.* **567**, 189–196.

23. Brigelius, R., Lenzen, R., and Sies, H. (1982) Increase in hepatic mixed disulphide and glutathione disulphide levels elicited by paraquat. *Biochem. Pharmacol.* **31**, 1637–1641.

24. Brigelius, R., Muckel, C., Akerboom, T.P., and Sies, H. (1983) Identification and quantitation of glutathione in hepatic protein mixed disulfides and its relationship to glutathione disulfide. *Biochem. Pharmacol.* **32**, 2529–2534.

25. Lash, L.H. and Jones, D.P. (1985) Distribution of oxidized and reduced forms of glutathione and cysteine in rat plasma. *Arch. Biochem. Biophys.* **240**, 583–592.

26. Go, Y.M., Ziegler, T.R., Johnson, J.M., Gu, L., Hansen, J.M., and Jones, D.P. (2007) Selective protection of nuclear thioredoxin-1 and glutathione redox systems against oxidation during glucose and glutamine deficiency in human colonic epithelial cells. *Free Radic. Biol. Med.* **42**, 363–370.

27. Sethuraman, M., McComb, M.E., Huang, H., Huang, S., Heibeck, T., Costello, C.E., and Cohen, R.A. (2004) Isotope-coded affinity tag (ICAT) approach to redox proteomics: identification and quantitation of oxidant-sensitive cysteine thiols in complex protein mixtures. *J. Proteome Res.* **3**, 1228–1233.

28. Halvey, P.J., Watson, W.H., Hansen, J.M., Go, Y.M., Samali, A., and Jones, D.P. (2005) Compartmental oxidation of thiol-disulphide redox couples during epidermal growth factor signalling. *Biochem. J.* **386**, 215–219.

Part V
Protein Dynamics in the Nucleus

Chapter 18
Fluorescence Correlation Spectroscopy to Assess the Mobility of Nuclear Proteins

Stefanie Weidtkamp-Peters, Klaus Weisshart, Lars Schmiedeberg, and Peter Hemmerich

Keywords Cell nucleus; Obstructed diffusion; Kinetic microscopy; Living cell; Diffusion coefficient; Enhanced green fluorescent protein; Bleaching

Abstract Recent developments in cell biology and microscopy techniques enable us to observe macromolecular assemblies in their natural setting: the living cell. These emerging technologies have revealed novel concepts in nuclear cell biology. In order to further elucidate the biochemistry of gene expression, replication, and genome maintenance, the major challenge is now to precisely determine the dynamics of nuclear proteins in the context of the structural organization of the nucleus. Fluorescence correlation spectroscopy (FCS) is an attractive alternative to photobleaching and photoactivation techniques for the analysis of protein dynamics at single-molecule resolution. Here we describe how FCS can be applied to retrieve biophysical parameters of nuclear proteins in living cells.

1 Introduction

1.1 Dynamic Organization of the Nucleus

The eukaryotic nucleus is responsible for the storage (DNA compaction), propagation (DNA replication), maintenance (DNA repair), and expression (RNA transcription) of the genetic material it contains. Therefore, a highly organized machinery is required for these processes to take place in an extremely condensed nuclear environment with protein concentrations ranging between 50 and 400 µg/µL. DNA in the form of chromatin is organized in distinct chromosome territories composed of transcriptionally active (euchromatin) and inactive domains (heterochromatin) (see Chap. 15 in Vol. 1 by M. Cremer et al.). Active DNA and RNA metabolism such as transcription and replication occurs within numerous focal sites throughout the nuclear volume, sometimes referred to as "factories" (Fig. 18.1).

R. Hancock (ed.) *The Nucleus: Volume 2: Chromatin, Transcription, Envelope,*
Proteins, Dynamics, and Imaging,
© Humana Press 2008

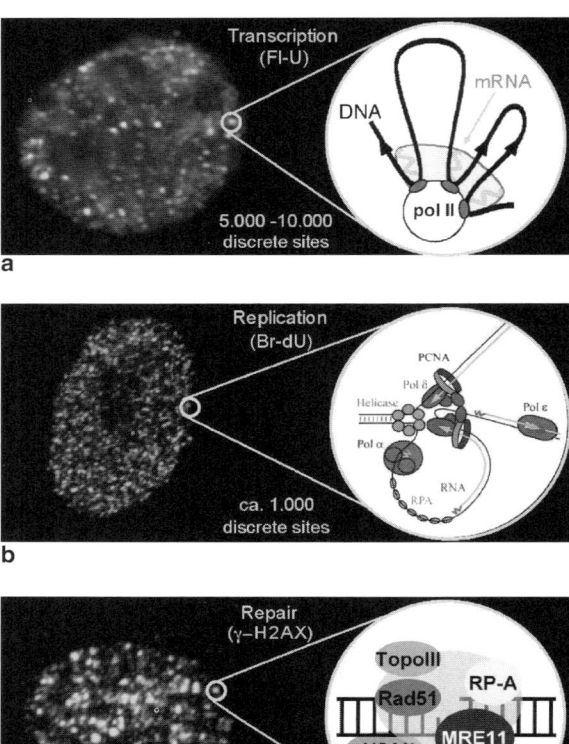

Fig. 18.1 "Factories" in the cell nucleus. **a** Transcription can be visualized by exposure of living cells to the nucleotide analog fluorouridine (Fl-U) for short periods of times. Nascent RNA transcripts are then detected by indirect immunostaining of the Fl-U epitope. Depending on the imaging technique, several hundred (confocal microscopy) to more than 10,000 (electron microscopy) discrete transcription sites can be detected throughout the entire nuclear volume of proliferating mammalian cells. **b** If bromodeoxyuridine (Br-dU) is used in an incorporation assay, nascent DNA can be visualized in cells that are in S-phase. The pattern of Br-dU incorporation changes throughout S-phase with strikingly similar dynamics in all mammalian cell nuclei. **c** Genotoxic stress, such as irradiation or radiomimetic drugs, can induce a variety of DNA damage that needs to be repaired. Common to all lesions within genomic DNA is an immediate phosphorylation of the histone variant H2AX. Using antibodies against this phosphorylated epitope (γ-H2AX), the sites of DNA damage can be identified. During the repair process, many DNA repair factors are recruited in a timely and organized fashion

Another hallmark of the mammalian cell nucleus is the presence of visually defined structural compartments. Most of these structures participate in the synthesis, processing, and modification of RNA, and form in response to gene expression, replication, and DNA repair. These subnuclear structures are macromolecular

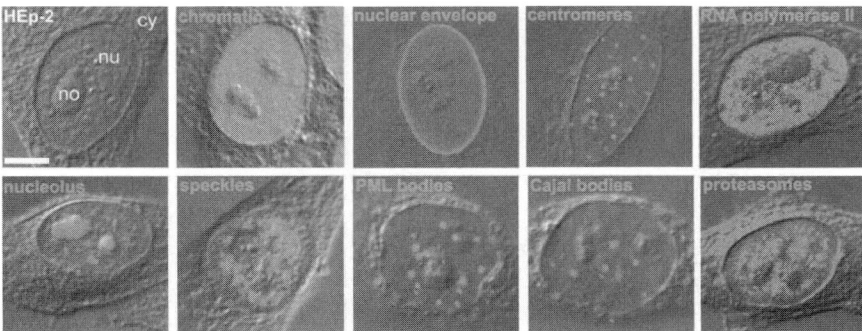

Fig. 18.2 Subnuclear structures. Confocal microscopy images of HEp-2 cells stained with specific antibodies decorating the indicated nuclear substructures (*white*). Differential interference contrast (DIC) images (*dark grey*) were taken from each cell at the same time and merged with the respective immunofluorescence image. DIC reveals the cytoplasm (cy), nucleoplasm (nu), and nucleoli (no). Subnuclear structures were detected with antibodies against histone H2A (chromatin), lamin A/C (nuclear envelope), the hyperphosphorylated form of the largest subunit of RNA polymerase, fibrillarin (nucleolus), splicing factor SC35 (speckles), the promyelocytic leukemia protein (PML) bodies), p80 coilin (Cajal bodies), and the 20 S proteasome. Bar, 5 μm. To view this figure in color, See COLOR PLATE 9

complexes that consist of membraneless accumulations of specific sets of functionally related molecules. For example, components of the ribosome biogenesis pathway are predominantly confined to the nucleolus, while proteins and ribonucleoprotein complexes involved in mRNA metabolism occupy the interchromatin space. Compartments identified in the interchromatin space include active factories (Fig. 18.1), nuclear speckles containing spliceosomal components, Cajal bodies involved in small nuclear ribonucleoprotein (snRNP) biogenesis, promyelocytic leukemia protein (PML) nuclear bodies that are enriched with proteins of various functions, and focal regions with proteasome activity (Fig. 18.2) (*(1)*; see also Chap. 14 in Vol. 1 by von Mikecz et al.).

The definition of specific biochemical interactions among nuclear proteins in distinct compartments has led to an image of structural continuity and functional stability within the nucleus. For years, nuclear pathways have been exhaustively examined using biochemical and molecular approaches without much consideration of the special restrictions presented by the nuclear architecture. The development of in vivo microscopy techniques using genetically encoded fluorescent tags, such as the green fluorescent protein (GFP), has opened the door to probe nuclear architecture and function in living cells. These powerful methods have recently been combined with photobleaching techniques: fluorescence recovery after photobleaching (FRAP), photoactivation (PA), fluorescence correlation spectroscopy (FCS), fluorescence resonance energy transfer (FRET), and single particle tracking, allowing one for the first time not only to visualize protein dynamics and interactions but also to quantitatively determine biophysical properties of proteins in intact cells (Fig. 18.3) (see also Chap. 19 by Siebrasse and Kubitscheck, and Chap. 20 by van

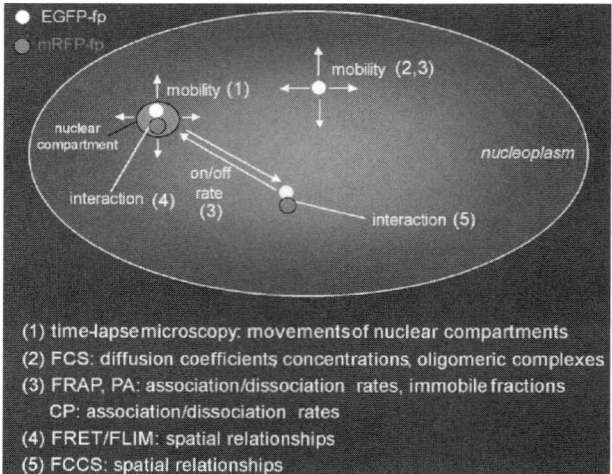

Fig. 18.3 Live-cell imaging tools to assess nuclear dynamics. Nuclear proteins can be introduced into living cells as fusions proteins (*fp*) tagged with fluorescent proteins such as enhanced green fluorescent protein (*EGFP*) or monomeric red fluorescent protein (*mRFP*). Depending on its current function, the nuclear protein(s) may occur diffusely in the nucleoplasm, in a factory, or in a specific subnuclear compartment. Mobility and interactions can be assessed by time-lapse microscopy, fluorescence correlation spectroscopy (*FCS*), fluorescence recovery (or redistribution) after photobleaching (*FRAP*), photoactivation (*PA*), continuous photobleaching (*CP*), fluorescence resonance energy transfer (*FRET*), fluorescence lifetime imaging (*FLIM*), and fluorescence cross-correlation spectroscopy (*FCCS*)

Royen et al. in this volume, and Chap. 9 in Vol. 1 by Louvet et al.). These studies indicate that nuclear compartments have their component parts in a notable state of constant flux *(2)*.

The structural instability of subnuclear compartments such as the nucleolus and nuclear bodies seems to allow functional flexibility, but this flexibility comes at a price: many nuclear proteins are known to interact dynamically with one or more of these compartments, and disruption of the specific organization of nuclear protein assemblies results in defects of cell functions *(3)*. In order to gain insight into such subnuclear pathologies, it is important to study the mechanisms of nuclear substructure formation, maintenance, and disassembly by assessing the mobility of their component parts in living cells. In a simplified model of nuclear protein dynamics, we can assume at least three different protein pools: nuclear proteins at "factories" are actively involved in biochemical reactions (pool 1); another population exists in a soluble pool in the nucleoplasm (pool 2); but there is also a subset in a more insoluble fraction associated with distinct intranuclear compartments or chromatin (pool 3). A major challenge in nuclear cell biology is now to precisely determine the dynamics of nuclear proteins within these pools and the spatio-temporal relationship between them in order to fully understand the molecular details of nuclear functions.

1.2   Fluorescence Correlation Spectroscopy

FCS is a method to analyze diffusing particles in solution or in living cells, introduced in the early 1970s by Elson, Magde, and Webb *(4)*. In FCS, fluorescent
molecules or particles diffuse by Brownian motion in-and-out of a space-limited
detection volume. This detection volume represents a diffraction-limited small
illumination ellipsoid created by a laser beam that is focussed through a high
numerical aperture objective (Fig. 18.4A, B). Photons emitted from the fluorescent particles are counted continuously through the same optics over time.
The intensity fluctuations reflect the photophysical and hydrodynamic properties
of the diffusing particles (Fig. 18.4C). While the fluctuation amplitude depends
on particle concentration and brightness, its frequency contains information on
the diffusion times of the particles. For quantitative evaluation, the fluctuation

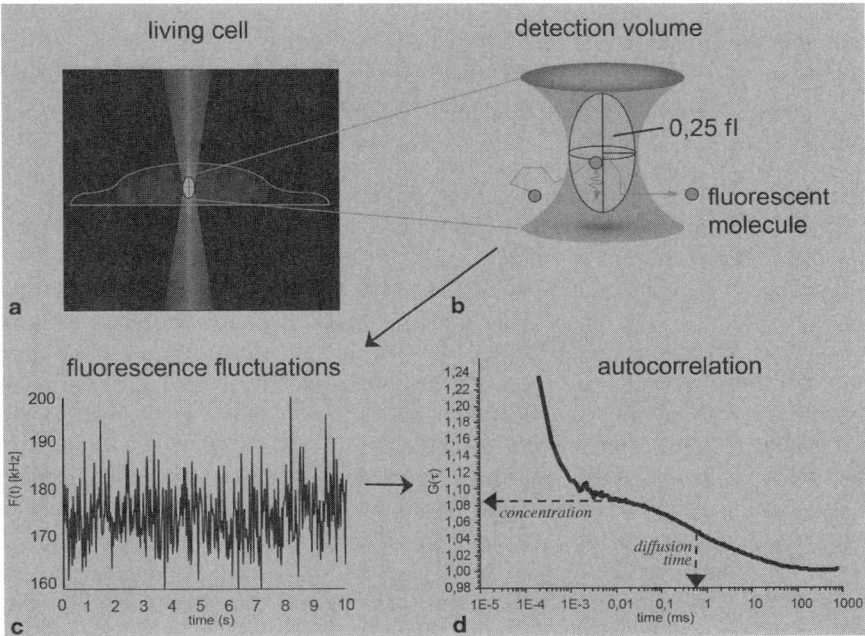

Fig. 18.4 FCS analysis procedure. **a** Schematic side view of a living cell with the FCS laser beam
focussed to a position within the nucleus (*blue*). **b** The beam focus creates a laser light-illuminated
confocal volume with an ellipsoidal shape (~1×0.3×0.3 μm in *z*, *x*, *y*). Fluorescent molecules only
emit light (*red arrow*) when they diffuse through this detection volume. The emitted photons are
monitored over time yielding a fluorescence fluctuation plot (**c**). These statistical fluctuations are
mathematically processed using an autocorrelation algorithm (**d**). By fitting the measured data
points to appropriate diffusion models, one can extract from the reciprocal of the amplitude and
the decay half-time value the number of particles in the detection volume (*concentration*) and the
diffusion time, respectively. To view this figure in color, see COLOR PLATE 10

frequency is correlated with a time-shifted replica of itself (autocorrelation) at different time values (Fig. 18.4D). A detailed description of this method can be found elsewhere *(5)*. By fitting theoretical model functions to the measured auto-correlation curves, the diffusion coefficient and the concentration of the diffusing species can be extracted.

The full potential of FCS was only realized in the 1990s when new lasers (especially two-photon techniques), single-photon detectors (avalanche photodiodes), and new microscope objectives with high numerical apertures became available. The combination of a confocal microscope with a FCS unit complemented live cell images with biophysical information *(6)*. The new developments provided reduction of the detection volume into the femtoliter range and short measurement times. Considering a particle concentration of $1\,nM$, the FCS detection volume contains less than one particle at any given time, thus providing single molecule resolution. This advantage of FCS is at the same time its drawback: it works only properly within a limited concentration range, typically between $10\,nM$ and $1\,\mu M$. Due to its high sensitivity, FCS is subject to certain artefacts that must be carefully controlled, such as photobleaching, cellular autofluorescence, intramolecular dynamics of the fluorophore, laser beam polarization effects, the refractive index of the objective's immersion medium, pinhole mis-adjustment, cover slide thickness, and optical saturation. Further information on these issues can be found elsewhere *(7, 8)*. In recently developed commercial FCS devices, these potential dangers are mostly eliminated. However, the first-time user should start under professional guidance until they feel safe to evaluate the data properly. In the Methods and Notes sections (**Sections 3** and **4**), we provide advice on how to avoid FCS artefacts.

The time scale of resolution of FCS is in the femtoseconds to seconds range, making it a powerful tool to study biological processes, particularly in living cells, which complements related techniques such as FRAP, PA, or single-particle tracking, which are described in other chapters of these volumes. Applications of both FRAP and FCS on the same molecules allows determination of the full spectrum of dynamics of a nuclear protein *(9)*. Typical FCS measurements performed in the nucleus are shown in Fig. 18.5. FCS is non-invasive and gentle for the biological specimen, since the high detection sensitivity requires only small laser intensities for fluorophore excitation. The probe can be monitored repeatedly (i.e. during the cell cycle) and remains accessible for manipulations (i.e. drug application). Important extensions of FCS include dual-color cross-correlation spectroscopy (FCCS) which allows analysis of protein–protein interactions *(10)*, continuous fluorescence photobleaching (CP) FCS that enables determination of binding residence times of proteins *(11)*, and photon-counting histogram (PCH) analysis, which has the potential to resolve protein oligomerization in living cells *(12)*. In this chapter, we focus on the determination of diffusion coefficients and concentrations of fluorophore-tagged nuclear proteins using the Zeiss LSM/ConfoCor system (Carl Zeiss MicroImaging GmbH, Jena, Germany).

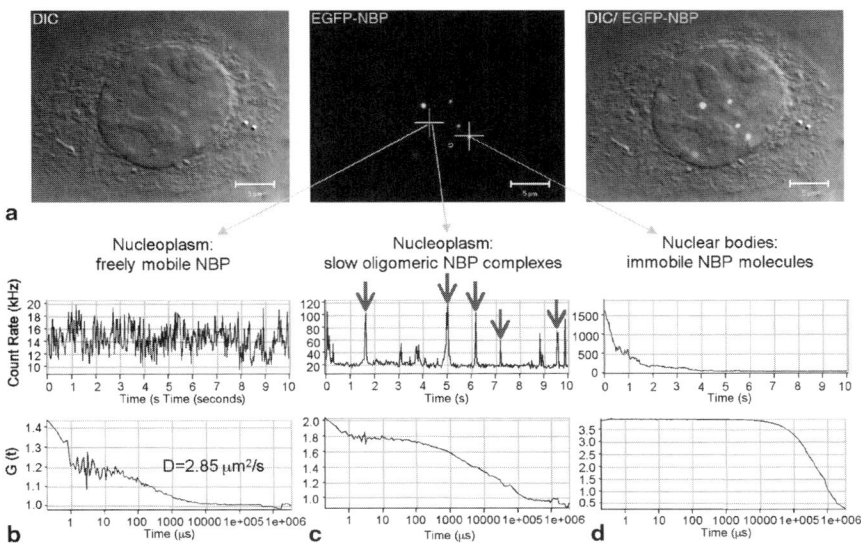

Fig. 18.5 FCS measurements of a nuclear protein in the nucleus. **a** Confocal image of a HEp-2 cell expressing an EGFP-tagged nuclear body protein (*GFP-NBP*). A differential interference contrast image (*left*) and the GFP fluorescence (*middle*) of the same cell were recorded separately before the FCS measurement. The image on the right shows a merge of these two images. Then, FCS measurements were performed at different positions of the nucleus (*crosses*). Different count rate traces and the respective autocorrelation curves of FCS measurements are obtained within the nucleoplasm (**b**, **c**). The kinetics of soluble NBP molecules (**b**) are readily accessible by FCS measurements, yielding a diffusion coefficient for GFP-NBP of $D = 2.85 \, \mu m^2 \, sec^{-1}$. Spikes in (**c**) correspond to large mobile structures containing several GFP-NBP molecules, which migrate slower due to their increased size and/or increased binding to chromatin. They are more sensitive to bleaching then dimeric or monomeric molecules since they stay longer in the illuminated spot. In (**d**) the laser beam was positioned in *x*, *y*, and *z* to hit a nuclear body. The initial fluctuation decay within the first seconds is indicative of an immobile or very slow fraction of NBP within the nuclear body structure. After this initial "bleaching" period, a count rate trace can be recorded that results in autocorrelation curves similar to those obtained for mobile NBP

2 Materials

2.1 Plasmid Transfection and Live Cell Imaging

1. Dulbecco's modified Eagle's medium (DMEM) without serum (PAA, Pasching, Austria).
2. Transfection reagent: FuGene6 or FuGeneHD (Roche, Penzberg, Germany), stored at 4°C.
3. Plasmid DNA, stored at −20°C.

2.2 FCS Measurements in Nuclei of Living Cells and Data Evaluation

1. Sterile glass coverslips (42-mm diameter, 0.17-mm thick; Saur, Reutlingen, Germany).
2. Sterile glass coverslips (e.g. 18×18×0.17 mm; Roth, Karlsruhe, Germany).
3. DMEM without phenol red, with HEPES buffer (Gibco/BRL, Bethesda, MD, USA).
4. Live cell chamber (PeCon, Eberbach-Bach, Germany).
5. Alexa 488 dye solution for pinhole adjustment (Invitrogen/Molecular Probes, Heidelberg, Germany).
6. Labware: No. 1 LabTek chambers (Nalge Nunc, Naperville, IL, USA) or glass-bottom culture dishes (MatTek, Ashland, MA, USA) for cell culture growth.
7. Fluorescence-free water as immersion medium (e.g. Sigma-Aldrich, St. Louis, MO, USA) (*see* **Note 4**).
8. Origin data analysis and visualization software (ADDITIVE, Friedrichsdorf, Germany)
9. FCS instrument: combined imaging and spectroscopic analysis is realized on an inverted microscope to which are attached a confocal laser scanning microscope (LSM) and a confocal fluorescence correlation spectrometer detection unit (Fig. 18.6). We will focus in this article on the LSM 510–ConfoCor 2 system (Carl Zeiss MicroImaging GmbH) *(13, 14)*, since this is the most widely used commercial system (*see* **Note 1**). However, most of the principles outlined will apply to other systems as well, and we will refer to major differences if necessary.
10. Correlation: FCS measures the stochastic intensity fluctuations δI of fluorescently labeled molecules caused by diffusion or photophysical processes by the so-called autocorrelation function G(τ), which is defined as:

$$G(\tau) = \frac{\langle I(t) \cdot I(t+\tau) \rangle}{\langle I(t) \rangle^2} = 1 + \frac{\langle \delta I(t) \cdot \delta I(t+\tau) \rangle}{\langle I(t) \rangle^2},$$

(18.1)

where ( ) denotes the time average and $\delta I(t) = I(t) - I(t)$ describes the fluctuations around the mean intensity *(15)*. In this way the correlation describes the chances of still seeing a molecule that has been detected at a certain time t after a certain lag or correlation time τ later (Fig. 18.7). By fitting the data to appropriate models, the number of molecules N, which is inversely proportional to the autocorrelation amplitude G(0), and the diffusion times $\tau_{d,i}$ and fractions f_i of each species i can be determined *(16, 17)* (*see* **Note 2**). Photophysical processes like triplet states will give an extra shoulder in the autocorrelation function and are accommodated for by an exponential function, which describes the fraction of molecules in the triplet state T and the triplet time τ_T *(5)*. For a free three-dimensional translational diffusion with up to three components and triplet state, the autocorrelation function can be described as:

$$G(\tau) = 1 + d + (1 - \frac{I_b}{I_t})^2 \cdot \frac{\gamma}{N} \cdot (1 + \frac{T \cdot e^{-\tau/\tau_T}}{1 - T}) \cdot$$

$$\sum_{i=1}^{3} \frac{f_i}{\left(1 + \left(\tau/\tau_{D,i}\right)\right) \cdot \left(1 + \left(\tau/\tau_{D,i}\right) \cdot 1/S^2\right)^{1/2}} \cdot \qquad (18.2)$$

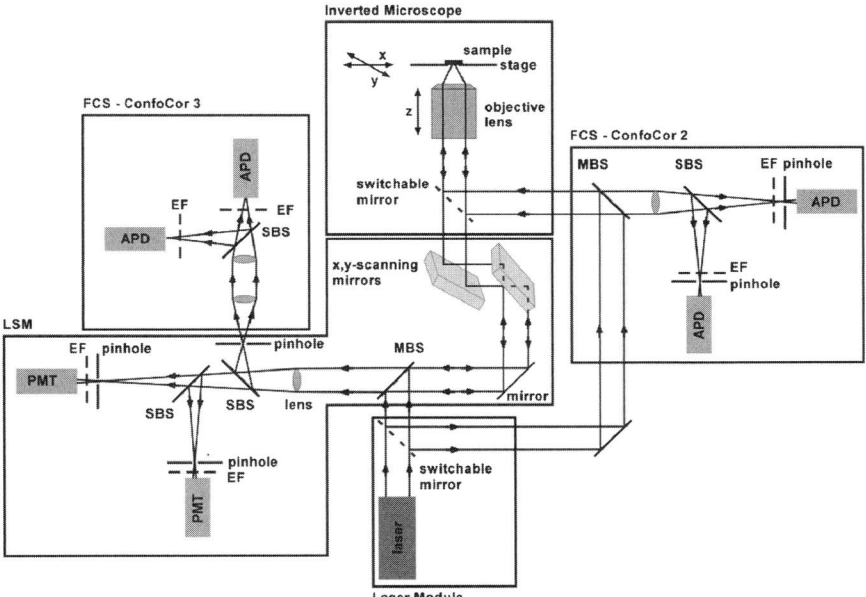

Fig. 18.6 Principle confocal setups for a combination of imaging and FCS. The set-up consists of a laser module, an inverted microscope, a confocal laser scanning microscope (*LSM*), and a confocal fluorescence correlation spectroscopy (*FCS*) spectrometer. The FCS unit can be attached separately to the microscope, as in the ConfoCor 2, or to a channel of the LSM as in the ConfoCor 3. A ConfoCor 3-like setup is also realized in the Leica TCS SP2–FCS2 system with the exception that one pinhole is used for all detectors. In the Picoquant Microtime 200 system, the ConfoCor 2 principle is used without an LSM, and avalanche photodiodes (*APDs*) serve for imaging as well. In the ISS ALBA system, the APDs and photomultiplier tubes (*PMTs*) are housed in the same detection unit and can be interchanged. The laser line is usually reflected via the major dichroic beam splitter (*MBS*) and the scanning mirrors through the objective lens into the sample. In the ConfoCor 2-like setups the scanning mirrors are bypassed; the emitted light returns through the same objective and due to its higher wavelength, it passes the MBS. By way of secondary beam splitters (*SBS*), the light can be spectrally resolved and pass through the pinholes and emission filters (*EF*), which serve to block remaining excitation light onto the detectors. APDs run in counting mode, and PMTs in integration mode

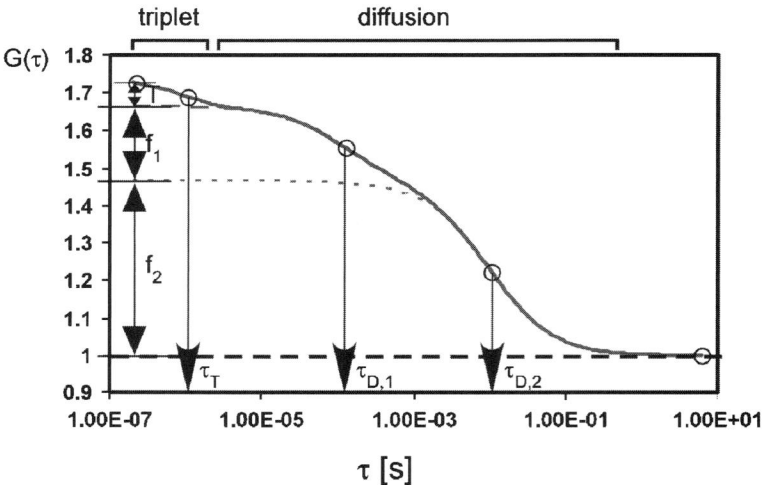

Fig. 18.7 Autocorrelation functions. A simulated curve with the following parameter settings: triplet fraction T, 10%; triplet decay time τ_T, 1 msec; number of particles N, 1.5; fraction of 1st component f_1, 0.3; fraction of 2nd component f_2, 0.7; diffusion time of first component $\tau_{D,1}$, 120 µsec; diffusion time of 2nd component $t_{D,2}$, 10 msec; and structural parameter S, 5. The curve decays sigmoidally for the diffusion process, where different components add up. A correlation with only the slower component is indicated by the *dotted line* for comparison. The triplet state, which is represented by an exponential decay, adds a shoulder to the autocorrelation function (*straight line*). For comparison, the *stippled curve* shows the correlation function without triplet. The total number of particles can be calculated from G(0), which is obtained by extrapolation of the correlation time to 0; each fraction can be derived from the single amplitudes. Decay and diffusion times can be determined from the inflection points of the curve

The geometric factor γ takes into account the Gauss intensity profile of the exciting laser beam. For a three-dimensional Gauss profile, this factor equals 0.35 *(18)*. The offset d is needed if the correlation function converges to a value above 1; otherwise it is set to 1. The background correction $(1 - I_b/I_t)^2$ corrects for a decrease in the amplitude if a non-correlating background intensity I_b contributes to the total intensity I_t *(19)*. If the background is negligible (below 10% of the total intensity), the factor can be set to 1. The structural parameter S is an instrumental parameter and defined as the ratio of the $1/e^2$ axial ω_z and radial ω_r extensions of the confocal volume, which is approximated by an ellipsoid of volume V *(8)* (*see* **Note 3**).

$$S = \frac{\omega_z}{\omega_r}, \qquad (18.3)$$

$$V = \pi^{3/2} \cdot \omega_r^2 \cdot \omega_z. \tag{18.4}$$

The diffusion times $\tau_{D,i}$ and number of molecules N_i can be transformed into diffusion coefficients D_i and concentrations c_i by the following relations:

$$D_i = \frac{\omega_r^2}{4 \cdot \tau_{D,i}} \text{ and} \tag{18.5}$$

$$c_i = \frac{N_i}{V \cdot N_A}, \tag{18.6}$$

where $N_A = 6.023 \times 10^{23} \text{ mol}^{-1}$ represents Avogadro's number and $N_i = N \times f_i$ equals the number of diffusing particles of species i.

In case of obstructed movement due to molecular crowding effects or barriers, an anomalous three-dimensional translational diffusion model with up to two components and triplet state is more appropriate, where the magnitude of obstruction is described by the anomaly parameter α_i for each species:

$$G(\tau) = 1 + d +$$

$$(1 - \frac{I_b}{I_t})^2 \cdot \frac{1}{N} \cdot (1 + \frac{T \cdot e^{-\tau/\tau_T}}{1 - T}) \cdot \sum_{i=1}^{2} \frac{f_i}{\left(1 + \left(\tau/\tau_{d,i}\right)^{\alpha_i}\right) \cdot \left(1 + \left(\tau/\tau_{d,i}\right)^{\alpha_i} \cdot 1/s^2\right)^{1/2}} \tag{18.7}$$

Under conditions of free diffusion, α equals 1, and for sub-diffusion it drops below 1. The transport coefficient Γ_i can be calculated from the anomalous diffusion times $\tau_{A,i}$ according to:

$$\Gamma = \frac{\omega_{xy}^2}{\tau_A^\alpha}. \tag{18.8}$$

For anomalous diffusion, the diffusion coefficient D_i is time-dependent and related to the transport coefficient Γ_i in the following way:

$$D_i(t) = \frac{\Gamma_i \cdot t^{\alpha-1}}{4}. \tag{18.9}$$

3 Methods

3.1 Plasmid Transfection and Live Cell Imaging

1. For live cell imaging, a 60-mm dish is prepared as follows: a small coverslip (e.g. 18×18×0.17 mm) is sterilized by dipping it into 95% ethanol and passing through the flame of a Bunsen burner. The hot coverslip is placed on the bottom of the plastic dish and allowed to stick to it, then a sterile 42-mm coverslip is placed on the smaller one; this facilitates the later removal of the fragile 42-mm coverslip (42-mm coverslips should be sterilized by washing for 1–2 h in 70% ethanol followed by 1–2 h in 100% ethanol, then placed between sheets of filter paper in a large glass Petri dish, autoclaved, and dried). Cells are seeded on the coverslip in a total volume of 5 mL.
2. On the day of transfection the cells should have ~60% confluency.
3. DMEM without serum is pre-warmed for the preparation of the transfection mix.
4. For transfection of cells on a 42-mm coverslip, 200 μL of medium without FCS is aliquoted in a tube. If necessary, a mastermix for the transfection of more then one dish with the same plasmid DNA can be prepared.
5. 2 μg of plasmid DNA is added and mixture is stirred with the tip of the pipette.
6. 6 μL of FuGene6 are added and mixed by vortexing for a second.
7. Allow transfection complex formation by incubation for 15 min at room temperature.
8. Finally the transfection mix is pipetted drop-wise onto the cells and the dish is swirled carefully. Change of culture medium is not necessary.
9. Cells are cultured for 16–48 h before FCS measurements.

3.2 FCS Measurements in Nuclei of Living Cells and Data Evaluation

3.2.1 Calibration of the Confocal Volume

1. Prepare 10^{-6} M stock solutions of Rhodamine 6 green (Rh6G from Sigma-Aldrich, St. Louis, MO, USA) and Cy 5 (GE Healthcare Life Sciences, Piscataway, NJ, USA) in water. Dilute to 10^{-7} M and 10^{-8} M in water.
2. Fill 400 μL of the 10^{-8} M dye solution into a well of an 8-well LabTek chamber and mount on the microscope.
3. Orient the sample carrier as outlined by the manufacturer. This will allow changing wells automatically if a scanning stage is available.
4. Find the upper coverslip's glass surface by moving the objective cautiously upward until the reflex images of the upper and lower surfaces are detected by

the FCS camera or in a line scan. Stop at the second reflex and focus at least 10 μm into the solution.

5. Adjust the count rate with a laser power of approximately 40 kHz and perform 10× 10-sec FCS measurements. Fit the average correlation function using Eq. 18.2 with one component. Calculate ω_r from Eq. 18.5 with the fitted τ_D and S values and the known diffusion coefficient D of the dye. Use ω_r to convert τ_D values into D values of your measurement using Eq. 18.5.

3.2.2   Offset Determination Between Stage and Scan Mirrors

1. Prepare a 10^{-6} M solution of Rhodamine 6 green (RhG) in ethanol.
2. Fill 200 μL of the 10^{-6} M dye solution into a well of an 8-well LabTek chamber and let the ethanol evaporate overnight with the lid open, creating a dye layer on the coverslip. Alternatively, use a layer of thin fluorescent glue such as glue #44 (Carl Zeiss MicroImaging) squeezed between the glass slide and a coverslip.
3. Focus on the dye layer using the LSM.
4. Either perform one FCS measurement for 2 sec with a closed pinhole and a stationary laser using high laser power to bleach a hole into the dye layer, or select four positions in the image and bleach holes at every one of them using stage movement.
5. Rescan and determine where the stationary laser points are in the image. Use different zooms and record coordinates. Alternatively, if stage positioning is required, determine offsets between the four selected positions and actual bleach sites using the LSM software. Enter the average of the offset values for offset correction into the Software according to the manufacturer (*see* **Note 5**).

3.2.3   Pinhole Alignment

1. Define the beam path configuration to be used in your experiment.
2. Prepare 10^{-6} M, 10^{-7} M, and 10^{-8} M solutions in water of a dye suitable for excitation and detection for your beam path, e.g. Rhodamine 6 Green (RhG) or Alexa 488 for excitation with 458, 488, or 514 nm, Alexa 543 for 543 nm, or Cy 5 or Alexa 633 for 633 nm.
3. Fill 400 μL of the 10^{-6} M, 10^{-7} M, and 10^{-8} M dye solutions into wells of an 8-well LabTek chamber and place on the microscope.
4. Define or load the appropriate beam path configuration. Set the pinhole to 1 Airy unit.
5. Adjust the count rate to 100–400 kHz by tuning the laser power and changing to the most appropriate dye concentration.

6. Perform pinhole alignment in x, y, and z directions (first coarse, than fine for x and y, only coarse for z) and store the positions according to the manufacturers protocol (*see* **Note 6**).
7. Switch to the 10^{-8} M dye solution, adjust the count rate to approximately 40 kHz, and turn the correction ring while monitoring the count rate. Adjust to maximum count rate and repeat the pinhole alignment.
8. Adjust all other beampaths to be used in the same manner.

3.2.4   Determination of Background Fluorescence

1. Seed mock-transfected cells onto LabTek chambers or MatTek dishes to 80% confluency. Alternatively, use non-expressing cells of a transiently transfected cell line (*see* **Note 7**).
2. Take a transmitted light image and place the FCS measurement position into the nucleus of a non-expressing cell (stage positioning) or position the nucleus manually using a cross mark to indicate the site of the laser beam.
3. Load the respective beam path configuration.
4. Perform several 1× 10-sec measurements using a range of laser powers expected to be used in your experiments. Record the count rate obtained at each laser power (*see* **Note 7**).

3.2.5   Measuring the Correlation Function

1. Seed transfected cells into LabTek chambers or MatTek dishes to 80% confluency, or perform transfection directly in these supports as described in **Section 3.1.** In the latter case, remove the glass coverslip with the transfected cells from the culture dish and place into the PeCon live cell chamber. Change the medium for pre-warmed medium without phenol red buffered with HEPES.
2. Place the chamber on the microscope stage. If temperature control is required, use an appropriate heated microscope stage with heated objective. The system should be allowed to adjust to 37°C for at least 15 min.
3. Determine the transfection efficiency and screen the expression levels of cells, either in wide-field illumination or by raster scanning (*see* **Note 8**). Select cells with very low expression levels of fluorescently tagged proteins within the nucleoplasm. In LSM it can be advisable to open the pinhole to be able to detect low expression.
4. Move the nucleoplasm of a selected cell manually to the defined position of the FCS measurement as determined in **Section 3.2.2,** or select measurement positions in the nucleus for stage or mirror positioning. Use fast scanning mode in the LSM and zoom factors up to four to minimize laser exposure. An LSM image with a higher resolution of the cell before the measurement should be stored in an image database.

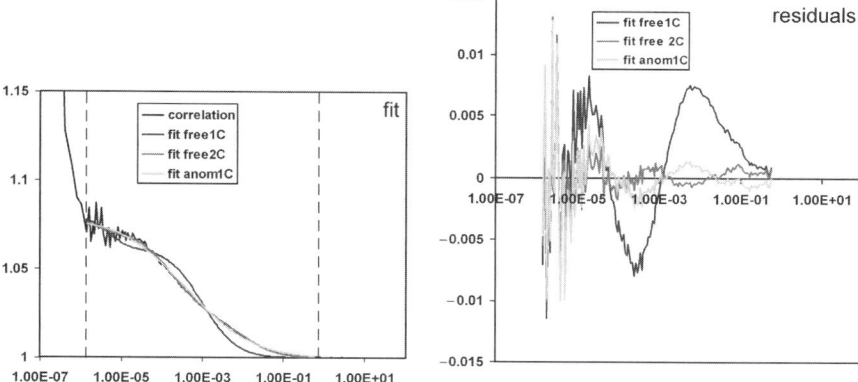

Fig. 18.8 Fitting an autocorrelation function to model equations. The *left panel* shows a correlation function of GFP in the nucleus (*black line*) and fits using a free three-dimensional diffusion model with triplet and one (*blue line*) or two (*red line*) components, as well as an anomalous three-dimensional diffusion model with triplet and one component (*yellow line*). The fit limits are indicated by the *left* (start value) and *right* (end value) *vertical stippled bars*. The start bar on the left is set to cut out the afterpulsing contribution of the detector, which is obvious as a steep increase at shorter correlation times. The *right panel* displays residual plots from the fit. Note that the free one-component model leads to huge systematic deviations of the fit compared with the correlation, and should be discarded as an appropriate model. The free two-component and the anomalous one-component models yield satisfactory fits with only minor systematic deviations. To view this figure in color, see COLOR PLATE 11

5. Adjust laser power for a reasonable count rate, preferentially 10× the background count rate and check the obtained counts per molecule (cpm) per second (*see* **Note 8**).
6. Perform 10× 10-sec measurements and save the data in the FCS database. Name the files appropriately to link to the image. Rescan the cell after the measurement to make sure that it remained unaffected by the measurement and to evaluate the extent of bleaching. Select the next cell and repeat for 50–100 cells (*see* **Note 9**).
7. Screen each measurement and discard those where the measured autocorrelation curve drops below 1 or that show other artefacts like aggregation (*see* **Note 10**). Take the average curve for fitting.

3.2.6 Fitting the Data to Model Equations

1. Load the correlation function into a fitting software, either that supplied by the manufacturer or an external program like Origin (*see* **Note 11**).
2. Check if the model requires accommodation of geometric factor, offset, and background correction and omit these parameters if possible (*see* **Note 12**).

3. Start with a one-component fit for free diffusion. Fix as many known parameters as possible. Set the fit limits to exclude afterpulsing and bleaching. Within the correlation curves, these regions are represented at the start (maximum) and at the end (minimum) of the correlation function, respectively (*see* **Note 13**).

4. Advance to more components or to an anomalous model only if the fit does not yield a satisfactory result (Fig. 18.8) (*see* **Note 14**).

5. Try to fit each curve from one data set (normally 10 measurements) to obtain similar parameter readings. Depending on the software package options, include globally linking a parameter *(20)*, defining start values for a parameter, altering the fit range, and defining limits for the parameter values (*see* **Note 15**).

6. Fit 50–100 data sets and built a histogram of diffusion times versus their frequency (*see* **Note 15**).

7. If required, convert diffusion times and number of molecules to diffusion coefficients, transport coefficients, and concentrations using Eqs. 18.5, 18.6, and 18.8 (*see* **Note 16**).

4   Notes

1. Suppliers of complete systems include Carl Zeiss MicroImaging GmbH, Jena, Germany; Leica Microsystems, Mannheim, Germany; PicoQuant GmbH, Berlin, Germany; Becker & Hickl GmbH, Berlin, Germany; and ISS, Champaign, IL, USA. For best performance, water-immersion lenses of high numerical aperture and with correction rings should be used.

2. The correlation function decays to 0 or 1 depending on which expression of Eq. 18.1, intensity or intensity fluctuation, is used for data analysis. In the Carl Zeiss and PicoQuant software, the correlation decays to 1, and in the ISS software also used by Leica to 0. Processes ranging between $1\,\mu sec$ to $100\,msec$ can be accessed accurately by FCS. To display such a wide range, correlation functions are normally displayed on a logarithmic scale for the lag times. In order to distinguish two species by the autocorrelation function, their diffusion times have to differ by a factor of at least 1.6, but in cases of low signal to noise, this can increase to 10 and higher. In these cases a two-component fit might not be justified later on in data analysis.

3. The structural parameter theoretically amounts to 2 to 3 for high numerical aperture objectives, but due to imperfect optics, it is in the range of 5 to 8. It is recommended to fix the structural parameter to 5 (for excitation with the 488 nm line) and 8 (for excitation with the 633 nm line) in the ConfoCor 2 and ConfoCor 3 for fitting the data. To determine the $1/e^2$ radial ω_r and axial ω_z extensions, dyes with a known diffusion constant like Rh6G ($D = 2.8\times10^{-10}$ m^2/sec) and Cy5 ($D = 3.16\times10^{-10}$ m^2/sec) with high quantum yield and low photophysics should be used for calibration. With the fitted τ_D and S values from Eq. 18.2, using a one component model ω_r and ω_z can be calculated from Eqs. 18.3

and 18.5. For the ConfoCor 2 and ConfoCor 3 using the C-Apochromat ×40 NA 1.2-W objective, the $1/e^2$ radial and axial extensions of the confocal volume amount to approximately 0.15 μm and 0.75 μm, respectively, for 488 nm excitation light, yielding a confocal volume of approximately 0.1 fL. The volumes for other laser lines scale with the third power of the wavelength ratios.

4. Most water objectives from commercial suppliers with correction rings are corrected for coverslip thicknesses between 0.13 and 0.19 mm, corresponding to No. 0 and No. 1 coverslips. If the glass is too thin or thick, chromatic aberrations will reduce signal to noise ratios in FCS, or in the latter case the working distance of the objective might be too small.

5. This calibration is only necessary for systems where the LSM and FCS detection units are mounted on different ports of the microscope and FCS positioning is done by a stage. Note that stage positioning has an accuracy of at best 1 μm. Using a Piezo stage like that in the PicoQuant MicroTime 200 or manual positioning, accuracies in the nanometer range can be obtained and therefore these should be used if higher precision is necessary. In systems where positioning is done by scan mirrors like the ConfoCor 3 (Carl Zeiss), the TCS SP2-FCS2 (Leica) or the ALBA (ISS), the calibration step is not necessary and position accuracies are normally in the nanometer range.

6. Pinhole sizes should be set to 1 Airy unit to guarantee confocality. In the ConfoCor 2 the optimal positions of the pinholes for given beam paths can be stored and are automatically reloaded when the beam path is used. For the ConfoCor 3, the pinhole is adjusted only in x and y. In the Leica TCS SP2–FCS2 the pinhole is fixed and no adjustment needs to be done. In the PicoQuant Time Harp the position of the pinhole is manually adjusted only once.

7. Background can arise due to light scattering or autofluorescence. In these cases the background fluorescence does not correlate, but will lead to a decrease in the correlation amplitude and hence an overestimation of the number of molecules; the diffusion time is generally not affected. If background is below 10% it can be neglected, and background between 10–30% can be corrected for. Backgrounds over 30% should be avoided and another cell line should be considered. Background can also be due to fluorescent chemicals like phenol red in the culture medium; this background is correlated and data have to be fitted to two components, complicating data analysis. Hence, phenol red-free medium or phosphate-buffered saline (PBS) should be used in the experiments. A range of laser powers should be tested since the power needed for measurements of fluorescent protein-expressing cells will vary depending on the expression level.

8. FCS can assess concentrations between picomolar and micromolar. However, the micromolar range is often exceeded in highly expressing cells, especially when using strong promotors, and fluctuations become so low that no correlation can be measured. Using endogenous or inducible promotors can provide some control towards lower expression. Over-expressing cells should be avoided in general due to the unphysiologically high expression levels of protein, which will lead to artificial binding to membranes and organelle structures. In general, the lower the expression the more pronounced are the fluctuations

and the better is the signal-to-noise ratio (SNR). The SNR can be determined by the counts per molecule (cpm) per second, which is a measure of how many photons are obtained by one molecule. The higher that number, the better the SNR. Normally the cpm will reach a maximum with a certain laser power, and increasing beyond that point will lead to a reduction due to saturation effects in the dye caused by photophysical processes. However, optimal laser powers can often not be used in a cellular environment due to bleaching problems and one often has to compromise for lower cpm values. Also if background is an issue, it will contribute more to the total intensity for low expression levels, reducing cpm values. In these cases it can be even beneficial to use stronger-expressing cells to reduce background levels below 10% with a concomitant rise in the cpm.

9. If molecules become slow, e.g. by binding to nuclear structures or chromatin, one has to ensure that measurement times are long enough to capture the process (as a rule of thumb 1,000× the diffusion time). Ten seconds seems to be a good compromise between obtaining a good SNR (which is proportional to the measurement time and the square of dye brightness) and photobleaching of the sample due to continuous illumination. Therefore it can be beneficial to measure repeatedly for shorter time intervals and to average the signal, to obtain statistics nearly as good as one long measurement will provide. Very slow and immobile molecules cannot be assessed by classical FCS methods, but rather are bleached during data acquisition, detected by a decreasing count rate trace and a correlation function that diverges below 1. If the molecules do not bleach, on the other hand, they cause an offset above 1 to which the correlation function converges. The contribution of very slow or immobile molecules can be removed by a short pre-bleach prior to the measurement. Measured diffusion times depend very much on the cellular environment that can change dynamically over time. Therefore, in FCS, which is a spot technology, there often is not one consistent diffusion time measured, but rather a distribution of diffusion times. To obtain sufficient statistics, many cells should be sampled and the data displayed in histograms where the probability is plotted versus a certain diffusion time range.

10. The first one to three of a set of ten measurements might be affected by bleaching with curves diverging below 1, especially for highly expressing cells, and these curves should be discarded. Bleaching is due to very slow or immobile fractions, which might be meaningful, and their dynamics has to be estimated by other techniques like FRAP. They might also be pure artifacts due to proteins unspecifically attached to cellular structures. Also, aggregates can sometimes be observed as huge peaks in the count rate trace (Fig. 18.5C). Since molecules contribute with the square of their brightness, huge aggregates dominate the correlation function and should either be removed from the raw data trace if available in the software, or the correlations discarded. For fitting, the average curve should be taken. For software that does not allow averaging correlation functions, averaging of the parameter values obtained in single measurements can alternatively be performed, however with less statistical significance.

11. Most manufacturers provide basic software for data analysis, and enhanced packages can be purchased additionally. Zeiss (with extra software modules to their basic package not available for older release versions), PicoQuant, and ISS, which is also used by Leica, provide most models needed for cell work. Data can be exported and mathematical programs like Origin, Igor, SigmaPlot, or MatLab can be used to fit the data. This requires, however, some knowledge of programming.

12. For diffusion measurements or if only relative concentrations have to be measured, the geometrical factor can be set to 1. If background is below 10% and curves converge to 1, the expressions for the offset and the background corrections can be omitted.

13. Afterpulsing of detectors occurs in the nanosecond time scale and leads to a steep increase at the start of the correlation function, especially for measurements with low cpm. It is due to secondary electrons released from the APD in addition to the electron triggered by the photon. Cellular movement or bleach artefacts can lead to a decrease or increase of the correlation function above or below 1 at longer correlation times. If the curve has converged to 1 for a certain time, it can still be fitted.

14. As an iron rule, one should start with as simple a model as possible. This is normally a free three-dimensional diffusion model with one component and triplet state. If a curve can be fitted to one component, the data do not justify fitting to more components. Only if the fit is not satisfactory can one go to a more-component fit or to an anomalous diffusion model. Any parameters that are known beforehand should be fixed, like the structural parameter. A three-component fit for free diffusion or a two-component fit of anomalous diffusion is only advisable if at least one parameter describing a component can be fixed. The more parameters that are floating, the less reliable will be the fit. Often both a two-component free and a one-component anomalous fit can yield similarly satisfactory results. In this case, prior knowledge or higher consistency for fitting different data sets can be used to bias the choice in favour of a certain model.

15. In global analysis, a parameter linked for the different measurements will be fitted to the same value for each of them. This immediately shows if one diffusion time is suitable for each of the measurements. Runaway measurements must be discarded, since they will lead to unsatisfactory fits. Some software packages provide initial guesses for fitting. Especially in noisier correlation curves, initial guesses might be incorrect and the fit will end in a local rather than in the global minimum. Due to the so-called multiple-time algorithms, correlation functions are noisier for short correlation times and the position of the start value for fitting can influence the initial guess. In these cases, changing the fit start can alter the initial guesses. Alternatively, amplitude and diffusion times can be obtained by visual inspection of the correlation function and provided as start values for the fit. Also, if one knows the value range for a parameter, like 1 to $5\,\mu$sec for the triplet time, defining the parameter limits can be beneficial for excluding local minima.

16. Diffusion times and numbers of particles depend on the dimensions of the confocal volume that are different for different excitation wavelengths. Diffusion coefficients and concentrations are normalized for the dimensions and are therefore independent of the sizes of the confocal detection volume. If only relative changes are necessary, just displaying the curves can be sufficient. An increase in the number of molecules is associated with a decrease in the correlation amplitude G(0). The amplitude represents the number of diffusing particles only if the triplet state is normalized by division of the term $(1 - T)$ as in Eqs. 18.2 and 18.7; otherwise it will represent the total number of diffusing particles and those in the triplet state. Diffusion times are best compared if the correlation functions are normalized to 2. The higher the diffusion time, the more right-shifted the correlation curve will be.

Acknowledgments This work was supported by grant HE 2484/3-1 within the priority program "Optical Analysis of the Structure and Dynamics of Supramolecular Biological Complexes" from the Deutsche Forschungsgemeinschaft awarded to P.H.

References

1. Hemmerich, P. and Diekmann S. (eds.) (2005) *Visions of the cell nucleus*. American Scientific Publishers, CA, USA.
2. Misteli T. (2001) Protein dynamics: implications for nuclear architecture and gene expression. *Science* **291**, 843–847.
3. von Mikecz A. and Hemmerich, P. (2005) Subnuclear pathology. In: *Visions of the cell nucleus* (Diekamnn, S. and Hemmerich, P. eds.), American Scientific Publishers, CA, USA, pp. 184–203.
4. Magde, D., Elson, E., and Webb, W.W. (1972) Thermodynamic fluctuations in a reacting system—measurement by fluorescence correlation spectroscopy. *Phys. Rev. Lett.* **29**, 705–708.
5. Widengren, J. and Mets, Ü. (2002) Conceptual basis of fluorescence correlation spectroscopy and related techniques as tools in bioscience. In: *Single-molecule detection in solution—methods and applications* (Zander, C., Enderlein, J., and Keller, R. A., eds.) Wiley-VCH, Berlin.
6. Rigler, R. and Widengren, J. (1990) Ultrasensitive detection of single molecules by fluorescence correlation spectroscopy. *Biosciences* **3**, 180–183.
7. Enderlein, J., Gregor, I., Patra, D., and Fitter J. (2004) Art and artefacts of fluorescence correlation spectroscopy. *Curr. Pharm. Biotechnol.* **2**, 155–161.
8. Bacia, K. and Schwille, P. (2003) A dynamic view of cellular processes by in vivo fluorescence auto- and cross-correlation spectroscopy. *Methods* **29**, 74–85.
9. Schmiedeberg, L., Weisshart, K., Diekmann, S., Meyer Zu Hoerste, G., and Hemmerich, P. (2004) High and low mobility populations of HP1 in heterochromatin of mammalian cells. *Mol. Biol. Cell.* **15**, 2819–2833.
10. Schwille, P., Meyer-Almes, F.-R., and Rigler, R. (1997) Dual-color fluorescence cross-correlation spectroscopy for multicomponent diffusional behaviour in solution. *Biophys. J.* **72**, 1878–1886.
11. Wachsmuth, M., Weidemann, T., Muller, G., Hoffmann-Rohrer, U.W., Knoch, T.A., Waldeck, W., and Langowski, J. (2003) Analyzing intracellular binding and diffusion with continuous fluorescence photobleaching. *Biophys. J.* **84**, 3353–3363.
12. Chen, Y., Müller, J. D., Ruan, Q., and Gratton, E. (2002) Molecular brightness characterization of EGFP in vivo by fluorescence fluctuation spectroscopy. *Biophys. J.* **82**, 133–144.
13. Jankowski, T. and Janka, R. (2001) in *Fluorescence correlation spectroscopy: theory and applications* (Rigler, R., and Elson, E. L., Eds.), Vol. 65, Springer, Berlin Heidelberg. pp. 331–345.

14. Weisshart, K., Jungel, V., and Briddon S. J. (2004) The LSM 510 META - ConfoCor 2 system: an integrated imaging and spectroscopic platform for single-molecule detection. *Curr. Pharm. Biotechnol.* **2**, 135–154.
15. Magde, D., Elson, E. L., and Webb, W. W. (1974) Fluorescence correlation spectroscopy. II. An experimental realization. *Biopolymers* **1**, 29–61.
16. Meseth, U., Wohland, T., Rigler, R., and Vogel, H. (1999) Resolution of fluorescence correlation measurements. *Biophys. J.* **3**, 1619–1631.
17. Haustein E. and Schwille P. (2003) Ultrasensitive investigations of biological systems by fluorescence correlation spectroscopy. *Methods* **2**, 153–156.
18. Muller, J. D., Chen, Y., and Gratton, E. (2003) Fluorescence correlation spectroscopy. *Methods Enzymol.* **361**, 69–92.
19. Hink, M. A., Borst, J. W., and Visser A. J. (2003) Fluorescence correlation spectroscopy of GFP fusion proteins in living plant cells. *Methods Enzymol.* **361**, 93–112.
20. Skakun, V.V., Hink, M. A., Digris, A. V., Engel, R., Novikov, E. G., Apanasovich, V. V, and Visser. A. J. (2005) Global analysis of fluorescence fluctuation data. *Eur. Biophys J.* **4**, 323–334. (Erratum in *Eur. Biophys. J.* (2005) **7**, 972).

Chapter 19
Single Molecule Tracking for Studying Nucleocytoplasmic Transport and Intranuclear Dynamics

Jan Peter Siebrasse and Ulrich Kubitscheck

Keywords Single-molecule fluorescence microscopy; Single-molecule tracking; Light microscopy; Nucleocytoplasmic transport; Fluorescence labelling

Abstract Microscopic imaging of single fluorescent molecules within cells provides a molecular, real-time view of physiological processes in vivo. Single fluorescent molecules produce diffraction-limited light spots in the image plane, which can be localised with a very high precision. In single-molecule fluorescence microscopy (SMF) the achievable localisation precision depends only on the signal-to-noise ratio (SNR) and the stability of the optical setup. Typically values between 20 and 40 nm can be achieved. Highly dynamic processes and Brownian motion characterised by diffusion coefficients $<20 \mu m^2/\text{sec}$ can be followed by high-speed imaging, hence the method is an ideal tool to study intranuclear protein or ribonucleoprotein particle mobility. In contrast to conventional techniques, different forms of mobility in a heterogeneous system may well be distinguished from each other. Furthermore, specific binding and bimolecular interaction events can be followed at the single molecule level. A prominent example of an application is the study of nucleocytoplasmic transport one molecule at a time. In this case, the high localisation precision allows to analyse the binding site distribution of single molecules at the nuclear pore complex, and the high time resolution allows determination of the binding duration of soluble receptors and transport substrates.

1 Introduction

In the recent few years the development of new light microscopy techniques made great progress, and exciting new methods were developed that allow a more and more accurate observation of cellular structures in the sub-micrometer range. Since the first theoretical formulation of optical resolution by Ernst Abbe in 1873, physicists have tried to overcome the diffraction-limited spatial resolution in light microscopy. A very explicit trend in more recent developments is the use of different illumination modes like in stimulated emission depletion (STED) microscopy or nonlinear structured illumination microscopy, where optical resolutions down to 20–40 nm are now within reach *(1, 2)*. However, there are still some drawbacks. To transform these advanced concepts into working microscopes, usually extremely expensive set-ups have to be realised. Such high-end microscopes require well-trained experts to ensure optimal performance and results.

In single-molecule fluorescence microscopy (SMF), detection of fluorescent molecules is still diffraction-limited. A single fluorescent molecule produces a blurred light spot in the image plane, whose extension does not correspond to the physical dimension of the originating fluorescent source but only to the wavelength of the emitted light and the numerical aperture of the objective employed. The resulting intensity pattern is theoretically described by an Airy function, which may be well approximated by a two-dimensional Gaussian function with a full width at half maximum (FWHM) of approximately 250 nm, when light with a wavelength of 500 nm and an objective lens with a numerical aperture (NA) of 1.4 is used for imaging. A two-dimensional fit of such a Gaussian can provide the centre position of the blurred light spot, which corresponds to the actual position of the fluorescent molecule. The achievable precision in this localisation process is only dependent on the signal-to-noise ratio (SNR) of the peak, and the mechanical stability of the microscope. Instead of an increased optical resolution as in the methods mentioned above, a precision of localisation for single molecules in the nanometre range is achieved. When the motion of single molecules is followed over time by video microscopy, single-molecule trajectories can be constructed that report on the dynamics of the observed entities. This approach is especially well suited for the analysis of fast motions or diffusional processes with a diffusion coefficient below 20 μm²/sec. Furthermore, by using multicolour labelling, binding events between single molecules and further intracellular structures can be followed over time with interacting times as low as 2 to 10 msec. Thus, the technique is ideally suited to study nucleocytoplasmic transport of single fluorescent cargo molecules. In that case, GFP-conjugated nuclear pore complexes can be used as markers for the position of the nuclear envelope, which can be localised with a very high precision *(3, 4)*. The technique is also very valuable for the study of the intranuclear mobility of specific intranuclear proteins, which may be imaged in relationship to defined intranuclear structures such as replication foci, nucleoli, or speckles *(5)*. In contrast to conventional techniques for the analysis of molecular mobility such as fluorescence correlation spectroscopy (FCS) or photobleaching techniques like fluorescence recovery after photo bleaching (FRAP), multiple forms of mobility in heterogeneous systems may be well discriminated by SMF microscopy. Prerequisites for intracellular single molecule detection are the reduction of background fluorescence, optimisation of light transmission in the microscope, utilisation of laser light sources for excitation, and high-speed CCD camera systems for signal detection in combination with digital image processing *(6)*.

2   Materials

2.1   *Single-Molecule Fluorescence Microscopy*

2.1.1   Microscope

The central component of the custom-built microscope set-up is a conventional wide-field epi-fluorescence microscope from Zeiss (Axiovert 200, Carl Zeiss, Jena, Germany; Fig. 19.1). Three lasers are available for illumination: a Sapphire 100 DPSS-laser with 488 nm (Coherent, Santa Clara, CA, USA), a frequency-doubled Nd-YAG laser, and a HeNe-Laser emitting at 532 and 635 nm, respectively. The laser beams are coupled into an optical mono-mode fibre (Pointsource, Hamble, UK) after passage through an acousto-optical tuneable filter (AOTF; AA Opto-Electronic, Orsay, France). The collimated output of the fibre illuminates a conjugated image plane of the microscope, and the magnification is chosen such that an object field with a diameter of ~10 μm is illuminated. In order to alternatively illuminate the complete field of view, a high-pressure HBO lamp can be used for illumination. The light emitted by the HBO lamp can be coupled into the excitation beam path by a motor-driven mirror, which allows for fast and comfortable switching

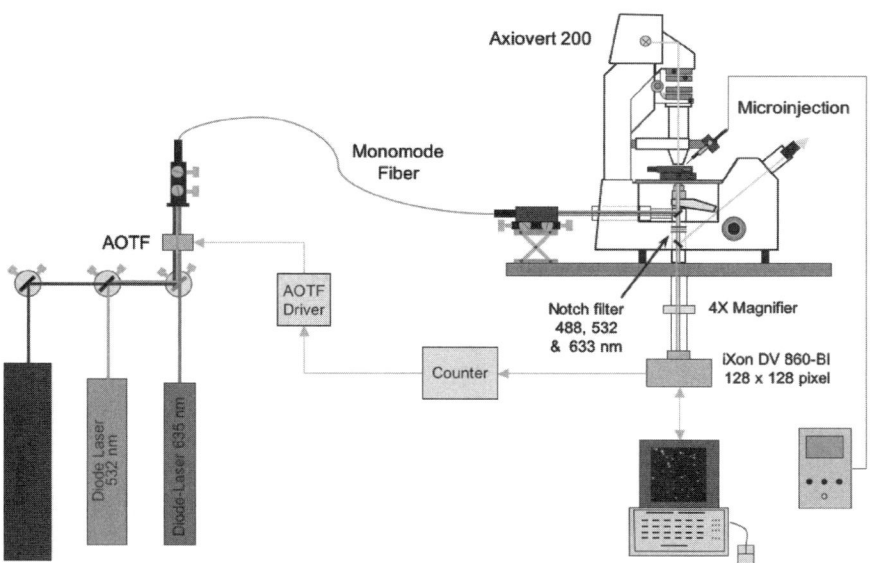

Fig. 19.1 Microscope setup for single-molecule fluorescence microscope. Details are given in the text

between the high intensity, narrow-field laser and wide-field HBO illumination. The emitted fluorescence light is separated from the excitation light by a triple dichromatic beam splitter that reflects the three excitation lines onto the sample, while the three fluorescence bands are transmitted to the detection system via three Notch filters extinguishing remaining excitation light (Semrock, Rochester, NY, USA). The microscope is mounted on a custom-made steel scaffold fixed to the optical table, which allows connecting of the video camera to the base port of the microscope.

2.1.2 Image Acquisition

An electron-multiplying CCD camera (EMCCD) is used (Andor iXon DV 860-BI; Andor Technology, Belfast, Northern Ireland) which features a 128×128 chip with a pixel size of 24×24 μm^2. The small chip size guarantees optimal imaging rates. In combination with a ×4 magnifier in front of the camera the pixel size in the object plane is 95.2 and 150 nm for a ×63 and a ×40 objective lens, respectively. Hence, the total observation field has a size of $(12.2\,\mu m)^2$ or $(19.2\,\mu m)^2$ depending on the objective lens employed. By means of a laboratory chiller attached to the internal Peltier cooling element, the chip is cooled to −90°C to minimise thermal noise.

Laser illumination can be regulated with a custom-built, programmable electronic counter, which is triggered by the camera and switches sequentially between the different lasers at will. It also allows to conveniently regulate the laser power via the AOTF driver. In this manner, the laser illumination and camera detection are synchronised and bleaching of the sample is minimised, because the laser illumination is only directed onto the sample if the camera is integrating.

Nanolocalisation in SMF requires the highest possible stability of the optical setup and this includes the mounted sample. For mounting the specimen onto the microscope stage, a special sample holder and mounting chamber were developed. This chamber is firmly attached to the microscope stage and comprises two micrometer screws for moving the sample laterally. The microscope stage itself is rendered unmovable and fixed for image acquisition to increase overall stability of the sample.

2.1.3 Software

1. Andor SOLIS for Imaging (Andor Technologies, Belfast, Northern Ireland).
2. DiaTrack for Windows 3.01 (SemaSopht, North Epping, Australia).
3. Plug-in for ImageJ (http://rsb.info.nih.gov/ij/) specifically developed to meet our demands (information available from the first author).

2.2   Fluorescence Labelling of Protein Molecules

1. Phosphate-buffered saline (PBS) without Mg^{2+} and Ca^{2+}: freshly prepared from a commercially available stock solution (Biochrom AG, Berlin, Germany). One vial is dissolved in 1 L of deionised water. The pH must be adjusted to 7.4 and the buffer stored at 4°C.
2. DTT stock solution: dissolve 7.71 g of dithiothreitol (DTT) in 50 mL of water. Prepare 1-mL aliquots and store at −20°C.
3. Maleimide dye stock solution (*see* **Note 1**): Dissolve 1 mg of the maleimide dye (Invitrogen, Karlsruhe, Germany) in anhydrous DMSO (Sigma-Aldrich, St. Louis, MO, USA) at a final concentration of 20 mM and prepare 10-μL aliquots.
4. Tris-(2-carboxyethyl)phosphine (TCEP) stock solution: dissolve TCEP (Invitrogen, Karlsruhe, Germany) in water at a final concentration of 10 mM and stored in 100 μL aliquots at −20°C.
5. Sephadex G-25 (Sigma-Aldrich) or BioRad-P6 desalting gel (Bio-Rad, Hercules, CA, USA) (*see* **Section 3.1.1.1**).
6. Chromatography column: glass, 150-mm long with a diameter of ~10 mm.
7. Equipment for separation of proteins by SDS-PAGE.
8. Amicon filter devices for concentrating (Millipore, Billerica, MA, USA) (*see* **Section 3.1.1.9**).
9. UV-VIS absorption spectrophotometer.

2.3   Digitonin-Permeabilisation of Cells

1. Transport buffer: this buffer was initially formulated by Adam et al. (7) and contains 110 mM K acetate, 5 mM Na acetate, 2 mM Mg acetate, 1 mM EGTA, and 2 mM DTT. The buffer is prepared as 10× stock solution and diluted with water prior to use. Adjust to pH 7.3 with 2 M KOH and add 2 mL of 1 M DTT stock solution to 1 L of buffer. The 1× transport buffer with DTT is stable for at least 1 week.
2. Digitonin solution: digitonin (Merck, Nottingham, UK) is dissolved in anhydrous DMSO at a final concentration of 20 mg/mL and stored in small aliquots at −20°C. For permeabilising cells, the stock solution is diluted 500-fold in ice-cold transport buffer resulting in a final concentration of 40 μg/mL.
3. Cell culture dishes with cover slips embedded in the bottom (Mattek, Ashland, MA, USA) (optional, *see* **Section 3.3**).

2.4   Microinjection of Fluorescent Probes
   (Optional: See Section 3.4)

1. Cell culture medium for microinjection: prepare a stock solution of 100 mM HEPES and adjust the pH to 7.4 with HCl, sterilise by filtration through a 0.22-μm

filter, and store at 4°C. To prepare the microinjection medium, add 2 mL of HEPES stock to 8 mL of Dulbecco's modified Eagle medium (DMEM) (Invitrogen) with 10% fetal calf serum (FCS) and warm it to 37°C.

2. Eppendorf InjectMan combined with a FemtoJet (Eppendorf, Hamburg, Germany).

3. FemtotipsII microinjection needles and a pipette equipped with a microloader tip (Eppendorf).

3   Methods

3.1   Fluorescence Labelling of Proteins for SMF Microscopy

There are some general considerations for selecting the optimal fluorescent dye (*see* **Note 2**). Covalent crosslinking of recombinant proteins or antibodies with fluorescent dyes is usually accomplished on either their lysine or cysteine residues. Succidinimidyl esters react with the amino-terminal NH_2-group and with the ε-amino groups of lysines within the polypeptide chain. Most proteins contain numerous lysines, and due to their hydrophilicity, these are often located at the protein surface. Labelling of lysine residues often results in high numbers of conjugated dye molecules, which produces bright probes. However, it is not always compatible with protein function, and proteins might also become sticky or even insoluble. Protein functionality may be influenced by the label, e.g. classic nuclear localisation signals contain essential lysines that are sensitive to crosslinking and loose import capability after lysine labelling. In such cases maleimide-conjugated dyes should be used. Maleimide-coupled dyes are reactive to the sulfhydryl group of cysteine residues, forming a stable ether bond. However, often labelling via cysteines is not straightforward. Most proteins contain only a few cysteines, which are mostly located within the protein. Furthermore, their sulfhydryl group is often masked by disulfide bridges. To render these sulfhydryl groups reactive the disulfide bonds must be reduced. Standard reducing reagents like β-mercaptoethanol or DTT can be used, but would quench the crosslinking reaction when not removed by dialysis prior to labelling. A milder reducing reagent is TCEP, which is often sufficient for reducing disulfides and does not interfere with the labelling reaction. Sometimes the protein of interest has no available cysteine or is sensitive to reduction of intra-protein disulfide bonds. If recombinant proteins are used, additional cysteines may be introduced genetically, e.g. placed between the tag for affinity purification and the protein coding sequence. If the protein of interest is to be labelled for the first time, then both labelling strategies should be tested. It is of utmost importance to check the functionality of the fluorescent protein carefully, e.g., by testing for an in vitro interaction with a known interaction partner in a pull-down assay. The dye-conjugated protein should perform identically compared with the unlabeled species.

3.1.1 Lysine Labelling with Succinimidyl Esters

1. Prepare the gel matrix for the chromatography column by hydrating the gel powder in the buffer used for separation. For green dyes, Sephadex G-25 can be used but it is not recommended for red dyes like Cy-5 or Alexa-633 (*see* **Note 3**). For separation of red dyes, BioRad-P6 desalting gel should be used instead (Bio-Rad, Hercules, CA, USA). The molecular cut off is 6 kDa, and it is well suited for separation of unbound dye molecules of ~800–1,200 Da. Let the gel swell for several hours or overnight at room temperature and pour it into a clean glass column. For proper resolution and separation, the gel bed height should be at least 150 mm with a diameter of ~10 mm.

2. The protein should be dissolved in an inorganic buffer like PBS or bicarbonate buffer, because organic buffers like Tris or HEPES contain free amino groups and will quench the reaction. The reactivity of succinimidyl esters is strongly pH-dependent, and therefore the pH of the reaction should be adjusted to 8.5 to 9.5 for optimal labelling of the lysine residues.

3. Dissolve the dye in either anhydrous DMSO or dimethylformamide (DMF; Sigma-Aldrich). Dissolved dye can not be stored for a long period of time without losing reactivity, so the dye solution must always be prepared directly before use. Labelling efficiency is depending on the molar dye to protein ratio. To avoid excess labelling of the protein, add only a twofold to threefold molar excess of dye to the protein.

4. Incubate the reaction vial for 2 h at room temperature or overnight on ice. Flip the vial occasionally but strictly avoid foaming. It is important to protect the vial from light at all times.

5. Centrifuge the solution at 20,000×g at 4°C for 20 min and keep the supernatant.

6. To load the column, remove the buffer from the gel surface with a Pasteur pipet and place the complete labelling solution on the column. Avoid disturbing the gel surface and let the sample enter the column by gravity flow. Slight suction might be applied to the exit of the column to move the sample into the gel matrix, but the column must not become dry.

7. Connect the column to a peristaltic pump and adjust the flow rate to 0.5–0.75 mL/min. If no pump is available, apply the buffer by gravity flow or manually using a syringe, but be careful not to compress the gel matrix by applying too high pressure.

8. Unbound dye and labelled protein should separate on the column into at least two distinct bands. Sometimes the dye band splits further into two bands representing dye aggregates. Due to the size exclusion limit of the gel, the protein will move faster through the column and elute first. Protein-containing fractions should be collected and checked by SDS-PAGE for complete removal of free dye. Unbound dye will run in the buffer front of the gel, and therefore the gel run should be stopped before the buffer front reaches the edge of the gel. The fluorescent dye can then be visualised by placing the gel directly after the run on an UV illuminator without fixation and staining. Red dyes are usually more difficult to detect due to their inefficient excitation by the UV light, but if a gel documentation system is used this can

be overcome by a longer integration time of the camera. If an unlabelled protein sample is added in an adjacent lane, the degree of labelling can be roughly estimated from the increase in the apparent molecular weight of the labelled protein compared with the unlabelled form after the gel has to been fixed and stained.

9. Protein fractions that do not contain unbound dye should be pooled and further concentrated with an Amicon filter device (Millipore, Billerica, MA, USA). Chose the molecular weight cut-off according to the protein size and transfer the solution into the filter device, which is then placed in a 12-mL or 50-mL tube and centrifuge for 10 min at 4,000×*g*. Mix the protein solution carefully before recovering it from the filter device, since a density gradient is usually formed during concentration and the solution will be more concentrated at the bottom of the device.

10. The concentrated protein should be aliquoted, snap frozen in liquid nitrogen, and stored at −80°C for long-term storage. Take one aliquot to determine the concentration and the degree of labelling of the protein (*see* **Section 3.2**).

3.1.2   Cysteine Labelling with Maleimide Dyes

1. Prepare the gel matrix for the column as described above.
2. Add TCEP stock solution to the protein solution to a final concentration of 2 m*M* and incubate for 45 to 60 min at room temperature to allow for reduction of the disulfides.
3. Dissolve the maleimide dye in anhydrous DMSO or thaw an aliquot of the stock solution (*see* **Section 2.2**). Add enough dye to the protein solution to get an equimolar to twofold molar excess of dye compared with free SH-groups. The volume of added DMSO should not exceed ~1/10 of the overall volume to avoid precipitation of the protein. Incubate the mixture 2 h at room temperature or overnight on ice. Protect the vial from light at all times.
4. Separate the protein from unbound dye and process the protein fractions as described above (*see* **Section 3.1.1.5**).

3.2.   Determining the Concentration and Labelling Degree of Fluorescent Proteins

3.2.1   Bradford Assay

The protein concentration can be determined by standard dye-based procedures like the Bradford or the bicinchoninic acid (BCA) method or directly by the protein absorption at 280 nm. We will describe the method according to Bradford; stabilised Bradford reagent is available from Sigma-Aldrich and can be used directly. For the standard curve, a dilution series of a commercial BSA stock solution should be prepared using at least four different BSA concentrations. Use different amounts of your protein solution e.g. 1, 2, and 5 µL to make sure that at least one measurement will be in the best concentration range of the test method.

1. Prepare standards by diluting the BSA stock solution in 50 µL of buffer to get the appropriate concentrations.
2. Mix the standards and the samples with the labelled protein with 1 mL of Bradford solution in a 2-mL vial and vortex the vials thoroughly.
3. Incubate for at least 5 min at room temperature and transfer the solution into cuvettes.
4. Start the spectrophotometer and set the wavelength to 595 nm. Let the excitation lamp warm up for 30 min.
5. Measure the absorbance of the blank standard and set it to zero, and measure the absorbance of the BSA standards and determine the standard curve.
6. Measure the absorbance of the samples and calculate the protein concentration from the standard curve.

3.2.2 Labelling Ratio

If the molar concentration of the fluorescence-labelled protein and the molar extinction coefficient of the fluorescence dye at its maximum absorption wavelength (ε_{dye}) are known, then the mean number of dye molecules per protein molecule can be determined. A labelling ratio above 2 is desirable in order to avoid problems due to dye blinking and to yield a high fluorescence signal. However, excessive labelling certainly abolishes any protein function and might also cause non-specific binding. For best results, the measurement of the labelling ratio of fluorescent proteins should be performed in an UV-VIS absorption spectrophotometer.

1. Place a cuvette with the buffer in the spectrophotometer and measure it as blank sample at the two wavelengths used.
2. Measure the absorbance of the fluorescent protein at the excitation maximum of the attached dye (A_{max}). To save material, a dilution of the protein should be used but make sure that the values obtained are in the measuring range of the spectrophotometer.
3. The labelling ratio of the protein can be roughly estimated according to the equation:

$$\text{Moles of dye per mole of protein} = \frac{A_{max} \text{ of the labeled protein}}{\varepsilon_{dye} * \text{protein concentration (M)}} \times Dilution factor$$

3.3 Using Digitonin Permeabilisation for Delivery of Molecules into Cells

Permeabilisation by digitonin is a well-established method for delivering macro-molecules into mammalian cells. It is especially well suited for analysis of nucleo-cytoplasmic transport, since the soluble part of the transport machinery that is lost after permeabilisation can easily be replaced by recombinant proteins or cell lysate (7). The main disadvantage of this approach is that it is not a live-cell experiment. The cell membrane is permeabilised by incubation in transport buffer containing 40 μg/mL of digitonin for 4–6 min depending on the cell type and the confluence of the cells. Digitonin is a natural glycoside obtained from the plant *Digitalis purpurea*, and can specifically precipitate the cholesterol in the plasma membrane of eukaryotic cells. Thus treating mammalian cells with digitonin will create holes in the plasma membrane, while the nuclear envelope is left unaffected (*see* **Note 4**). Large molecules from the surrounding medium can diffuse freely into the cytosol of permeabilised cells, but will be excluded from the nucleus as long as they are not actively imported. As a control, the nuclear uptake of an inert molecule like fluorescent BSA or dextrans with a molecular weight ≥70 kDa should be monitored. Since these molecules cannot pass the nuclear pore complex due to its size exclusion limit of 40–60 kDa, any increase in nuclear fluorescence indicates leakage of the nuclear envelope. To maintain the nuclear integrity in transport experiments, initial experiments for determining the optimal permeabilisation time should be performed. The solution containing the fluorescent probe molecules should be prepared prior to cell permeabilisation. A crucial aspect of preparing the solution containing the molecules that will be tracked, both for digitonin-permeabilised cells and for microinjection, is to choose the optimal dilution of the fluorescent molecules. If they are highly concentrated, the SNR will be diminished by out-of-focus fluorescence. Furthermore, the tracks of different single molecules might be confused in the analysis if the signal density is too high. More detailed considerations are given below (*see* **Section 3.5**).

1. If no custom-made sample holder for cover slips is available, cell culture dishes with cover slips embedded in their bottom are very convenient (Mattek, Ashland, MA, USA).
2. Prepare the cells 24–48 h before the experiment. Trypsinise the cells with 1 mL of trypsin/EGTA solution (0.25% v/v in PBS) for 2–5 min at 37°C, add 9 mL of DMEM with 10% FCS and transfer the cell suspension into a sterile tube. Centrifuge the cells at 200×g for 10 min at 4°C and aspirate the supernatant.
3. Resuspend the cells in fresh DMEM medium with 10% FCS. If HeLa POM121-GFP cells are used (*see* Fig. 19.2), 50 μg/mL of gentamycin (Biochrom, Berlin, Germany) should be added.
4. Seed the cells in fresh dishes containing cover slips. Since most cell lines grow only slowly on untreated glass surfaces compared to cell culture dishes, the cells should not be too diluted. HeLa cells can be split 1:2–1:3 and are ~70% confluent after 24–48 h. Transfer 10 mL of cell suspension to a 100-mm dish or 2.5 mL to a Mattek dish.
5. For permeabilisation of cells on coverslips, a 6-well cell culture plate is practical. Place the culture plate on ice and transfer a cover slip with cells to a well filled with 2 mL of cold transport buffer.
6. Wash the cells 3× with ice-cold transport buffer (*see* **Note 5**), add 2 mL of transport buffer with 40 μg/mL of digitonin, and incubate for about 5 min (the optimal time must be determined beforehand as stated above).

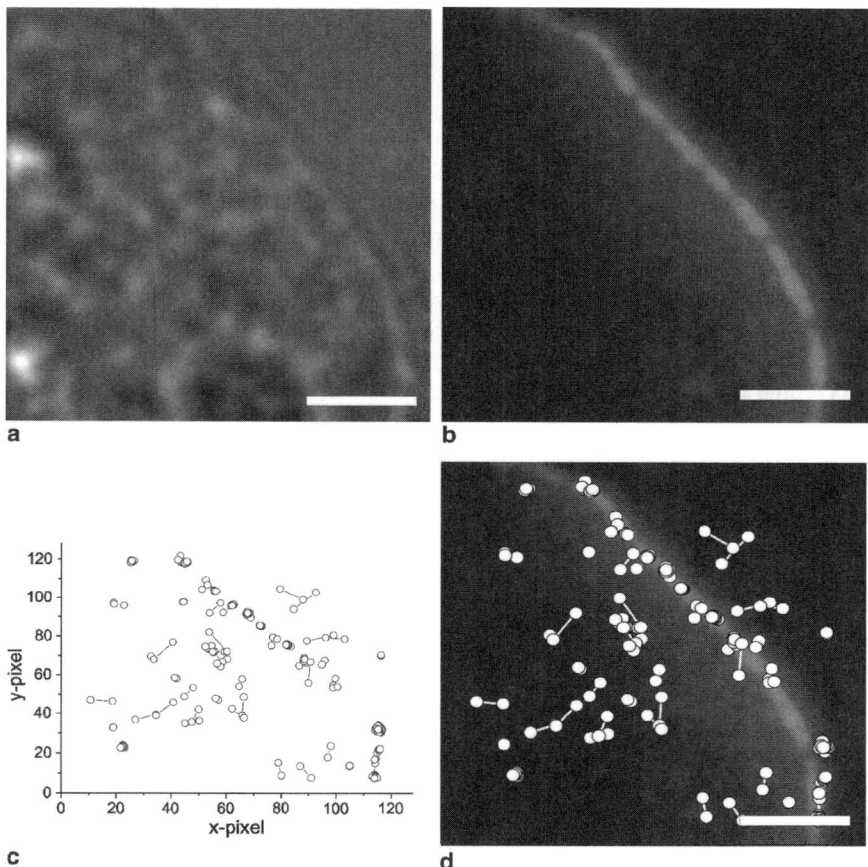

Fig. 19.2 Import of single molecules of human NTF2/p10 (a 14-kDa protein implicated in NLS-mediated nuclear import) in digitonin-permeabilised HeLa cells stably expressing POM121-GFP, a GFP conjugate of the nucleoporin POM121 (*see* **ref.** *(3)*). **a** Bright-field image of the nucleus. Cells were seeded in a Mattek-culture dish 24 h before the experiment. Permeabilisation was achieved with 40 μg/mL of digitonin in Transport Buffer. After the last wash step the cells were covered with import mix containing NTF2/p10-Alexa488 and placed on the microscope (×63 objective, NA 1.4, size bar, 2.5 μm). **b** POM121-GFP-staining of the nuclear envelope: The HeLa cells stably expressed POM121-GFP, which was used for focussing on the equatorial plane of the nucleus. The GFP signal was bleached by repeated image series at low laser power until the fluorescence of single import molecules could be perceived. **c** After bleaching of the POM121-GFP, the cytosolic and nuclear movements of single NTF2/p10-Alexa488 molecules were recorded and their positions were plotted. Their locations are marked by *open circles*. Trajectories of single NTF2/p10 molecules are indicated by *full lines*. **d** Overlay of the POM121-GFP staining with NTF2/p10 trajectories. Obviously, most observations were made directly at the nuclear envelope as indicated by the numerous circles along the line of GFP staining. To view this figure in color, see COLOR PLATE 12

7. Remove the digitonin solution quickly and wash the cells 3× with ice-cold transport buffer.

8. The cells are now ready to be loaded with the import mix. The latter contains the labelled protein probe at the appropriate concentration and, in the case of substrates that are translocated across the nuclear pore complex by signal-mediated transport, also an energy regenerating system and either reticulocyte lysate as source of transport factors, or purified soluble transport receptors (3). If glass-bottom dishes are used, they can be placed directly on ice and otherwise be treated accordingly.

9. Transfer the cover slip with the cells or the Mattek chamber, respectively, to the sample holder and fix it in place. Cover the cells with 200 µL of import mix and start a timer. Carefully dry the bottom side of the coverslip from residual medium and place the sample holder on the microscope.

10. Employ a marker grid in the eyepiece of the microscope to move a cell into the field of view of the camera, and adjust the focal plane. For the study of nucleocytoplasmic transport, this corresponds to the equator of the nucleus (*see* **Note 6**).

11. If the fluorescence emission spectrum of the labelled probe molecule overlaps the fluorescence spectrum of the marker employed, the latter has to be bleached before single molecule observation because it dominates the weak single molecule signals. Bleaching may be performed by illuminating with high laser power, but this might severely harm cell functionality and viability. To avoid this, bleaching can be achieved by acquisition of repeated image series at low laser power until the marker fluorescence is bleached and the fluorescence of the single probe molecules can be perceived. Prerequisites for this approach are that the marker molecules are immobile, and that the probe molecules can move into the illuminated observation field from an unbleached region of the cell.

12. Now movies of transport events or intranuclear molecular motion can be acquired. Important considerations on data acquisition are given below in Section 3.5. Most importantly, the laser power and image acquisition frame rate should be set such that optimal SNRs of the fluorescent probe molecules are obtained. A SNR ≥ 3 is recommendable.

3.4 Using Microinjection for Delivery of Molecules into Living Cells

Microinjection is a powerful technique for direct delivery of molecules into the cellular interior. The most important advantage compared with the digitonin method is that single live cells can be analysed, and that only the proteins of interest have to be brought into the cells with no need for exogenous lysate or soluble transport factors. Furthermore, proteins may be directly injected into the cell nucleus with no need for nuclear import, but impairment of nuclear integrity or structure should be considered. We use an Eppendorf InjectMan combined with a FemtoJet for microinjection. If no CO_2 incubation chamber is used, the cells should be kept in properly buffered medium to avoid an increase in pH. Depending on the cell type used, the medium can be supplemented with $20\,mM$ HEPES (pH 7.4) to stabilise the pH, or a commercially available buffered medium can be used.

1. All solutions used for microinjection should be centrifuged prior to use to remove any aggregates or precipitated protein, which would otherwise clog the injection capillary. Prepare the injection mix and centrifuge it for $30\,min$ at $\geq 40,000 \times g$ and $4°C$ or $10,000 \times g$ if larger nanoparticles such as quantum dots are used. Recover the supernatant, transfer it to a fresh vial, and store it on ice until needed.

2. Prepare the cells for microinjection. Place the cells on the microscope and cover them with $200\,\mu L$ of microinjection medium pre-warmed to $37°C$. Carefully dry the bottom side of the cover slip from residual medium and transfer the sample holder to the microscope stage.

3. Use the marker grid to move a cell into the field of view and adjust the focal plane as described in **Section 3.3.6**.

4. Start the microinjection pressure supply and micromanipulator control. We use FemtotipsII injection needles and a pipette equipped with a microloader tip to fill the microinjection needle with 2–$3\,\mu L$ of injection solution. It is important to avoid any air bubbles in the capillary. After loading, remove the protection tip from the needle by carefully screwing it from the plastic holder (*see* **Note 7**).

5. Mount the microinjection needle in the holder, connect the pressure tube to the pressure supply, and quickly dip the needle into the medium covering the cells to avoid it drying up.

6. Localising the needle with the microscope requires some practice because only a limited depth of focus is available. Start with a smaller magnification using a ×10 to ×20 objective lens to localise the needle and move the micro-capillary carefully when it approaches the cover slip, switching the capillary mover from coarse to fine movement mode. It is important to avoid any direct contact between the capillary and the cover slip surface. The holding pressure should be set such that some fluid is pumped out of the needle constantly to avoid medium entering the needle (~10–$20\,hPa$).

7. The Eppendorf micromanipulator offers two injection modes for automatic injection. Either the needle is simply moved down along the vertical direction into the cell, or with a combined vertical–horizontal movement when the needle punctures the membrane more perpendicularly. For both injection modes, the bottom limit has to be carefully adjusted. Chose the injection pressure and time such that the interior of the cell is not visibly disturbed by the injection process, e.g. avoid settings that will pump up the cell since harsh injection conditions harm cells and might severely modify the intracellular mobility of the injected molecules. Adherent cells unfortunately have a very flat shape, and the only region where the needle can penetrate the plasma membrane without touching the glass surface below the cell is the region very close to the nucleus.

8. For cytosolic injection, the microinjection needle should be aligned tangentially to the nuclear envelope to inject directly the cytosol adjacent to the nucleus.

9. For nuclear injection, a slightly different strategy should be employed. The holding pressure should be increased such that the microinjection solution is continuously pumped out of the needle (~20–40 hPa). The nucleus should then be penetrated carefully by moving the needle manually with the micromanipulator. Take care not to pump up the nucleus by the injection, since this will most probably destroy it.

10. After injection, move the capillary out of the field of view and allow the cells to recover from injection for some minutes. It is advisable to move the needle always to the same parking position because it can be more easily relocated for subsequent injections.

11. Switch to fluorescence detection and start data acquisition. For choosing the appropriate camera settings, considerations similar to those for digitonin-permeabilised cells are valid.

3.5 Data Acquisition

Several parameters have a strong impact on the successful imaging and tracking of single molecules. SMF microscopy requires the maximum stability of the optical setup, including the mounted sample. The complete microscopic setup has to be aligned carefully to achieve the highest available sensitivity and maximum SNR. The analyte solution itself also contributes to the data quality. As mentioned above, the fluorescent dye employed should be as stable and as bright as possible, and the protein of interest must be highly labelled without impairing its biological function. Also, the concentration of the fluorescent probe in the sample is of great importance. To estimate the useful limiting concentration we proceed as follows. Our single molecule microscope has a lateral field of view of $12.2 \times 12.2 \, \mu m^2$. Given a depth of field of $1 \, \mu m$, the total focal observation volume is $V_{obs} \approx 150 \, \mu m^3$. Usually a number of 10 molecules in this volume is the upper limit that can be tolerated; otherwise the moving molecules cannot be distinguished from each other any more. This corresponds to a concentration $C_{max} \approx 110 \, pM$. Therefore, we consider a concentration in the range of $100 \, pM$ as an optimal concentration. One problem is distinguishing the probe molecules from each other, and a second is of course that higher probe concentrations lead to higher out-of-focus fluorescence, severely reducing the SNR (*see* **Note 8**).

The total number of distinguishable objects in the field depends on the optical resolution and the total field of view. A more general approach is therefore to consider the number of focal volume elements V_f within the field of view. V_f is also designated as resolution element ("resel"), and is given by the area of the Airy disk multiplied by the depth of field. The latter corresponds to the twofold axial resolution. For the high numerical aperture objective lenses that are mandatory in single molecule microscopy, $V_f \approx \pi \, 0.25^2 \times 1 \, \mu m^3 \approx 0.2 \, \mu m^3$. A rule of thumb is that there should be at maximum one object per 100 resels corresponding to a concentration of $1/(20\,\mu m^3)$, which is about 80 pM. On the other hand, if the probe is too diluted then too few events will be detected per measurement, which in turn significantly increases the amount of data that has to be analysed.

Beside these aspects, the optimal settings for data acquisition are critical for single molecule detection. Here we will describe the different settings for acquiring data using the software *Andor SOLIS for Imaging* and the EMCCD camera iXon DV-860-BI (Andor Technologies).

1. Directly after starting the software, the temperature control under *Hardware* should be switched on, and the CCD chip of the camera should be cooled to −90°C. It is important that the camera is connected to the water cooler and that water cooling is running. The cooling fan should be turned off because it is not needed when using water cooling, and it produces only unwanted vibrations.

2. The image acquisition setup can be found under the menu item *acquisition* or started directly by pressing the screw-wrench symbol. In the first tabulator *Setup CCD* the basic parameter settings are defined. For taking image series the *Acquisition Mode* has to be set to *Kinetic*, *Triggering* is set to *Internal*, and *Read Out mode* should be *Image*. Under the menu item *Timings*, the image series can now be specified. First the *Frame Transfer* menu box should be checked, then *Exposure Time*, *Kinetic Series Length*, and *Number of prescans* can be inscribed. The number of prescans can be left at zero.

3. The length of the image series (*Kinetic Series Length*) depends on the frequency of analysable events, but the maximum series length should not exceed ~2,000 frames since at 14-bit colour depth (16-bit stored) and with full resolution of 128×128 pixels, such a stack is already ~128 MB in size. The handling, and in particular the evaluation, of larger series becomes increasingly inconvenient but obviously depends on the available computing power.

4. The adequate exposure time is directly dependent on the mobility and brightness of the probe molecules. Together with the read out settings for the CCD chip, the *Exposure Time* will define the overall time resolution which is reported as the *Kinetic Cycle Time*. The chosen time resolution has to be high enough to follow the trajectories of single protein molecules and depends on the velocity and the diffusion coefficient of the probe molecules. Using probes labelled by single dye molecules, we find a 5 msec exposure time to be the lower limit to obtain images of sufficient SNR at irradiances in the range of 1 to 2 kW/cm². Higher irradiances cause excessive bleaching of most dyes.

5. The overall achievable image read out rate depends on the *Vertical Pixel Shift* and *Horizontal Pixel Shift*, which are connected with the chip architecture. High-speed cameras contain a frame transfer chip, in which the light-sensitive portion of the array is used for signal acquisition from where the accumulated charge is rapidly shifted into a light-protected storage region (*Vertical Pixel Shift*) for transfer to the serial output register (*Horizontal Pixel Shift*). Reduction of the pixel shift time will improve time resolution, but also increase the noise level. As a trade-off, the *Shift Speed* can be set to 0.9 μsec and the *Vertical Clock*

Voltage Amplitude to normal. The *Readout Rate* in the *Horizontal Pixel Shift* menu should be chosen as low as possible and a *5.3× Pre-Amplifier Gain* is usually sufficient.

6. One method to improve time resolution while even improving data quality is the so-called pixel binning. Thereby the signals from several adjacent pixels on the chip are combined before readout, which can significantly increases the achievable frame rate because less data have to be digitised. However, the effective pixel size in the object field is scaled up accordingly, which may render nanolocalisation impossible because the required details of the point spread function are missing. Another means to increase the imaging rate is to read out only a certain portion of the chip as a *Sub Image*. This improves the frame rate at the cost of the size of the field of view, but preserves the overall imaging resolution.

7. Usually it is also helpful to adjust the *Image Orientation* so that the displayed digital image corresponds to the eyepiece view. The exact settings here depend on the orientation of the camera with regard to the optical axis of the microscope. For our instrument, the image must be rotated by 90° anti-clockwise and flipped horizontally in order to obtain the original orientation as seen visually in the eyepiece.

3.6   Data Evaluation

The Andor software itself offers several features for data evaluation, but for an in-depth analysis we usually take advantage of additional software. For image analysis and single molecule tracking we either use *DiaTrack for Windows 3.01* or a plug-in for ImageJ (http://rsb.info.nih.gov/ij/) which was specifically developed to meet our demands (*see* **Section 2.1.3**). While the data acquired with the Andor software should be stored initially in the Source Input Format (SIF) to preserve essential information on acquisition parameters, for evaluation they should be transformed to the Tagged Image File format (TIF) with ImageJ. To this end, the image stacks are imported into ImageJ using a SIF import plug-in. If DiaTrack is used later on, the colour depth must be reduced to 8 or 16 bit and the stack has to be saved as multiplane TIF file. Contrast and brightness can be adjusted if necessary, but non-linear image modifications like gamma correction should be avoided since it might introduce artefacts. To speed up the evaluation process the series can be truncated and frames without analysable events discarded. For the detailed analysis of the movement of single molecules or particles within a cell, a cellular compartment, or in solution, one has to follow their pathways over time. For this purpose the position of the molecule has to be determined in single, subsequent frames of an image series. Based on the position at a given time, complete trajectories can be determined. By such measurements intracellular transport pathways, mobility restrictions, and intranuclear binding processes can be documented, and different forms of mobility in heterogeneous systems may well be identified. Given below is the detailed tracking procedure with DiaTrack 3.01.

1. For opening the TIF stack use the *File/Load Data* command and by default DiaTrack will open the *Windows/Temp* directory. The program asks whether the TIF stack contains 4D data, which should be denied. A principle decision to be taken is whether a two-dimensional Gaussian function will be used for fitting the particle position, or the centre of gravity approach which is more robust but not as exact. For using the Gaussian in the *Options/Particle Production* menu, the *High Precision* option must be checked, otherwise the program will use the centre of gravity method for finding particle positions. With *Options/Particle Production/High Speed* marked, the software will not store any temporarily files and the program speed is significantly higher. This can be used to speed up routine evaluations, but in the beginning the temporarily stored images files might be helpful in manually controlling the tracking results. The basic parameters for evaluation are given on the left side of the screen. For background subtraction the corresponding menu box must be marked, and then a value for filtering can be entered or set with the *Filter Data* control. Start with a mild filtering option (~1.1) and try not to exceed ~1.7.

2. Next the *Find Particles* button has to be pushed and then a region of interest can be defined with the *Select Region* tool. With the left mouse button you can define nodes in the outline of the region, the right mouse button will end the selection. We routinely define a region of interest to avoid molecules or particles that will appear close to the image borders. DiaTrack offers four ways of predefining particles that should be considered in the analysis, which are denoted as *Trash dim*, *Trash bright*, *Trash blurred*, and *Trash unrounded*. The best way is using the *Trash dim* option, which will sort out particles that fall under a certain intensity threshold. Start with a *Trash dim* value of ~130 and increase this value until no more bogus particles are detected.

3. With the *Process next frames* button pushed, the software will now search for particles frame by frame based on the parameters given. This may take some time dependent on number of events present and the overall image stack size. When all frames are processed, the final tracking can be started with the *Track!* button and the software asks for the maximum jump distance in pixels that will be considered for connecting the trajectories. Therefore a theoretical estimation of the expected maximum jump distance has to be made first. We use the following estimate for the maximum jump distance r for a given molecule with a known diffusion coefficient D at a given time interval Δt:

$$r\max = 3 \cdot RMS = 3 \cdot \sqrt{4D\Delta t}. \tag{19.1}$$

If the maximum jump distance is set to 3RMS (root mean square displacement), then 99.7% of all theoretically expected jumps will be tracked. The tracking process is usually much faster than the preceding image processing and the trajectories can then be saved with *File/Export/Trajectories*.

4. Use Option b in the dialog window to export the trajectories column by column. The resulting TXT file can then be used to transfer the trajectories to any software for further analysis.

4 Notes

1. Maleimide dyes are usually not very stable in aqueous solutions, and therefore aqueous stock solutions are not recommended for long-term storage. In our hands, stock solutions prepared in anhydrous DMSO are clearly more stable, and remain reactive for several months when stored at −20°C. Succidinimidyl ester solutions should be prepared freshly and not stored as stock solutions, since the dye will rapidly lose reactivity.

2. It is essential to use fluorescence labels of highest photostability. Dyes emitting in the far red spectrum of visible light tend to be more photostable than green fluorescent molecules. Another advantage of using red fluorescent dyes is the almost complete absence of cellular autofluorescence in this part of the spectrum. Usually mammalian cells exhibit a strong cytoplasmic autofluorescence when illuminated with laser light at 488 nm, which can severely diminish the achievable SNR. Thus a strong autofluorescence can render tracking of green-emitting molecules impossible, but using fluorescence in the far-red spectrum can circumvent this problem. In addition, the autofluorescence within the cell nucleus is usually significantly lower than in the cytoplasm.

3. Red fluorescent dyes like Cy-5 or Alexa-633 interact strongly with the Sephadex matrix, and bands will be smeared during separation. A complete separation is impossible.

4. Permeabilisation efficiency is often influenced by cell density, and confluent cultures may need a longer permeabilisation time.

5. This can most easily be accomplished by adding the buffer with a pipette to the well, and removing it with a suction pump.

6. In case molecular events at other fluorescently labelled intranuclear sites are examined, the focus should be adjusted to these accordingly. If no fluorescent reference structure is employed the bright field image of the sample has to be used for focusing. Keep in mind that the visual impression of the optimal focal plane in the bright field mode might differ from the optimal plane in the fluorescent image.

7. Try not to remove the protection tip by hand, instead just let it fall off after untightening it. Otherwise it will touch the needle and most probably break it.

8. It is advisable to begin single molecule observation by visual inspection. This, however, works only with green-labelled probes, because the human eye is not sensitive enough in the red part of the spectrum. In case you cannot make out single molecules by eye or by the camera, it is most probable that the concentration is too high. It must not exceed 5 nM, because then there is a molecule in almost every focal volume and they cannot be distinguished any more.

References

1. Klar, T. A., Jakobs, S., Dyba, M., Egner, A., and Hell, S. W. (2000). Fluorescence microscopy with diffraction resolution barrier broken by stimulated emission. *Proc. Natl. Acad. Sci. USA* **97**, 8206–8210.
2. Gustafsson, M. G. (2005). Nonlinear structured-illumination microscopy: wide-field fluorescence imaging with theoretically unlimited resolution. *Proc. Natl. Acad. Sci. USA* **102**, 13081–13086.
3. Kubitscheck, U., Grunwald, D., Hoekstra, A., Rohleder, D., Kues, T., Siebrasse, J. P., and Peters, R. (2005). Nuclear transport of single molecules: dwell times at the nuclear pore complex. *J. Cell Biol.* **168**, 233–243.
4. Yang, W. and Musser, S. M. (2006). Visualizing single molecules interacting with nuclear pore complexes by narrow-field epifluorescence microscopy. *Methods* **39**, 316–328.
5. Grunwald, D., Spottke, B., Buschmann, V., and Kubitscheck, U. (2006). Intranuclear binding kinetics and mobility of single U1 snRNP particles in living cells. *Mol. Biol. Cell* **17**, 5017–5027.
6. Siebrasse, J. P., Grunwald, D., and Kubitscheck, U. (2007). Single-molecule tracking in eukaryotic cell nuclei. *Anal. Bioanal. Chem.* **387**, 41–44.
7. Adam, S. A., Marr, R. S., and Gerace, L. (1990). Nuclear protein import in permeabilized mammalian cells requires soluble cytoplasmic factors. *J. Cell Biol.* **111**, 807–816.

Chapter 20
Fluorescence Recovery After Photobleaching (FRAP) to Study Nuclear Protein Dynamics in Living Cells

Martin E. van Royen, Pascal Farla, Karin A. Mattern, Bart Geverts, Jan Trapman, and Adriaan B. Houtsmuller

Keywords Fluorescence recovery after photobleaching; FRAP; Protein mobility; Confocal microscopy; Androgen receptor; Fluorescent proteins

Abstract Proteins involved in chromatin-interacting processes, like gene transcription, DNA replication, and DNA repair, bind directly or indirectly to DNA, leading to their immobilisation. However, to reach their target sites in the DNA the proteins have to somehow move through the nucleus. Fluorescence recovery after photobleaching (FRAP) has been shown to be a strong approach to study exactly these properties, i.e. mobility and (transient) immobilisation of the proteins under investigation. Here, we provide and discuss detailed protocols for some of the FRAP procedures that we have used to study protein behaviour in living cell nuclei. In addition, we provide examples of their application in the investigation of the androgen receptor (AR), a hormone-inducible transcription factor, and of two DNA-maintenance factors, the telomere binding proteins TRF1 and TRF2. We also provide protocols for qualitative FRAP analysis and a general scheme for computer modelling of the presented FRAP procedures that can be used to quantitatively analyse experimental FRAP curves.

1 Introduction

1.1 Green Fluorescent Protein

In the past decade, genetic labeling with fluorescent proteins has caused a revolution in molecular cell biological research. Ever since they became available, green fluorescent protein (GFP) and its color variants have been used at a tremendous scale to study the dynamic behavior of proteins in their most natural environment, the living cell. GFP was derived from the jellyfish *Aequorea Victoria* (*1*). By mutagenesis of wild-type GFP, enhanced versions, such as enhanced GFP (EGFP) with improved brightness and expression properties have been developed (reviewed in ref. *(2)*). In addition, currently a large array of color variants has been generated (reviewed in refs. *(3, 4)*). GFP and its color variants provide minimally invasive tools, not only to determine the dynamic intracellular localization using e.g. confocal time-lapse imaging, but also to study the dynamic behavior of proteins in living cells (reviewed in ref. *(5)*).

1.2   Fluorescence Recovery After Photobleaching (FRAP)

Proteins involved in DNA-interacting processes, like DNA repair, DNA replication, and transcription, bind directly or indirectly to DNA to exert their function. To reach their target sites in the DNA, either DNA damage, replication origins, or transcription sites, nuclear proteins have to move through the nucleus. Fluorescence recovery after photobleaching (FRAP) has proven to be a strong approach to qualitatively or quantitatively study exactly these properties, i.e. the mobility and (transient) immobilization of molecules in living cells (*(6–12)*, and reviewed in ref. *(13)*). FRAP was developed in the 1970s by Axelrod and coworkers. Early FRAP investigations were focused on the mobility of fluorescently labelled constituents of the cell membrane *(14)*. The development in the 1980s and 1990s of confocal microscopy and GFP technology enormously enhanced the applicability of FRAP. Currently, by far the most FRAP studies use confocal microscopes, although wide-field systems are also becoming increasingly available (e.g. ref. *(15)*).

In a typical FRAP experiment, a small defined region within a larger volume (for instance the cell nucleus) is shortly illuminated at high laser intensity (Fig. 20.1a) *(13)*. Immediately after the bleach-pulse the majority of the GFP-tagged proteins within the region have irreversibly lost their fluorescent properties, a process referred to as photobleaching. In a situation where all GFP-tagged proteins are mobile, proteins from outside will diffuse into the bleached region resulting in an increase of the fluorescent signal in the region until the signal inside the bleached region is equal to the signal outside the bleached region. In contrast, if permanently immobile proteins are present, these will not diffuse into the strip, resulting in an incomplete recovery of the fluorescent signal inside the bleached region relative to the remainder of the nucleus (Fig. 20.1b) (*see* **Note 1**). Transient immobilization, as was observed for many active nuclear proteins, including the AR, results in a delayed, secondary fluorescence recovery in the bleached region because a fraction of immobilized proteins will release and become mobile during the FRAP experiment, and then contribute to fluorescence recovery but later than the mobile fraction (Fig. 20.1) *(13)*. Summarizing, FRAP experiments yield information on essentially three mobility parameters: diffusion coefficient, immobile fraction, and the time spent in the immobile state. Assuming elementary binding kinetics, the size of the immobile fraction and the duration of immobilisation are determined by the on- and off-rates of the investigated protein to and from immobile complexes (*see* **Note 2**).

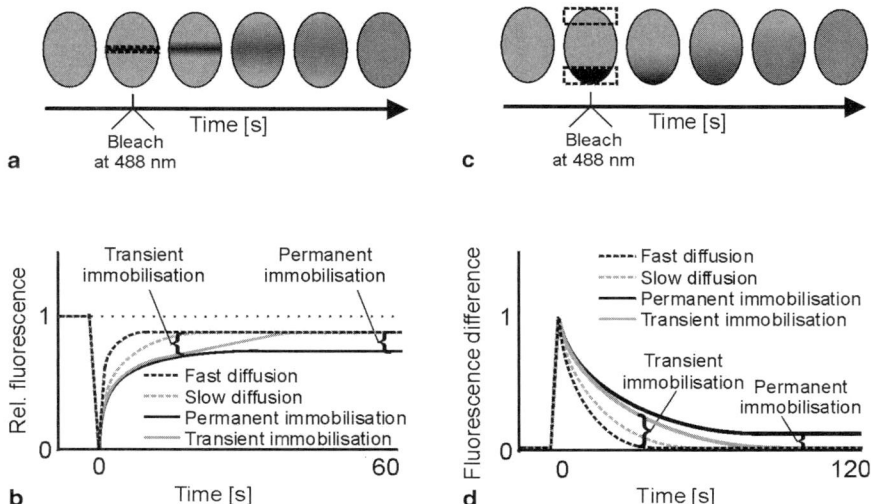

Fig. 20.1 Schematic representation of the principles of strip-FRAP and combined FLIP-FRAP. **a** In strip-FRAP, the recovery of GFP fluorescence is recorded in time after shortly bleaching a small strip spanning the nucleus. **b** Strip-FRAP curves representing different scenarios are expressed relative to pre-bleach values and the intensity directly after bleaching. Permanent immobilization of GF-tagged proteins (*solid curve*) can be identified by an incomplete recovery compared to FRAP curves of molecules that are freely mobile (*dotted curve* with fast diffusion and *dotted grey curve* with slow diffusion). Transient immobilization leads to a secondary recovery of fluorescence (*solid grey curve*). **c** In FLIP-FRAP experiments, the fluorescence in the bleached region and in the region at the opposite nuclear pole are recorded in time after photobleaching until a new steady state is reached. **d** The differences in fluorescence between the two opposite poles identify the different scenarios. Similar to strip-FRAP, a permanent immobilisation results is an incomplete redistribution and thus a constant difference between both signals in the two measured regions

Several variants of FRAP have been developed, including spot-FRAP, strip-FRAP, fluorescence loss in photobleaching (FLIP) *(16)*, combined FLIP-FRAP *(6, 7, 17, 18)* and inverted FRAP (iFRAP) *(19)*. Spot-FRAP is based on photobleaching of a small spot whereas in strip-FRAP, a larger region, for instance a narrow strip spanning the nucleus, is bleached. The latter method is favourable when signals are very low, e.g. due to low expression of the GFP-tagged protein. In FLIP, a very common variant of FRAP, the loss of fluorescence in a region or structure distant from the bleached region is monitored. FRAP and FLIP can also be combined (FLIP-FRAP): two regions at two poles of an ellipsoid nucleus are monitored simultaneously after bleaching only one of them. FLIP and combined FLIP/FRAP are specifically useful to determine the residence time of proteins inside subnuclear structures, such as telomeres, repair foci, or speckles. In iFRAP the entire nucleus is bleached with exception of a structure of interest. Immediately after bleaching, loss of fluorescence in the structure fully represents the off-rate of the associated protein (*see* **Note 3**).

1.3 Application of FRAP to Proteins in the Living Cell Nucleus

Application of FRAP to investigate the dynamic behaviour of nuclear proteins has provided new insights in nuclear protein function. The first FRAP studies revealed an unexpected high mobility of many nuclear factors, including components of the nucleotide excision repair (NER) machinery, which removes certain types of single-strand DNA damage *(20–24)*, transcription factors *(9)*, and RNA-splicing factors *(25)*. It was shown that in the absence of DNA damage the NER-factors ERCC1/XPF *(20)*, XPA *(21)*, PCNA *(22, 23)*, and recently XPG *(24)* were highly mobile in living cells and bound transiently to DNA damage. Similarly, high mobility and transient immobilisation due to DNA-binding were found for steroid receptors *(6, 9)* and for many more nuclear proteins with roles in a diversity of other processes *(26)*, including double-strand break repair *(27–30)*, DNA replication *(22, 23, 31)*, chromatin structure *(18, 32–34)*, and RNA processing and transcription *(19, 25, 34–36)*.

1.4 Examples of FRAP Applications: Androgen Receptors and Telomere-Binding Proteins

In this chapter, we provide detailed procedures for two types of FRAP, strip-FRAP and combined FLIP-FRAP, to study the dynamic behavior of the androgen receptor (AR) and give an example of a investigation of the telomere-binding proteins TRF1 and TRF2. We also provide methods to qualitatively analyze FRAP curves (Fig. 20.2) as well as an elementary modelling algorithm to generate FRAP curves with varying mobility parameters to fit and quantify the experimental data (Fig. 20.3).

The AR is a hormone-induced transcription factor and a member of the nuclear receptor (NR) superfamily. The AR is involved in the development and maintenance of the male phenotype and also plays a crucial role in the development and progression of prostate cancer *(37, 38)*. Like all NRs the AR consists of three domains: a conserved DNA-binding domain (DBD), a C-terminal ligand-binding domain (LBD), and a more variable N-terminal domain (NTD) *(39)*. Ligand-activated ARs translocate to the nucleus where they exert their activity by binding to specific androgen response elements (AREs) in promoter and enhancer sequences of AR-regulated genes *(40, 41)*. Several FRAP studies show a high mobility and transient immobilization not only of the AR but also of other NRs (Fig. 20.4) *(6–9, 11, 42, 43)*.

This immobilization is lost in a non-DNA-binding AR mutant (AR A573D) *(6, 44)* and in wild-type AR in the presence of antagonists *(7)*. In FRAP experiments direct comparison of active versus inactive states, like DNA repair proteins in the presence or absence of DNA damage or non-DNA binding transcription factor mutants in transcription, greatly simplifies the interpretation of the generated data. Computer simulation-aided analysis of combined experimental strip-FRAP and FLIP-FRAP data (Fig. 20.4a–d) showed that the wild-type AR kinetics could not be described by a model of freely diffusing molecules only (Fig. 20.4e, f). A model

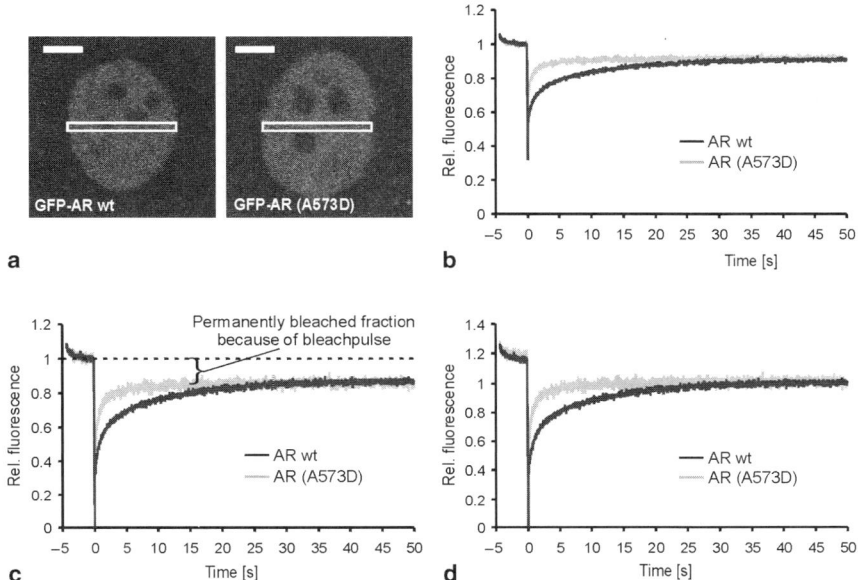

Fig. 20.2 Strip-FRAP applied to wild-type and mutant ARs. **a** and **b** Confocal images of Hep3B cells stably expressing GFP-AR (wild type) or the DNA binding-deficient mutant, GFP-AR (A573D) (**a**) and their strip-FRAP curves normalised by the three different normalisation procedures given in **Section 3.4.3**. **b–d** Bar, 5 μm. The recovery of fluorescence in a small strip spanning the nucleus (*white box* in **a**) is recorded in time after shortly photobleaching the fluorescence. **b** The most straightforward normalisation procedure is to normalise the data relative to pre-bleach fluorescence intensities. Comparing normalised data of active versus inactive proteins like wild-type AR and AR (A573D) used here enables identification of transient and permanent immobilisation. The wild-type AR shows a slower total recovery of fluorescence compared with the non-DNA-binding mutant, AR (A573D), due to transient immobilisation of wild-type AR. The difference in intensity of the DNA binding-deficient AR (A573D) before and after complete recovery does not reflect a permanent immobilisation, but is caused by the permanent bleaching of a fraction of protein by the bleach pulse. Curves represent data of at least 10 cells. *Rel*, relative. **c** Applying a second normalisation procedure, correcting possible variations in bleach-depths, data is expressed relative to pre-bleach intensities and the intensity directly after bleaching. The difference between pre-bleach intensity and the intensity after complete redistribution of AR (A573D) reflects the fraction permanently bleached due to the bleach pulse. **d** A third way of normalization yields a curve running from 0 directly after bleaching, to 1 after complete recovery, allowing also quantitative analysis by fitting the data to any equation that represents the diffusion process (and transient immobilization) *(13)*. After this normalization, it is no longer possible to extract information on permanent immobilization (**Section 3.4.3**)

of freely diffusing molecules together with a transiently immobilized fraction fitted to both strip-FRAP and FLIP-FRAP curves (Fig. 20.4g, h) *(7)*.

A second example of FRAP application concerns the investigation of telomere-binding proteins. Telomeres are nucleoprotein structures at chromosome ends. Telomere-binding proteins play a key role in the regulation of the length of the telomeric DNA tract. In addition these proteins prevent end-to-end fusion of

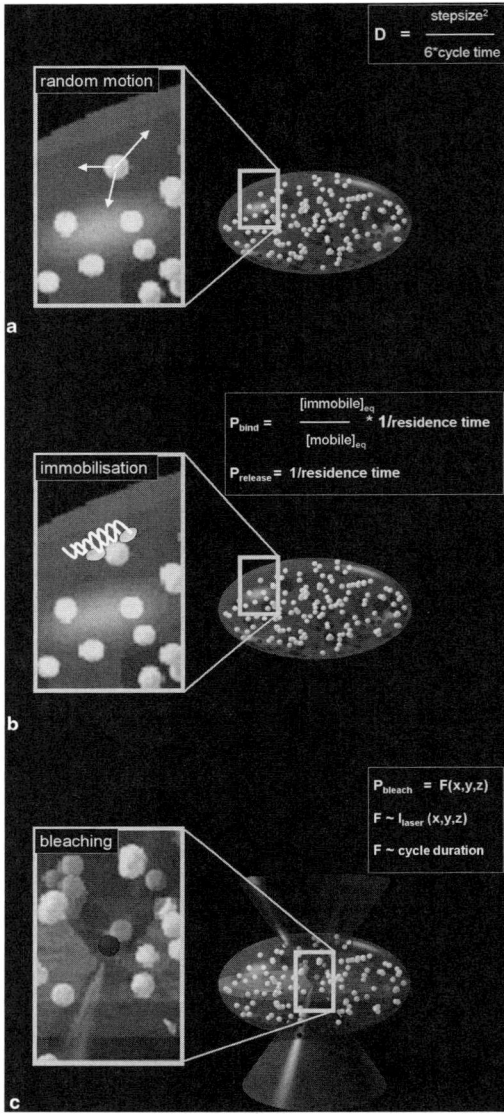

Fig. 20.3 General scheme for Monte Carlo simulation of FRAP on nuclear proteins. **a** Schematic drawing of a cell nucleus (*ellipsoid*) containing randomly distributed GFP-tagged proteins (*spheres*). Random Brownian motion (*inset*) is simulated on the basis of the Einstein-Smoluchowsky equation D = stepsize^2/6* cycle time (*see* **Section 3.4.4**). **b** Simulation of binding to randomly distributed immobile target sites in the DNA (*inset*) is simulated by evaluating a chance to bind or to release based on simple binding kinetics, where the ratio between on- and off-rates (defined by k_{on} and k_{off}, *see* **Section 3.4.4**) equals the ratio between the number of immobile and mobile molecules. **c** Photobleaching is simulated by evaluating a chance to get bleached based on the intensity profile of the laser beam. This profile can be obtained experimentally by illuminating a paraformaldehyde-fixed nucleus with a stationary laser beam at different intensities and collecting 3-D image stacks afterwards. Also GFP-blinking can be simulated (*see* **Note 25**)

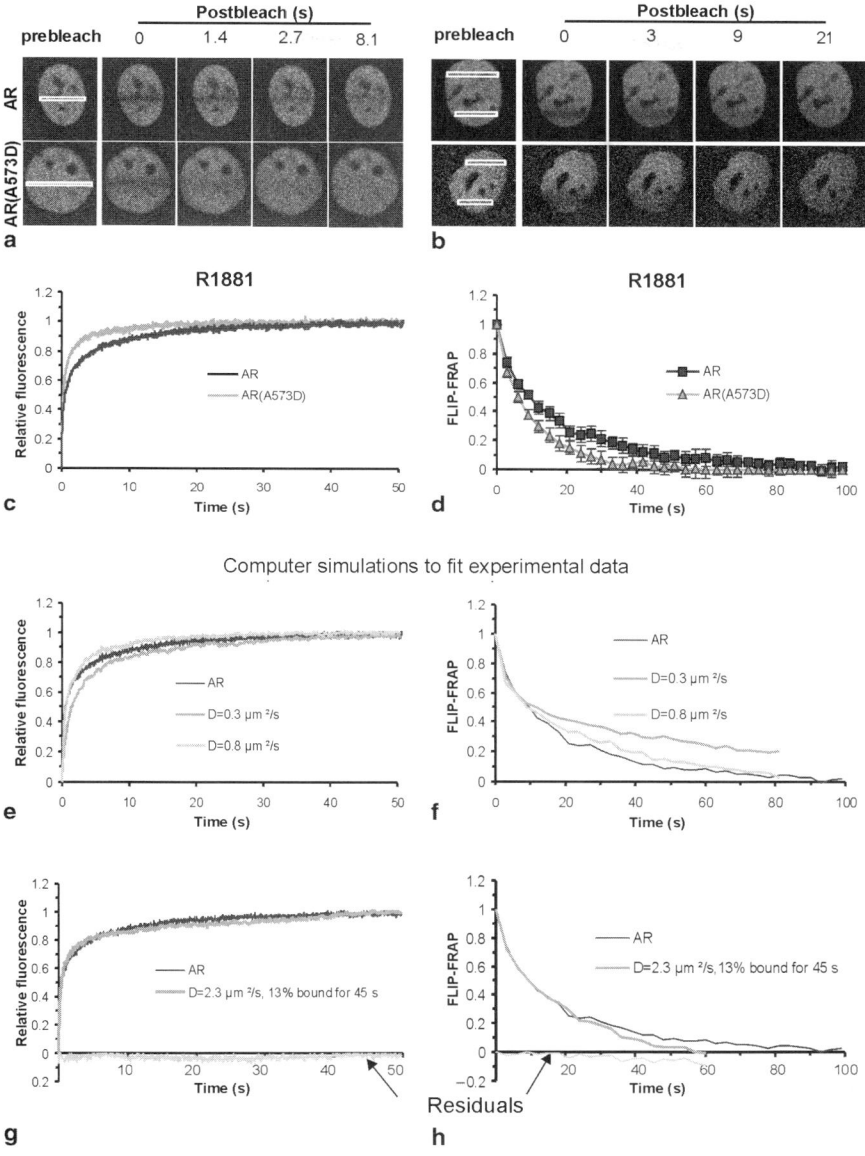

Fig. 20.4 Combined strip-FRAP and FLIP-FRAP reveal that a fraction of agonist-liganded GFP-ARs is transiently immobilized. **a** In the strip-FRAP procedure a strip in the centre of a nucleus is bleached (*rectangle*) at high laser power. Subsequently, fluorescence in the strip is measured at regular time intervals. **b** In the FLIP-FRAP procedure a strip at one pole of the nucleus was bleached for a relatively long period. The difference between fluorescence signals in the bleached region (FRAP, *lower rectangle*) and a distal region at 10 μm from the bleached region of the nucleus (FLIP, *upper rectangle*) was determined at regular time intervals. **c, d** Strip-FRAP and FLIP-FRAP experiments of GFP-AR or the non-DNA-binding mutant GFP-AR(A573D) in the

chromosomes. Application of FRAP to two telomeric proteins (TRF1 and TRF2) revealed that telomere binding occurs in a complex dynamic fashion (Fig. 20.5).

2   Materials

2.1   Constructs

1. Standard EGFP, EYFP, and ECFP vectors for cloning (Clontech, Palo Alto, CA, USA) (*see* **Note 4**).
2. pAR0, expressing human full-length wild-type AR *(39)*, is used to fuse the AR with the fluorescent proteins.
3. pGEM-T-Easy (Promega, Madison, WI, USA).

2.2   Cell Culture and Transfection

1. Hep3B human hepatocellular carcinoma cell line (ATCC HB-8064; LGC Promochem, Teddington, UK) (*see* **Note 5**).
2. Alpha Minimal Essential Medium (αMEM) (Bio-Whittaker/Cambrex, Verviers, Belgium) supplemented with 2 m*M* L-glutamine (Bio-Whittaker/Cambrex), 100 U/mL of penicillin, 100 μg/mL of streptomycin (Bio-Whittaker/Cambrex) and 5% v/v triply 0.1-μm sterile-filtered fetal bovine serum (FBS) (HyClone, South Logan, UT, USA). Store at 2–8°C.

Fig. 20.4 (continued) presence of an agonistic ligand (10^{-9} *M* R1881). **c** Graph showing fluorescence intensities relative to complete redistribution of the non-DNA-binding mutant GFP-AR(A573D) in the presence of R1881 plotted as a function of time. Mean values of at least 10 cells are plotted. All experiments were performed at least three times. **d** Graph showing the difference between fluorescence intensity in the FLIP and FRAP regions (*rectangles* in **b**) relative to the difference directly after bleaching, plotted against time. Mean values ±2× the SEM of two independent experiments on at least ten cells are plotted. **e, f** Computer simulations of strip-FRAP and FLIP-FRAP of freely diffusing molecules do not explain the experimental FRAP data obtained with both methods. Experimental strip-FRAP data on wild-type GFP-AR lies in between curves representing the indicated scenarios of free diffusion (**e**), whereas experimental FLIP-FRAP data on wild-type GFP-AR lies outside these boundaries (**f**). **g, h** Computer simulations representing a model where, next to freely diffusing molecules, a fraction is transiently immobilized, fitted both strip-FRAP and FLIP-FRAP experimental curves on wild-type GFP-AR. Computer simulations correspond to the average of best fits of FRAP and FLIP-FRAP experiments respectively, so are not necessarily the best fits of the individual experiments. Absolute value of residuals of the computer simulation fit and the experimental data on each time point are plotted below the *x*-axis. (Figure adapted from ref. *(7)*)

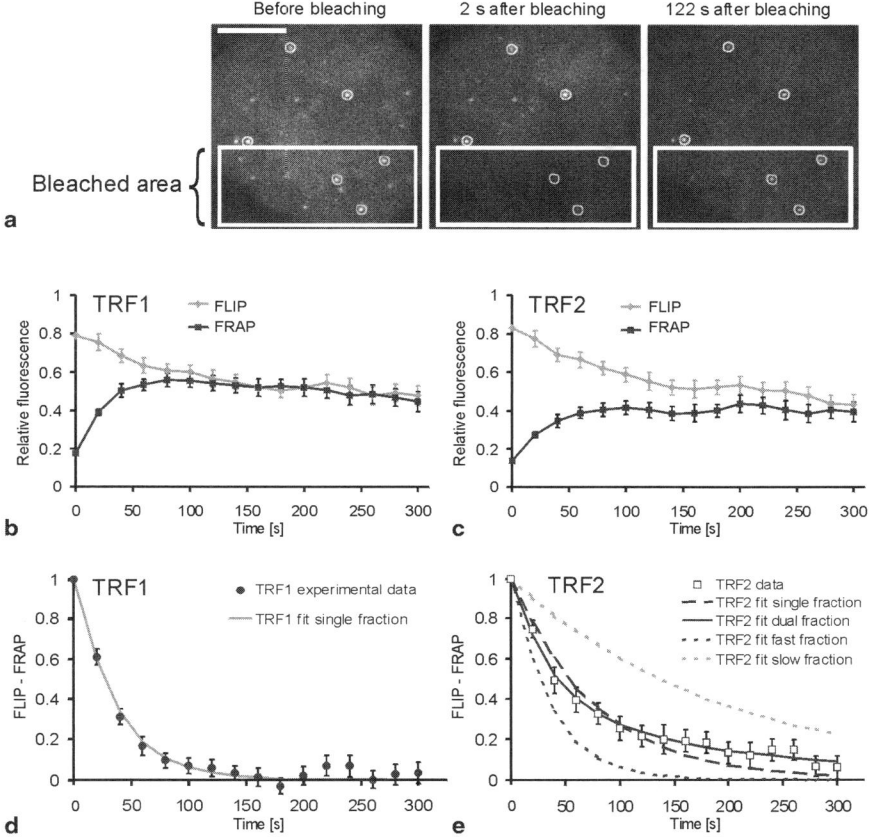

Fig. 20.5 Simultaneous FLIP-FRAP of telomere-bound proteins TRF1 and TRF2. **a** FLIP-FRAP on living HeLa cells expressing GFP-TRF1. Cells are photobleached over a region covering about one half of the nucleus (indicated by a *white box*). The images were acquired before bleaching and at 20-sec intervals after bleaching, starting at 2 sec. The *circles* in the bleached area and unbleached area indicate the regions that are used to calculate fluorescence redistribution. Scale bar, 5 μm. **b**, **c** Quantitative analysis of redistribution of GFP-TRF1 (**b**) and GFP-TRF2 (**c**) at telomeres separately in bleached (in *white box*) and unbleached (*upper*) half of the nucleus. Values are means ± SEM from at least 40 cells. (**d**) Difference (Δ) in telomere intensity in bleached and unbleached parts of the cell, calculated from the data shown in **b** (TRF1) and **c** (TRF2). **e** A fitting analysis of the experimental data in **d** to the equation $\Delta I_{\mathrm{rel}}(t) = f_1 e^{-k_1 t} + f_2 e^{-k_2 t}$ indicated a good fit with the single binding kinetics of GFP-TRF1 (*solid grey line*). In contrast, GFP-TRF2 redistribution does not fit with single binding kinetics (*dotted black line*) but does fit with dual binding kinetics (*solid black line*). Note the similarity between the fitted curves of the fast fraction of TRF2 and of TRF1. (Figure adapted from ref. (18))

3. G418 sulfate (Geneticin) (HyClone): working solution is 100 mg/mL active concentration in PBS. Final concentration in culture medium is 0.6 mg/mL of G418 (*see* **Note 6**).

4. Methyltrienolone (R1881; NEN DuPont, Boston, USA): dissolve in EtOH to 1 mM stock solution. The stock solution is diluted stepwise (1:10) in EtOH down to 1 nM R1881 to generate an array of working solutions. For our experiments we used the 1 μM working solution to obtain a final concentration of 1 nM of hormone in our culture medium. R1881 is light-sensitive and stored at −18°C.

5. Charcoal-stripped serum (Hyclone, or can be prepared by a protocol such as that described at www.sigmaaldrich.com/sigma/usage/f2442use.pdf).

6. Trypsin-EDTA (Bio-Whittaker/Cambrex): contains 200 mg/mL of Na$_4$EDTA, 500 mg/L of trypsin 1:250. Sterile filtered. Store the stock solution at −10°C and the working solution at 2–8°C.

7. Cover slips: 24-mm diameter, 0.13–0.16 mm thickness (Menzer-Gläser/Menzel Gerhard, Braunschweig, Germany) (*see* **Note 7**).

8. Polystyrene 6-well Cell Culture Cluster (Corning B.V. Life Sciences, Schiphol-Rijk, Netherlands).

9. FuGENE6 transfection medium (Roche Molecular Biochemicals, Indianapolis, IN, USA). Store at 2–8°C.

2.3   Generation of Stable Cell Lines

1. Polystyrene tissue culture dishes 100×20 mm (Falcon; BDBiosciences, Alphen aan den Rijn, Netherlands).

2. Polystyrene 6-well cell culture cluster (Corning B.V. Life Sciences).

3. HyQ G418 sulfate (HyClone): working solution is 100 mg/mL active concentration in PBS. Final concentration in culture medium is 0.6 mg/mL (*see* **Note 6**).

2.4   Fluorescence Recovery After Photobleaching

1. All the quantitative FRAP procedures are performed on a Zeiss Confocal Laser Scanning Microscope LSM510 META equipped with a ×40/1.3A NA oil immersion objective, a Lasos LGK 7812 ML-4 Laser Class 3 B Argon laser (30 mW) with the excitation laser lines 458 and 488, and an acousto-optical tunable filter (AOTF) (Carl Zeiss MicroImaging, Jena, Germany) (*see* **Note 8**).

2. The filter set to specifically image GFP is shown in Table 20.1 (*see* **Note 4**).

3. Temperature is controlled by a heatable stage and a lens-heating device, both developed in our laboratory (*see* **Note 9**).

4. The LSM 5 software, Version 3.2 controls the microscope, the scanning and laser modules, and the image acquisition process. This software is also used to analyze the images.

Table 20.1 Filter sets used in FRAP experiments

Fluorophore	Excitation	Main beam splitter	Secondary beam splitter	Emission filter
EGFP	488 nm	HFT 488	Mirror	BP 505–530
ECFP	458 nm	HFT 458/514	NFT 515	BP 470–500
EYFP	514 nm	HFT 458/514	NFT 515	LP 560

3 Methods

3.1 Constructs

1. The GFP-AR coding construct is generated by performing PCR on pAR0 (39) using a sense primer (5′-GCAGA AGATCT GCAGGTGCTGGAGCAGGTGCT GGAGCAGGTGCTGGAGAAGTGCAGTTAG-3′) to introduce a *Bgl*II restriction site and a flexible (GlyAla)$_6$ spacer sequence and an anti-sense primer in the AR cDNA overlapping a *Sma*I site (5′-TTGCTGTTCCTCATCCAGGA-3′) (*see* **Note 10**).
2. The PCR product is cloned in pGEM-T-Easy and the sequence is verified.
3. The *Bgl*II-*Sma*I fragment is inserted in the corresponding sites of pEGFP-C1.
4. Next, the *Sma*I fragment from pAR0 is inserted into the *Sma*I site to generate pGFP-(GlyAla)$_6$-AR (further referred to as GFP-AR).
5. The non-DNA-binding mutant is obtained by exchanging the *Asp*718I-*Sca*I fragment from pAR(A573D) in GFP-AR.

3.2 Cell Culture and Cell Transfection (see Note 11)

1. Hep3B cells are grown in αMEM supplemented with L-glutamine, penicillin, streptomycin, and 5% FCS at 37°C and 5% CO_2 and passaged when approaching confluence (every 3–4 days) with trypsin/EDTA to provide experimental cultures.
2. Two days before confocal microscopy, Hep3B cells are seeded on a coverslip in a 6-well plate at a concentration of ~3×10^5 cells per well in 2 mL of αMEM with 5% FCS (*see* **Note 12**). This concentration will provide near-confluent cultures at the time of the experiment and a sufficient amount of cells at the time of transfection. The cells are grown overnight at 37°C with 5% CO_2.
3. Between 24 and 32 h before confocal microscopy, the medium is replaced by 1 mL of αMEM supplemented with 5% charcoal-stripped serum, L-glutamine and antibiotics, without washing the cells.
4. After 2 h the transfection mix is prepared for transfection of 1 µg of GFP-AR coding vector. Three microliters of FuGENE6 per microgram of DNA to be

transfected is added to 100 μL of serum-free αMEM. After 5 min incubation, the DNA is added. The transfection mix is then gently mixed by pipetting up and down, and left at room temperature for at least 30 min (*see* **Note 13**).

5. Four hours after medium replacement, the transfection mix is gently added to the cells under gentle mixing. The cells are then incubated for 4 h at 37°C and 5% CO_2.

6. Four hours after transfection the medium is replaced again by 2 mL of αMEM supplemented with charcoal-stripped serum with or without 1 nM R1881. The cells are further incubated overnight at 37°C and 5% CO_2.

3.3 Generation of Stable Cell Lines

To avoid transfections before each experiment and to simplify the selection of cells with physiologically relevant expression level of the tagged protein, a cell line stably expressing a receptor with fluorescent label(s) can be generated. It is often stated that the use of cells stably expressing fusion proteins is essential to avoid overexpression, but microscopic approaches enable preselection of cells with a physiological expression level also in transient transfected cells. With this in mind, also transiently transfected cells can be studied when a sufficient number of cells in the required expression range can be found (*see* **Note 11**).

1. Hep3B cells are seeded in 6 wells of a 6-well plate in αMEM with 5% FCS as described in Section 3.2, and incubated overnight at 37°C and 5% CO_2.

2. After overnight incubation, the cells in all 6 wells are transfected with the same expression vector (1 μg of DNA for each well) as described in Section 3.2.

3. Four hours after transfection, the medium is replaced by 2 mL of αMEM/5% FCS and the cells are incubated overnight at 37°C and 5% CO_2.

4. After overnight incubation the cells are trypsinised and an array of dilutions ranging from 9:10 to 1:1,000 are seeded in 10-cm-diameter dishes. The Hep3B cells are incubated again overnight at 37°C and 5% CO_2.

5. After another overnight incubation the medium is replaced by αMEM with 5% FCS supplemented with 0.6 mg/mL of G418 (active concentration). The cells are further incubated at 37°C/5% CO_2. Twice a week, the medium is replaced by fresh medium with G418.

6. When the clones grown from single cells consist of 100–200 cells, the locations of the colonies are indicated with a fine marker underneath the bottom of the dish. Using a 200-μL pipette, each clone is carefully scraped while slowly collecting the cells by pipetting.

7. The collected cells are seeded in a glass-bottomed 24-well dish in 1 mL of αMEM with G418 and incubated for 1 or 2 weeks.

8. When the clones have grown, the expression of GFP-AR is judged by confocal microscopy. The clones with sufficient expression are passaged and used for experiments (*see* **Note 14**).

3.4 Fluorescence Recovery After Photobleaching

3.4.1 Strip-FRAP

In the most elementary FRAP experiment, a small circular area that is illuminated by a stationary laser beam (the diffraction-limited spot) is bleached and the recovery of fluorescence inside the spot is monitored in time *(14)*. The use of a confocal scanning microscope makes it possible to bleach larger areas by scanning at high intensity. In addition, not only the recovery of fluorescence in the bleached area, but also the fluorescence in the entire nucleus can be monitored.

There may be several potential drawbacks of bleaching a small area. First, the recovery of fluorescence is very fast, possibly too fast for the imaging system, especially when the entire nucleus is imaged. Second, the amount of fluorescence emitted by the limited number of molecules in a single spot is relatively low, leading to low signal-to-noise levels. Third, the relative location of the bleached spot inside the nucleus may influence the recovery curve. Therefore, we provide a FRAP protocol, designated strip-FRAP, in which a narrow is strip is bleached with a width of ~700 nm (corresponding to 10 pixels at zoom 6 on a Zeiss LSM 510 microscope) and spanning the entire nucleus. To assure a sufficient time resolution to also follow fast diffusion processes (like diffusion of free GFP), only the fluorescence in the bleached strip is monitored after photobleaching (Figs. 20.1a and 20.4a, c).

1. These instructions assume use of a Zeiss CLSM 510 confocal laser scanning microscope. For GFP-imaging in general, an argon laser is used, adjusted to 6.1-A tube current and allowed to prewarm for at least 15 min.
2. For live cell imaging, an inverted microscope is used. Cells are grown on a coverslip that forms the bottom of a container enabling the addition of medium on top of the cells. In the examples given here we investigated Hep3B cells expressing GFP-tagged androgen receptor (GFP-AR). The glass-bottom container containing the cells and medium is transferred into a heatable stage holder at 37°C (*see* **Note 9**), which is mounted on a motorized scan stage.
3. GFP-fluorescence is detected using 488-nm excitation, usually by an argon laser, using a 488-nm beam splitter and a band-pass emission filter passing GFP emission at 505–530 nm (Table 20.1) (*see* **Note 4**). Cells are imaged using a ×40/1.3 NA oil objective. The confocal pinhole is adjusted such that the estimated optical slice thickness is ~2 μm (corresponding to 2.48 Airy units).
4. Scanning is performed unidirectional with scan speed 9 to enable fast recording of the fluorescent signal in the strip. Laser intensity is attenuated to approximately 1% of the maximum output using an AOTF for scanning (~0.5–0.8 μW). Detector gain is set on 900, the amplifier offset on 0, and the amplifier gain on 1. Fluorescent signals are recorded with an 8-bit data depth (*see* **Notes 15–17**).
5. A nucleus expressing GFP-AR at physiologically relevant expression level (see **Note 18**) is selected at zoom 1 (*see* **Note 19**). To limit potential effects of nuclear shape, all nuclei are oriented in the same way relative to the bleaching strip (in our experiments we make sure that the longer axis of the more-or-less ellipsoid-shaped

nucleus is perpendicular to the strip. When the nucleus is oriented correctly, zoom is adjusted to 6 corresponding to a lateral pixel distance of 70 nm. A 10-pixel-wide (700-nm) strip spanning the nucleus is selected in the Edit ROI panel, for recording the recovery of the signal (Fig. 20.1a) (*see* **Note 20**).

6. The fluorescent signal is monitored at 21-msec intervals by scanning the region of interest (ROI) for 4000 iterations (~80 sec) prior to bleaching (Time Series Control) at low excitation (*see* **Note 21**). After 200 scans, GFP is bleached locally inside the ROI using a scan (1 or 2 iterations) at maximum intensity (Fig. 20.1a). The time-series are initiated using the Mean ROI option in the time-series control (*see* **Notes 22** and **23**).

7. After the scan the data can be copied directly to Excel or the file can be saved as a data-file for later analysis.

3.4.2 FLIP-FRAP

Next to the comparison of FRAP curves obtained from proteins under different experimental conditions, it may be useful to apply two different FRAP variants to a protein under the same conditions. For instance, the combination of strip-FRAP (Fig. 20.1a, b) and combined FLIP-FRAP (Fig. 20.1c, d) can be used to investigate transient immobilization *(6, 13, 16)*. This approach is based on the fact that often more than one scenario (with different diffusion coefficient, immobile fraction, and time of immobilization) fits the experimental data. To differentiate between different scenarios that fit well to the results of a strip-FRAP experiment, it is possible to perform a complementary combined FLIP-FRAP experiment (Fig. 20.1c, d) since two different scenarios that result in similar curves in a strip-FRAP experiment give clearly different curves in a complementary FLIP-FRAP experiment *(13)*. In FLIP-FRAP, the recovery of the fluorescence in a region at one end of the nucleus after bleaching for a relative long period (FRAP) is measured in parallel with the decrease of the signal in a similar region at the other pole of the nucleus (FLIP) (Fig. 20.1c, d). As stated above, the combination of strip-FRAP and FLIP-FRAP is especially applicable for studying transient immobilization as found for proteins (Fig. 20.4) *(6, 7)*.

1. Confocal settings are similar to those used in the strip-FRAP procedure (steps 1–3) except for the Detector Gain, which is set on 1,000.

2. A 20-pixel-wide ROI spanning the nucleus at one pole is selected in the Edit ROI panel, for recording the recovery of the signal after bleaching. Using the Define Region option of the Bleach Control panel, the same ROI is selected to locally bleach GFP. A second ROI of similar width is selected spanning the nucleus at the other pole where the decrease of fluorescence due to redistribution of the proteins from the bleached area is measured (Fig. 20.1c). The distance between the two regions should be kept constant in different cells.

3. The fluorescent signal is monitored by scanning the two regions of interest for 35 iterations (approx. 100 sec) at a low excitation level with a 3-sec time interval (Time Series Control) (*see* **Note 21**). After the first scan the GFP is bleached locally inside the lowest ROI using 10 iterations of 488 nm laser light at maximum voltage (Fig. 20.1c). The time-series are initiated using the mean ROI option in the time-series control (*see* **Notes 22** and **23**).

4. After the scan the data can be copied directly to Excel or the file can be saved as a data-file for later analysis.

3.4.3 Qualitative FRAP Analysis

FRAP data can initially be qualitatively analyzed by comparing data from experiments applied when the protein under investigation is active or inactive. In the case of the study of steroid receptors like the AR, comparison of non-DNA-binding mutant versions and the wild type is possible where the differences observed between these can be ascribed to DNA binding. In our studies, direct comparison of a wild-type or mutant AR to a non-DNA-binding mutant (e.g. AR [A573D]) greatly simplifies the interpretation of the data, because this mutant is not immobilized (Fig. 20.2) *(6, 7, 44)*.

In strip-FRAP the recovery of fluorescence in a narrow strip spanning the nucleus is scanned at low laser intensity after shortly bleaching the fluorescence inside this region (Fig. 20.1a, b). The qualitative comparison between FRAP curves from different experimental conditions requires some kind of normalization of the data. There are several ways to normalize FRAP data, each of which enables the extraction of different parameters (Fig. 20.2b–d) *(13)*.

1. The most straightforward normalization of FRAP data is to express intensities relative to the average of a sufficient number of measurements before bleaching ($I_{prebleach}$): $I_{norm,t} = (I_t - I_{background})/(I_{prebleach} - I_{background})$, where $I_{background}$ is the background signal (Fig. 20.2b) (*see* **Note 24**).

2. Straightforward normalization can be extended with the normalization to the fluorescence intensity directly after bleaching (I_0), by expressing intensity values relative to the intensity directly after bleaching as well as to the average prebleach intensity: $I_{norm,t} = (I_t - I_0)/(I_{pre} - I_0)$. This way of normalization enables the quick visual estimation of the size of a potentially present immobilized fraction (Fig. 20.2c) (*see* **Note 1**).

3. A third way of normalization is to express intensity values relative to the fluorescence after complete recovery (I_∞), and the intensity directly after bleaching (I_0): $I_{norm,t} = (I_t - I_0)/(I_\infty - I_0)$. This yields a curve running from 0 directly after bleaching, to 1 after final recovery, allowing direct comparison of apparent diffusion rates irrespective of a potentially present immobile fraction, since different final recoveries due to different immobile fraction are not visible. In addition, this normalization is also often used for quantitative analysis by fitting the data

to any equation that represents the diffusion process (and transient immobilization) (Fig. 20.2d) *(13)*.

4. Combined FLIP-FRAP data can be analyzed by first calculating the fluorescence intensity difference between the FLIP-region and the FRAP-region: $I_{\text{FLIP-FRAP}} = (I_{\text{FLIP-ROI}} - I_{\text{FRAP-ROI}})$. Subsequently the intensity differences can be expressed relative to the highest difference, i.e. immediately after bleaching (Fig. 20.1d and Fig. 20.4d).

3.4.4 Quantitative FRAP Analysis

Quantitative analysis of FRAP is mostly performed by fitting experimental data to mathematical models. Many different models have been brought forward ranging from simplified models based on 1-D diffusion *(20, 45)* to very sophisticated 3-D models incorporating as many aspects of the FRAP experiment as possible *(46–52)*. These approaches have shown to be very useful for quantitative FRAP analysis, a slight drawback being that for instance diffusion in ellipsoid volumes or conically shaped laser beams are relatively difficult to solve analytically. Another possible approach to quantitatively analyse FRAP results is to generate FRAP curves by Monte Carlo simulation of diffusion of individual molecules and their binding to immobile elements (representing chromatin binding) in an ellipsoidal volume (representing the nucleus) as well as the shape and intensity distribution of the applied laser beam (Fig. 20.3) *(6, 17, 20)*. The strength of Monte Carlo simulation is that it generates the highly complicated outcome of a set of relatively simple mathematically definable rules, such as diffusion of single particles in an ellipsoid volume, the presence of nucleoli, or the typical shape and intensity distribution of a focused laser beam, which are hard to solve analytically. The drawback of Monte Carlo simulations is that they are very time-consuming.

1. In our simulations, the size of the ellipsoid representing the nuclear volume is based on the experimentally determined average size of the investigated nuclei.
2. At the start of a Monte Carlo simulation, all tagged proteins start at a random position inside an ellipsoid volume representing the nucleus (Fig. 20.3a). Subsequently, the simulation goes through cycles representing a minimum time span. Typically in our simulations the time step was 21 msec, corresponding to the time it takes to scan a 10-pixels-wide strip spanning the width of the nucleus one time.
3. Diffusion of single particles (representing GFP-tagged proteins) is simulated using the strikingly simple Einstein-Smoluchowski relationship for 3-D Brownian motion: $D = s^2/6\,T$, where D is the diffusion coefficient, s is the average distance moved by the particles, and T is the time span in which the particles move, in our strip-FRAP simulations, 20 msec (Fig. 20.3a). Simulation of Brownian motion of individual particles only involves the additional consideration that the equation predicts the average movement of all particles, whereas single particles will travel

over slightly different distances. Therefore, diffusion of each particle is simulated by displacing it over a certain distance (corresponding to a certain diffusion coefficient) plus or minus a small variation defined by a Gaussian with standard deviation σ, which represents the stepsize per unit time and is related to the diffusion coefficient according to the above equation: $\sigma = \sqrt{6 D \Delta t}$. In practice the step to be made by the molecule is split in three steps, in x-, y-, and z-direction, respectively: at each new time $t + \Delta t$, a new position $(x_{t+\Delta t}, y_{t+\Delta t}, z_{t+\Delta t})$ is derived for all mobile molecules from their current position (x_t, y_t, z_t) by $x_{t+\Delta t} = x_t + G(r_1)$, $y_{t+\Delta t} = y_t + G(r_2)$, and $z_{t+\Delta t} = z_t + G(r_3)$, where r_i is a random number ($0 \le r_i \le 1$) chosen from a uniform distribution, and $G(r_i)$ is an inversed cumulative Gaussian distribution with $\mu = 0$ and $\sigma^2 = 6 D \Delta t$, where D is the diffusion coefficient. Note that the latter follows directly from the Einstein-Smulochowski equation, the average stepsize being equal to the standard deviation of the Gaussian distribution.

4. Binding to and releasing from immobile elements in the nucleus (like chromatin) can also be simulated by simple mathematical expressions (Fig. 20.3b). The probability for an immobile molecule to release is the most easy to explain since, in simple binding kinetics, this only depends on the affinity of the protein, yielding an equation defining the chance of a molecule to release: $P_{mobilise} = k_{off} = 1/T_{imm}$, where T_{imm} is the time spent in the immobile state expressed in number of time steps. If for instance the average immobilisation time is three time steps (60 msec in our typical strip-FRAP simulation), an immobile molecule will in each step have a chance of 1/3 to release, leading to an expected residence time of three steps. From this simple equation, the chance for a free molecule to become immobile can be calculated when one takes into account the fact that the ratio of immobile and mobile molecules is equal to the ratio between k_{on} and k_{off} (law of mass action): $k_{on}/k_{off} = F_{imm}/F_{mob}$, where F_{mob} is the number of mobile molecules and F_{imm} is the number of immobile molecules. The probability for each particle to become immobilized (representing chromatin binding) then is $P_{immobilise} = k_{on} = k_{off} \cdot F_{imm}/F_{mob}$. Since above we saw that $k_{off} = 1/T_{imm}$, the chance to immobilise can be expressed fully in terms of immobile fraction and immobilisation time, the two typical mobility parameters obtained from FRAP experiments: $P_{immobilise} = k_{on} = (F_{imm})/(T_{imm} \cdot F_{mob})$

5. For bleaching simulation we use experimentally derived 3-D laser intensity profiles, determining the probability for each molecule to get bleached considering their 3-D position relative to the laser beam (Fig. 20.3c). The 3-D laser intensity profile is derived from the bleach pattern in confocal image stacks of chemically fixed nuclei containing GFP that were exposed to a stationary laser beam at various intensities and varying exposure times.

6. For quantitative analysis of the FRAP data, raw FRAP curves are normalized to pre-bleach values and the best-fitting curve (by ordinary least squares) is selected from a large set of FRAP curves generated as described above in which three parameters representing mobility properties were varied: diffusion rate (ranging from, e.g. 0.04 to 25 μm^2/sec), k_{on} and k_{off} (corresponding to immobile fractions of, e.g. 0, 10, 20, …, 90%) and time spent in the immobile state (e.g. 2, 4, 8, 16, 32, 64, 128, 256, 512, 1,024, ∞ sec).

4   Notes

1. In FRAP, a considerable fraction of the fluorescent proteins inside a nucleus will be irreversibly bleached during the bleach pulse, resulting in incomplete recovery of the fluorescent signal independent of any immobile fraction.

2. Next to a diffusion coefficient, an immobile fraction and a residence time in the immobile state can be derived from FRAP curves. It is also possible to describe the dynamic binding and release from immobile components of the nucleus in terms of immobilisation (k_{on}) and release rate (k_{off}). The ratio between k_{on} and k_{off} then is equal to the ratio between mobile and immobile fraction. The residence time is equal to the inverse of the release rate k_{off} (*see also* **Section 3.4.4**).

3. Analysis of FRAP or FLIP curves is more complicated, since the recovery or loss of fluorescence is a result of dissociation and association of tagged proteins *(13)*.

4. The examples given here are all based on EGFP-tagged proteins. If ECFP, EYFP, or other fluorescent dyes are used, imaging protocols should be modified according to the fluorescent properties of the chosen dye (see Table 20.1).

5. Hep3B cells do not express endogenous AR, are easy to transfect, and are relatively large, enhancing microscopic analysis. When using other cell types the expression of endogenous nuclear receptors needs to be taken into account. Although many GFP-tagged proteins are functional, their functionality will usually be less than the wild type. Endogenously expressed proteins compete with the tagged version, thereby limiting its activity compared with when no endogenous protein is present. A potential threat to the use of cell lines lacking endogenous expression is the potential absence of specific cofactors that modulate or enhance the activity of the studied protein.

6. Antibiotics are potentially harmful and should always be treated with proper personal protection.

7. Coverslips should not be thicker than 0.16 mm, because of the high numerical aperture of some of the lenses of the confocal microscope.

8. The LSM 510 is a laser hazard class 3 B instrument and is marked as such. This moderate-risk class includes medium-power lasers. You must take care not to expose yourself to the radiation of such lasers.

9. We use temperature-control equipment developed in our laboratory. Alternatively, commercially available equipment can be used varying from heatable plates up to complete incubators mounted on the microscope.

10. The insertion of a flexible stretch, for instance a short glycine–alanine repeat, between the protein of interest and the fluorescent label may decrease the potential (negative) effect of the label on the functionality of the protein of interest. For tagged ARs, most often reporter assays using androgen-regulated luciferase reporter genes are used to verify the functionality (6).

11. Although the use of cell lines stably expressing the tagged protein has a number of advantages, it may under circumstances be more convenient to investigate transiently transfected cells. In these cases it is important to only select cells that do not overexpress the protein. This can be achieved by, for example,

selecting cells at the same settings as used for stably expressing cells. Using these settings, fluorescence levels should be similar to cells stably expressing the protein at physiological levels. Fluorescence of overexpressing cells will be at the maximum in the entire nucleus (each pixel will have maximum intensity, in 8-bit imaging corresponding to a value of 255).

12. Coverslips in 6-well plates are sterilized by 45 min of UV treatment. Alternatively, they are submerged in ethanol and flamed before placing in the 6-well plate.

13. Do not vortex the transfection mix (refer to the FuGENE6 Transfection Reagent Instruction Manual). Contact between the undiluted FuGENE6 and any plastic surface (except for the pipette tip) should be avoided.

14. Alternatively, modern FACS-sorters could be used to select single cells on the basis of their fluorescent level, representing the expression level of the tagged protein. Each cell can be deposited in a separate well of, for instance, a 96-well plate. After incubation, the clones can be selected that have the appropriate expression level (step 7–8 above).

15. Although a higher detector gain (DG) is favorable to obtain higher signals, it increases noise. Therefore, a trade-off of settings to reduce both noise (e.g. lower DG, averaging) and monitor bleaching (e.g. lower excitation level, rapid scanning) but still giving high enough signal in low-expressing cells is necessary to optimize the experimental setup (e.g. wider pin-hole, higher DG, higher excitation level). These settings may also depend on the level or pattern of expression of the protein of interest. In general, for FRAP excitation laser intensity should be as low as possible, to avoid monitor bleaching, which is hard to correct for *(13)* (*see* **Note 16**).

16. The rate at which monitor bleaching occurs can be determined in a non-bleached area of the nucleus or by performing a FRAP experiment without applying the bleach pulse. Subsequently, experimental FRAP curves can be corrected according to the observed monitor bleach rate. However, when a significant immobile fraction is present, this fraction will contribute more to the monitor bleaching in the control measurements, than the mobile fraction, since immobile molecules stay constantly in the illuminated area, whereas mobile ones move in and out. In a FRAP experiment this immobile fraction is largely bleached, leading to less monitor bleaching after photobleaching than estimated on the basis of the control curves. Thus, the presence of a (transiently) immobile fraction will lead to overcorrection of the experimental curve and subsequent underestimation or failure to detect this fraction. Therefore, it is important to avoid monitor bleaching or to limit it as much as possible.

17. Settings for imaging in the LSM510 software must be selected in several panels; Scan Control, Edit ROI, Time Series Control, and Bleach Control. Settings for imaging are selected using the Scan and Time Series control menus.

18. For all the approaches discussed, it is essential to select cells with physiologically relevant expression levels. Overexpression can lead to aggregation and artificial immobilization of the receptors (53). For quantitative measurements like FRAP described here, high-resolution imaging is not essential.

19. To speed up the procedure it is advisable to use software to put the center of the nucleus in the middle of the scanning area ("center" macro in the macro-directory).

20. It is not essential to use the exact values given here, but the values chosen should be kept constant (similar) over sets of experiments to allow comparison between curves of different proteins/conditions.

21. It is important to limit monitor bleaching by applying excitation at a low laser power. Although monitor bleaching correction procedures are designed for FRAP experiments, it is favorable to limit the monitor bleaching during data collection.

22. By selecting the Mean ROI option in the time series control panel, only the mean intensity inside the ROI is plotted. In contrast, by selecting StartB (or StartT) all scans of the ROI are saved for later analysis.

23. In the configuration, the monitor diode (ChM) can be selected also to monitor potential fluctuations of laser intensity during scanning, and used afterwards for correction. Note that lasers become less stable towards the end of their life.

24. In data normalisation where the intensity after bleaching is set to 0, some information is lost, since the depth of the bleach pulse is not only determined by laser intensity but also by protein mobility and size of the immobile fraction: the slower the protein and the larger the immobile fraction, the deeper the bleach pulse will be. If experimental curves are fit to models that take into account bleach depth, the estimates of D, k_{on}, and k_{off} (or immobile fraction and residence time in the immobile state) may be more accurate than when bleach depth is not taken into account.

25. Although in many FRAP analyses it is assumed that proteins are irreversibly photobleached, it has been shown that a fraction of proteins will regain their fluorescence, a process often referred to as blinking. Blinking not only occurs when GFP is illuminated at high intensity for photobleaching, but also at lower monitor intensities, although the time spent in the fluorescent state (on-time) of blinking GFPs is shorter at higher intensities. In contrast, the off-times are independent of excitation intensity *(54)*. Therefore, in a FRAP-experiment the fraction of GFPs in the reversible off-state will increase during the high intensity bleach pulse, and decrease again during subsequent monitoring at low intensity, leading to a recovery of fluorescence that is not related to protein mobility. Although the contribution of this will be limited if molecules are relatively mobile, in cases where the majority or all of the investigated proteins are immobile (for instance core histones) the effect will be substantial, leading to underestimation of the immobile fraction.

References

1. Tsien, R. Y. (1998) The green fluorescent protein. *Ann. Rev. Biochem.* **67**, 509–544.
2. Lippincott-Schwartz, J. and Patterson, G. H. (2003) Development and use of fluorescent protein markers in living cells. *Science* **300**, 87–91.
3. Shaner, N. C., Steinbach, P. A., and Tsien, R. Y. (2005) A guide to choosing fluorescent proteins. *Nat. Meth.* **2**, 905–909.

4. Heim, R. and Tsien, R. Y. (1996) Engineering green fluorescent protein for improved brightness, longer wavelengths and fluorescence resonance energy transfer. *Curr. Biol.* **6**, 178–182.

5. Giepmans, B. N. G., Adams, S. R., Ellisman, M. H., and Tsien, R. Y. (2006) The fluorescent toolbox for assessing protein location and function. *Science* **312**, 217–224.

6. Farla, P., Hersmus, R., Geverts, B., Mari, P. O., Nigg, A. L., Dubbink, H. J., Trapman, J., and Houtsmuller, A. B. (2004) The androgen receptor ligand-binding domain stabilizes DNA binding in living cells. *J. Struct. Biol.* **147**, 50–61.

7. Farla, P., Hersmus, R., Trapman, J., and Houtsmuller, A. B. (2005) Antiandrogens prevent stable DNA-binding of the androgen receptor. *J. Cell Sci.* **118**, 4187–4198.

8. Rayasam, G. V., Elbi, C., Walker, D. A., Wolford, R., Fletcher, T. M., Edwards, D. P., and Hager, G. L. (2005) Ligand-specific dynamics of the progesterone receptor in living cells and during chromatin remodeling in vitro. *Mol. Cell. Biol.* **25**, 2406–2418.

9. McNally, J. G., Müller, W. G., Walker, D., Wolford, R., and Hager, G. L. (2000) The glucocorticoid receptor: rapid exchange with regulatory sites in living cells. *Science* **287**, 1262–1265.

10. Schaaf, M. J. and Cidlowski, J. A. (2003) Molecular determinants of glucocorticoid receptor mobility in living cells: the importance of ligand affinity. *Mol. Cell. Biol.* **23**, 1922–1934.

11. Stenoien, D. L., Patel, K., Mancini, M. G., Dutertre, M., Smith, C. L., O'Malley, B. W., and Mancini, M. A. (2001) FRAP reveals that mobility of oestrogen receptor-alpha is ligand- and proteasome-dependent. *Nat. Cell Biol.* **3**, 15–23.

12. Agresti, A., Scaffidi, P., Riva, A., Caiolfa, V. R., and Bianchi, M. E. (2005) GR and HMGB1 interact only within chromatin and influence each other's residence time. *Mol. Cell* **18**, 109–121.

13. Houtsmuller, A. B. (2005) Fluorescence recovery after photobleaching: application to nuclear proteins. In: *Advances in biochemical engineering* (Rietdorf, J., Ed.), Vol. 95, Springer-Verlag, Berlin, pp. 177–199.

14. Axelrod D, K. D., Schlessinger J, Elson E, and Webb WW. (1976) Mobility measurement by analysis of fluorescence photobleaching recovery kinetics. *Biophys. J.* **16**, 1055–1069.

15. Fukano, T., Hama, H., and Miyawaki, A. (2004) Similar diffusibility of membrane proteins across the axon-soma and dendrite-soma boundaries revealed by a novel FRAP technique. *J. Struct. Biol.* **147**, 12–18.

16. Houtsmuller, A. B. and Vermeulen, W. (2001) Macromolecular dynamics in living cell nuclei revealed by fluorescence redistribution after photobleaching. *Histochem. Cell Biol.* **115**, 13–21.

17. Hoogstraten, D., Nigg, A. L., Heath, H., Mullenders, L. H. F., van Driel, R., Hoeijmakers, J. H. J., Vermeulen, W., and Houtsmuller, A. B. (2002) Rapid switching of TFIIH between RNA polymerase I and II transcription and DNA repair in vivo. *Mol. Cell* **10**, 1163–1174.

18. Mattern, K. A., Swiggers, S. J., Nigg, A. L., Lowenberg, B., Houtsmuller, A. B., and Zijlmans, J. M. (2004) Dynamics of protein binding to telomeres in living cells: implications for telomere structure and function. *Mol. Cell. Biol.* **24**, 5587–5594.

19. Dundr, M., Hoffmann-Rohrer, U., Hu, Q., Grummt, I., Rothblum, L. I., Phair, R. D., and Misteli, T. (2002) A kinetic framework for a mammalian RNA polymerase in vivo. *Science* **298**, 1623–1626.

20. Houtsmuller, A. B., Rademakers, S., Nigg, A. L., Hoogstraten, D., Hoeijmakers, J. H. J., and Vermeulen, W. (1999) Action of DNA repair endonuclease ERCC1/XPF in living cells. *Science* **284**, 958–961.

21. Rademakers, S., Volker, M., Hoogstraten, D., Nigg, A. L., Mone, M. J., van Zeeland, A. A., Hoeijmakers, J. H. J., Houtsmuller, A. B., and Vermeulen, W. (2003) Xeroderma pigmentosum group A protein loads as a separate factor onto DNA lesions. *Mol. Cell. Biol.* **23**, 5755–5767.

22. Sporbert, A., Gahl, A., Ankerhold, R., Leonhardt, H., and Cardoso, M. C. (2002) DNA polymerase clamp shows little turnover at established replication sites but sequential de novo assembly at adjacent origin clusters. *Mol. Cell* **10**, 1355–1365.

23. Essers, J., Theil, A. F., Baldeyron, C., van Cappellen, W. A., Houtsmuller, A. B., Kanaar, R., and Vermeulen, W. (2005) Nuclear dynamics of PCNA in DNA replication and repair. *Mol. Cell. Biol.* **25**, 9350–9359.

24. Zotter, A., Luijsterburg, M. S., Warmerdam, D. O., Ibrahim, S., Nigg, A., van Cappellen, W. A., Hoeijmakers, J. H. J., van Driel, R., Vermeulen, W., and Houtsmuller, A. B. (2006)

Recruitment of the nucleotide excision repair endonuclease XPG to sites of UV-induced DNA damage depends on functional TFIIH. *Mol. Cell. Biol.* **26**, 8868–8879.

25. Phair, R. D. and Misteli, T. (2000) High mobility of proteins in the mammalian cell nucleus. *Nature* **404**, 604–609.

26. Phair, R. D., Scaffidi, P., Elbi, C., Vecerova, J., Dey, A., Ozato, K., Brown, D. T., Hager, G., Bustin, M., and Misteli, T. (2004) Global nature of dynamic protein-chromatin interactions in vivo: three-dimensional genome scanning and dynamic interaction networks of chromatin proteins. *Mol. Cell. Biol.* **24**, 6393–6402.

27. Essers, J, Houtsmuller, A. B., van Veelen, L., Paulusma, C., Nigg, A. L., Pastink, A., Vermeulen, W., Hoeijmakers, J. H., and Kanaar, R. (2002) Nuclear dynamics of RAD52 group homologous recombination proteins in response to DNA damage. *EMBO J.* **21**, 2030–2037.

28. Lukas, C., Falck, J., Bartkova, J., Bartek, J., and Lukas, J. (2003) Distinct spatiotemporal dynamics of mammalian checkpoint regulators induced by DNA damage. *Nat. Cell Biol.* **5**, 255–260.

29. Lukas, C., Melander, F., Stucki, M., Falck, J., Bekker-Jensen, S., Goldberg, M., Lerenthal, Y., Jackson, S., Bartek, J., and Lukas, J. (2004) Mdc1 couples DNA double-strand break recognition by Nbs1 with its H2AX-dependent chromatin retention. *EMBO J.* **23**, 2674–2683.

30. Bekker-Jensen, S., Lukas, C., Melander, F., Bartek, J., and Lukas, J. (2005) Dynamic assembly and sustained retention of 53BP1 at the sites of DNA damage are coordinated by Mdc1/NFBD1. *J. Cell Biol.* **170**, 201–211.

31. Leonhardt, H., Rahn, H.-P., Weinzierl, P., Sporbert, A., Cremer, T., Zink, D., and Cardoso, M. C. (2000) Dynamics of DNA replication factories in living cells. *J. Cell Biol.* **149**, 271–280.

32. Kimura, H. and Cook, P. R. (2001) Kinetics of core histones in living human cells: little exchange of H3 and H4 and some rapid exchange of H2B. *J. Cell Biol.* **153**, 1341–1354.

33. Kimura, H. (2005) Histone dynamics in living cells revealed by photobleaching. *DNA Rep. (Amst.)* **4**, 939–950.

34. Chen, D., Dundr, M., Wang, C., Leung, A., Lamond, A., Misteli, T., and Huang, S. (2005) Condensed mitotic chromatin is accessible to transcription factors and chromatin structural proteins. *J. Cell Biol.* **168**, 41–54.

35. Kruhlak, M. J., Lever, M. A., Fischle, W., Verdin, E., Bazett-Jones, D. P., and Hendzel, M. J. (2000) Reduced mobility of the alternate splicing factor (ASF) through the nucleoplasm and steady state speckle compartments. *J. Cell Biol.* **150**, 41–52.

36. Kimura, H., Sugaya, K., and Cook, P. R. (2002) The transcription cycle of RNA polymerase II in living cells. *J. Cell Biol.* **159**, 777–782.

37. Trapman, J. (2001) Molecular mechanisms of prostate cancer. *Eur. J. Cancer* **37**, S119–125.

38. Feldman, B. J. and Feldman, D. (2001) The development of androgen-independent prostate cancer. *Nat. Rev. Cancer* **1**, 34–45.

39. Brinkmann, A. O., Faber, P. W., van Rooij, H. C. J., Kuiper, G. G. J. M., Ris, C., Klaassen, P., van der Korput, J. A. G. M., Voorhorst, M. M., van Laar, J. H., Mulder, E., and Trapman, J. (1989) The human androgen receptor: domain structure, genomic organization and regulation of expression. *J. Steroid Biochem.* **34**, 307–310.

40. Claessens, F., Verrijdt, G., Schoenmakers, E., Haelens, A., Peeters, B., Verhoeven, G., and Rombauts, W. (2001) Selective DNA binding by the androgen receptor as a mechanism for hormone-specific gene regulation. *J. Steroid Biochem. Mol. Biol.* **76**, 23–30.

41. Cleutjens, K. B. J. M., van der Korput, J. A. G. M., van Eekelen, C. C. E. M., van Rooij, H. C. J., Faber, P. W., and Trapman, J. (1997) An androgen response element in a far upstream enhancer region is essential for high, androgen-regulated activity of the prostate-specific antigen promoter. *Mol. Endocrinol.* **11**, 148–161.

42. Schaaf, M. J. M., Lewis-Tuffin, L. J., and Cidlowski, J. A. (2005) Ligand-selective targeting of the glucocorticoid receptor to nuclear subdomains is associated with decreased receptor mobility. *Mol. Endocrinol.* **19**, 1501–1515.

43. Van Royen, M. E., Cunha, S. M., Brink, M. C., Mattern, K. A., Nigg, A. L., Dubbink, H. J., Verschure, P. J., Trapman, J., and Houtsmuller, A. B. (2007) Compartmentalization of androgen receptor protein-protein interactions in living cells. *J. Cell Biol.* **177**, 63–72.
44. Bruggenwirth, H. T., Boehmer, A. L. M., Lobaccaro, J. M., Chiche, L., Sultan, C., Trapman, J., and Brinkmann, A. O. (1998) Substitution of Ala564 in the first zinc cluster of the deoxyribonucleic acid (DNA)-binding domain of the androgen receptor by Asp, Asn, or Leu exerts differential effects on DNA binding. *Endocrinology* **139**, 103–110.
45. Ellenberg, J., Siggia, E. D., Moreira, J. E., Smith, C. L., Presley, J. F., Worman, H. J., and Lippincott-Schwartz, J. (1997) Nuclear membrane dynamics and reassembly in living cells: targeting of an inner nuclear membrane protein in interphase and mitosis. *J. Cell Biol.* **138**, 1193–1206.
46. Blonk, J. C. G., A. Don, H. Van Aalst, and J. J. Birmingham (1993) Fluorescence photobleaching recovery in the confocal scanning light microscope. *J. Micros.* **169**, 363–374.
47. Braga, J., Desterro, J., and Carmo-Fonseca, M. (2004) Intracellular macromolecular mobility measured by fluorescence recovery after photobleaching with confocal laser scanning microscopes. *Mol. Biol. Cell* **15**, 4749–4760.
48. Braga, J., McNally, J. G., and Carmo-Fonseca, M. (2007) A reaction-diffusion model to study RNA motion by quantitative fluorescence recovery after photobleaching. *Biophys. J.* **92**, 2694–2703.
49. Sprague, B. L., and McNally, J. G. (2005) FRAP analysis of binding: proper and fitting. *Trends Cell Biol.* **15**, 84–91.
50. Sprague, B. L., Pego, R. L., Stavreva, D. A., and McNally, J. G. (2004) Analysis of binding reactions by fluorescence recovery after photobleaching. *Biophys. J.* **86**, 3473–3495.
51. Braeckmans, K., Peeters, L., Sanders, N. N., De Smedt, S. C., and Demeester, J. (2003) Three-dimensional fluorescence recovery after photobleaching with the confocal scanning laser microscope. *Biophys. J.* **85**, 2240–2252.
52. Carrero, G., McDonald, D., Crawford, E., de Vries, G., and Hendzel, M. J. (2003) Using FRAP and mathematical modeling to determine the in vivo kinetics of nuclear proteins. *Methods* **29**, 14–28.
53. Marcelli, M., Stenoien, D. L., Szafran, A. T., Simeoni, S., Agoulnik, I. U., Weigel, N. L., Moran, T., Mikic, I., Price, J. H., and Mancini, M. A. (2006) Quantifying effects of ligands on androgen receptor nuclear translocation, intranuclear dynamics, and solubility. *J. Cell. Biochem.* **98**, 770–788.
54. Garcia-Parajo, M. F., Segers-Nolten, G. M. J., Veerman, J.-A., Greve, J., and van Hulst, N. F. (2000) Real-time light-driven dynamics of the fluorescence emission in single green fluorescent protein molecules. *Proc. Natl. Acad. Sci. USA* **97**, 7237–7242.

Part VI
Imaging Methods

Chapter 21
Nanosizing by Spatially Modulated Illumination (SMI) Microscopy and Applications to the Nucleus

Udo J. Birk, David Baddeley, and Christoph Cremer

Keywords SMI nanosizing; Structured illumination; Sizes and distances; Transcription factories; Fluorescence microscopy

Abstract In this chapter we present the method of spatially modulated illumination (SMI) microscopy, a (far-field) fluorescence microscopy technique featuring structured illumination obtained via a standing wave field laser excitation pattern. While this method does not provide higher optical resolution, it has been proven a highly valuable tool to access structural parameters of fluorescently labeled macromolecular structures in cells. SMI microscopy has been used to measure relative positions with a reproducibility of <2 nm between fluorescing objects. Among others, we have measured size distributions of protein clusters with an accuracy much better than the resolution achievable e.g. in confocal microscopy. The advantages of the SMI microscope over other (ultra-)high resolution light microscopes are its easy sample preparation and microscope handling as well as the comparably fast acquisition times and large fields of view.

1 Introduction

The spatial architecture not only of the genome, but of many protein complexes with respect to their functions and dynamics in the cell nucleus, have emerged as a major topic of biomedical research. Describing the complex nuclear genome structure requires the combination of advanced experimental and theoretical approaches *(1)*. However, precise information on specific chromatin nanostructures is still scarce. For example, compaction of specific gene domains is generally regarded to be closely correlated to the transcription potential of a gene or cluster of genes; recent biocomputing approaches *(2)* have indicated that compaction and other nanostructural features of specific gene domains might play a decisive role in the accessibility of transcription regulating factors to specific domain sites. If so, gene domain nanostructure would constitute an additional level of gene regulation. Another example for the requirement for improved knowledge about nuclear genome nanostructures is the induction of cancer-related genome instability and epigenetic reprogramming. These processes are regulated to a significant part at the chromatin level, involving complex changes in histone tail modifications as well as three-dimensional (3D) chromatin folding. Although ultrastructural methods (e.g. electron microscopy) provide optical resolution down to the nanometer range and have provided most important contributions, they are limited in terms of detecting multiple differently labeled molecules in the same cell, preserving cellular nanostructure, and relaying structural information back to defined genetic sequence information.

In recent years methods to narrow the microscopic point spread function (PSF), and hence to overcome the conventional light optical resolution limit of about 200 nm laterally and 600 nm axially (in the direction of viewing), have been developed. These methods have been employed in novel types of microscopes such as the 4Pi microscope *(3, 4)* improving axial optical resolution to about 100 nm, or the stimulated emission depletion (STED) microscope *(5)*, which is currently capable of pushing the lateral optical resolution below 20 nm in the focal plane *(6)*. An increase in the optical resolution is also possible using non-focusing techniques *(7–11)*. An additional approach to the study of cellular nanostructures is given by spectral precision distance microscopy (SPDM), which measures the distance between spectrally distinct small fluorescent structures or even single molecules, allowing a distance resolution in the range of few tens of nanometers *(12–14)*. This method is based on high-precision localization of such structures. Recently, Betzig et al. *(15)* and Hess et al. *(16)* described another method of localization microscopy; this technique, termed photoactivated localization microscopy (PALM), allows the requirement for multiple spectral signatures to be overcome by detecting individual fluorophores after sparse photoactivation.

All the techniques mentioned above, as well as the SMI method described in this chapter, rely on fluorescent labeling. The procedures of fluorescence in situ hybridization (FISH) to specifically label gene regions and immunostaining protocols to specifically target structures in fixed cell preparations are well established. With the urgent need to also stain live cells, "in vivo" labels based on fluorescent proteins were developed. With the SMI microscope described below, all of these methods are equally suitable. In contrast to many other novel fluorescence microscopic procedures, most of the widely used fluorophores yield in the SMI microscope a signal that is by far sufficient in combination with short acquisition times and in compliancy with commonly followed cell culture/cell preparation protocols.

In the case when the fluorescently labeled objects are sparsely distributed ("optical isolation"), important structural features can be analyzed using structured illumination achieved by an interference pattern in the object space *(17)*. In many applications, it is sufficient to know the size and the position of subwavelength-sized objects, a realm where SMI microscopy *(18–20)* excels, at least in transparent specimens. Possible biological applications comprise the analysis of chromatin complexes such as the following: topological analysis of genomic control elements regulating gene activity such as silencing/activating complexes together with their genomic accessibility; conformational influence of drugs and other environmental agents or radiation; experimental validation of quantitative modeling and simulation of the localization of supramolecular complexes in vivo; nanosizing of gene duplications/deletions in individual cell nuclei; nanoscopical analysis of the structure of genetically active/inactive chromosomal vectors in adult stem cells and in animal embryonic stem cells; and also questions relating to conformational changes of chromatin in gene expression due to aging or diseases (e.g. cancer).

We have previously shown that SMI approaches can be used to measure distances with an outstanding precision *(18)* and to measure the diameter of individual fluorescent targets down to a few tens of nanometers *(19)*, i.e. for objects the volumes of which are several orders of magnitude smaller than the observation volume of the point spread function of confocal laser scanning microscopy (CLSM) *(21)*.

A presumed reorganization of the chromatin induced by transcriptional activity may lead to a difference in dimension and compaction of DNA in the vicinity of the respective gene domain when compared with its transcriptionally inactive state, a change which may be detected with appropriate high resolution microscopy. First steps towards this goal have been reported using SMI microscopy by extracting, with high precision, size- and compaction-related parameters of the 7q22 gene region *(22)* and of other specific gene regions *(23)* in 3D conserved cells after FISH. In general, we have shown that the SMI microscope can be applied to address the structural parameters of not only specific gene domains, but also of macromolecular objects involved in and undergoing other structural changes, e.g. the accumulation of nuclear proteins during transcription-related processes.

Here we present an application of SMI microscopy to the measurement of the size of transcription factories as an example of the performance range of this special type of microscopy. A thorough description of this study can be found in ref. *(24)*. In the nucleus of an actively growing human (HeLa) cell, three different types of RNA polymerase (Pol I, II, and III) are localized in approximately 10,000 biomolecular complexes termed transcription factories *(25)*; see also *(26)*. The three different Pols are defined by different drug sensitivities, nuclear localization, and the genes they transcribe. Since in mammalian cells each transcription unit is transcribed by only one (or a few) Pol II enzymes *(27, 28)*, one would expect that each Pol II transcription factory consisting of about eight active Pols *(25)* is simultaneously transcribing multiple transcription units. Due to the size of these transcription factories, a direct verification of co-localization of at least two different nascent transcripts within the same factory is difficult, because of the limited resolution and sensitivity of current imaging systems. In the following an example of SMI size determination is presented in measuring the axial extent of fluorescently labeled Pol II factories in fixed cells. With the present SMI setup, first measurements have shown that acquisitions of live cells are also possible. However, the data illustrated here are based on fixed cell experiments.

2 Materials

2.1 The SMI Microscope

The setup of the SMI microscope is depicted in Fig. 21.1. The fluorescence light is collected by an objective lens (×100, oil immersion, 1.4 NA) and detected on a CCD camera, after passing an appropriate filter set (beam splitter and emission filter) similar to a wide field fluorescence microscope. In contrast to a widefield microscope, an illumination light pattern is used in the SMI microscope to vary the excitation intensity in the object space, i.e. at the position of the object slide between the two objectives (see setup scheme). This structured illumination pattern, or standing wave field, is achieved by the interference of two laser light beams in the object space, yielding so called fringes or light sheets stacked along the optical axis (z), i.e. a modulation pattern where the intensity varies according to a $\sin^2(z)$ function (see Fig. 21.2).

The cells containing the fluorescent objects are mounted between a conventional object slide and a conventional cover slip. The sample preparation is brought into the space between the two opposite objective lenses, then the sample is moved along the z-axis (typical step size 40 nm) through the focus of the detection objective

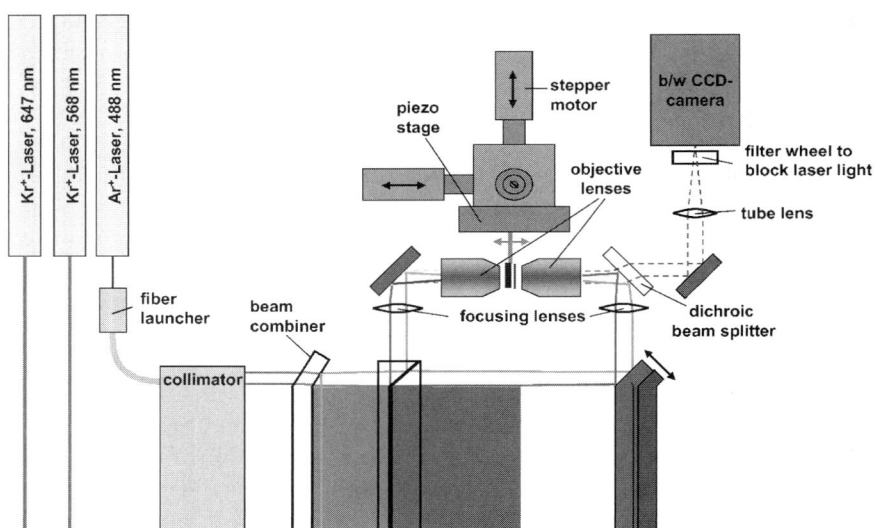

Fig. 21.1 Setup of the present SMI microscope. Here, only one combination of the possible sets of excitation wavelengths is shown. After the combination of the different excitation laser lines, the illumination light is split by the 50:50 beam splitter into two beams of equal intensities. Both beams pass a focusing lens and an objective lens each, before the two counter propagating waves are brought to interference yielding a standing wave field in the object space between the two objective lenses. Only the right objective lens is used for detection of the fluorescence light

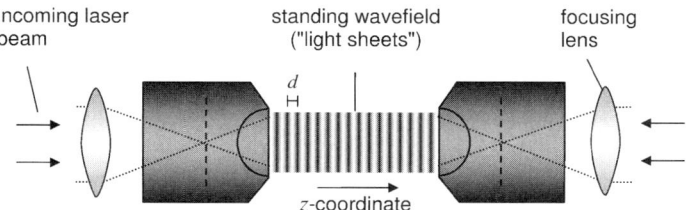

incoming laser beam

standing wavefield ("light sheets")

focusing lens

z-coordinate

Fig. 21.2 Along the optical axis (z), the excitation light is modulated (standing wave field) by the interference of the two laser beams from both sides according to a $\sin^2$, with a modulation wavelength $d \approx 163\,nm$ when exciting fluorophores (in a medium with refractive index $n \approx 1.5$) using the 488-nm laser line. The illumination intensity perpendicular to the optical axis is considered to be constant (i.e. the illumination intensity for a single image in the 3D recording process)

lens to obtain a 3D image data stack with the SMI microscope. During this movement, the fluorophores are excited and thus emit fluorescence light according to the excitation light pattern described above.

As an alternative to the stage scan, the object can remain stationary and the wave field can be shifted axially through the object space by actuation of the piezo-controllable mirror (see Fig. 21.1). The piezoelectric object stage itself is attached to a stepper motor-actuated object stage for the macroscopic movement of the object slide; this provides lateral translation for the search of suitable objects and for moving the object slide in or out of the space between the objective lenses.

2.2 Cell Culture and Preparation

As with any of the novel microscopic techniques, good quality of the sample is necessary to obtain trustworthy results. In particular, take special care to remove all unspecific fluorescence and to obtain the lowest possible autofluorescence. Culture and prepare cells as you would for high quality confocal imaging. For the transcription factories studied here, the following protocol was used:

1. HeLa cells are fixed in 4% and then in 8% paraformaldehyde in 250 mM Hepes, pH 7.2, and frozen.
2. Cryosections (~140-nm thick as deduced from interference color) are cut from the frozen blocks using an UltraCut UCT 52 ultracryomicrotome (Leica, Milton Keynes, UK), captured on drops of 2.1 M sucrose in PBS, and transferred to cover slips.
3. Pol II sites were indirectly immunolabelled using the mouse monoclonal antibody H5 against the hyperphosphorylated C-terminal domain (CTD) of the largest subunit of Pol II (phosphorylated at Ser^2 of the heptad repeat); BAbCO/Covance, Richmond, CA, USA) diluted 1:2,000 in PBS, for 2 h.
4. For CLSM and SMI microscopy, sections were incubated for 1 h with Alexa Fluor 488-conjugated antibodies raised against mouse IgG (1:1,500; Molecular

Probes, Eugene, OR, USA) or, for a three-layer detection protocol, incubated for
1 h with antibodies raised in rabbit against mouse Ig (1:200; Cappel; MP
Biomedicals, Illkirch, France) and then for 1 h with Alexa Fluor 488-conjugated
antibodies raised in goat against rabbit IgG (1:4,000; Molecular Probes).
5. Sections were washed in 0.1% (v/v) Tween-20 in PBS before nucleic acids were
counterstained for 45 min in 20 μM TOTO-3 (Molecular Probes) in PBS-Tween,
washed successively in PBS-Tween and PBS (5×), and mounted in VectaShield
(Vector Laboratories, Peterborough, UK).

2.3 Beads

Fluorescent microspheres (beads) with a diameter of 100 nm are used to calibrate
the standing wave field of the microscope interferometer. Green and orange fluorescent beads are available from Duke Scientific (Microgenics Corp., Fremont, CA,
USA; http://www.dukescientific.com) or from Molecular Probes (Invitrogen Corp.,
Carlsbad, CA; http://probes.invitrogen.com), which also supplies red fluorescent
beads. Other suppliers are Bangs Laboratories (Bangs Laboratories, Fishers, IN,
USA; http://www.bangslabs.com) and Polysciences (Polysciences, Warrington, PA,
USA; http://www.polysciences.com). The beads have to be selected such that their
emission spectra match those of the fluorophores with which the cellular objects
were labeled. For multi-color distance measurements, multi-spectral beads (e.g.
TetraSpeck; Bangs Laboratories) are used to simultaneously correct for chromatic
aberrations of the microscope optics.

3 Methods

The SMI microscope is a robust system that in contrast to other fluorescence microscopes does not feature visual detection through an ocular, but only via the highly
sensitive CCD camera. In our experience, a Ph.D. student in cell biology with
understanding of light microscopy will be able to use the instrument independently
after a few days of guidance.

3.1 Sample Preparation

1. For SMI microscopy sample preparations, conventional glass object slides and
cover slips (BK7-glass, 170-μm thickness) can be used (see **Note 1**).
2. Cells are prepared preferably on a cover slip.
3. Since the SMI microscope has a wide field detection, all the out-of-focus fluorescence contributes to the detected intensity. It is important to optimize the
samples with respect to autofluorescence and unspecific labeling.

4. For densely populated objects as in the case of transcription factories, additional measures have to be taken when the objects are no longer optically isolated. For this particular problem, two options are available. First, the cells can be cut in thin sections as would be done for electron microscopy preparations. Second, the cells can be artificially flattened by hypotonic treatment so that they spread out on the cover slip, and the objects are laterally resolved.

5. Prepare a 1:1,000 dilution of the 100-nm beads (*see* **Note 2**), fluorescing in the same colour as your samples. Use multi-spectral beads if necessary.

6. Add beads to the object slide and let them dry so that they stick to the glass surface (*see* **Note 2**).

3.2 Starting the Microscope

1. Turn on water cooling for the lasers (*see* **Note 3**).

2. Turn on lasers (for each laser): Switch power on for the high-voltage power supply. Turn the key on the laser launcher. Ensure that light and current dials are set to low values. Set lasers to "Current Control". Press "Power on". After warm-up time (~90 sec), press "Start". Change to "Light Control", and adjust to the desired stable light output power.

3. Switch on all electrical devices: stepper motors, piezo-electrical control unit, and camera.

4. If required, switch on the white light illumination (*see* **Note 4**).

5. Switch on the computer and monitor.

6. Start the SMI control software *PySMI. NB: A warning will automatically appear "remove the object mounting"*. Make sure the object holder is removed before continuing, because the stepper motor stages will now be initialized. Wait for initialization movements to finish.

7. Place the object slide in the object holder with the labeled end facing towards the detection objective lens and pointing up.

8. Attach the slide holder on top of the stepper motor stage.

9. In the control software, switch joystick ON, and increase the joystick speed to "5" to conveniently move the slide in between the two objective lenses. Be careful not to let the slide or the slide holder touch the objective lenses.

10. Apply a small drop of immersion media (oil) on both sides between the sample and the objective lenses.

11. Reduce the joystick speed to "1".

3.3 Data Acquisition

1. At the beginning and end of each measurement day, the wavefield of the SMI microscope must be calibrated. This is done by acquiring a few 3D stacks with

beads of known diameters as test specimen (calibration measurement, *see* **Section 2.3** and **Note 2**). Data acquisition is the same as for cell preparations.

2. Focus the cells (beads) by using the 3D stepper motor stage.
3. If the focal plane is difficult to find in biological samples, e.g. when the cells are sparsely distributed or weakly labeled, use the white light illumination. This helps to prevent photobleaching prior to the measurement.
4. In the camera properties dialog, select the color channels you want to acquire. For high-resolution images, ensure that binning is switched off.
5. Before starting the acquisition, select the *z*-scanning range in the *PySMI* acquisition dialog. A typical stack consists of about 100 images spaced by 40 nm in the axial direction, spanning 4 μm in the object space. For thick samples, the number of images may be increased, but not the axial stepsize.
6. Save the data immediately after it is displayed.

3.4 Data Evaluation

The SMI data is evaluated offline on a different computer. For data evaluation, the MATLAB environment and the DipImage toolbox are used. A toolbox named *sviewer* has been written to evaluate the SMI data stacks, which is accessed via the MATLAB command window. LaTeX needs to be installed on the machine in order to automatically produce reports from the results.

1. When using the *sviewer* toolbox, the parameters for automated object detection and fitting are stored in a database. Connect to the database with the command "connectToDB". This will open a browser window, where additional explanations/help on the various parameters are given (*see* **Note 5**).
2. By typing "svlauncher" into the command line, the *sviewer* toolbox is initialized. Here, the SMI data stack can be read from file before starting the *sviewer* user interface (see Fig. 21.3). In this dialog window, the appropriate color channel can be selected.
3. In a first evaluation run, the file with the acquired beads data will be used to determine the quality of the modulation of the standing wavefield i.e. the unmodulated part (parameter UMOD) of the excitation pattern. As the modulation is used to measure the objects' axial extensions, good statistics with several hundred beads is required. The size of the beads is entered as a parameter in the REF_BEADS field. The following steps are the same for analyzing data from beads and/or biological samples. Repeat **step 4–6** from **Section 3.4** for each acquired data file. **Step 7** from **Section 3.4** is only necessary for bead measurements.
4. For the analysis of both the beads and the cells, an automated object-finding routine is used to detect the fluorescently labeled objects in the 3D image stacks (*see* **Note 6**). To initialize this, type "ofindroi" in the command line, and select the region of interest by outlining the cell with the mouse.
5. All identified objects can now be analyzed. Start the automated object-fitting evaluation with the command "dofits".

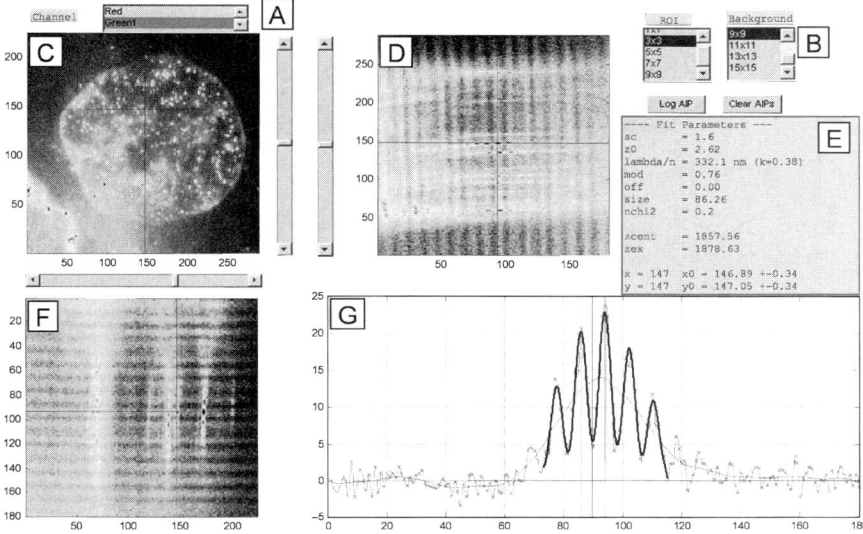

Fig. 21.3 The *sviewer* user interface. Here, the green color channel has been selected (**a**). Values for ROI and BACKGOUND (**b**) are taken from the database. An object is selected in the *z*-section (**c**), for which the respective *z-y*-section (**d**) and *x-z*-section (**f**) are displayed. For the selected object, the axial intensity profile (**g**) and a summary of the results of the SMI nanosizing procedure (**e**) are shown. For this polymerase II complex, a size of 86 nm was found

6. Results are visualized with the command "`showres`" (see Fig. 21.4), which will display a maximum intensity projection (or alternatively the intensities averaged over the depth of focus), a histogram of the measured sizes and a figure, where the detected objects are drawn as small spheres colored according to the size extracted from the automated SMI nanosizing analysis. Objects that could not be fitted either due to a poor signal-to-noise ratio or a size outside the measurable range are displayed in purple or black. The results can be saved to the hard disk and a PS/PDF report generated with the command "`savefits_rep`". More results can be displayed with the commands "`show_more_res`" and "`show_lots_more_res`".

7. In the first step, you have now analyzed the 100-nm beads for the calibration of the wavefield. The unmodulated excitation intensity can now be extracted from the variable "`umods`", by typing "`median(umods)`". Enter this value into the field UMOD of the database's dialog for the parameters, then save. Using the thus calibrated wavefield, the data from the biological samples can be analyzed similarly starting from **Section 3.4, step 4**.

As argued in this chapter, the SMI microscope is a reliable and easy-to-use instrument providing structural information on the nanoscale even without an

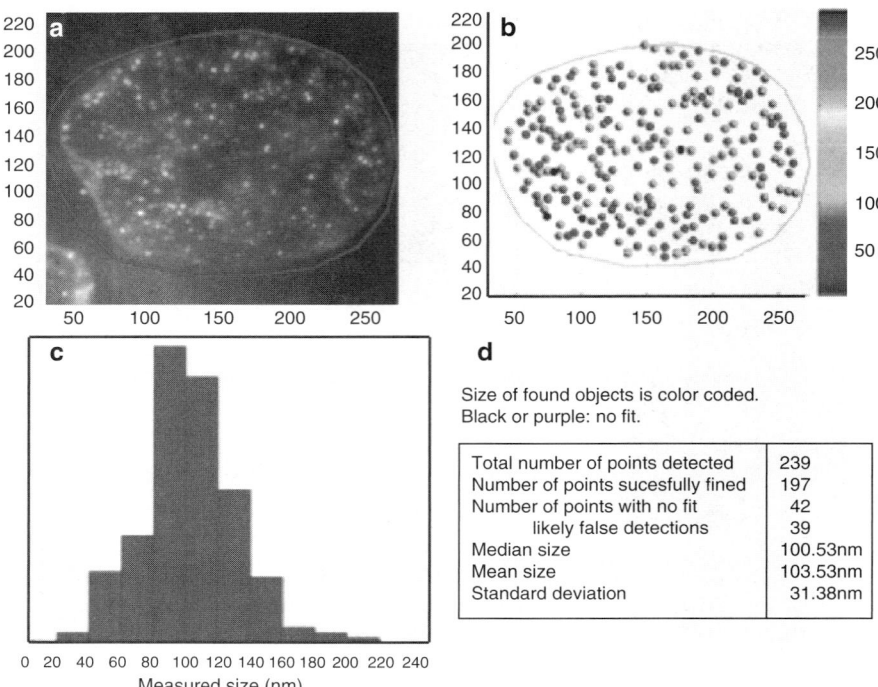

Fig. 21.4 The results as saved by the command "savefis_rep." **a** Maximum intensity projection; unprocessed image data. **b** Automatically detected and analyzed objects. The object color corresponds to the extracted SMI size (nanometers). Purple and black indicate objects for which no fit was obtained (e.g. the signal-to-noise ratio was too poor to analyze or the objects' sizes are above the size range for which SMI nanosizing is feasible). **c** Histogram of the extracted object sizes. **d** Summary of the results for this cell. The analysis of the transcription factories yielded an average size of 100 nm for this cell. Note: The system is optimized for axial size and position determination. The lateral image (**a**) corresponds to that of a conventional wide field fluorescence microscope. To view this figure in color, see COLOR PLATE 13

increased optical resolution in its classical sense. By the use of this method, valuable insights into the architecture of not only cell nuclei can be gained. Even though the SMI microscope in its present layout is limited to access the z-direction only, current work on the setup promises to overcome this impasse by the implementation of a 3D illumination structure. As a result, this microscope is going to be an even more promising device for future biological and biomedical research applications.

4   Notes

1. The accessible range is limited by the stepper motor stage, and the objects have to be centered on the whole object slide (not on the transparent part of the slide).

2. For best results, the cells should be placed on the cover slip and the beads on the object slide. Beads can be diluted in ethanol, water, or PBS. Especially when preparing the beads on the cover slip, dilution in ethanol is preferable because then the beads are distributed more evenly. Typical batches from commercial suppliers are shipped with 1% solids. These need to be diluted 1:1,000 to obtain good statistics for the calibration of the wave field.

3. Lasers should be switched on 2–3 h before the experiment. Valve 1 (Ventil 1) is used to switch the water cooling of the blue-green Argon Ion Laser (488-nm excitation) and the dark red Lexel Krypton Ion Laser (647 nm). Valve 2 (Ventil 2) does the same thing for the yellow Lexel Krypton Ion Laser (568-nm excitation).

4. A super-bright LED connected to a lab power supply serves as a white light illumination. This can be used to find the focal plane, and to visually inspect the cell morphology. The images achieved with this illumination are similar to phase contrast.

5. Different types of parameters are required for different steps in the data evaluation. For data extraction, setting the parameters ROI = 2 for averaging and BACKGROUND = 5 for background subtraction are recommended. During the evaluation, the intensity data is fitted to a model function. Here, the axial STEPSIZE (=40 typically) needs to be entered (in nanometers), as well as the effective modulation wavelength MED_K_EFF = $(2\pi n/\lambda)\cdot$STEPSIZE, which should be 0.75, when using λ = 488-nm excitation and a specimen with refractive index n = 1.44.

6. The parameters OFIND_BLUR (=2 typically) and OFIND_THRESH can be varied to change the sensitivity of the automated object-detection routine. Lower values for OFIND_THRESH will result in the detection of more objects.

Acknowledgments The authors thank Dr. Sonya Martin and Dr. Ana Pombo from the MRC London. This work was supported by the German Research Foundation, Projects DFG CR 60/16-1–3 and DFG CR 60/23-1–2. The authors gratefully acknowledge initial funding for the SMI microscope setup by the BMBF (Bundesministerium für Bildung und Forschung). Udo Birk gratefully acknowledges Funding through a Marie-Curie Intra-European Fellowship (MEIF-CT-2006-041827)

References

1. O'Brien, T. P., Bult, C. J., Cremer, C., Grunze, M., Knowles, B. B., Langowski, J., McNally, J., Pederson, T., Politz, J. C., Pombo, A., Schmahl, G., Spatz, J. P., and van Driel, R. (2003) Genome function and nuclear architecture: from gene expression to nanoscience. *Genome Res.* **13**, 1029–1041.

2. Odenheimer, J., Kreth, G., and Heermann, D. W. (2005) Dynamic simulation of active/inactive chromatin domains. *J. Biol. Phys.* **31**, 351–363.

3. Egner, A. and Hell, S. W. (2005) Fluorescence microscopy with super-resolved optical sections. *Trends Cell Biol.* **15**, 207–215.

4. Hell, S. W., Lindek, S., Cremer, C., and Stelzer, E. H. K. (1994) Measurement of the 4Pi-confocal point spread function proves 75 nm resolution. *Appl. Phys. Lett.* **64**, 1335–1337.

5. Willig, K. I., Rizzoli, S. O., Westphal, V., Jahn, R., and Hell, S. W. (2006) Nanoscale resolution in GFP-based microscopy. *Nature* **440**, 935–939.

6. Westphal, V. and Hell, S. W. (2005) Nanoscale resolution in the focal plane of an optical microscope. *Phys. Rev. Lett.* **94**, 143903.

7. Gustafsson, M. G. L., Agard, D. A., and Sedat, J. W. (1995) Seven-fold improvement of axial resolution in 3-D widefield microscopy using two objective lenses. *Proc. SPIE* **2412**, 147–156.

8. Gustafsson, M. G. (2005) Nonlinear structured-illumination microscopy: wide-field fluorescence imaging with theoretically unlimited resolution. *Proc. Natl. Acad. Sci. USA* **102**, 13081–13086.

9. Heintzmann, R., Jovin, T. M., and Cremer, C. (2002) Saturated patterned excitation microscopy (SPEM)—a novel concept for optical resolution improvement. *J. Opt. Soc. Am. A. Opt. Image Sci. Vis.* **19**, 1599–1609.

10. Frohn, J. T., Knapp, H. F., and Stemmer, A. (2001) Three-dimensional resolution enhancement in fluorescence microscopy by harmonic excitation. *Opt. Lett.* **26**, 828–830.

11. Frohn, J. T., Knapp, H. F., and Stemmer, A. (2000) True optical resolution beyond the Rayleigh limit achieved by standing wave illumination. *Proc. Natl. Acad. Sci. USA* **97**, 7232–7236.

12. Bornfleth, H., Sätzler, K., Eils, R., and Cremer, C. (1998) High precision distance measurements and volume-conserving segmentation of objects near and below the resolution limit in three-dimensional confocal fluorescence microscopy. *J. Microsc.* **189**, 118–136.

13. Cremer, C., Edelmann, P., Bornfleth, H., Luz, H., Kreth, G., Münch, H., and Hausmann, M. (1999) in *Handbook of computer vision and applications, Vol. 3* (Jahne, B., Haußecker, H. and Geißler, P., Eds.), pp. 839–857, Academic Press, San Diego, New York.

14. Heilemann, M., Herten, D. P., Heintzmann, R., Cremer, C., Muller, C., Tinnefeld, P., Weston, K. D., Wolfrum, J., and Sauer, M. (2002) High-resolution colocalization of single dye molecules by fluorescence lifetime imaging microscopy. *Anal. Chem.* **74**, 3511–3517.

15. Betzig, E., Patterson, G. H., Sougrat, R., Lindwasser, O. W., Olenych, S., Bonifacino, J. S., Davidson, M. W., Lippincott-Schwartz, J., and Hess, H. F. (2006) Imaging intracellular fluorescent proteins at nanometer resolution. *Science* **313**, 1642–1645.

16. Hess, S. T., Girirajan, T. P., and Mason, M. D. (2006) Ultra-high resolution imaging by fluorescence photoactivation localization microscopy. *Biophys. J.* **91**, 4258–4272.

17. Bailey, B., Farkas, D. L., Taylor, D. L., and Lanni, F. (1993) Enhancement of axial resolution in fluorescence microscopy by standing-wave excitation. *Nature* **366**, 44–48.

18. Albrecht, B., Schweitzer, A., Failla, A. V., Edelmann, P., and Cremer, C. (2002) Spatially modulated illumination (SMI) microscopy allows axial distance resolution in the nanometer range. *Appl. Opt.* **41**, 80–87.

19. Failla, A. V., Spoeri, U., Albrecht, B., Kroll, A., and Cremer, C. (2002) Nanosizing of fluorescent objects by spatially modulated illumination microscopy. *Appl. Opt.* **41**, 7275–7283.

20. Failla, A. V., Albrecht, B., Spöri, U., Schweitzer, A., Kroll, A., Hildenbrand, G., Bach, M., and Cremer, C. (2003) Nanostructure analysis using spatially modulated illumination microscopy. *Com-PlexUs* **1**, 77–88.

21. Spöri, U., Failla, A. V., and Cremer, C. (2004) Superresolution size determination in fluorescence microscopy: A comparison between spatially modulated illumination and confocal laser scanning microscopy. *J. Appl. Phys.* **95**, 8436–8443.

22. Mathée, H., Baddeley, D., Wotzlaw, C., Fandrey, J., Cremer, C., and Birk, U. (2006) Nanostructure of specific chromatin regions and nuclear complexes. *Histochem. Cell Biol.* **125**, 75–82.

23. Hildenbrand, G., Rapp, A., Spoeri, U., Wagner, C., Cremer, C., and Hausmann, M. Nano-sizing of specific gene domains in intact human cell nuclei by spatially modulated illumination light microscopy. (2005) *Biophys. J.* **88**, 4312–4318.

24. Martin, S., Failla, A. V., Spori, U., Cremer, C., and Pombo, A. (2004) Measuring the size of biological nanostructures with spatially modulated illumination microscopy. *Mol. Biol. Cell* **15**, 2449–2455.

25. Pombo, A., Jackson, D. A., Hollinshead, M., Wang, Z., Roeder, R. G., and Cook, P. R. (1999) Regional specialization in human nuclei: visualization of discrete sites of transcription by RNA polymerase III. *EMBO J.* **18**, 2241–2253.
26. Wansink, D. G., Sibon, O. C., Cremers, F. F., van Driel, R., and de Jong, L. (1996) Ultrastructural localization of active genes in nuclei of A431 cells. *J. Cell. Biochem.* **62**, 10–18.
27. Miller, O. L., Jr. and Bakken, A. H. (1972) Morphological studies of transcription. *Acta Endocrinol. Suppl.* **168**, 155–177.
28. Jackson, D. A., Iborra, F. J., Manders, E. M., and Cook, P. R. (1998) Numbers and organization of RNA polymerases, nascent transcripts, and transcription units in HeLa nuclei. *Mol. Biol. Cell* **9**, 1523–1536.

Chapter 22
Visualisation of RNA by Electron Microscopic In Situ Hybridisation

Jacques Rouquette, Karl-Henning Kalland, and Stanislav Fakan

Keywords Electron microscopic in situ hybridisation; RNA probes; Ultrastructural analysis; EDTA staining

Abstract Visualisation of RNA at an ultrastructural level represents a major approach to study organisation and function of the cell nucleus. In addition to methods allowing one to visualise a general distribution of RNA-containing structural constituents, in situ hybridisation (ISH) is a powerful tool for revealing specific RNA sequences or species. In this chapter we describe a method for detecting RNA by electron microscopic in situ hybridisation (EMISH) using anti-sense RNAs as probes. We first present the protocol for preparation of anti-sense RNA probes labeled with different markers, and then describe how such probes are applied to ultrathin sections by a method of ultrastructural ISH. The great advantage of this method is that it does not require denaturing either the specimen or the probe, thus allowing nuclear fine structure to be well preserved. The presence of the marker in the probe can be detected by immunoelectron microscopy using colloidal gold-conjugated antibodies, offering the possibility to evaluate the signal quantitatively. The method can also be combined with cytochemical techniques such as EDTA staining for preferential visualisation of ribonucleoprotein-containing nuclear structural components.

1 Introduction

Ribonucleoproteins are major components of the cell. They contain different types of RNA associated with specific proteins and are involved in many biological processes such as gene expression and regulation, formation of messenger RNA (mRNA), and ribosome subunit assembly in the nucleus, as well as translation of mRNAs in the cytoplasm. Moreover, there are nuclear RNA–protein complexes that are essential for pre-mRNA processing or those containing RNA species restricted to the nucleus and playing an as yet unknown role in nuclear metabolism and functional architecture.

The intranuclear distribution of ribonucleoprotein-carrying structural constituents has been extensively investigated during the last four decades, using various methods of ultrastructural cytochemistry. The introduction of EDTA staining, which preferentially contrasts RNP-containing nuclear structural domains while leaving chromatin weakly contrasted *(1)*, allowed substantial progress providing a general view of the intranuclear distribution of RNP structures. It could be combined with high-resolution autoradiography *(2–4)* and later also with ultrastructural immunocytochemistry, using both resin sections and cryosections *(5)*. More recently, a specific staining method for RNA has been developed, offering the possibility of visualising the general distribution pattern of RNA in the nucleus *(6)*. However, these approaches enable an overall visualisation of RNA or RNP-containing structural components in the nucleus in situ, but cannot provide information about the distribution pattern of specific RNA sequences.

While the protein moieties of ribonucleoproteins can be detected by means of specific antibodies and immunoelectron microscopy, RNA species can be identified by techniques of ultrastructural molecular in situ hybridisation (ISH) using probes constituted by nucleotide sequences complementary to the target RNA molecules. The probes contain markers, such as digoxigenin or biotin, that can be identified by appropriate antibodies coupled to colloidal gold particles. ISH consists of the detection of cellular RNA by a specific probe synthesised in vitro. This probe can be marked by incorporation of digoxigenin or biotin labelled-NTPs all along the probe.

Following the introduction of molecular ISH methods at the light microscope level *(7)*, different technical variants of this approach have been developed for the application at an ultrastructural level. These extend from the use as a target of ultrathin cryosections to ultrathin sections of samples previously embedded in different sorts of synthetic resins (reviewed in, e.g. refs. *(8–10)*). The use of resin sections, even though giving rise to a relatively low hybridisation efficiency restricted by the limited number of target sites made available on the surface of the section by "uncovering" during the cutting process, offers a better stabilisation of the fine cellular structure thanks to the degree of polymerisation of the resin. Acrylic resins, in particular those that are partially hydrosoluble *(11)*, seem to be especially suitable for ISH.

In this chapter we concentrate on the ultrastructural in situ visualisation of RNA at the cellular level using labeled RNAs as probes. This approach, when combined with ultrathin resin sections, is at present the best for specific and high-resolution visualisation of different RNA species. Due to the fact that anti-sense RNAs are the probes of choice (Fig. 22.1), the reaction does not require denaturing either the specimen or the probe, thus offering the most favorable conditions to provide a sufficient signal (*see*, e.g. **refs.** *(12, 13)* and Fig. 22.3). Moreover, the fact that the final detection is carried out by means of antibodies conjugated to colloidal gold particles makes it possible to evaluate the labelling in different nuclear domains or compartments quantitatively and to express the signal in terms of labeling density, i.e. the number of gold particles per unit surface.

We choose to use RNA probes synthesised by in vitro transcription for two main reasons. The first is that their use avoids the utilisation of denaturing agents such as formamide and so favours better structural preservation of the sample. The second

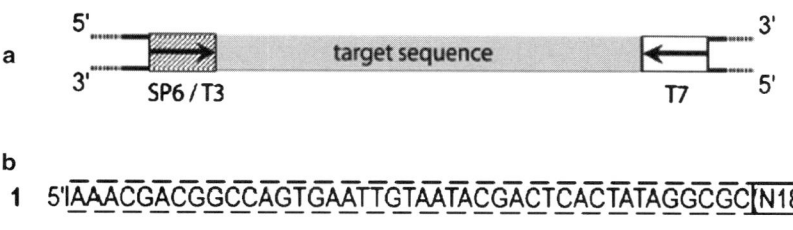

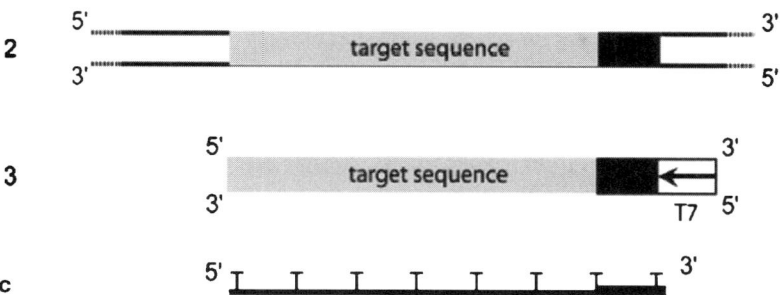

Fig. 22.1 Schematic drawing of steps for in vitro transcription. **a** The target sequence inserted into a transcription vector is flanked by bacteriophage promoters. **b** Generation of PCR fragments as template for in vitro transcription. **b1** The reverse primer used for PCR to generate anti-sense probes; the T7 RNA polymerase promoter sequence is in the *dashed box* and the nucleotide flanking the sequence of interest in the *full box*. **b2** The appropriate template sequence for PCR containing the target sequence. Flanking sequence is indicated by the black box. **b3** The amplified fragment used as template for in vitro transcription. **c** Labelled riboprobes, Flanking sequence is indicated by the black box. *T* represents a digoxygenin- or biotin-NTP

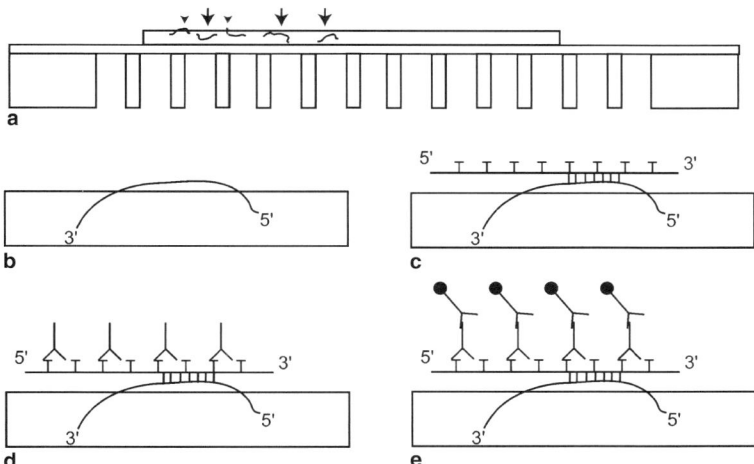

Fig. 22.2 Steps of the hybridisation protocol. **a** Section on formvar/carbon coated grid. *Arrows* correspond to RNA embedded within the section thickness; *arrowheads* represent RNA molecules made accessible to probes by sectioning. **b** Enlargement of RNA exposed on the surface of the section. **c** RNA recognition by a specific labeled probe; *T* corresponds to Tag-coupled UTP. **d** Interaction between primary antibodies and the labeled probe. **e** Interaction between gold-labeled secondary antibodies and primary antibodies. Note that more than one secondary antibody can interact with one primary antibody

is that in vitro transcription allows synthesis of long RNAs in which digoxigenin- or biotin-labelled NTPs are incorporated all along the probe, increasing the potential sensitivity of their detection. The length of the probe is an important aspect; RNA within a resin section is only accessible on its surface (Fig. 22.2a) so that a long probe offers the advantage that even when only a short region is annealed, the long region remaining unhybridised is accessible for secondary detection by antibodies and so the specific signal is amplified.

2 Materials

2.1 Synthesis of Probes

1. The desired sequence cloned in a transcription vector, e.g. pGEM (Promega, Wallisellen, Switzerland) or pBluescript (Stratagene, La Jolla, CA, USA).
2. Restriction enzymes and buffer for vector linearisation.
3. Nuclease-free H_2O.
4. Phenol:chloroform:isoamyl alcohol (25:24:1 v/v, pH 8) (Invitrogen, Carlsbad, CA, USA).
5. Phase Lock Heavy tubes, 1.5 mL (Eppendorf, Hamburg, Germany; cat. 0032 007.953).
6. UltraPure Glycogen (Invitrogen; cat. 10814-010).
7. Absolute ethanol p.a.
8. Ammonium acetate (Sigma-Aldrich, St. Louis, MO, USA) or Ambion RNase-free buffer kit (Applied Biosystems, Foster City, CA, USA; cat. AM9010).
9. Equipment for agarose gel electrophoresis or a 2100 Bioanalyser (Agilent Technologies, Santa Clara, CA, USA).
10. UV spectrophotometer or equivalent equipment for nucleic acid quantitation.
11. In vitro transcription kit: MEGAScript T7 (Ambion; cat. 1334, Applied Biosystems).
12. For digoxigenin labelling: 200 nmol of DIG-11-UTP (Roche, Indianapolis, IN, USA; cat. 03359247-910).
13. For biotin labelling: 250 nmol each of biotin-11-CTP (10 mM) and biotin-16-UTP (10 mM) (both Enzo Life Sciences, Farmingdale, NY, USA; cat. 42818 and 42814).
14. RNeasy mini RNA cleanup kit (Qiagen, Hilden, Germany; cat. 74106).
15. $ZnCl_2$ or Ambion RNA fragmentation reagent (Applied Biosystems; cat. 8740).
16. Taq polymerase, buffer, and dNTPs for polymerase chain reaction (PCR).

2.2 In Situ Hybridisation

1. Ultrathin resin sections on Formvar and carbon-coated nickel grids.
2. Digoxigenin- or biotin-labeled probe.

3. SSC solutions: 20× SSC (sodium citrate saline) stock solution (1× SSC is 0.15 M NaCl, 0.015 M sodium citrate, pH 7.0) and freshly diluted 1×, 2×, and 4× solutions.

4. Hybridisation buffer (HB): 20× SSC diluted 3:10 to give 6× SSC, with 0.7 mg/mL of yeast transfer RNA (tRNA) (Sigma-Aldrich).

5. Tris-buffered saline–Tween-20 (TBS-T): TBS is 150 mM NaCl, and 10 mM Tris-HCl, pH 7.5. Add 0.05% v/v Tween-20 (Sigma-Aldrich).

6. TBS-BR: TBS containing 0.5% blocking reagent (Roche).

7. Plastic box for incubation in a humid atmosphere.

8. Primary antibody: mouse monoclonal anti-digoxigenin or anti-biotin (Roche).

9. Secondary antibody: goat anti-mouse conjugated with colloidal gold (6–15-nm diameter (Jackson Immunoresearch, West Grove, PA, USA; or Orion, Wageningen, NL).

3 Methods

3.1 Preparation of Probes

RNA probes (riboprobes) offer several advantages for the visualisation of nucleic acids using ISH. They are single stranded ribonucleic acids with a sense or antisense orientation and can be labelled to a high specific activity by enzymatic incorporation of different modified ribonucleotides with covalently coupled groups such as biotin, digoxigenin (DIG), or aminoallyl. Microgram amounts of labelled riboprobes can be transcribed in a reaction volume of 20 µL. Most commonly, the sequence of interest is cloned into a plasmid with flanking promoters for T7, SP6, or T3 RNA polymerase. Such plasmids, called transcription vectors, are exemplified by the commercially available vectors pGEM and pBluescript. Prior to in vitro transcription, the plasmid is linearised using a restriction enzyme with its cutting site at the opposite flank of the cloned sequence from the promoter sequence (Fig. 22.1).

3.1.1 Preparation of Plasmid DNA

1. Linearise 2–10 µg of the appropriate transcription vector containing the target sequence flanked by bacteriophage promoters, using the appropriate restriction enzyme and buffer in a 50-µL reaction at the optimal temperature for 1 h (Fig. 22.1a).

2. Check an aliquot (~100 ng DNA) to verify complete linearisation by electrophoresis in a 1% agarose gel and ethidium bromide visualisation under UV light.

3. Add 150 µL of phenol:chloroform:isoamylalcohol (50:49:1) and 100 µL of nuclease-free H_2O to the restriction enzyme digest, vortex for 10 sec, transfer the mixture to a precentrifuged 1.5-mL Eppendorf PLG Heavy tube and centrifuge for 5 min at 15,000×g.

4. Transfer the upper water phase to a clean tube and adjust to $2M$ ammonium acetate. Mix in $10\,\mu g$ of glycogen and 2.5 volumes of absolute ethanol.
5. Following at least 30 min in the freezer, the plasmid is pelleted by centrifugation at $15,000\times g$ for 30 min. The pellet is overlaid by $500\,\mu L$ of 70 % ethanol and recentrifuged for 5 min before all ethanol is carefully removed.
6. Following air drying for 15 min, the pelleted plasmid is dissolved to $200\,ng$ DNA/μL in nuclease-free H_2O.

3.1.2 Transcription and Purification of Riboprobes

1. Thaw the components of the Ambion MegaScript kit and prepare the reaction at room temperature, not on ice.
2. Add to give a final reaction volume of $20\,\mu L$: $1\,\mu g$ (~0.5 picomoles) of linearised transcription vector, transcription buffer, ribonucleotides (final concentrations $1.5\,mM$ each ATP, CTP, and GTP; $1\,mM$ UTP; and $0.5\,mM$ DIG-UTP) (*see* **Note 1**), $2\,\mu L$ of enzyme mix (RNA polymerase and RNAse inhibitor), and nuclease-free H_2O, and mix.
3. Incubate for 6 h at 37°C for T7 and T3 RNA polymerases, or at 40°C for SP6 RNA polymerase.
4. (Optional) To remove the DNA template, add $1\,\mu L$ of DNAse 1 from the kit and incubate for 1 h at 37°C. At this step the reaction mixture can be stored frozen before proceeding to the purification step.
5. Clean up the labelled riboprobe using a Qiagen RNeasy minikit, following the supplier's instructions.
6. Measure the OD_{260} and OD_{280} using $2\,\mu L$ of DIG-cRNA + $38\,\mu L$ of dH_2O (dilution 2:40) in a spectrophotometer that can read 384-well plates. The quality of the labelled RNA can be checked by electrophoresis of $100\,ng$ in a 1% agarose gel and ethidium bromide staining.
7. (Optional) To achieve shorter RNA fragments to improve the hybridisation kinetics of labelled riboprobes, $ZnCl_2$ can be added to $10\,mM$ and the riboprobes incubated for 15 min at 60°C. Alternatively, use the Ambion fragmentation reagents according to the supplier's instructions.
8. Store the DIG-cRNA at −20°C (short-term) or at −80°C (long-term) (Fig. 22.1c).

3.1.3 Generation of PCR Fragments Containing a Flanking T7 RNA Polymerase Promoter

If the sequence of interest is not available in a transcription plasmid, the labelled riboprobe can be obtained as follows:
1. Order a 60-mer synthetic oligonucleotide containing the 40-nucleotide T7 RNA polymerase promoter sequence fused to an 18-nucleotide primer (N_{18}) flanking the sequence of interest AAACGACGGCCAGTGAATTGTAATACGACTCAC-TATAGGCGC(N_{18}). For the generation of antisense riboprobes, the T7 RNA

polymerase promoter sequence is fused to the specific reverse primer (Fig. 22.1, b1). For the generation of sense riboprobes, the T7 RNA polymerase promoter is fused to the specific forward primer.

2. The PCR reaction is done in a volume of 25 μL containing the appropriate template sequence (Fig. 22.1, b2), buffer, dNTPs, 1 μM of each primer and the thermoprofile according to the Tm of the primers for 40 cycles.

3. The amplified fragment (~0.2 picomoles) (Fig. 22.1, b3) can be substituted for linearised plasmid as described in **Section 3.1.2**.

3.2 Electron Microscope In Situ Hybridisation (EMISH)

ISH assays consist of six steps: pre-hybridisation, probe linearisation and disaggregation, hybridisation, post-hybridisation washes, detection with primary and secondary antibodies, and final washes (Fig. 22.2).

1. Pre-hybridisation reduces non-specific interactions between the probe and the specimen. Grids are placed on a ~10-μL drop of HB buffer on a plastic plate covered with Parafilm in a humid chamber containing moistened paper to humidify the atmosphere and prevent the solution from drying, which is sealed with Parafilm and incubated for 1 h at 55°C.

2. Probe linearisation and disaggregation (to disperse RNA fragments and denature any that may be associated). Probes are diluted in HB buffer to 5–30 ng/μL and incubated for 5 min in a water bath at 70°C. Mix by vortexing and repeat the incubation. Finally, keep the probes on ice at least 5 min before hybridisation (*see* **Note 2**).

3. For hybridisation, grids with sections are placed on 10-μL drops of probe solution (*see* **Section 3.1.2**) in a humid chamber with 6× SSC. The chamber is sealed with Parafilm and incubated for 4 h at 55°C (Fig. 22.2c).

4. The grids are washed at room temperature by placing them on drops on a sheet of Parafilm. First, they are floated for 2 and 5 min on 4× SSC, followed by four washes of ~5 min each on 2× SSC, 1× SSC, TBS-T, and TBS-BR.

5. Antibodies for the detection of hybrids depend on the marker used for probe labelling. Typically, use monoclonal primary antibodies directed against digoxygenin-UTP (MAD) or biotin-UTP (MAB). Grids are incubated on a drop of antibody diluted in TBS-BR (usual dilution is 1:10) for at least 30 min at room temperature in the humid chamber (Fig. 22.2d).

6. Grids are washed 4× 5 min at room temperature on drops of TBS-T placed on Parafilm.

7. To localise the primary antibodies, we use goat anti-mouse antibodies coupled with colloidal gold particles of various dimensions, diluted in TBS-BR according to the manufacturer's indication. Grids are incubated on drops in the humid chamber containing 2× SSC at room temperature for at least 30 min (Fig. 22.2e).

8. Grids are washed 3× 5 min on TBS-T, then on distilled or ultrapure H_2O for 5 min.
9. Finally, grids are rinsed dropwise in dH_2O from a wash bottle and air dried (see **Note 3**). All washes are performed at room temperature.

3.3   *Increasing Contrast*

To increase the contrast of cellular structures, a double staining with uranyl acetate and lead citrate can be used. All steps are performed at room temperature.

1. Uranyl acetate: grids are incubated on uranyl acetate solution (2.5% w/v in dH_2O) drops for 5 min, rinsed dropwise in dH_2O from a wash bottle (see **Note 3**), and air dried.
2. Lead citrate: grids are incubated on drops of lead citrate solution for 3 min for LRWhite and 1.5 min for Lowicryl K4M resin sections in a chamber containing NaOH pellets to bind CO_2 from the air, rinsed dropwise in dH_2O, and air dried.

3.4   *Cytochemical Reaction: EDTA Staining*

The EDTA regressive staining procedure (1) was developed with the aim of preferentially visualising ribonucleoprotein-containing nuclear structural constituents, while chromatin remains light grey. The procedure consists of first staining ultrathin sections with uranyl acetate, which binds to both the deoxyribonucleoproteins (chromatin) and ribonucleoproteins. The sections are then treated with a neutral solution of EDTA, which chelates uranyl ions bound to DNA–protein complexes, thus giving rise to a relatively higher contrast of RNP structures. Finally, lead from lead citrate stain binds to the uranyl ions remaining on RNPs and enhances their contrast. This results in an image of the cell nucleus exhibiting bleached chromatin and well-contrasted RNP-containing structural domains. Although the exact mechanism of the staining reaction has not been clearly elucidated, the method has been widely used for preferential visualisation of nuclear RNP constituents in combination with other cytochemical methods. When applied to ultrathin acrylic resin sections previously submitted to ISH, the following protocol is recommended:

1. Prepare $0.2 M$ EDTA solution: for 100 mL, put 7.44 g of EDTA into 50 mL of ultrapure or distilled H_2O. Adjusted the pH to 7 by adding very slowly 1 N NaOH. At this point, the EDTA should be completely dissolved. Finally, complete to 100 mL with H_2O and filter *(1)*.
2. Treat grids with sections for 1 min at room temperature with saturated uranyl acetate in aqueous solution (about 4.7%).
3. Wash with dH_2O and air dry.
4. Treat with 1:10 EDTA solution, for 3 min.
5. Wash with dH_2O and air dry.
6. Post-stain with lead citrate solution *(14)* for 1 min, wash and air dry.

3.5   Controls of Specificity

Among the best controls for revealing a specific hybridisation signal on an RNA target, treatment of ultrathin sections with RNase prior to ISH ensures the RNA nature of the reactive sites in the specimen:

1. Typically, sections are floated on 0.2–0.5% w/v RNase in a buffer (e.g. triethanolamine, pH 7.3) or on dH_2O, for 8 to 12 h at 37°C, then washed with dH_2O and submitted to ISH.
2. Another negative control is to use a sense-RNA probe labeled with the same marker, which should result in a background level of signal.

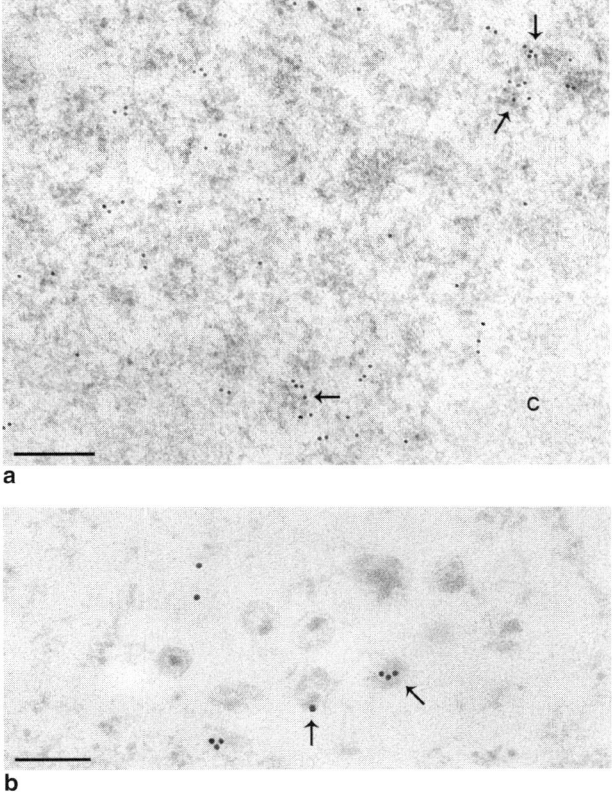

Fig. 22.3 Ultrathin sections of HIV-1-transfected COS cells embedded in Lowicryl K4M resin. The distribution of viral RNA is revealed by ISH with a digoxigenin-labeled anti-sense RNA probe followed by an anti-digoxigenin antibody complex with colloidal gold. **a** The signal represents 12-nm gold particles mainly associated with ribonucleoprotein-containing fibrils (some indicated by *arrows*) in the nucleoplasm; **c** indicates condensed chromatin. **b** Some viral particles are labeled in the extracellular space (*arrows*). Gold grains in this experiment were 15 nm in diameter. Micrographs courtesy of Dr. D. Cmarko. For more details see ref. (13). Bars, 0.2 μm

3. The usual control of the final immunocytochemical reaction consists of omitting the primary antibody recognising the marker on the probe.

3.6 Observations

Observations are made with a conventional transmission electron microscope, typically at 80 kV using a medium-size objective aperture. Micrographs are recorded either by a high-resolution digital camera or on negative film (Fig. 22.3).

4 Notes

1. For biotin-labelled probes, use the following final concentrations of rNTPs instead: 1 mM ATP and GTP, and 0.5 mM each of CTP, UTP, Bio-11-CTP, and Bio-16-UTP.
2. If you prepare only the exact final volume and not more, remember to centrifuge your probe solution in order to collect the liquid spread on the walls of the tube, thus keeping the correct volume.
3. To wash with a wash bottle, direct a weak flow on the tweezers and not directly on the grids.

Acknowledgments We thank Dr. D. Cmarko for kindly providing his micrographs, and we are indebted to Mr. W. Blanchard for photographic assistance. The work in the authors' laboratories has been supported by the Swiss National Science Foundation and by the Research Council of Norway and Helse Vest.

References

1. Bernhard, W. (1969) A new staining procedure for electron microscopical cytology. *J. Ultrastruct. Res.* **27**, 250–265.
2. Fakan, S. and Bernhard, W. (1971) Localisation of rapidly and slowly labelled nuclear RNA as visualized by high resolution autoradiography. *Exp. Cell. Res.* **67**, 129–141.
3. Nash, R. E., Puvion, E., and Bernhard, W. (1975) Perichromatin fibrils as components of rapidly labeled extranucleolar RNA. *J. Ultrastruct. Res.* **53**, 395–405.
4. Fakan, S., Puvion, E., and Spohr, G. (1976) Localization and characterization of newly synthesized nuclear RNA in isolate rat hepatocytes. *Exp. Cell Res.* **99**, 155–164.
5. Fakan, S., Leser, G., and Martin, T. E. (1984) Ultrastructural distribution of nuclear ribonucleoproteins as visualized by immunocytochemistry on thin sections. *J. Cell Biol.* **98**, 358–163.
6. Biggiogera, M. and Fakan, S. (1998) Fine structural specific visualization of RNA on ultrathin sections. *J. Histochem. Cytochem.* **46**, 389–395.
7. Gall, J. G. and Pardue, M. L. (1969) Formation and detection of RNA-DNA hybrid molecules in cytological preparations. *Proc. Natl. Acad. Sci. U S A* **63**, 378–383.
8. Morel, G. (1993) *Hybridization techniques for electron microscopy.* CRC Press.

9. Puvion-Dutilleul, F. and Puvion, E. (1996) Non-isotopic electron microscope in situ hybridization for studying the functional sub-compartmentalization of the cell nucleus. *Histochem. Cell Biol.* **106**, 59–78.
10. Cmarko, D. and Koberna, K. (2007) Electron microscopy in situ hybridization. In: *Electron microscopy: Methods and protocols* (Kuo, J., Ed.), Humana, Totowa, NJ, pp. 213–228.
11. Carlemalm, E., Garavito, R. M., and Villiger, W., (1982) Resin development for electron microscopy and an analysis of embedding at low temperature. *J. Microsc.* **126**, 123–143.
12. Fischer, D., Weisenberger, D., and Scheer. U. (1996) In situ hybridization of DIG-labeled rRNA probes to mouse liver ultrathin sections. In: *Nonradioactive in situ hybridization. Application manual, 2nd ed.* Boehringer, Mannheim, Germany, pp. 148–151.
13. Cmarko, D., Boe, S. O., Scassellati, C., Szilvay, A. M., Davanger, S., Fu, X. D., Haukenes, G., Kalland, K. H., and Fakan, S. (2002) Rev inhibition strongly affects intracellular distribution of human immunodeficiency virus type 1 RNAs. *J. Virol.* **76**, 10473–10484.
14. Reynolds, E. S. (1963) The use of lead citrate at high pH as an electron-opaque stain in electron microscopy. *J. Cell Biol.* **17**, 208–212.

Chapter 23
Electron Spectroscopic Imaging of the Nuclear Landscape

Kashif Ahmed, Ren Li, and David P. Bazett-Jones

Keywords Electron spectroscopic imaging; Correlative microscopy; Promyelocytic leukemia nuclear bodies; Nucleus; Chromatin; Nuclear structure

Abstract Our understanding of sub-nuclear organisation is largely based on fluorescence and electron microscopy methods. Conventional electron microscopy, which depends on heavy atom contrast agents, provides excellent contrast of condensed chromatin and some sub-nuclear structures such as the nucleolus. Unfortunately, other components, 10-nm chromatin fibres for example, do not contrast well. Electron spectroscopic imaging partially overcomes this limitation. In particular, phosphorus and nitrogen mapping provide sufficient contrast and resolution to visualise 10-nm chromatin fibres, while providing an opportunity to distinguish protein-based from nucleic acid-based supramolecular structures, such as the cores of nuclear bodies. Electron spectroscopic imaging, therefore, offers an approach to address many questions related to the functional organisation of the interior of the cell nucleus, which is illustrated in this chapter.

1 Introduction

The eukaryotic nucleus is perhaps the most complex compartment in the cell, which until recently was viewed as a storage compartment for chromatin. Our understanding of the nucleus and nuclear compartmentalisation has been expanded profoundly mainly due to improved cell imaging techniques and functional biochemical assays *(1)*. These methods have provided the basis for detailed description of nuclear structure now available. We now know that many sub-nuclear compartments exist within the nucleus, such as interchromatin granule clusters (IGCs), Cajal bodies, polycomb bodies, and promyelocytic leukemia (PML) nuclear bodies (NB) *(2)*. DNA within the nucleus is complexed with histones to form chromatin, which plays an important role in regulation of transcription, replication, and repair of DNA.

Microscopical techniques have played a central role in our understanding of chromatin and other sub-nuclear compartments. Several microscopy techniques

R. Hancock (ed.) *The Nucleus: Volume 2: Chromatin, Transcription, Envelope, Proteins, Dynamics, and Imaging,*
© Humana Press 2008

have been described that either map specific molecules in situ, in fixed cells by indirect immunofluorescence, or in living cells by expressing specific proteins fused with fluorescent tags such as green fluorescent protein (GFP) *(3)*. Despite the power and the enormous value of these light microscopy methods, it is recognised that higher-resolution techniques are required to define the ultrastructure responsible for the nuclear complexity observed with the light microscope.

Electron microscopy (EM) has played a significant role in defining nuclear ultrastructure. In particular, the discovery of the nucleosomal subunit by both biochemical and EM data *(4)* and the imaging of actively transcribing ribosomal genes *(5)* have together greatly affected our concept of how DNA is organised and transcribed within the nucleus. Conventional EM is facilitated by the use of molecular stains that provide contrast between relatively thick chromatin fibre and the nucleoplasm. But during interphase, the majority of chromatin exists in a decondensed state and as a result has little contrast in conventional electron micrographs prepared by the typical fixation and staining procedures using uranyl acetate and/or lead citrate. The use of energy-filtered transmission electron microscopy (EFTEM), also referred to as electron spectroscopic imaging (ESI), can overcome this problem when imaging interphase chromatin by deriving contrast in electron micrographs from the elemental content (nitrogen and phosphorus) of the chromatin itself *(6, 7)*.

ESI is based on the principle of electron energy-loss spectroscopy. When a specimen is bombarded with electrons, the elements within the specimen become ionised and the energy for an ionisation event is equal to the energy lost by the responsible incident electron. An imaging electron spectrometer produces an energy-loss spectrum reflecting the elemental content of the sample and is capable of reconstituting an image with electrons that have lost a particular amount of energy by interaction with a particular element. Nitrogen in the nucleus can be used as an "endogenous stain" for both protein and nucleic acid, and phosphorus as a specific marker for nucleic acids. Thus, chromatin can be imaged within the nucleus without the use of heavy atom stains that might obscure ultrastructural detail or otherwise create ambiguities because of the non-uniform staining characteristics of different biochemical components in situ. The advantage of ESI, based on element-specific imaging, is that the distribution of nucleic acid can be resolved and delineated from the protein component of macromolecular complexes. High-resolution images of the phosphorus distribution, when compared with the total mass image or the nitrogen distribution, can be used to delineate the nucleic acid component from the protein-based components in a complex. Quantification of such elemental signals provides stoichiometric relationships of protein and nucleic acid, an important supplement to the structural information provided.

Some features of nuclear organisation that ESI can provide are illustrated in Fig. 23.1. A fluorescence microscope image of a physical section (Fig. 23.1a) can be used to identify a region of interest to image at high resolution in the electron microscope (Fig. 23.1b–h). A net nitrogen (Fig. 23.1b) and net phosphorus map (Fig. 23.1c) reveal biochemical differences in the various structures. The core of the PML NB, for example, is nitrogen-rich but phosphorus-depleted. Segmenting of the phosphorus and nitrogen signals can provide further information. For example, the phosphorus map can be subtracted from the nitrogen map (N–P), and represented in shades of blue. When the phosphorus map, after representation in shades of yellow, is superimposed on the N–P image (Fig. 23.1d, g, h), chromatin fibres (Ch) are visualised as yellow fibres, the protein-based core of the PML NB is represented as a blue structure, and the cluster of ribonucleoprotein (RNP) granules is a mixture of yellow and blue shades. Single DNA fibres, comprising 10-nm chromatin, can be visualised (arrowheads). Silver enhancement of ultra-small gold-tagged secondary antibodies can reveal the location of a protein of interest (detected with an anti-PML antibody). The silver signal contrasts well in the nitrogen maps (Fig. 23.1b, e) and can be superimposed on the composite images.

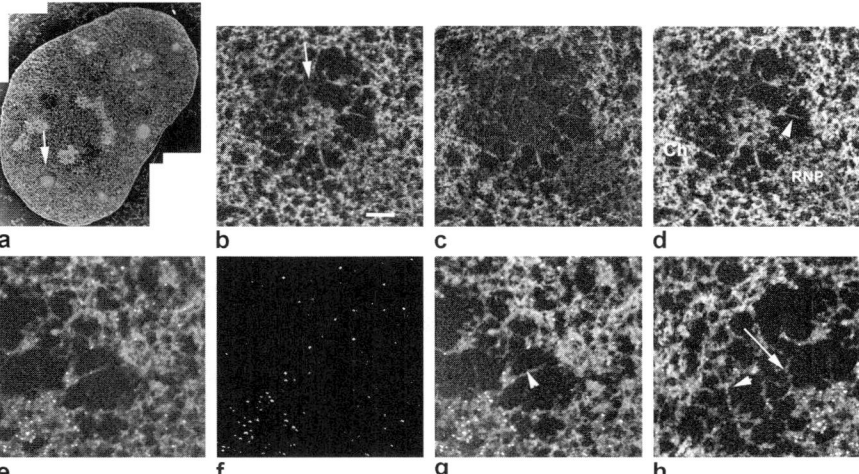

Fig. 23.1 Correlative and fluorescence micrographs of a 70-nm-thick section of an SK-N-SH neuroblastoma cell nucleus. **a** An immunofluorescence image of the section of a cell labelled for PML protein is overlaid onto a low-magnification EM image of the same cell. The PML NB (*arrow*) is shown at high magnification in the electron micrographs (**b–h**). The orientation of the *arrow* indicates the rotation of the images between magnification steps. **b** Nitrogen map. **c** Phosphorus map. **d** Phosphorus map subtracted from the nitrogen map of the field shown in (**b**) and (**c**). *RNP* signifies a region of ribonucleoprotein granules. **e** Higher magnification of the nitrogen map shown in (**b**). **f** Silver-enhanced signal of ultra-small secondary antibody. Silver is contrasted due to an ionisation edge that overlaps with the nitrogen energy-loss ionisation edge. **g, h** Higher magnification composites of the nitrogen and phosphorus information seen in (**d**). *Arrowheads* in (**d**), (**g**), and (**h**) indicate single DNA molecules comprising 10-nm chromatin fibres. Scale bar, 150 nm in **b–d** and 75 nm in **e–h**. Ch, chromatin; RNP, ribonucleoprotein. To view this figure in color, see COLOR PLATE 14

2 Materials

2.1 Cell Culture and Sample Preparation

1. Dulbecco's modified Eagle's medium (DMEM) (Wisent, Québec, Canada) supplemented with 10% fetal bovine serum (FBS, Hyclone, Logan, UT, USA).
2. Phosphate-buffered saline (PBS), 10× stock: 1.37 M NaCl, 27 mM KCl, 100 mM Na$_2$HPO$_4$, and 18 mM KH$_2$PO$_4$, adjust to pH 7.4 with HCl and autoclave before storage at room temperature. Prepare a working solution by dilution of one part with nine parts of water (see **Note 1**).
3. Parafilm M (Alcan Packaging, Menasha, WI, USA).
4. Paraformaldehyde (PFA): 16% stock solution (Electron Microscopy Sciences, Hatfield, PA, USA), store at 4°C. Dilute with PBS to the working concentration just before use.
5. Glutaraldehyde: 25% stock solution (Electron Microscopy Sciences), store at 4°C. Dilute with PBS to the working concentration just before use.
6. Permeabilisation solution: 0.5% (v/v) Triton X-100 in PBS.
7. Blocking buffer: 5% donkey serum (Jackson ImmunoResearch, West Grove, PA, USA) in PBS.
8. Primary antibody: rabbit anti-PML (cat. 1370; Chemicon, Temecula, CA).
9. Secondary antibodies: donkey anti-rabbit Aurion Ultra-small Immunogold (Electron Microscopy Sciences); donkey anti-rabbit Cy3 (Jackson ImmunoResearch).

2.2 Dehydration, Silver Enhancement, and Embedding

1. Anhydrous ethyl alcohol.
2. Silver enhancement kit: Aurion R-Gent SE-EM (cat. 25521; Electron Microscopy Sciences).
3. Quetol 651 epoxy resin (cat. 20440) and Quetol 651-NSA Kit (cat. 14640; Electron Microscopy Sciences) (store Quetol 651 resin at 4°C).
4. 60-mm Petri dishes (Permanox; Nalge Nunc International, Rochester, NY, USA) or glass Petri dishes.

2.3 Sectioning, Grids, and Carbon Film

1. Ultracut UCT ultramicrotome (Leica Microsystems, Richmond Hill, ON, Canada).
2. Vacuum coater (Edwards E306) (BOC Edwards, Mississauga, ON, Canada) and an electron beam evaporator system (Cressington Scientific Instruments, Ltd., Watford, UK).

3. 400-mesh copper electron microscope grids with parallel bars containing a single perpendicular centre bar (Electron Microscopy Sciences; cat. G400PB-Cu).

3 Methods

A key question in molecular cell biology is what role sub-nuclear organisation plays in regulating gene activity *(8, 9)*. Nuclear compartments are assembled and maintained in a dynamic fashion, reflecting their function and involvement in nuclear processes *(10)*. Moreover, each chromosome occupies a discrete volume and is located non-randomly in the nucleus *(8, 11, 12)*. Furthermore, in concert with cellular differentiation, movement of specific loci from the nuclear periphery to the interior, or from the interior to the exterior of a chromosome territory, are also observed *(13, 14)*. These correlations of gene activity with location indicate that sub-nuclear environments must be considered as a potential contributing factor in gene regulation.

PML NBs represent one of several sub-nuclear "organelles" that contribute to the compartmentalisation of nuclear proteins. About 5–30 PML NBs ranging from 0.2 to 1 μm in diameter are present in an average nucleus. PML NBs are implicated in a number of cellular processes including transcriptional regulation, apoptosis, DNA repair, and replication of both viral and cellular DNA *(15, 16)*. Using biochemical and microscopical techniques, including ESI, to study the structure, biochemical composition, and dynamics of PML NBs, we have shown that the sub-nuclear positioning of PML NBs as well as their structural integrity is dependent on chromatin contacts *(17, 18)*. Immunolabelling of the PML protein is used to identify PML NBs by correlative fluorescence microscopy and ESI of physical sections, or by immuno-gold and silver enhancement methods.

3.1 Cell Culture and Immunolabelling

1. Cells are cultured on sterilised glass coverslips in 60-mm dishes or 6-well plates, and grown to the desired confluency using standard procedures.
2. Cells are removed from the medium and rinsed rapidly with PBS.
3. Cells are then fixed with 2% paraformaldehyde in PBS for 10 min at room temperature (*see* **Note 2**).
4. Paraformaldehyde is discarded into a hazardous waste container, and the samples are washed 3× 5 min with PBS.
5. The cells are permeabilised by incubation in PBS containing 0.5% Triton X-100 for 5 min at room temperature and washed 3× 5 min with PBS.
6. Samples are blocked by incubation in blocking buffer for ~30 min at room temperature.
7. The blocking buffer is removed and replaced with the primary antibody diluted 1:500 in blocking buffer for 1 h at room temperature (*see* **Note 3**).

8. The primary antibody is removed and the sample washed 3× 5 min with PBS.
9. Ultra-small immunogold secondary antibody diluted 1:100 in blocking buffer is then added for 2 h at room temperature or overnight at 4°C, and samples are washed 3× 5 min with PBS. If performing correlative microscopy, the gold-conjugated antibody is combined with a Cy3-labelled antibody at 1:100 and 1:500 dilutions, respectively.

3.2 Silver Enhancement, Dehydration, and Embedding

1. The samples are post-fixed with 2% glutaraldehyde in PBS for 5 min at room temperature and washed 3× 5 min with PBS.
2. The cells are washed 3× 5–10 min with ultra-pure, deionised water (*see* **Note 4**).
3. Samples are then subjected to silver enhancement. In a 1.5-mL tube, mix 10 μL of initiator and 390 μL of activator (1:40; mix 1). Take 20 μL of mix 1 and add 380 μL of enhancer (1:20; mix 2). Leave mix 2 at room temperature for 10 min, place drops of 50–100 μL on Parafilm, and place the coverslips on the drops with the cells facing down. Incubate at 30°C for 15 min (*see* **Note 5**).
4. Remove the coverslips from the Parafilm and wash the cells 3× with ultra-pure distilled water for 10 min.
5. Cells are dehydrated with a series of graded ethanol steps at 30, 50, 70, 90, and 100%, respectively, with incubation on a shaker for 30–60 min at each step (*see* **Note 6**).
6. Discard the ethanol, add enough Quetol 651 resin to cover the cells sufficiently (3–4 mL in a 60-mm Permanox dish), and incubate with shaking for 2–3 h at room temperature. Prepare the Quetol-mix during this incubation (**step 7**).
7. To prepare 100 mL of Quetol-mix, add 35 mL of Quetol 651, 54 mL of NSA, 11 mL of NMA, and 1.5–2 mL of DMP-30 (components of the Quetol 651-NSA kit). Stir the contents using a magnet stirrer for 2 h (*see* **Note 7**).
8. Exchange the Quetol 651 for Quetol-mix and incubate with shaking for 2–3 h at room temperature. It is important to remove as much Quetol 651 resin as possible before the addition of Quetol-mix.
9. Change the Quetol-mix and incubate further for 2–3 h at room temperature with shaking.
10. Change the Quetol-mix once more and add enough new Quetol-mix to cover the cells to a depth of 1–2 mm (~3 mL in a 60-mm dish). Incubate at 65–70°C for at least 24 h.

3.3 Sectioning and Carbon Coating

1. Remove the dish from the incubator and let it cool to room temperature.

2. Mark the edge of the coverslip on the hardened Quetol-mix. Place the Permanox dish on a hot plate with the temperature set at about 90°C and cut the block using a sharp razor. Heating at 90°C makes the Quetol-mix soft, making it easier to cut. Avoid the breaking of coverslip or block.

3. Carefully peel off the block from the coverslip.

4. Trim the block on the hot plate into small cubes, and glue the cubes with the cell side facing up onto resin "bullets".

5. The trimmed block, glued onto bullets, is sectioned with the ultramicrotome. Generally a ribbon of 5–15 serial sections of 60- to 90-nm thickness is obtained on the surface of water.

6. A 400-mesh copper electron microscope grid with parallel bars containing a single perpendicular centre bar is used to pick up the sections. As the ribbon of sections is picked up, a region of interest must be carefully positioned so that it is located over a grid opening and not on a grid bar.

7. The grids are then checked for integrity of the sections and to identify structures of interest (e.g. PML NBs) by epifluorescence microscopy. A grid with serial sections is placed on a glass slide, covered with a glass coverslip, and secured with transparent tape and images are recorded at low magnification; these images reveal both the integrity of the section as well as the relative orientation and position of the section with respect to the grid centre. Images are recorded and displayed on a computer at the electron microscope, in order to locate the same sections and regions of interest in the electron microscope. Cells of interest can be marked on the image and their position noted with respect to the grid centre, which facilitates pinpointing the cell's position when examined in the electron microscope.

8. Sections between 60 and 90 nm in thickness embedded in Quetol 651 become resistant to deformation in the electron beam if they are coated with a very thin carbon film 3–4 nm in thickness. When preparing these films initially, their thickness is estimated by comparison of the optical density in dark-field energy-loss images of a double or triple layer of the film. These optical densities are compared with those of an internal mass standard such as pure plasmid DNA or tobacco mosaic virus (TMV). Subsequently, the acceptable range of carbon film thickness can be estimated by the reflectivity of the films by eye when they are floated on the surface of water. The carbon films are prepared by evaporating carbon onto the surface of freshly cleaved mica by electron beam evaporation in a vacuum evaporator operated at approximately 10^{-6} Torr. The films are then floated onto distilled water and picked up on the specimen grids with the section side facing up.

3.4 Imaging

1. After addition of a carbon film, the regions of interest are imaged with a Tecnai 20 transmission electron microscope (FEI/Philips Electron Optics, Eindhoven, Netherlands) equipped with an electron imaging spectrometer (Gatan, Pleasanton, CA, USA), and operated at 200 kV.

2. Serial sections are first stabilised to the beam by imaging with a low flux of electrons at a low magnification. Stabilisation helps to prevent deformation, stretching, or rupture of thin sections during imaging.

3. Net phosphorus ratio maps are produced from pre- and post-edge images recorded at 120 and 155 eV ($L_{II,III}$ edge), and net nitrogen ratio maps are produced from pre- and post-edge images recorded at 385 and 415 eV (K edge).

4. Elemental maps are generated by dividing the element-enhanced post-edge image by the pre-edge image following alignment by cross-correlation (the software used is Digital Micrograph; Gatan). The recording times required to obtain the pre-edge and post-edge images are in the range of 10 to 30 sec, and result in electron exposures of the specimen to approximately 10^5 electrons/nm^2.

4 Notes

1. All solutions should be prepared in deionised and filtered water that has a resistivity of 18.2 MΩ-cm or better. Water purity is most critical in washing steps before and after silver enhancement (Section 3.3).

2. Cells can be fixed in a range of 1–4% paraformaldehyde for 5–10 min. By varying the fixation conditions, it is possible to vary the contrast of certain subnuclear structures visualised in net nitrogen images. For instance, high-contrast images of chromatin can be obtained by gentle fixation with 1% paraformaldehyde followed by extraction of soluble nuclear proteins and nucleic acids by the Triton X-100 treatment. The choice of fixation conditions is even more important for immuno-EM, as a balance must always be found between the preservation of ultrastructure and the preservation of antigenicity. Individual conditions for a particular antibody–protein antigen pair must be determined empirically.

3. For economy, only 50–100 μL of antibody per sample is required at this step. Place a sheet of Parafilm onto a clean glass plate and place a drop of 50–100 μL of antibody solution on the Parafilm; we normally use 50 μL. Carefully pick up the coverslip using a forceps and place it on the drop of antibody with cells side facing down. After incubation, add ~200 μL of PBS to the corner of the coverslip using a pipette. This will float the coverslip on the buffer and it can be picked up easily using forceps. The same procedure can be performed for the secondary antibody.

4. The water washes before and after silver enhancement steps are critical to achieve good enhancement with low background. The highest quality of water available should be used at these steps. We generally use distilled water from Gibco (cat. 15230; Invitrogen, Grand Island, NY, USA) for these steps.

5. Perform this step under reduced light, as some components of the kit are light-sensitive. Subsequent steps can be performed at normal light levels.

6. A longer first step for 1–2 h at 30% ethanol is desirable. Subsequent steps are at least 30 min in duration, with a longer incubation at 100% ethanol. At least two changes of 100% for 10 min each followed by a longer incubation for 1–2 h at room temperature or overnight at 4°C give satisfactory results. If performing overnight dehydration, on the next day change 100% ethanol for 3× 10 min.

7. It is important to keep the contents mixing at all times after addition of DMP-30. For convenience, Quetol-mix without DMP-30 can be prepared in larger quantities and stored at 4°C. When needed, remove the desired amount and add DMP-30 accordingly (i.e. 2 mL per 100 mL) and mix at room temperature.

Acknowledgments This work was funded by operating grants from the Natural Sciences and Engineering Research Council and the Canadian Institutes of Health Research. DPB-J holds the Canada Research Chair in Molecular and Cellular Imaging.

References

1. Cremer, T. and Cremer, C. (2001) Chromosome territories, nuclear architecture and gene regulation in mammalian cells. *Nat. Rev. Genet.* **2**, 292–301.
2. Lamond, A. I. and Earnshaw, W. C. (1998) Structure and function in the nucleus. *Science* **280**, 547–553.
3. Kruhlak, M. J., Lever, M. A., Fischle, W., Verdin, E., Bazett-Jones, D. P., and Hendzel, M. J. (2000) Reduced mobility of the alternate splicing factor (ASF) through the nucleoplasm and steady state speckle compartments. *J. Cell. Biol.* **150**, 41–51.
4. Olins, A. L. and Olins, D. E. (1974) Spheroid chromatin units (nu bodies). *Science* **183**, 330–332.
5. Miller, O. L., Jr. and Beatty, B. R. (1969) Visualization of nucleolar genes. *Science* **164**, 955–957.
6. Bazett-Jones, D. P. and Hendzel, M. J. (1999) Electron spectroscopic imaging of chromatin. *Methods* **17**, 188–200.
7. Dehghani, H., Dellaire, G., and Bazett-Jones, D. P. (2005) Organization of chromatin in the interphase mammalian cell. *Micron* **36**, 95–108.
8. Bolzer, A., Kreth, G., Solovei, I., Koehler, D., Saracoglu, K., Fauth, C., Muller, S., Eils, R., Cremer, C., Speicher, M. R., and Cremer, T. (2005) Three-dimensional maps of all chromosomes in human male fibroblast nuclei and prometaphase rosettes. *PLoS Biol.* **3**, e157.
9. van Driel, R., Fransz, P. F., and Verschure, P. J. (2003) The eukaryotic genome: a system regulated at different hierarchical levels. *J. Cell Sci.* **116**, 4067–4075.
10. Dundr, M. and Misteli, T. (2001) Functional architecture in the cell nucleus. *Biochem. J.* **356**, 297–310.
11. Boyle, S., Gilchrist, S., Bridger, J. M., Mahy, N. L., Ellis, J. A., and Bickmore, W. A. (2001) The spatial organization of human chromosomes within the nuclei of normal and emerin-mutant cells. *Hum. Mol. Genet.* **10**, 211–219.
12. Misteli, T. (2004) Spatial positioning: A new dimension in genome function. *Cell* **119**, 153–156.
13. Osborne, C. S., Chakalova, L., Brown, K. E., Carter, D., Horton, A., Debrand, E., Goyenechea, B., Mitchell, J. A., Lopes, S., Reik, W., and Fraser, P. (2004) Active genes dynamically colocalize to shared sites of ongoing transcription. *Nat. Genet.* **36**, 1065–1071.
14. Chambeyron, S. and Bickmore, W. A. (2004) Chromatin decondensation and nuclear reorganization of the HoxB locus upon induction of transcription. *Genes Dev.* **18**, 1119–1130.
15. Borden, K. L. (2002) Pondering the promyelocytic leukemia protein (PML) puzzle: possible functions for PML nuclear bodies. *Mol. Cell. Biol.* **22**, 5259–5269.
16. Dellaire, G. and Bazett-Jones, D. P. (2004) PML nuclear bodies: Dynamic sensors of cellular stress and DNA damage. *Bioessays* **26**, 963–977.
17. Eskiw, C. H., Dellaire, G., and Bazett-Jones, D. P. (2004) Chromatin contributes to structural integrity of promyelocytic leukemia bodies through a SUMO-1-independent mechanism. *J. Biol. Chem.* **279**, 9577–9585.
18. Block, G. J., Eskiw, C. H., Dellaire, G., and Bazett-Jones, D. P. (2006) Transcriptional regulation is affected by subnuclear targeting of reporter plasmids to PML nuclear bodies. *Mol. Cell. Biol.* **26**, 8814–8825.

Chapter 24
Cryoelectron Microscopy of Vitreous Sections: A Step Further Towards the Native State

Cedric Bouchet-Marquis and Stanislav Fakan

Keywords Cell nucleus; Cryoelectron microscopy; Cryofixation; Vitrified sections; Cryoelectron microscopy of vitreous sections; CEMOVIS

Abstract Nuclear architecture has been investigated intensively by various electron microscopy (EM) methods. Most of these require chemical fixation of the sample, although cryofixation has also been used in combination with cryosubstitution and resin embedding. This approach allowed one to considerably increase the knowledge about the structural features of different nuclear domains and their involvement in nuclear functions. Cryoelectron microscopy of vitreous sections (CEMOVIS) has added a new dimension to the ultrastructural analysis of the cell nucleus, especially thanks to the possibility of observing the specimen in its hydrated state. In this way one can analyse, at high resolution, cellular structures as close as possible to their native state. In this chapter we describe in detail the different steps of the CEMOVIS method, which should allow an electron microscopist to perform cryosectioning and cryoelectron microscopy of vitrified biological material.

1 Introduction

Conventional electron microscopy (EM) has been, for a long time, the technique of choice for high-resolution analysis of biological specimens. It consists of chemical fixation, mostly with aldehydes, followed by dehydration and resin embedding of the sample. The resin-embedded sample is cut into ultrathin sections and stained with electron dense contrasting agents such as uranyl acetate and lead citrate in order to be observed in a transmission electron microscope. Electron microscopy opens ultrastructural studies of biological samples down to dimensions in the range of 10 Å, from which conclusions about structural features with these dimensions can be drawn. However, at this resolution sample preparation is critical.

The functional architecture of the nucleus has been analysed since the early 1960s by means of different methods of ultrastructural cytochemistry (*1*). These included various techniques of preferential or specific contrasting of nuclear constituents such as DNA or RNA, often combined with labelling by radioactive or halogenated precursors. Nuclear domains are also analysed by means of specific immunocytochemical or molecular probes and high-resolution autoradiography or immuno-EM. These different approaches have been applied on chemically fixed or cryofixed specimens ultimately embedded into different epoxy or acrylic resins (for review, see ref. *(2)*), or on cryosections of aldehyde-fixed samples *(3)*. In order to observe specimens in the best conditions of preservation, cryoelectron microscopy of vitreous sections (CEMOVIS) has been developed over the last three decades.

CEMOVIS was designed with the aim of performing high-resolution imaging of biological material cut into ultrathin vitreous sections *(4–11)*. The goal is to observe the specimen as close as possible to its native state. High-pressure freezing is the technique of choice for vitrification of specimens up to about 250 μm in thickness. The advantage of CEMOVIS is that the specimens are observed in their hydrated state, thus preventing as much as possible any rearrangements or aggregation of fine biological structures *(8, 12)*. Such artefacts may occur during conventional sample preparation, probably due to aldehydic bridging and/or dehydration process. After vitrification the sample is cut into ultrathin vitreous sections, allowing one to subsequently perform high-resolution cryo-imaging.

In this chapter, the preparation and imaging of fully hydrated cultured Rat Hepatoma (HTC) cells is described to illustrate the different steps of the CEMOVIS procedure. The method outlines 1) preparation and high pressure freezing of the sample; 2) sectioning of the vitrified sample into ultrathin cryosections; and 3) observation of the cells in vitreous sections at low temperature in the transmission cryoelectron microscope.

2 Materials

1. Monolayer of cells cultured in a 50-mL flask, 70 to 80% confluent.
2. Cell scraper.
3. Eppendorf micropipettes with small yellow tips.
4. Cryoprotectant medium: 20% (w/v) dextran (40 kDa, e.g. Sigma-Aldrich, St. Louis, MO, USA) in PBS (140 mMNaCl, 2.7 mM KCl, 8.1 mM Na$_2$HPO$_4$, and 1.5 mM KH$_2$PO$_4$, pH 7.2–7.4).
5. Parafilm, hydrophobic paper.
6. Decon 90 detergent solution (Decon Laboratories, Hove, UK).
7. Carbon-coated copper electron microscope grids, 1,000-mesh size.
8. Wooden rods with an eyelash glued to the end using nail varnish.
9. Small 4-chamber cryo-grid boxes to store cryosections in liquid nitrogen (Gatan, Abingdon, UK).
10. A centrifuge suitable for pelleting cells using glass centrifuge tubes.

11. A high-pressure freezing apparatus. The description given here is for a Leica EM PACT1 (Leica Microsystems, Wetzlar, Germany) with the following accessories: Freezing specimen carriers: copper tubes.
    Aluminium supports in which to mount the copper tubes.
    A piece of metal wire of the same diameter as the inner diameter of the tubes.
12. A cryoultramicrotome. We use a Leica UCT FCS placed in an air-conditioned room with controlled humidity and O_2 level, with a cryotool kit comprising:

    (a) A polished metal plate with a polished metal puncher.
    (b) A grid holder.
    (c) An antistatic device.
    (d) Thermally isolated tweezers.
    (e) A cryotrimming knife with 45° oriented cutting side edges or with a simple 45° straight cutting angle.
    (f) A cutting knife with 45° or 35° cutting angles (Diatome, Bienne, Switzerland.

13. A transmission cryoelectron microscope equipped with:

    (a) A cryostage.
    (b) A cryoholder.
    (c) An anti-contaminator coupled to the column of the microscope.
    (d) A CCD camera or film acquisition system.

3   Methods

3.1   Preparation of Cells for Cryofixation

1. Detach the cells from the surface of the flask using a cell scraper (*see* **Note 1**). Cells floating in the culture medium, sometimes in clusters, are separated by gently pipetting three or four times using a micropipette with a yellow tip; this small aperture makes separation of the cells more efficient so that most cell clusters will be disassembled.
2. Cells are pelleted for 5 min in a conical glass centrifuge tube.
3. Remove the supernatant completely and resuspend the pellet in 300 µL of cryo-protectant solution. Again, homogenise the suspension using a pipet with a yellow tip to separate the cells as much as possible. It is recommended that the suspension be checked in the light microscope before using the suspension to ensure that the cells are well dispersed.
4. Five minutes after addition of the cryoprotectant medium, place a drop of the cell suspension on a piece of Parafilm (*see* **Note 2**). Time is crucial at this step; the cell suspension should not be used after more than 10 min of incubation in the cryoprotectant medium. In fact, our experiments have revealed that cells cannot survive in such a viscous medium for more than 30–45 min (data not shown).

3.2 Cryofixation

1. Fix the holder for cell suspensions of the Leica high-pressure freezer, a copper tube with an outer diameter of 500 μm and an inner diameter of 250 μm (Fig. 24.1a), into an aluminium support (Fig. 24.1b) in order to hold it and to allow its insertion into the cryo-chamber of the machine (*see* **Note 3**).
2. Introduce the cell suspension into the copper tube, using a piece of wire, which drives the cells surrounded by the cryoprotectant medium into the hole of the tube (Fig. 24.1c). For this, first introduce the wire inside the tube, plunge its tip into a droplet of cell suspension, and remove the wire gently from the tube. This piece of wire should have approximately the same diameter as the inner diameter of the tube to ensure total filling of the tube and to avoid the presence of air bubbles in the volume of the cylinder (*see* **Note 4**).
3. The copper tube is then introduced into the cryo-chamber of the high-pressure freezer. The procedure consists of raising the pressure in the cryo-chamber to 2,100 bars using the main hydraulic pump and pressure accumulator *(13)*. Ten milliseconds later, a jet of liquid nitrogen is applied on both sides of the tube to freeze the specimen with a speed approaching 15,000 K/sec (Fig. 24.2). Under such a pressure, according to the phase diagram of water, ice crystals do not form at 0°C but only at −22°C *(14–16)*. Applying a jet of liquid nitrogen on the sample at this stage allows one to reach a freezing speed higher than that needed to form nucleation centres from which ice crystals are initiated, and sufficient to obtain the vitreous state.

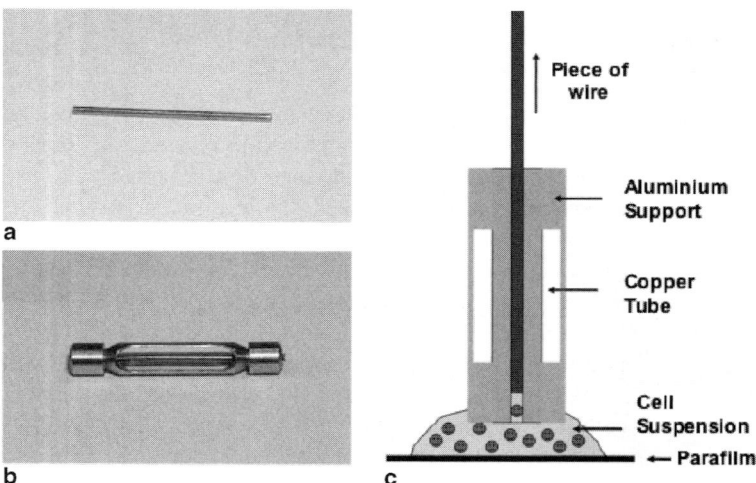

Fig. 24.1 **a** Hollow copper tube with an inner diameter of 250 μm. **b** The copper tube is first inserted into its aluminium support which has two open sides. This shape allows for spraying liquid nitrogen directly on each side of the tube during the vitrification process. **c** Scheme representing a droplet of cells suspended in PBS/20% dextran solution, deposited on a piece of Parafilm, and sucked into the hole of the copper tube using a piece of wire as a piston

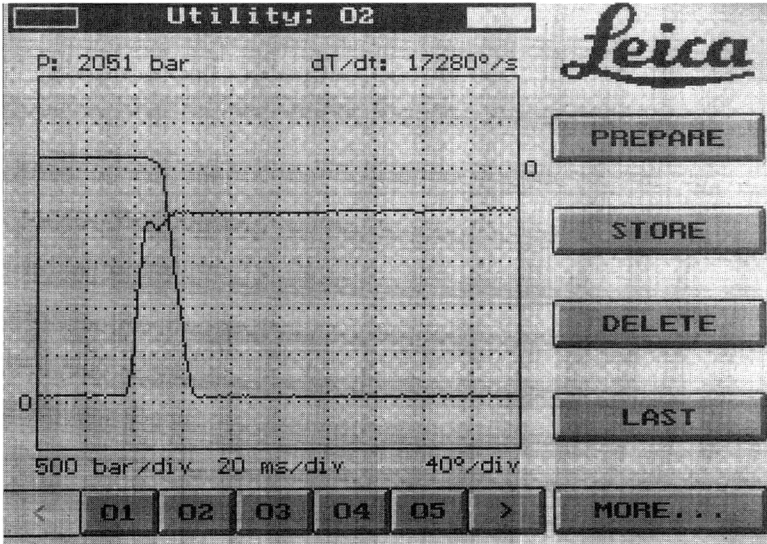

Fig. 24.2 Graph displaying the vitrification process at the stage of the cryo-chamber of the Leica high-pressure freezing apparatus. One curve on the graph displays the increase in pressure applied to the sample, the other displays the decrease in temperature in the cryo-chamber. First, the pressure is raised to 2,000 bars and 10 msec later a jet of liquid nitrogen slams the copper tube to cool it at about 17,000 K/sec, sufficient to vitrify the whole volume of the tube

4. At the end of the procedure, the tube in its aluminium support is automatically released from the cryo-chamber into a bath of liquid nitrogen by using the residual pressure. The sample can be either stored in liquid nitrogen or directly processed for cryosectioning.

3.3   Cryosectioning

Since the cells are now considered as vitrified, one must be very careful during further handling; indeed, from this step the sample temperature should not rise above −135°C (the devitrification temperature) to keep the specimen in its vitreous state (Fig. 24.3) (*see* **Note 5**).

Working with clean knives is essential to produce good quality cryosections.

3.3.1   Knife Cleaning

1. Bevel a wooden rod to an angle of approximately 45° using an oil-free razor blade, and then dip it into a 5% Decon solution and remove the excess solution while checking under the stereomicroscope.

Fig. 24.3 Electron-diffraction patterns.
a Typical pattern of a vitreous sample. The
presence of two diffuse rings (*arrowheads*) is
the only real proof of the amorphous state of
the water contained in the cryosectioned
specimen. **b** If the temperature rises above
−135°C, cubic ice crystals will form providing a characteristic pattern where diffuse
rings turn into two full and sharp rings
(*arrows*). **c** In the case of a higher temperature of the specimen, larger hexagonal crystals will form giving rise to a hatched circular
pattern (*arrows*) representative of the size and
orientation of the crystals

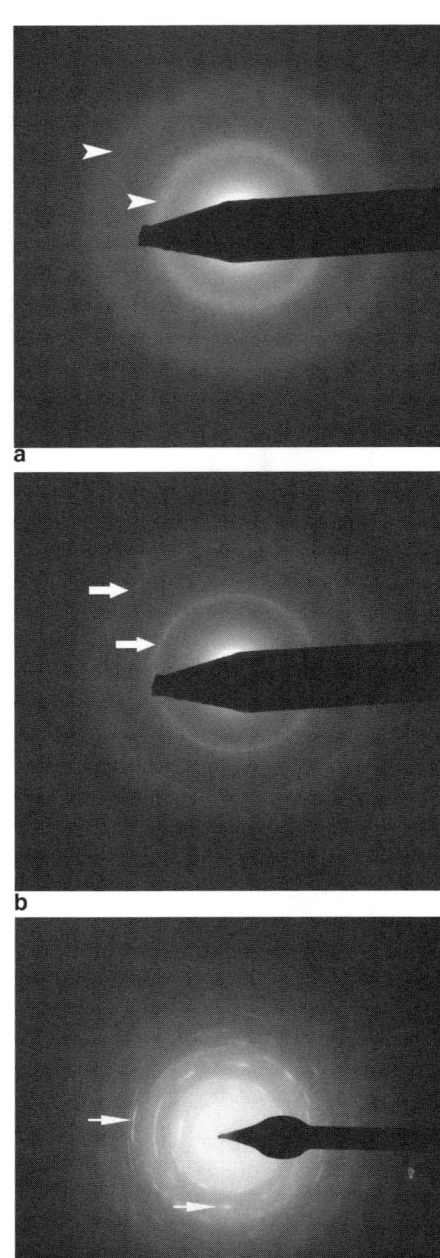

2. Pass the rod over the cutting edge of the knife without applying pressure.
3. Remove any Decon by pouring distilled water on the surface of the knife.
4. Pour 100% ethanol on the cutting edge and dry it using an air blower.
5. Do not hesitate to repeat the procedure until the edge and the surface of the knife are perfectly clean.
6. At the end of each cutting session, rinse the knife directly from the cryo-chamber under tap water.

3.3.2 Mounting the Specimen

1. Using a home-made cutting tool, cut out the central section of the copper tube (~half the original length) containing the vitreous sample, leaving the extremities attached to the aluminium support.
2. Precool the cryo-chamber of the microtome containing trimming and cutting knives to −140°C in order to maintain the temperature below the devitrification point (*see* **Note 6**).
3. Transfer the copper tube rapidly into the chamber of the microtome using pre-cooled tweezers and clamped into the specific chuck (provided by Leica, allowing for a reduction of the original chuck's diameter). At this stage, it is still difficult to know if the sample is vitreous or not, since ice contamination remains at the extremities of the tube.

3.3.3 Trimming the Specimen

1. Approach the trimming knife to the copper tube, using the light at the bottom of the microtome chamber to illuminate from underneath. At a certain point during the approach, the light reflected by the back of the knife is visible on the surface of the sample when moving it back and forth at the level of the cutting edge. The cutting edge is reached when the light vanishes, and the trimming process can start.
2. The first trimming consists in removing a substantial portion of the specimen at the cut extremities of the copper tube. One should trim the overall surface to remove a depth of ~200 to 300 μm (*see* **Notes 7** and **8**) with a cutting speed between 500 and 1,000 mm/sec (*see* **Note 9**) until obtaining a nice smooth and shiny surface.
3. One must adjust the light at the bottom of the chamber, looking at the shiny surface of the specimen with the light on but also trying to localise any ice crystals possibly present in the depth of the vitrified specimen with the light off. If ice crystals are present, they appear like small white dust surrounded by a black background of vitreous material when the light is turned off. A smooth and shiny surface and black or transparent material are two criteria prerequisites indicating, with a high probability, the amorphous state of the specimen.

4. At this point, a pyramid is trimmed on the top of the tube (Fig. 24.4). The right edge of the knife is aligned with the left side of the tube (Fig. 24.4a). From the first copper shavings, one should trim 200 μm deeply through the specimen (the trimming depth can be adjusted as a function of the desired final top surface of the pyramid).

5. The same conditions should then be applied to trim the three other sides of the pyramid. When trimming the first two sides, the first 150 μm can easily be trimmed with a speed of 500 mm/sec because the presence of the copper reinforces the sample and should prevent its breakage. The last 50 μm are trimmed with a speed of 200 mm/sec to ensure maintaining sharp edges, which are of crucial importance if one wants to produce good-quality cryosections. After a rotation of 90° (Fig. 24.4b), the other two sides of the pyramid are trimmed with a cutting speed between 100 and 200 mm/sec (Fig. 24.4c, d). At this point the copper support is no longer present, making the pyramid more fragile, and the sample is now ready for cryosectioning.

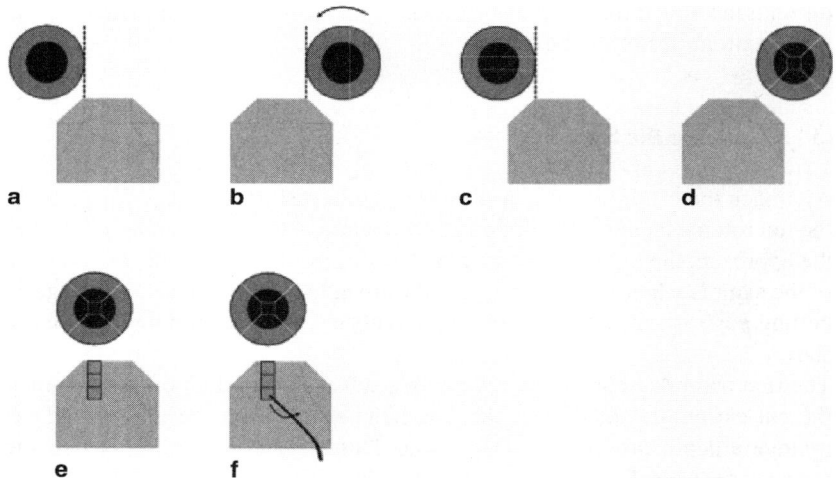

Fig. 24.4 Cryo-trimming and picking up of cryosections. From **a** to **d**, the different steps of the trimming process are shown. Based on the achievement of a black/transparent smooth and shiny surface of the vitrified suspension, the tube is first trimmed on one side, aligning its border with the side of the knife edge. Then, the opposite side is trimmed in the same way. After rotating the tube by 90° the same procedure is followed. At the end, a pyramid with a truncated tip is obtained at the top of the tube, allowing one to perform cryosections. In **e** and **f**, the ribbons of cryosections obtained are picked up using the tip of an eyelash and deposited on the surface of a 1,000-mesh carbon-coated copper grid

3.3.4   Cryosectioning

1. The trimming knife is displaced to the side or out of the chamber, and the cutting knife is moved into its place. Make sure that this does not stay for a too long period in the microtome chamber in order to prevent ice crystal deposits on the edge; this is especially important when you work in relatively high humidity conditions (*see* **Note 10**).

2. When the knife is cold, the antistatic device is maintained in the cryo-chamber, at first far from the knife and the pyramid until its temperature reaches that of the cryo-chamber. In this way, heating and devitrification of the sample are avoided. Then the antistatic device is placed at about 2 cm from the edge of the knife during the cutting procedure (*see* **Note 11**); this distance can be modified if the operator has the impression that cutting is not optimal.

3. One edge of the pyramid is oriented parallel to the cutting edge of the knife and the knife is then gently approached to the top of the pyramid using the reflected light, as during the trimming procedure. It is rare that the orientation of the cutting knife exactly matches that of the pyramid's surface. Since the pyramid touches the cutting edge of the knife, it is necessary to start making small sections in order to enter the specimen before obtaining full-surface sections, which should glide nicely on the surface of the knife.

4. As soon as ultrathin cryosections start to be produced, the cutting speed is adjusted to 0.4 mm/sec and the nominal feed is set to 50 nm (*see* **Note 12**). Care must be taken to use a specific region of the knife edge only once. Even if a nice ribbon of sections is produced at one place, or if for whatever reason cutting is interrupted, the knife should be retracted and another place chosen along its edge (*see* **Note 13**).

5. If the sections do not glide properly on the surface of the knife, or if it is not possible to produce more than one section at each run, gliding may be improved by using the eyelash tool (**Section 2.8**). During cutting or between the production of two sections, try to pull gently on the section to favour its gliding on the surface of the knife; this is not easy at first, but can be essential in some cases. It is important to understand that vitreous specimens cannot be considered as a rigid material, but rather as a material which exhibits an extremely high viscosity and which can dilate with time when kept at the microtome cryo-chamber temperature. If the sample is not retracted and readjusted at the edge of the knife after a break between two cutting series, most of the time the specimen will dilate. As a consequence, when making the next section the pyramid will break because it will have to endure forces that even its high viscosity are not able to absorb.

6. Finally, when the whole knife edge is covered by nice ribbons of sections (Fig. 24.4e) they are collected on a carbon-coated copper grid (or more than one grid if there are many sections).

3.3.5   Collecting Sections on the Grid

1. A 1,000-mesh carbon-coated copper grid is introduced into the cryo-chamber of the cryoultramicrotome and maintained in the specific holder.
2. After 5 min for the temperature to equilibrate, the grid is brought as close as possible to the surface of the knife, underneath the ribbons of sections but without touching the knife (see **Note 14**).
3. The sections are collected using an eyelash tool (Fig. 24.4f); this must touch the ribbon on one side just enough to stick to it but not too much, otherwise the ribbon can easily be disrupted. The ribbon is gently pulled up and to one side to detach it from the edge of the knife. When the ribbons stick to the eyelash, they are gently deposited near the centre of the grid.
4. The grid is then released onto a polished metal surface and pressed using a polished metal punch, in order to make the ribbon of sections attach strongly to the surface of the grid.
5. The grids are placed into the grid box, and can either be stored in liquid nitrogen or observed directly in the transmission cryoelectron microscope (*see* **Note 15**).

3.4   Cryo-observation

CEMOVIS makes it possible to observe the specimen in its hydrated state. Consequently, extreme care should be taken with the electron beam illumination conditions when imaging cryospecimens. It is very important to maintain the electron dose as low as possible, both when looking for structures of interest and when imaging them at higher magnification. In general, observations are carried out at 80 kV.

3.4.1   Cryotransfer of the Grid into the Microscope Cryoholder

1. Once the cryoholder, cryotransfer device, and insulated tweezers have been pre-cooled in liquid nitrogen, the cryoholder rod is inserted inside the cryotransfer apparatus.
2. Then, the grid box is introduced into the cryotransfer chamber filled with liquid nitrogen.
3. At the tip of the rod, the clamp of the cryoholder is removed and kept in liquid nitrogen. At this stage, the grid box is opened and the liquid nitrogen level is lowered down the cryoholder's rod. One must be sure that no bubbling of liquid nitrogen reaches the place where the grid will be placed, otherwise it will be moved away before having time to be clamped.
4. A grid is mounted, sections upward, at the tip of the cryoholder's rod and subsequently clamped onto it. Liquid nitrogen is immediately poured into the transfer chamber in order to prevent warming of the rod and the specimen.
5. The cryoholder is inserted inside the transmission cryoelectron microscope, and the cryoholder's Dewar flask is filled with a small amount of liquid nitrogen to

prevent drift during image acquisition due to bubbling of nitrogen in it. The liquid nitrogen allows for stabilisation of the Dewar's temperature at about −180°C throughout observations.

3.4.2 Image Formation in the Transmission Cryoelectron Microscope When Observing Vitrified Specimens

In conventional EM, an image is formed due to the density differences between regions in the specimen due to amplitude contrast. However, in fully hydrated and unstained specimens at zero defocus, particles such as ribosomes are practically invisible. Fortunately, an electron beam, like light, has wave properties and is able to form an image by phase contrast (17). Imaging of vitreous sections relies nearly exclusively on phase contrast, which requires a selection of the defocus adapted to the desired dimensions one wants to observe.

Each individual biological structure, particle, or macromolecule is characterised by its own density and compaction. All such fine details are present in a thin section and will also be present in the image, but not all of them will be recognized. With CEMOVIS observations (Fig. 24.5) the major limitation arises not from the lack of contrast of unstained structures in water, but rather from the plethora of overlapping information within the thickness of the section.

3.4.3 Checking the Specimen's Vitreous State

Start each set of observations by checking the electron diffraction pattern produced by the specimen (15), which, together with the radius of the rings measured on it, is the only way to prove that a specimen is vitreous (Fig. 24.3a) or not (Fig. 24.3b, c). To do this:

1. Insert the diffraction aperture inside the column and centre it, then reduce the beam intensity, remove the objective aperture, and set the camera length to 210 mm.
2. Focus is then adjusted to obtain the narrowest spot at the centre of the diffractogram. The characteristic diffraction pattern of vitreous water reveals two diffuse rings (Fig. 24.3a).

3.4.4 The First Look at a Cryosection

Deformation of the sample under the local high pressure generated by the cutting process is inherent to vitreous sections (for reviews see refs. (18, 19)). Sections will show parallel lines at their surface, also called knife marks, attributed to imperfections and contamination at the knife's cutting edge. Banana-shaped fractures, most of the time homogeneously located on the overall surface of the section, can be observed. When the force applied to a section by the knife is too high, it forms a field of fractures termed crevasses, which penetrate to varying depths into the section (*see* **Note 16**). Finally, sections can exhibit a form of non-homogeneous

Fig. 24.5 a The organisation of an HTC
cell nucleus, separated from the cytoplasm (*Cyt*) by the nuclear envelope
(*Ne*), is shown by CEMOVIS. The process precludes any use of contrasting
agents such as heavy metal salts used in
more conventional methods, and structures display a relatively low contrast.
Nevertheless, condensed chromatin (*Cc*)
domains can be distinguished thanks to
their characteristic granular texture. The
nucleolus (*No*) is mainly seen thanks to
its granular component. At the border of
the condensed chromatin domains, thin
fibro-granular structures can be observed
that probably represent perichromatin
fibrils (*white arrowheads*) and granules
(*black arrowheads*). The rest of the
nucleoplasm is mainly composed of
interchromatin space (*IS*) crowded with
less-ordered material. Sometimes, vesicles (*V*) can be observed in the nucleoplasm (see ref. *(11)* for more details). In
b, a section of an HTC cell was submitted to beam irradiation, inducing bubbling (*arrow*) of the biological structures
within the thickness of the cryosection.
Note that not all structures are bubbling
at the same time; some are more responsive to beam irradiation than others.
Scale bars, 500 nm

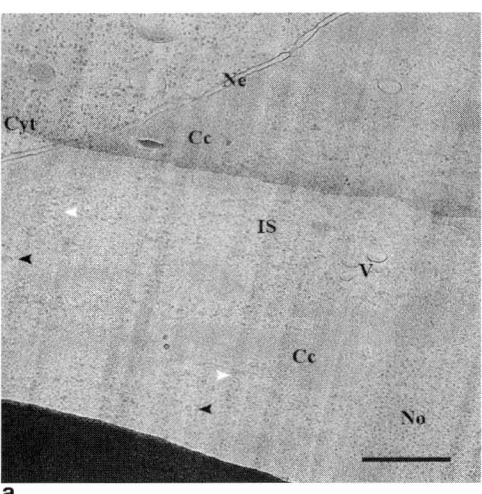

a

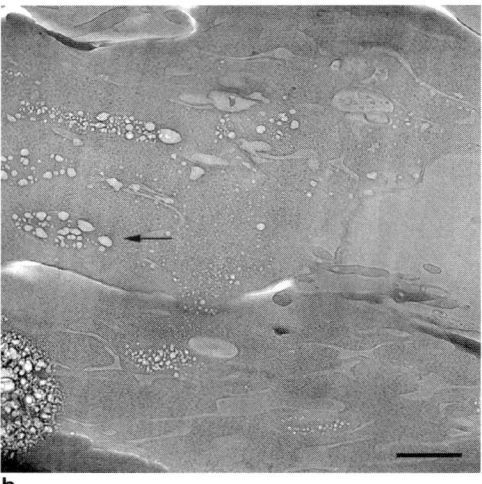

b

compression termed chatter, defined as an oscillatory variation of section thickness
along the cutting direction and due to the irregular friction of the forming section
against the knife surface, which produces variable compression and causes section
thickness to vary accordingly.

3.4.5 Image Acquisition in Low-Dose Mode

As previously mentioned, the control of the electron dose is important during imaging of cryosections *(20)*. Low-dose mode must be used when looking for the area
of interest at a relatively low magnification.

1. The exposure illumination is set to 2.5 sec and the time to 1 sec far from the area of interest, to be sure of avoiding damage to the specimen before or during the acquisition and to minimise troubleshooting of beam-induced drift.
2. The ideal region is then located in search mode at ~5,000× magnification using the lowest possible illumination conditions. Care must be taken to avoid irradiation of the specimen after finding the area of interest; in order to do so, the microscope's beam blank control that shuts down the beam when desired is used.
3. Then, the focus mode is chosen to set the defocus on an area outside the area of interest, using the magnification chosen for final acquisition. A hole is made through the section by concentrating the beam on it, the sides of this hole are used to set the defocus adequately for the next acquisition, and then the beam is blanked out.
4. Finally, switch to exposure mode and acquire the image (Fig. 24.5a). One has to be aware that one area of interest will support only one exposure with such illumination conditions. If the irradiation is too high, material rearrangements can occur rapidly followed at a certain threshold by bubbling of the material within the section (Fig. 24.5b) (*see* **Note 17**).

4   Notes

1. Trypsin can also be used to detach the cells; however, it is unknown to what extent trypsin action modifies the native state of cells.
2. After the drop is placed on Parafilm, the freezing process including loading of the sample into the tube and cryofixation itself must be fast. With time, cells tend to sediment onto the surface of the Parafilm, significantly reducing their concentration. After 2 min the drop has to be removed and replaced by a new one.
3. When mounting the tube on the support, care must be taken to maintain the tube in a straight position. Otherwise, the tube can bend or even break during the freezing process.
4. If any air is present along the wall of the cylinder, it will interfere with reaching high pressure, which can lead to poor freezing of the sample.
5. If the specimen's temperature rises above −135°C, the amorphous state is lost and cubic ice crystals form, exhibiting a characteristic and easily recognisable electron diffraction pattern. Further heating gives rise to hexagonal ice crystals. At this stage, observation of the specimen is compromised.
6. Cutting temperature is set at −140°C (*18*) for the following reason: vitrified specimens are not rigid objects even when handled at very low temperatures, but rather like objects showing an extremely high viscosity. At −140°C the vitrified material appears "soft", whereas a specimen cut at −160°C will appear more "brittle" when sectioning, and cutting will produce sections that exhibit more severe cutting artefacts such as crevasses.
7. During retrieval of the central part of the tube, it is cut on both sides while loosing its extremities. The force applied by the knife at the cutting points can be sufficient to promote crystallisation at the extremities of the tube. Consequently, the specimen must be trimmed for another 200 μm to reach vitrified material.

8. The specimen is trimmed at one extremity first in order to see if one can obtain a nice smooth and shiny surface. If the surface does not appear smooth after 300 μm of trimming, it is not useful to continue further on the same extremity. Turn the tube and test the opposite end in the same way. If the other surface is not smooth and shiny either, change for another tube.

9. Trimming of the copper tube is probably the only operation where such a high cutting speed can be employed. The presence of the copper around the sample makes it readily resistant to forces applied during trimming, but this is only valid for the first steps of trimming.

10. The length of time that the cutting knife can stay in the microtome chamber depends on the ambient humidity; if this is high, 15 min may be the limit and in this case, the surface of the knife will become completely contaminated with ice crystals that will prevent obtaining clean cryosections homogeneous in size, thickness, and flatness. Moreover, if the cryosections are covered by contaminants additional charge will appear, and picking the cryosections up and attaching them to the grid surface will become very difficult. Most of the time, it is recommended to work with a freshly cooled cutting knife.

11. The antistatic device plays an extremely important role for achieving good sections. It neutralises the electrostatic charges at the surface and edge of the knife and therefore improves the gliding of the sections during cryosectioning.

12. Setting the microtome to 50 nm nominal thickness will produce sections with a real thickness of about 70 nm if the compression factor is taken into account. According to ref. *(8)*, compression makes the sections shorter along the cutting direction compensated by an increase in thickness, while the lateral dimensions remains unchanged; a compression factor of around 30 to 40% is common when cutting cryosections.

13. One side of the knife usually produces better sections than the other. The best part of the knife is used for sectioning after approach to its edge, and the first few sections are made on the poorer part.

14. When the knife edge is touched, charges remaining at its surface are conducted onto the grid, making deposition of sections difficult due to repulsive forces.

15. The best way is to start microscopic observation of cryosections right after their preparation. Over time, the risk of contamination of sections by ice crystals on their surface increases.

16. Knife marks as well as soft crevasses are melted away by electron irradiation, thus giving the impression of an artefact-free section. In fact, fractures caused by crevasses remain generally marked in biological structures such as membranes. Moreover, a high electron dose causes a general loss of high-resolution information. The formation of crevasses can be overcome by cutting sections in the range of 30 to 50 nm (real thicknesses), when the thin layer of vitreous material reacts differently with regard to cutting-induced forces compared with thicker sections.

17. Bubbling is a typical phenomenon observed on cryospecimens. When irradiating the specimen with the electron beam, the energy induces chemical reactions that produce gas, and bubbling is caused by release of this gas concentrated in the vitreous specimen *(17)*.

Acknowledgments CBM would like to thank particularly Jacques Dubochet for having taught him the technical details as well as the theoretical background of vitrification and cryo-electron microscopy. The authors thank Jacques Dubochet and Graham Knott for critical reading of the manuscript and corrections.

References

1. Monneron, A. and Bernhard, W. (1969) Fine structural organization of the interphase nucleus in some. mammalian cells. *J. Ultrastruct. Res.* **27**, 266–288.
2. Fakan, S. (2004) The functional architecture of the nucleus as analysed by ultrastructural cytochemistry. *Histochem. Cell Biol.* **122**, 83–93.
3. Tokuyasu, K. T. (1973) A technique for ultracryotomy of cell suspensions and tissues. *J. Cell Biol.* **57**, 551–565.
4. McDowall, A. W., Chang, J. J., Freeman, R., Lepault, J., Walter, C. A., and Dubochet, J. (1983) Electron microscopy of frozen hydrated sections of vitreous ice and vitrified biological samples. *J. Microsc.* **131**, 1–9.
5. McDowall, A. W., Hofmann, W., Lepault, J., Adrian, M., and Dubochet, J. (1984) Cryoelectron microscopy of vitrified insect flight muscle. *J. Mol. Biol.* **178**, 105–111.
6. McDowall, A. W., Smith, J. M., and Dubochet, J. (1986) Cryo-electron microscopy of vitrified chromosomes in situ. *EMBO J.* **5**, 1395–1402.
7. Sartori Blanc, N., Senn, A., Leforestier, A., Livolant, F., and Dubochet, J. (2001) DNA in human and stallion spermatozoa forms local hexagonal packing with twist and many defects *J. Struct. Biol.* **134**, 76–81.
8. Al-Amoudi, A., Chang, J. J., Leforestier, A., McDowall, A., Salamin, L. M., Norlen, L. P., Richter, K., Blanc, N. S., Studer, D., and Dubochet, J. (2004) Cryoelectron microscopy of vitreous sections. *EMBO J.* **23**, 3583–3588.
9. Zuber, B., Nikonenko, I., Klauser, P., Muller, D., and Dubochet, J. (2005) The mammalian central nervous synaptic cleft contains a high density of periodically organized complexes. *Proc. Natl. Acad. Sci. USA* **102**, 19192–19197.
10. Eltsov, M. and Dubochet, J. (2005) Fine structure of the Deinococcus radiodurans nucleoid revealed by cryoelectron microscopy of vitreous sections. *J. Bacteriol.* **187**, 8047–8054.
11. Bouchet-Marquis, C., Dubochet, J., and Fakan, S. (2006) Cryoelectron microscopy of vitrified sections: a new challenge for the analysis of functional nuclear architecture. *Histochem. Cell Biol.* **125**, 43–51.
12. Dubochet, J. and Sartori Blanc, N. (2001) The cell in absence of aggregation artifacts. *Micron* **32**, 91–99.
13. Studer, D., Graber, W., Al-Amoudi, A., and Eggli, P. (2001) A new approach for cryofixation by high-pressure freezing. *J. Microsc.* **203**, 285–294.
14. Eisenberg, D. and Kauzmann, W. (1969) *The Structure and Properties of Water.* Oxford University Press, Oxford.
15. Franks, F. (1982) Water: A comprehensive treatise. In: *Water and aqueous solutions at sub-zero temperatures* (Franks, F., Ed.), Plenum Press, New York, pp. 215–338.
16. Dubochet, J., Adrian, M., Chang, J. J., Homo, J. C., Lepault, J., McDowall, A. W., and Schultz, P. (1988) Cryo-electron microscopy of vitrified specimens. *Q. Rev. Biophys.* **21**, 129–228.
17. Dubochet, J., Al-Amoudi, A., Bouchet-Marquis, C., Eltsov, M., and Zuber, B. (in press) CEMOVIS. Cryo-electron microscopy of vitreous sections. In: *Modern Cryopreparation methods for electron microscopy* (Cavalier, A., Humbel, B. M., and Spehner, A., eds.), CRC.
18. Al-Amoudi, A., Studer, D., and Dubochet, J. (2005) Cutting artefacts and cutting process in vitreous sections for cryo-electron microscopy. *J. Struct. Biol.* **150**, 109–121.
19. Al-Amoudi, A., Dubochet, J., Gnaegi, H., Luthi, W., and Studer, D. (2003) An oscillating cryo-knife reduces cutting induced deformation of vitreous ultrathin sections. *J. Microsc.* **212**, 26–33.
20. Sartori Blanc, N., Studer, D., Ruhl, K., and Dubochet, J. (1998) Electron beam-induced changes in vitreous sections of biological samples. *J Microsc* **192**, 194–201.

Index

A

Antibodies
  primary
    for atomic force microscopy
      recognition imaging, 126–136
    BrdU, 87, 143
    digoxygenin, 87, 407, 411
    glutathione S transferase, 188,
      200–203
    nucleoporin 62, 4
    O-GlcNAc-modified proteins, 229,
      233–243, 247–250
    phosphoinositide monophosphate
      kinase I, 218
    phospholipase Cγ, 218
    PML, 417–418
    poly(ADP-ribose) (PAR), PAR
      polymerases, PAR glycohydrolase,
      269–270, 274–280
    redox factor-1, 305–306, 309
    RNA polymerase 2 largest subunit,
      hyperphosphorylated C-terminal
      domain, 393
    SUMO, 257, 259
    thioredoxin-1, 305–306, 309
    U2 snRNP protein U2 B", 36
  secondary, gold-labelled
    detection of PML bodies, 417, 418
    labelling pre-mRNPs, 35–36,
      39, 49
    resolution, 35
    RNA detection by electron microscopy,
      407–411
    silver enhancement, 49, 417–420

B

Balbiani rings, 29–50

C

Cajal bodies, visualisation by
  immunofluorescence, 323
Chromatin
  electroelution of fragments, 146–148, 156
  extraction of histones, 290
  fibres
    dynamics, 31
    visualisation by electron spectroscopic
      imaging, 417
  Ku protein in, 133
  loops
    mapping by crosslinking, 105–120
    measurement of length, 142–147
    transcription, 32–33, 36, 44
  membrane vesicle binding, 216
  removal from *Xenopus* oocytes, 169
  visualisation
    atomic force microsopy, 126,
      128, 134
    cryoelectron microscopy, 436
    EDTA regressive staining, 410
    immunofluorescence of histone
      H2A or H2B, 101, 323
Chromosomes
  contacts between, 105–106
  polytene, 29–50
  telomere proteins, FRAP, 371
  territories, 154
Confocal microscopy
  confocal volume measurement, 325,
    332–333
  coverslip thickness for, 380
  measurement of nuclear envelope
    permeability, 165, 171–174
  nuclear protein import assay, 198–200
  setup for fluorescence correlation
    spectroscopy, 329

D
Data processing and analysis (*see* Software)
DiIC18 (1,1'-dioctadecyl-3,3,3',3'-
    tetramethyl-indocarbocyanine
    perchlorate) for labelling nuclear
    membranes, 7, 8

E
Electron microscopy
    immunolabelling
        efficiency of antibodies, 48
        export of mRNPs, 39
        isolated polytene chromosomes, 46
        nascent transcript and splicing
            complexes, 36
        salivary gland nuclei, 44–45
    nucleoskeleton, 148–149
Electrophoresis
    chromatin fragments for measuring loop
        size, 145–147
    detection of poly(ADP-ribose)
        glycohydrolase activity by
        zymography, 281–282
    proteins involved in nuclear architecture,
        153–155

F
Fixation
    DNA damage by paraformaldehyde, 282
    influence on detection of poly
        (ADP-ribose) polymerases, 275–276
    influence on contrast for electron
        spectroscopic imaging, 422
Fluorescence
    bleaching
        in single-molecule tracking, 353
        minimising, 346
        problems, 338, 357
    blinking, 351, 382
    dextran for nuclear envelope permeability
        assays, 171–174
    dyes for visualising chromatin-bound
        membrane vesicles, 209, 215, 216
    fluorescence-free water, 328
    fluorescence-labelled proteins
        cargoes for nuclear protein import
            assays, 198–201
        CFP-(nuclear localisation signal x3),
            3–4, 8
        EGFP-lamin A, 3–4, 8
        EGFP-nuclear body protein, 327, 335
        GFP-androgen receptor, 366–369, 373

GFP-histone H2B, 99, 101
GFP-nucleoporin POM121, 353
GFP-telomere-binding proteins TRF1
    and TRF-2, 367, 371
GFP 96-mer, 93, 97–98
labelling with succinimidyl esters or
    maleimide dyes, 348–351
sequence insertion to decrease
    effects of a label on protein
    functionality
in situ hybridisation on combed DNA,
    probes, 83–86
measuring nuclear strain, 13–26
membrane labels, 4, 7–8, 209, 216
molecular beacons, 92, 98–99, 102
single molecules, 343–361
studying protein mobility and
    interactions, 324

G
Glycoproteins in the nucleus, 228

H
Histones
    extraction, 290
    histone H2B-GFP, 99, 101
    histone H3 antibodies for atomic force
        microscopy imaging, 134

I
Immunofluorescence visualisation
    chromosome territories, 154
    correlation with electron microscopy, 407
    isolated polytene chromosomes, 45
    newly-replicated DNA, 74, 83–88, 150
    nuclear pores, 4, 7
    nuclear protein import, 200–201
    poly(ADP-ribose) (PAR), PAR
        polymerases, PAR glycohydrolase,
        275–280
    salivary gland cells, 43
    splicing factors, 32
    subnuclear structures, 323

K
Ku protein, 133

L
Laminopathies, 3, 13

M

Maleimide dye labelling of protein
        cysteines, 350
Microinjection
    equipment, 42, 93, 348
    methods
        molecular beacons, 99
        salivary gland cells, 46–47
        single molecule studies, 354, 355
        Xenopus oocytes, 166–168

N

Nuclear lamina
    lamin mutations in muscular dystrophy,
        3, 13
    visualisation by immunofluorescence, 323
    visualisation with EGFP-lamin A, 4, 5, 8
Nuclear envelope
    as a solid–elastic shell, 4
    electrical conductance, 170–171
    in cryoelectron microscopy, 436
    permeability to dextrans, 174
    preparation from *Xenopus* oocytes, 169
    visualisation by atomic force microscopy,
        174–177
    visualisation by fluorescent lipid stains, 4,
        7–8, 216
    visualisation using GFP-nuclear pore
        proteins, 344, 353
Nuclear pores
    fluorescent labelling with
        anti-nucleoporin 62, 4, 7
    fluorescent labelling with GFP-
        nucleoporin POM121, 353
    glycoproteins, 228
    imaging by atomic force microscopy, 162,
        174–178
    transport, 161–178, 181–204
Nuclease
    DNase 1, preparation of nuclear proteins,
        155
    restriction nucleases
        chromatin conformation capture, 111,
            115, 117–118
        cleavage of chromatin loops, 145–148,
            156
        cloning tandem repeats, 96–97, 101
        fragments for atomic force microscopy
            recognition imaging, 134
Nucleolus
    in cryoelectron microscopy, 436
    visualisation by immunofluorescence, 323
    visualisation of PARP-1 and PARP-2, 276

        visualisation of rRNA genes in
            Saccharomyces cerevisiae, 57
Nucleoplasm
    cryoelectron microscopy, 436
    fluorescence correlation spectroscopy, 327
    injection in *Xenopus* oocytes, 167, 168
    protein pools, 324
    ribonucleoprotein-containing fibrils, 411
Nucleoskeleton, 148–149

P

PCR
    chromatin conformation capture, 107,
        113–114, 117–120
    generation of antisense RNA probes,
        404–405, 408–409
PML bodies
    visualisation by electron spectroscopic
        imaging, 417
    visualisation by immunofluroescence, 323
Probes
    hybridisation on combed DNA, 83–86
    molecular beacons, 92
    RNA antisense, 404–405, 408–409, 411
Proteases
    inhibitors, 5, 108, 188, 210, 234, 236, 257,
        304, 307
    SUMO, 256
Proteasomes, visualisation by immunofluores-
        cence, 323

R

Replication
    factories, 148, 152, 322
    labelling sites, 150–151
    using combed DNA, 71–90
RNA
    isolation from salivary gland cells, 47–48,
        50
    probes, 405
    structure prediction, 95
    visualisation of nascent transcripts, 57, 58
RNA polymerase
    visualisation, 36, 319

S

Software
    Andor SOLIS for Imaging (single-
        molecule fluorescence), 346, 357
    DiaTrack (single-molecule fluorescence),
        346, 358–359

Software (*cont.*)
    Digital Micrograph (electron spectroscopic
        imaging), 422
    for tracking single mRNAs, 95, 99
    ImageJ (image processing), 11, 358
    Jmeasure 2.2.4 (combed DNA), 79
    Matlab (image processing), 16–17, 99
    m-fold (RNA secondary structure), 95
    Origin (FCS data), 328
    PCLAMP7 (atomic force microscopy), 165
    SimplePCI (quantifying nuclear import), 201
    SmartCapture 2.1 (camera, filter, objective,
        and stage control), 79
    Turbosequest (mass spectrometry), 247
Succinimidyl ester labelling of protein lysines,
    349, 350
Telomere-binding proteins GFP-TRF1 and
    TRF-2, FRAP, 365–367, 371
Transcription
    factors
        HSF2 sumoylation, 251–261
        binding and nuclear redox state, 300
        androgen receptor, studies by FRAP,
            360–364, 367–373
    sites
        visualisation, 151–152, 318, 322

    sizing by spatially modulated
        illumination microscopy, 381–392
    synthesis of RNA probes, 395–399
    visualisation on polytene chromosomes,
        29–53
    visualisation on yeast chromatin, 55–69
Single particle tracking
    messenger ribonucleoproteins, 93–105
    proteins, 337–355
Splicing factors
    in pre-mRNP complexes, immunoelectron
        microscopy, 36
    in polytene nuclei and chromosomes,
        immunofluorescence, 32
    SC35 in speckles, visualisation by
        immunofluroescence, 319

T
Tandom repeat sequences, cloning, 98

Z
Zymography, detection of poly
        (ADP-ribose) glycohydrolase
        activity, 281–282

Printed in the United States of America